BASIC LINEAR PARTIAL DIFFERENTIAL EQUATIONS

Pure and Applied Mathematics

A Series of Monographs and Textbooks

Editors **Samuel Eilenberg and Hyman Bass**

Columbia University, New York

RECENT TITLES

JAMES W. VICK. Homology Theory: An Introduction to Algebraic Topology

E. R. KOLCHIN. Differential Algebra and Algebraic Groups

GERALD J. JANUSZ. Algebraic Number Fields

A. S. B. HOLLAND. Introduction to the Theory of Entire Functions

WAYNE ROBERTS AND DALE VARBERG. Convex Functions

A. M. OSTROWSKI. Solution of Equations in Euclidean and Banach Spaces, Third Edition of Solution of Equations and Systems of Equations

H. M. EDWARDS. Riemann's Zeta Function

SAMUEL EILENBERG. Automata, Languages, and Machines: Volumes A and B

MORRIS HIRSCH AND STEPHEN SMALE. Differential Equations, Dynamical Systems, and Linear Algebra

WILHELM MAGNUS. Noneuclidean Tesselations and Their Groups

FRANÇOIS TREVES. Basic Linear Partial Differential Equations

WILLIAM M. BOOTHBY. An Introduction to Differentiable Manifolds and Riemannian Geometry

BRAYTON GRAY. Homotopy Theory: An Introduction to Algebraic Topology

ROBERT A. ADAMS. Sobolev Spaces

JOHN J. BENEDETTO. Spectral Synthesis

D. V. WIDDER. The Heat Equation

IRVING EZRA SEGAL. Mathematical Cosmology and Extragalactic Astronomy

J. DIEUDONNÉ. Treatise on Analysis: Volume II, enlarged and corrected printing; Volume IV

WERNER GREUB, STEPHEN HALPERIN, AND RAY VANSTONE. Connections, Curvature, and Cohomology: Volume III, Cohomology of Principal Bundles and Homogeneous Spaces

I. MARTIN ISAACS. Character Theory of Finite Groups

JAMES R. BROWN. Ergodic Theory and Topological Dynamics

In preparation

CLIFFORD A. TRUESDELL. A First Course in Rational Continuum Mechanics: Volume 1, General Concepts

K. D. STROYAN AND W. A. J. LUXEMBURG. Introduction to the Theory of Infinitesimals

B. M. PUTTASWAMAIAH AND JOHN D. DIXON. Modular Representations of Finite Groups

MELVYN BERGER. Nonlinearity and Functional Analysis: Lectures on Nonlinear Problems in Mathematical Analysis

GEORGE GRATZER. Lattice Theory

BASIC LINEAR PARTIAL DIFFERENTIAL EQUATIONS

François Treves

DEPARTMENT OF MATHEMATICS
RUTGERS UNIVERSITY
NEW BRUNSWICK, NEW JERSEY

1975

ACADEMIC PRESS

New York • San Francisco • London

A Subsidiary of Harcourt Brace Jovanovich, Publishers

ACADEMIC PRESS, INC.
111 Fifth Avenue, New York, New York 10003

United Kingdom Edition published by
ACADEMIC PRESS, INC. (LONDON) LTD.
24/28 Oval Road, London NW1

Library of Congress Cataloging in Publication Data

Treves, François, (date)
Basic linear partial differential equations.

(Pure and applied mathematics v. 62)
Bibliography: p.
Includes indexes.
1. Differential equations, Partial. 2. Differential equations, Linear. I. Title. II. Series: Pure and applied mathematics; a series of monographs and textbooks; v. 62
QA3.P8 [QA374] 515'.353 74-10206
ISBN 0-12-699440-4

AMS (MOS) 1970 Subject Classifications: 35-02, 35J05, 35K05, 35L05

PRINTED IN THE UNITED STATES OF AMERICA

Contents

CHAPTER I

The Basic Examples of Linear PDEs and Their Fundamental Solutions

CHAPTER II

The Cauchy Problem

CHAPTER III

Boundary Value Problems

CHAPTER IV

Mixed Problems and Evolution Equations

Preface

The discrepancy between what is taught in a standard course on partial differential equations and what is needed to understand recent developments in the theory is now very wide. It is a fact that only a relatively small number of specialists, in a few universities, are able, these days, to teach a course that is truly introductory to those developments. Perhaps this is not much different from what has been happening in all active areas of mathematics. But it is also true, speaking of the best graduate students, as well as of professional mathematicians, that when they are said to be conversant in all aspects of mathematics, this often excludes substantial portions of analysis and most of partial differential equations.

The complementary facet of such a state of affairs is that many up-to-date expositions fail, frequently because of lack of time, to show the link with the older results, and give the erroneous impression that the modern theories have no roots and are cut off from a rich past. The truth, of course, is that progress comes not only from pushing further and further into new territory but also from frequent returns to the familiar grounds, from seeking an ever-deeper understanding of their nature, and finding there new inspiration and guidance.

The archetypes of linear partial differential equations (Laplace's, the wave and the heat equations) and the traditional problems (Dirichlet's and Cauchy's) are the main topic of this book. Most of the basic classical results can be found here. But the methods by which these are arrived at are definitely not traditional; the methods are, in practically every instance, applications of those now in favor at a higher level of abstraction. The aim of this approach is twofold: it is, on one hand, that of recalling the classical material to the modern analyst, in a language he can understand; on the other hand, that of exploiting the same material, with the wealth of examples it provides, as an introduction to the modern theories.

Developments toward greater generality have not been avoided when it was felt that they represented the natural "next step" and afforded a meaningful opening to the more advanced stages of the theory—provided, also, that they did not require more machinery than had been made available up to that point. Thus the reader will find a discussion of the Cauchy problem for first-order systems of hyperbolic equations with constant coefficients in Section 15, following the study of the same problem for the wave equation. Similarly, Gårding's inequality for strongly elliptic equations of any (even) order is established in Section 36, and, in a somewhat more philosophical vein, the meaning of the Lopatinski boundary conditions is explained in Section 38.

The approach to the classical Dirichlet problem calls for some comment. Because I felt committed to describe the classical results, it was out of the question to limit the discussion to the weak solution, or variational, method—even strengthened by the proof of regularity up to the boundary, when the latter is sufficiently smooth. After all, one might want to have the solution to the Dirichlet problem in a cube when the boundary value is continuous. Thus I was resigned to the dichotomy between the variational methods within the framework of the Sobolev spaces, and the Perron–Brelot method, tied to potential theory, until Guido Stampacchia indicated to me how to make the transition from the former to the latter, by way of his weak maximum principle (Section 28). I have followed his advice and adapted the argument of his article [2] (where, needless to say, more general second-order elliptic equations than Laplace's are studied). From there on, the classical potential theory can easily take off, as is succinctly indicated in Sections 29 and 30.

Like potential theory, many other important topics are very lightly touched upon: for example, the Dirac equations, random walks, the finite difference method, and continuous semigroups of operators. Here the book is truly introductory; its sole ambition is to give an idea of what these topics are all about and a taste for learning more. Thus it is not a treatise. Nor is it a classroom text, due to its size and the quantity of its contents, although it is true that it began as a set of lecture notes used at the University of Miami and at Rutgers University. Some readers might find that the writing shows too little regard for concision—for which I apologize. I have made a point, rather, of explicitly formulating some of the many thoughts that usually go unformulated while writing, especially while writing mathematics; and I hope not to have totally failed in this.

Today, distributions are the language of linear PDE theory, and I am certainly not of the school that would like to do without them. But knowing that not all students are seriously exposed to distributions, I have limited their use to their more mechanical aspects—convergence of sequences, differentiation, convolution; sometimes, but not often, the local representation of a distribution as a finite sum of derivatives of continuous functions is used to advantage. Fourier transformation of distributions, however, is used systematically; the student genuinely interested in PDE must make an effort to learn it. Not that much effort is needed, for it is such a smooth and simple theory: Excellent expositions are found in Chapter VII of L. Schwartz' book [TD], or in Chapter I of Hörmander's book [LPDO] (and in many other texts). In particular, the reader should be familiar with the Plancherel theorem and with the Paley–Wiener(–Schwartz) theorem. As far as linear functional analysis is concerned, the basic facts about Hilbert and Banach spaces must be known, but nothing much deeper—although, from the middle of the book on, an ever greater use is made of functions (and, later, distributions) with values in Banach spaces. Finally, it is presumed that the student has a fairly good knowledge of holomorphic functions of one complex variable, of real variable theory, mainly Lebesgue integration, and a smattering of measure theory. A bit of linear algebra will be of help, here and there.

There are 390 exercises, and several contain detailed information which should enable the reader to reconstruct the proofs of some important results: for example, the hypoellipticity of elliptic equations—of any order—with C^∞ coefficients, in Exercises 36.4 and 36.7, or the theorem of supports—in one variable—in Exercises 43.4, 43.5, and 43.6. Other exercises are simple variants or straightforward applications of the results and the methods in the text.

Notation

$\mathbf{R}^n$ product of n copies of the real line $\mathbf{R}$
$x^1, \ldots, x^n$ coordinates in $\mathbf{R}^n$; also $y^1, \ldots, y^n$, etc.
$x = (x^1, \ldots, x^n)$ the variable in $\mathbf{R}^n$; also y, etc.
$\mathbf{R}_n$ *dual* of $\mathbf{R}^n$
$\xi_1, \ldots, \xi_n$ coordinates in $\mathbf{R}_n$; also $\eta_1, \ldots, \eta_n$, etc.
$\xi = (\xi_1, \ldots, \xi_n)$ the variable in $\mathbf{R}_n$; also η, etc.
$x \cdot \xi = x^1\xi_1 + \cdots + x^n\xi_n$ the scalar product between a vector in $\mathbf{R}^n$ and a covector in $\mathbf{R}_n$; also $\langle x, \xi \rangle$
$\mathbf{Z}^n$ product of n copies of the set $\mathbf{Z}$ of integers (of all signs)
$\mathbf{Z}^n_+$ product of n copies of the set $\mathbf{Z}_+$ of nonnegative integers
$\mathbf{C}^n$ product of n copies of the complex plane $\mathbf{C}$
$z^1, \ldots, z^n$ complex coordinates in $\mathbf{C}^n$
$z = (z^1, \ldots, z^n)$ the variable in $\mathbf{C}^n$
$\mathrm{Re}\, z = (\mathrm{Re}\, z^1, \ldots, \mathrm{Re}\, z^n)$ the *real part* of the complex vector z
$\mathrm{Im}\, z = (\mathrm{Im}\, z^1, \ldots, \mathrm{Im}\, z^n)$ the *imaginary* part of z
$\mathbf{C}_n$ dual of $\mathbf{C}^n$
$\zeta_1, \ldots, \zeta_n$ coordinates in $\mathbf{C}_n$
$\zeta = (\zeta_1, \ldots, \zeta_n)$ the variable in $\mathbf{C}_n$
$z \cdot \zeta = z^1\zeta_1 + \cdots + z^n\zeta_n$ the (real) scalar product between $z \in \mathbf{C}^n$ and $\zeta \in \mathbf{C}_n$; also $\langle z, \zeta \rangle$
$|x| = \{(x^1)^2 + \cdots + (x^n)^2\}^{1/2}$ the *Euclidean norm* on $\mathbf{R}^n$
$|\xi|, |z|, |\zeta|$ the Euclidean norms on $\mathbf{R}_n$, $\mathbf{C}^n$, $\mathbf{C}_n$, respectively
$|\alpha| = \alpha_1 + \cdots + \alpha_n$ the *length* of the n-tuple $\alpha \in \mathbf{Z}^n_+$
$\alpha! = \alpha_1! \cdots \alpha_n!$, $\binom{\alpha}{p} = \binom{\alpha_1}{p_1} \cdots \binom{\alpha_n}{p_n}$, $\binom{\alpha_j}{p_j} = \alpha_j!/\beta_j!(\alpha_j - \beta_j)!$ where $\alpha, \beta \in \mathbf{Z}^n_+$ and $\alpha_j \geq \beta_j$ for every $j = 1, \ldots, n$
$d(x, A)$ Euclidean distance from the point x to the set A
$B_r(x)$ open ball centered at x, having radius r
sup *supremum*, or least upper bound, of a set of real numbers
inf *infimum*, or greatest lower bound, of a set of real numbers
ch A *convex hull* of a set A (contained in a linear space)
Ω an open subset of $\mathbf{R}^n$
$[a, b]$ a *closed* interval, with limit points a and b in the real line $\mathbf{R}^1$ (also when either $a = -\infty$ and/or $b = +\infty$)

$[a, b[$ semiclosed interval $a \le t < b$; $]a, b] = \{t \in \mathbf{R}^1; a < t \le b\}$
$]a, b[$ open interval $a < t < b$
$\complement A$ complement of A
$A \backslash B$ complement in A of the subset B of A
$a \mapsto f(a)$ mapping which to the object a assigns the value $f(a)$
$f: E \to F$ mapping f, defined in the set E, and valued in the set F
supp u the *support* of u (smallest closed set outside which $u \equiv 0$)
$\dfrac{\partial u}{\partial x^j}$ partial derivative of u with respect to x^j; also u_{x^j}, $\partial_{x^j} u$; also u' if there is only one variable

$D_j = -\sqrt{-1}\,\dfrac{\partial}{\partial x^j}$; also D_{x^j}

$(\partial/\partial x)^\alpha = (\partial/\partial x^1)^{\alpha_1} \cdots (\partial/\partial x^n)^{\alpha_n}$, $\quad (\alpha \in \mathbf{Z}^n_+)$
$D^\alpha = D_1^{\alpha_1} \cdots D_n^{\alpha_n}$
$dx = dx^1 \cdots dx^n$ the *Lebesgue measure* on $\mathbf{R}^n$; $d\xi = d\xi_1 \cdots d\xi_n$ the analog in $\mathbf{R}_n$ (both measures dx and $d\xi$ assign the volume 1 to the unit cube)
dz^1 the *line measure* in the complex plane (oriented counterclockwise)
$dz = dz^1 \cdots dz^n$ the product of n line measures in $\mathbf{C}^n$

$(f * g)(x) = \int_{\mathbf{R}^n} f(x - y)g(y)\, dy$ the value at x of the *convolution* of f and g

$\hat{u}(\xi) = \int_{\mathbf{R}^n} e^{-ix\cdot\xi} u(x)\, dx$ *Fourier transform* of u $\quad (i = \sqrt{-1})$

$\mathscr{F}u$ also Fourier transform of u

$\mathscr{F}^{-1}v(x) = (2\pi)^{-n} \int_{\mathbf{R}_n} e^{ix\cdot\xi} v(\xi)\, d\xi$ *inverse Fourier transform* of v

$P(x, \partial/\partial x) = \sum_{|\alpha| \le m} c_\alpha(x)(\partial/\partial x)^\alpha$ a linear partial differential operator of order m in Ω (generally with C^∞ coefficients c_α)

${}^tP(x, \partial/\partial x) = \sum_{|\alpha| \le m} (-\partial/\partial x)^\alpha c_\alpha(x)$ the (*formal*) *transpose* of $P(x, \partial/\partial x)$

$P(x, \partial x)^* = \sum_{|\alpha| \le m} (-\partial/\partial x)^\alpha \overline{c_\alpha(x)}$ the (*formal*) *adjoint* of $P(x, \partial/\partial x)$

$\Delta = \left(\dfrac{\partial}{\partial x^1}\right)^2 + \cdots + \left(\dfrac{\partial}{\partial x^n}\right)^2$ the Laplace operator (or Laplacian) in n variables

$\square = \left(\frac{\partial}{\partial t}\right)^2 - \Delta$ the wave operator (or *d'Alembertian*) in *n space* variables (t always denotes the *time* variable)

E, **F**, ... Hilbert spaces or Banach spaces over the complex numbers (sometimes over the real numbers)

x, **e**, ... elements of the Banach space **E**

$\|\mathbf{e}\|_{\mathbf{E}}$ *norm*, in the Banach space **E**, of the element **e**

$L(\mathbf{E}; \mathbf{F})$ space of bounded linear operators of **E** into **F**, equipped with the *operator norm*

$$\|A\| = \sup_{0 \neq \mathbf{e} \in \mathbf{E}} \|A\mathbf{e}\|_{\mathbf{F}}/\|\mathbf{e}\|_{\mathbf{E}}$$

$\mathbf{E}', \mathbf{E}^*$ *topological dual* of the Banach space **E**, the space of continuous linear functionals on **E** [equipped with the dual norm, which is the operator norm when one recalls that $\mathbf{E}' = L(\mathbf{E}; \mathbf{C})$]

$\langle \mathbf{e}^*, \mathbf{e} \rangle$ or $\langle \mathbf{e}, \mathbf{e}^* \rangle$ the *duality bracket* between $\mathbf{e} \in \mathbf{E}$ and $\mathbf{e}^* \in \mathbf{E}^*$

$\overline{\mathbf{E}}'$ *antidual* of the Banach space **E**, i.e., the space of continuous antilinear functionals on **E** (a mapping $u : \mathbf{E} \to \mathbf{F}$ is *antilinear* if $u(\lambda\mathbf{x} + \mu\mathbf{y}) = \bar{\lambda}u(\mathbf{x}) + \bar{\mu}u(\mathbf{y})$, $\forall \mathbf{x}, \mathbf{y} \in \mathbf{E}$, $\forall \lambda, \mu \in \mathbf{C}$)

$\langle \mathbf{x}^*, \mathbf{x} \rangle^-$ the bracket of the antiduality between **E** and $\overline{\mathbf{E}}'$

Main Spaces of Functions and Distributions

$C^0(\Omega)$ space of complex-valued functions, defined and continuous in the open set Ω, equipped with the topology of uniform convergence on the compact subsets of Ω

$C^0(\overline{\Omega})$ space of complex continuous functions on the closure $\overline{\Omega}$ (supposed to be compact), equipped with the maximum norm

$C^m(\Omega)$ space of complex functions, defined and m times continuously differentiable in Ω, equipped with the topology of uniform convergence on every compact subset of Ω, of the functions and of each one of their derivatives of order $< m + 1$ ($m \in \mathbf{Z}_+$ or $m = +\infty$)

$\mathscr{B}^m(\Omega)$ subspace of $C^m(\Omega)$ consisting of the functions having all their derivatives of order $< m + 1$ bounded in the whole of Ω, equipped with the topology of uniform convergence over Ω of all these derivatives

$C^m(\overline{\Omega})$ subspace of $\mathscr{B}^m(\Omega)$ consisting of the functions all of whose derivatives of order $< m + 1$ can be extended as continuous functions to the closure $\overline{\Omega}$ (supposed to be compact) of Ω, equipped with the topology induced by $\mathscr{B}^m(\Omega)$

$C_c^m(\Omega)$ subspace of $C^m(\Omega)$ consisting of the functions having *compact* support; elements of $C_c^\infty(\Omega)$ are often referred to as *test functions* in Ω

$C_c^m(K)$ subspace of $C_c^m(\mathbf{R}^n)$ whose support is contained in the compact set K, equipped with the topology induced by $C^m(\mathbf{R}^n)$

$L^p(\Omega)$ Lebesgue space of *measurable* functions f such that the pth power of the absolute value $|f|$ is integrable over Ω $(1 \le p < +\infty)$, equipped with the norm $\|f\|_{L^p(\Omega)} = (\int_\Omega |f(x)|^p\,dx)^{1/p}$ (actually f represents an equivalence class of functions equal almost everywhere)

$L^\infty(\Omega)$ Lebesgue space of (classes of) measurable functions f in Ω which are essentially bounded, equipped with the norm $\|f\|_{L^\infty(\Omega)}$, the *essential supremum* of f

$L^p_{\text{loc}}(\Omega)$ space of locally-L^p functions f in Ω [i.e., if K is any compact subset of Ω, the function f_K equal to f on K and to zero in $\Omega\backslash K$ belongs to $L^p(\Omega)$]

$\mathscr{D}'(\Omega)$ space of distributions in Ω

$\mathscr{E}'(\Omega)$ space of distributions with compact support in Ω

$H^{m,p}(\Omega)$ space of functions u in Ω such that $D^\alpha u \in L^p(\Omega)$ for all n-tuples $\alpha \in \mathbf{Z}_+$, $|\alpha| \leqq m$ (D^α denotes the distribution derivative); $H^{m,p}(\Omega)$ is the *Sobolev space*, equipped with the norm

$$\|u\|_{H^{m,p}(\Omega)} = \left\{\sum_{|\alpha|\le m} \|D^\alpha u\|^p_{L^p(\Omega)}\right\}^{1/p}$$

$H_0^{m,p}(\Omega)$ closure in $H^{m,p}(\Omega)$ of $C_c^\infty(\Omega)$

$H^{-m,p}(\Omega)$ space of distributions u in Ω which can be written as finite sums of derivatives of order $\le m$ of functions belonging to $L^p(\Omega)$ $(m \in \mathbf{Z}_+)$

$\mathscr{S}(\mathbf{R}^n)$ or $\mathscr{S}$ space of C^∞ functions u in $\mathbf{R}^n$ such that, for any pair of nonnegative integers k, M,

$$p_{k,M}(u) = \sup_{x\in\mathbf{R}^n}\left\{(1 + |x|^2)^k \sum_{|\alpha|\le M} |D^\alpha u(x)|\right\} < +\infty,$$

equipped with the topology defined by the seminorms $p_{k,M}$ ($\mathscr{S}$ is the space of C^∞ functions in $\mathbf{R}^n$ *rapidly decaying at infinity*)

$\mathscr{S}'(\mathbf{R}^n)$ or $\mathscr{S}'$ the dual of $\mathscr{S}$, also the space of *tempered distributions* in $\mathbf{R}^n$

$H^s(\mathbf{R}^n)$ or H^s the *Sobolev space of order* $s \in \mathbf{R}$ in $\mathbf{R}^n$, i.e., the space of tempered distributions u in $\mathbf{R}^n$ whose Fourier transform $\hat{u}$ is a measurable function such that

$$\|u\|_s = \left(\int_{\mathbf{R}_n} |\hat{u}(\xi)|^2 (1 + |\xi|^2)^s \frac{d\xi}{(2\pi)^n}\right)^{1/2} < +\infty,$$

equipped with the Hilbert space structure defined by the norm $\|\ \|_s$

$H^s(K)$	subspace of H^s consisting of the elements having their support contained in the compact set K
$H^s_c(\Omega)$	space of distributions in Ω which belong to the space $H^s(K)$ for some choice of the compact subset K of Ω
$H^s_{loc}(\Omega)$	space of distributions u in Ω such that $\alpha u \in H^s$ for every $\alpha \in C^\infty_c(\Omega)$
$\mathscr{M}(\Omega)$	space of *Radon measures* in $\Omega[\mathscr{M}(\Omega)$ is the dual of $C^0_c(\Omega)]$
$\mathscr{D}'_+(\mathbf{R}^1)$ or $\mathscr{D}'_+$	space of distributions on the real line which vanish identically in the open negative half-line

Note: When $\Omega = \mathbf{R}^n$, $(\mathbf{R}^n)$ will often be omitted, e.g., as in C^∞, $\mathscr{D}'$, H^s, L^p, etc.

BASIC
LINEAR PARTIAL
DIFFERENTIAL
EQUATIONS

CHAPTER I

The Basic Examples of Linear PDEs and Their Fundamental Solutions

I

The Basic Examples of Linear PDEs

The theory of linear PDEs stems from the intensive study of a few special equations, whose importance was recognized in the eighteenth and nineteenth centuries. These were the basic equations in mathematical physics (gravitation, electromagnetism, sound propagation, heat transfer, and quantum mechanics). After their introduction in applied mathematics, they were shown to play important roles in pure mathematics: For instance, the Laplace equation was first studied as the basic equation in the theory of Newton's potential and in electrostatics; later, suitably reinterpreted, it was used to study the geometry and topology of Riemannian manifolds. Similarly, the heat equation was studied by Fourier in the context of heat transfer. Later it was shown to be related to probability theory. One of the basic examples, which we describe below, does not seem to have originated in applications to physics: the Cauchy–Riemann operator, which is used to define analytic functions of a complex variable. But to my knowledge, all the remaining ones have their origin in applied mathematics. At any rate, the general theory of linear PDEs is an elaboration of the respective theories of these special operators. During the twentieth century it was recognized that many properties which had seemed to be the prerogative of the Laplace equation or of the wave equation could in fact be extended to wide classes of equations. These properties usually center around a question or a problem that only makes sense for one or the other equation: for instance, around the Dirichlet problem, which makes sense for the Laplace equation but not really for the wave equation, or the Cauchy problem, which is well posed for the latter but not for the former. The purpose of this introductory course is to help the student to understand some of these problems and some of their solutions—but always by staying very close to the special equation for which they were originally considered. It is therefore necessary that we have the nature of the basic examples clearly in mind.

1.1 The Laplace Equation in $n > 1$ Variables

Let us denote by $x = (x^1, \ldots, x^n)$ the variables in the Euclidean space $\mathbf{R}^n$. Usually the *Laplace operator* is

$$\Delta = \left(\frac{\partial}{\partial x^1}\right)^2 + \cdots + \left(\frac{\partial}{\partial x^n}\right)^2.$$

Some people call the Laplace operator that which in our notation would be $-\Delta$. They have very good reasons to do this; it is a pity that historical custom is not on their side, but they are gaining ground. Indeed, $-\Delta$ is a positive operator; its Fourier transform is the square of the norm of the variable in $\mathbf{R}_n$, $|\xi|^2$. The latter remark underlines the close relationship between the Laplace operator and the Euclidean norm, the spheres in the Euclidean space, the orthogonal transformations, and so on. Indeed, Δ is *invariant under orthogonal transformations*; that is, if T is any such transformation in $\mathbf{R}^n$ and f any infinitely differentiable function of x, then

$$(1.1) \qquad (\Delta f)(Tx) = \Delta\{f(Tx)\}, \qquad x \in \mathbf{R}^n.$$

This, of course, is a crucial symmetry property of the Laplace operator and is part of the reason for its role in the description of many phenomena in *isotropic* media.

As a matter of fact, any linear transformation T of $\mathbf{R}^n$ such that (1.1) holds for all C^∞ functions f must be orthogonal: The orthogonal transformations are exactly those which leave Δ invariant (i.e., which commute with Δ).

The functions that satisfy the homogeneous Laplace equation

$$(1.2) \qquad \Delta h = 0$$

are called *harmonic functions.*

1.2 The Wave Equation

For reasons which will become clear when we begin using the Fourier transformation, it is convenient to replace the partial differentiations $\partial/\partial x^j$ by purely imaginary variables $\sqrt{-1}\,\xi_j$ $(j = 1, \ldots, n)$. Thus the operator $-\Delta$ becomes

$$(1.3) \qquad |\xi|^2 = \xi_1^2 + \cdots + \xi_n^2$$

which is a positive-definite quadratic form. Its *signature* is $(n, 0)$: It has n positive eigenvalues and no nonpositive ones. We may also look at quadratic forms with different signatures. An important case is the form with all eigenvalues strictly positive except one which is strictly negative. For various reasons it is convenient to consider such a form on an $(n + 1)$-dimensional space $\mathbf{R}_{n+1}$ where the variables are denoted by $(\xi_1, \ldots, \xi_n, \tau)$. It is essentially the form

$$|\xi|^2 - \tau^2 = \xi_1^2 + \cdots + \xi_n^2 - \tau^2 \tag{1.4}$$

corresponding to the partial differential operator in $\mathbf{R}^{n+1}$ (where the variables are denoted by $x^1, \ldots, x^n, t$):

$$\Box = \frac{\partial^2}{\partial t^2} - \Delta_x = \frac{\partial^2}{\partial t^2} - \left(\frac{\partial}{\partial x^1}\right)^2 - \cdots - \left(\frac{\partial}{\partial x^n}\right)^2.$$

This is the *wave operator* (sometimes called the *d'Alembertian*): The x^j's are called the *space* variables, and t is the *time* variable. It is the operator used to describe oscillatory phenomena and wave propagation.

If we are interested in those linear transformations of $\mathbf{R}^{n+1}$ which commute with $\Box$, we will have no trouble in determining what they are. Of course, they are the same as the linear transformations in (the dual space) $\mathbf{R}_{n+1}$ which leave the quadratic form (1.4) invariant. They form a group much used in physics since the advent of relativity: the Lorentz group.

The solutions of the wave equation

$$\Box g = 0 \tag{1.5}$$

have properties that are radically different from those of the Laplace equation, as will become clear when we take a closer look at them.

1.3 The Heat Equation

The examples given in §1.1 and §1.2 are both *homogeneous second-order* differential operators, that is, differential operators which involve second-order partial differentiations and none of order $\neq 2$. The *heat operator* in $\mathbf{R}^{n+1}$,

$$\frac{\partial}{\partial t} - \Delta_x, \tag{1.6}$$

is not of this type. It is used to describe various transfer phenomena, like the transfer of heat in isotropic media. At first glance the heat and the wave equations look alike, and indeed they have some properties in common. But there are also very deep differences. No wave propagation phenomena are associated with the solutions of the heat equation; phenomena of the diffusion type are. As a matter of fact, there is some similarity with the Laplace equation. It should not come as a surprise: The *leading terms* in the heat equation, that is, the second-order partial derivatives, are the same as in the Laplace equation in space variables.

We have just seen what are probably the most important examples of linear partial differential equations. The Laplace equation is the archetype of a large class of equations, called the *elliptic* PDEs. The reason for this is obvious: If we look at the quadratic form

$$a_1\xi_1^2 + a_2\xi_2^2, \qquad a_1 > 0, \quad a_2 > 0,$$

it is equal, up to a change of scale, to the *symbol* (1.3) of the Laplace operator in *two* variables. It is also the function in $\mathbf{R}_2$ whose level curves are ellipses.

Similarly, the wave equation is the archetype of the *hyperbolic* PDEs: The level curves of the function $\xi^2 - \tau^2$ in $\mathbf{R}_2$ are the standard hyperbola. The heat equation is the archetype of the *parabolic* PDEs: Its symbol can be defined as being the function $\xi^2 - \tau$ in $\mathbf{R}_2$ whose level curves are the standard parabola. As a matter of fact, again in view of our use of the Fourier transformation, we prefer to define its symbol as $|\xi|^2 + i\tau$, replacing $\partial/\partial t$ by $i\tau$ rather than by $-\tau$.

This has been the classical way of categorizing partial differential equations, when only those of first and second order were studied by mathematicians. It is quite inadequate to classify *systems* of PDEs, higher order equations, or equations with complex coefficients. It turns out that some of the essential properties of the Laplace equation follow from the fact that its symbol (1.3) only vanishes at the origin—and not from the fact that it is a positive-definite quadratic form. In other words, these properties subsist in other equations which partake of the former characteristic but not of the latter. This is the case of the equation we study next.

1.4 The Cauchy–Riemann Equation

Let x, y denote the variables in the plane $\mathbf{R}^2$. The homogeneous Cauchy–Riemann equation reads

$$\frac{\partial f}{\partial x} + \sqrt{-1}\,\frac{\partial f}{\partial y} = 0. \tag{1.7}$$

Here $f = u + iv$ is a complex-valued differentiable function (u, v are real). Equation (1.7) is equivalent to the system

$$\frac{\partial u}{\partial x} = \frac{\partial v}{\partial y}, \qquad \frac{\partial v}{\partial x} = -\frac{\partial u}{\partial y}. \tag{1.8}$$

Let us set $z = x + iy$, $\bar{z} = x - iy$, or, equivalently, $x = \frac{1}{2}(z + \bar{z})$, $y = (1/2i)(z - \bar{z})$. Thus any function such as $f(x, y)$ in a subset of $\mathbf{R}^2$ can also be viewed as a function of $(z, \bar{z})$. Equation (1.7) can then be rewritten (by the chain rule of differentiation) as

$$\frac{\partial f}{\partial \bar{z}} = 0. \tag{1.9}$$

We have set

$$\frac{\partial}{\partial \bar{z}} = \frac{1}{2}\left(\frac{\partial}{\partial x} + \sqrt{-1}\,\frac{\partial}{\partial y}\right).$$

Roughly speaking, (1.9) tells us that f is "independent of $\bar{z}$"; more precisely, it states that f (supposed to be sufficiently smooth) is an analytic function of z, i.e., has a complex derivative [at every point where (1.9) holds]. It is convenient to introduce also the "anti-Cauchy–Riemann" operator

$$\frac{\partial}{\partial z} = \frac{1}{2}\left(\frac{\partial}{\partial x} - \sqrt{-1}\,\frac{\partial}{\partial y}\right).$$

Note that

$$4\,\frac{\partial^2}{\partial z\,\partial \bar{z}} = \Delta, \tag{1.10}$$

the Laplace operator in two variables. The identity (1.10) points to strong relations between the Laplace and the Cauchy–Riemann equations. These will be confirmed when we study them. The symbol of $\partial/\partial \bar{z}$ is

$$\frac{i}{2}(\xi + i\eta) \tag{1.11}$$

(we have denoted by ξ, η the variables in the dual plane $\mathbf{R}_2$). Note that, like the symbol of the Laplace operator, it only vanishes at the origin. This property will have important consequences. Because of it, in the modern terminology, the Cauchy–Riemann operator is also said to be *elliptic*.

1.5 The Schrödinger Equation

In the study of partial differential equations one is quickly taught to expect important and deep implications to follow from merely formal differences. This is confirmed by everything that follows and is well exemplified by the theories of the heat and of the Schrödinger equations. The Schrödinger operator *with constant coefficients* in n-space variables is

$$\frac{1}{i}\frac{\partial}{\partial t} - \Delta_x. \tag{1.12}$$

The only difference with the heat operator is the presence of the factor i^{-1} in front of $\partial/\partial t$. Yet the solutions of the two equations exhibit very different kinds of behavior, as will be seen later. The Schrödinger equation was originally introduced to describe the behavior of the electron and other elementary particles. It has the defect of not being Lorentz-invariant and therefore of not fitting in the relativistic formulation of quantum mechanics. It is still used as an approximation, but in a more rigorous setup, it has been replaced by *Dirac's equations*.

So far we have only looked at examples of a single, or *scalar*, linear partial differential equation. But there are many important (for mathematics and for physics) examples of *systems* of equations. This means that we are given $N_1 N_2$ linear partial differential operators P_{jk} ($j = 1, \ldots, N_1$, $k = 1, \ldots, N_2$) and that we consider the N_1 equations in N_2 unknown functions u^k

$$\sum_{k=1}^{N_2} P_{jk} u^k = f_j, \qquad j = 1, \ldots, N_1. \tag{1.13}$$

The system (1.13) is said to be *determined* if $N_1 = N_2$, that is, if there are exactly as many equations as there are unknowns; *overdetermined* if $N_1 > N_2$, that is, if there are strictly more equations than unknowns; and *underdetermined* if there are strictly fewer equations than unknowns. The theory of systems is more difficult than the theory of single equations, especially the theory of overdetermined systems. At this stage we shall content ourselves with some examples. The Maxwell equations, on which classical electromagnetism is based, constitute an example of a determined system, as are the Dirac equations, alluded to above. Both are *hyperbolic systems*. Without getting into the technicalities of the definition, let us say that hyperbolic systems have formal and nonformal properties closely related to those of the wave equation. We next give some examples of systems of linear PDEs which are not determined.

1.6 The Gradient

Let $n > 1$ denote the number of independent variables x^j. The gradient,

$$\operatorname{grad} = \left(\frac{\partial}{\partial x^1}, \ldots, \frac{\partial}{\partial x^n}\right), \tag{1.14}$$

is an overdetermined system of differential operators: The number N_1 above is equal to n, $N_2 = 1$. The system of equations (1.13) reads here

$$\frac{\partial u}{\partial x^j} = f_j, \qquad j = 1, \ldots, n. \tag{1.15}$$

The study of overdetermined systems is beset with difficulties which are absent in the study of single equations; they are algebraic in nature and show up even in cases as simple as (1.15). Indeed, if the equations (1.15) hold (assuming that the functions $u, f_1, \ldots, f_n$ are sufficiently smooth), we derive from them

$$\frac{\partial f_j}{\partial x^k} = \frac{\partial^2 u}{\partial x^j \, \partial x^k} = \frac{\partial f_k}{\partial x^j}.$$

In other words, if we regard $(f_1, \ldots, f_n)$ as a vector-valued function $\mathbf{f}$, we see that (1.15) implies

$$\operatorname{curl} \mathbf{f} = 0. \tag{1.16}$$

Equation (1.16) is a system of $n(n-1)/2$ equations, which are called the *compatibility conditions* for the system (1.15). It is not difficult to see that if (1.16) holds, one can in fact solve (1.15)—at least locally.

The gradient is an *elliptic* (nonscalar) differential operator. We cannot go into the exact meaning of this, but it points to a certain analogy with the Laplace operator (see below).

1.7 The Divergence

The divergence operator acts on functions which are defined in a subset of $\mathbf{R}^n$ and are valued in $\mathbf{C}^n$ (most of the time, we shall deal with complex-valued scalar functions). The system (1.13) defined by the divergence operator reads

$$\frac{\partial u^1}{\partial x^1} + \frac{\partial u^2}{\partial x^2} + \cdots + \frac{\partial u^n}{\partial x^n} = f \tag{1.17}$$

and $N_1 = 1$, $N_2 = n$. If we set $\mathbf{u} = (u^1, \ldots, u^n)$, the left-hand side of (1.17) is usually denoted by div $\mathbf{u}$. Reasoning very formally we see at once that (1.17) admits many solutions. Fix j arbitrarily and take u^j to be any primitive of f with respect to x^j while taking $u^k \equiv 0$ for $k \neq j$; this gives us a solution. This kind of procedure works for all underdetermined systems: by setting $N_2 - N_1$ unknowns equal to zero, we reduce it to a determined system. This is a familiar device of elementary linear algebra; it shows that underdetermined systems are much easier to study than overdetermined ones (and in most cases, than determined ones also); they are also much less interesting.

Let $\mathbf{u} = (u^1, \ldots, u^n)$ denote a C^∞ function defined in the open subset Ω of

$\mathbf{R}^n$ and valued in $\mathbf{C}^n$, v a C^∞ function with compact support in Ω, complex-valued. By integration by parts we see at once that

$$\int (\operatorname{div} \mathbf{u}) v \, dx = -\int \langle \mathbf{u}, \operatorname{grad} v \rangle \, dx \tag{1.18}$$

where $\langle \mathbf{u}, \operatorname{grad} v \rangle = \sum_{k=1}^{n} u^k \, \partial v/\partial x^k$ is the standard scalar product between vectors (more precisely, between vectors and covectors). Formula (1.18) can be stated by saying that the operators $-\operatorname{div}$ and grad are the *transpose* of each other. Another well-known formula is

$$\operatorname{div} \operatorname{grad} = \Delta, \tag{1.19}$$

which underlines the relationship between gradient, divergence, and Laplace operator.

There are other examples of overdetermined systems of differential operators which are important in mathematics and its applications: We have come across one of them above, the *curl*. Another one is the complex gradient $\bar{\partial}$:

$$\bar{\partial} = (\partial/\partial \bar{z}^1, \ldots, \partial/\partial \bar{z}^n).$$

The system of homogeneous equations

$$\bar{\partial} u = 0 \tag{1.20}$$

is called *the Cauchy–Riemann equations* in several variables. The solutions of (1.20) (where u is supposed to be once continuously differentiable) are the holomorphic functions of the n complex variables $z^1, \ldots, z^n$, whose theory has been vigorously developed in recent years.

In addition to the preceding examples we shall often refer to *ordinary differential equations* (ODEs) as a particular case of PDEs; of course, this is simply the case where the number of independent variables is $n = 1$. The theory of linear ODEs is known in much greater detail than that of PDEs. Some of the properties of ODEs can be generalized to PDEs; other properties can be exploited to construct solutions of certain PDEs. This will be seen in the subsequent discussions.

Exercises

1.1. Using polar coordinates, write the expression of the Laplace operator in two variables and that of the Cauchy–Riemann operator.

1.2. Write the expression of the Laplace operator in three variables in spherical and in cylindrical coordinates.

1.3. Consider a linear partial differential operator of order 1, with constant complex coefficients, in n variables:

$$L = \alpha^1 \frac{\partial}{\partial x^1} + \cdots + \alpha^n \frac{\partial}{\partial x^n}.$$

Prove the following statements:

(1) If the complex vector $\boldsymbol{\alpha} = (\alpha^1, \ldots, \alpha^n)$ is of the form $z\mathbf{a}$ where $\mathbf{a}$ is a real vector and z is some complex number, we may perform a linear change of variables $x \to y$ in $\mathbf{R}^n$ such that the expression of L in the y coordinates becomes $z\, \partial/\partial y^1$.

(2) Suppose now that the vectors $\operatorname{Re} \boldsymbol{\alpha}$ and $\operatorname{Im} \boldsymbol{\alpha}$ are linearly independent (this demands $n > 1$); show that we may now perform a linear change of variables $x \to y$ in $\mathbf{R}^n$ such that the expression of L in the y coordinates becomes

$$\frac{1}{2}\left(\frac{\partial}{\partial y^1} + \sqrt{-1}\, \frac{\partial}{\partial y^2}\right).$$

1.4. Consider a linear partial differential operator of order 1, with C^∞ *real* coefficients in a neighborhood of the origin in $\mathbf{R}^n$:

$$L = \sum_{j=1}^{n} \alpha^j(x) \frac{\partial}{\partial x^j}.$$

Suppose that at least one of the coefficients α^j does not vanish at the origin. Show that there is a C^∞ change of variables $x \to y$ in a neighborhood of the origin such that the expression of L in the y coordinates is $\omega(y)\, \partial/\partial y^1$ with $\omega(0) \neq 0$. Is it always possible to choose the coordinates y so as to have $\omega(y) \equiv 1$ near $y = 0$?

1.5. Let $\Box = (\partial^2/\partial x^2) - (\partial^2/\partial y^2)$ be the wave operator in the plane. Show that there is a linear change of variables in $\mathbf{R}^2$ transforming $\Box$ into $4\, \partial^2/\partial x'\, \partial y'$. Use this fact to prove that all the solutions of the wave equation in the plane, $\Box u(x, y) = 0$, are of the form $u(x, y) = f(x + y) + g(x - y)$, where f and g are functions on the real line (the student may assume that u is twice continuously differentiable).

Comparing with the property of the Laplacian in the plane, expressed in formula (1.10), is it true that every harmonic function in two variables is of the form $u(x, y) = f(x + iy) + g(x - iy)$, where f and g are holomorphic functions of one complex variable?

1.6. Prove that, if f is a C^1 function, defined in the closed unit ball $\bar{B}_1 = \{x \in \mathbf{R}^n;\ |x| \leq 1\}$, valued in $\mathbf{C}^n$, which satisfies (1.16), the system (1.15) always has a C^1 solution u, defined in $\bar{B}_1$ (and valued in $\mathbf{C}$). Show by an example that this is not so if $n = 2$ and if one replaces the closed unit disk

$\bar{B}_1 \subset \mathbf{R}^2$ by the semiclosed annulus $0 < x^2 + y^2 \leqq 1$. Relate this fact with the fact that the anti-Cauchy–Riemann equation $\partial u/\partial z = f$ does not always have a solution u which is a holomorphic function of z in that annulus, when f itself is holomorphic (say in the complement of the origin in the plane).

1.7. Consider the heat equation in one space variable:

$$\frac{\partial u}{\partial t} = \frac{\partial^2 u}{\partial x^2} \qquad (\text{in } \mathbf{R}^2). \tag{1.21}$$

Describe all solutions of the form $u(x, t) = v(x)w(t)$ (the student may assume that v and w are infinitely differentiable).

Apply the same method to the wave equation and to the Schrödinger equation (also in one space variable) and compare the results in the three cases.

1.8. Consider the wave equation in the plane:

$$\frac{\partial^2 u}{\partial t^2} = \frac{\partial^2 u}{\partial x^2}. \tag{1.22}$$

Let $u_0(x)$, $u_1(x)$ be two C^∞ functions in the real line, vanishing outside the interval $|x| \leqq 1$. By exploiting the description of all the solutions of (1.22) given in Exercise 1.5, show that there is a unique solution u of (1.22) such that

$$u(x, 0) = u_0(x), \qquad u_t(x, 0) = u_1(x). \tag{1.23}$$

Suppose that $u_1 \equiv 0$ and that $u_0(x) > 0$ for $|x| < 1$. What is the support of the solution u? What is the region in the plane in which the solution $u(x, t) > 0$?

1.9. Let $u_0(x)$ be a C^∞ *periodic* function on the real line. One may then write

$$u_0(x) = \sum_{p=-\infty}^{+\infty} u_{0,p} e^{ip\omega x}, \tag{1.24}$$

where $\omega = 2\pi/T$, T the period of u_0, with Fourier coefficients $u_{0,p}$ whose absolute value tends to zero, as $|p| \to \infty$, faster than any power of $1/|p|$. Admitting these facts (if he does not know them) the student should try to exploit the method of separation of variables indicated in Exercise 1.7, in order to solve the initial value problem for (1.21), that is, find a solution u of that equation, which satisfies $u(x, 0) = u_0(x)$.

1.10. Give necessary and sufficient conditions on the function $u_0(x)$, defined and C^∞ in the open interval $|x| < 1$, in order that there be a solution $u(x, t)$ to the initial value problem:

$$\frac{\partial u}{\partial t} = i\frac{\partial u}{\partial x} \qquad \text{in the region} \quad x^2 + t^2 < 1, \tag{1.25}$$

$$u(x, 0) = u_0(x) \qquad \text{in the interval} \quad |x| < 1. \tag{1.26}$$

1.11. Let Ω be an open subset of $\mathbf{R}^n$, $P(x, \partial/\partial x)$ a linear partial differential operator with complex coefficients, defined and C^∞ in Ω, Σ a space of distributions in Ω [that is, a linear subspace of $\mathscr{D}'(\Omega)$, equipped with a locally convex topology finer than the one induced by $\mathscr{D}'(\Omega)$]. Prove that the linear subspace $\mathrm{Ker}_\Sigma\, P$ consisting of the distributions, belonging to Σ, which satisfy the homogeneous equation $P(x, \partial/\partial x)h = 0$ in Ω, is closed in Σ.

1.12. Let Ω and $P(x, \partial/\partial x)$ be as in Exercise 1.11 and write

$$P(x, \partial/\partial x) = \sum_{|\alpha| \leq m} c_\alpha(x)(\partial/\partial x)^\alpha,$$

where $(\partial/\partial x)^\alpha = (\partial/\partial x^1)^{\alpha_1} \cdots (\partial/\partial x^n)^{\alpha_n}$ and $|\alpha| = \alpha_1 + \cdots + \alpha_n$. Let x_0 be a point in Ω. Prove the following statement:

LEMMA 1.1. *If there is an n-tuple α with length $|\alpha| = m$ such that $c_\alpha(x_0) \neq 0$, the homogeneous equation $P(x, \partial/\partial x)h = 0$ in Ω does not have any nonzero distribution solution whose support consists of the single point x_0.*

[*Hint*: Use the fact that any distribution u whose support is $\{x_0\}$ is a finite linear combination of derivatives of the Dirac measure at x_0:

$$u = \sum_{|\beta| \leq m'} a_\beta\, \delta^{(\beta)}(x - x_0).]$$

2

Existence and Smoothness of Solutions Not Submitted to Side Conditions

Since we now take a closer look at linear partial differential operators, it is time to adopt convenient notation and terminology. As it is now universally accepted, we adopt the multi-index notation. A linear partial differential operator in n independent variables $x^1, \ldots, x^n$, with complex coefficients defined in an open subset Ω of $\mathbf{R}^n$, is a polynomial in the partial differentiations and has the form

$$P(x, \partial/\partial x) = \sum_{|\alpha| \leqq m} c_\alpha(x)(\partial/\partial x)^\alpha. \tag{2.1}$$

Here α is a *multi-index*, that is, an n-tuple of integers $\alpha_j \geqq 0$; $|\alpha|$ denotes its *length* $\alpha_1 + \cdots + \alpha_n$. Also $(\partial/\partial x)^\alpha = (\partial/\partial x^1)^{\alpha_1} \cdots (\partial/\partial x^n)^{\alpha_n}$. The integer m is usually the *order* of the operator; this assumes that for some multi-index α with length $|\alpha| = m$, the coefficient $c_\alpha(x)$ is not identically equal to zero. If the coefficients c_α are constant throughout Ω, we write $P(\partial/\partial x)$ instead of $P(x, \partial/\partial x)$. Because of the usefulness of the Fourier transformation, one often prefers to deal with the elementary operators

$$D_j = -\sqrt{-1}\, \partial/\partial x^j, \qquad j = 1, \ldots, n,$$

rather than $\partial/\partial x^j$. We consider then differential operators of the form

$$P(x, D) = \sum_{|\alpha| \leqq m} c_\alpha(x) D^\alpha \tag{2.2}$$

rather than of the form (2.1). A typical differential operator with constant coefficients is denoted by $P(D)$.

When passing from the study of ordinary differential equations to partial differential equations, the first novelty one notices is that all the examples of

PDEs one encounters have infinitely many linearly independent solutions. Take, for instance, a homogeneous PDE with constant coefficients

$$P(D)u = 0. \tag{2.3}$$

For suitable complex vectors ζ, we may take $u = \exp(i\langle\zeta, x\rangle)$, where

$$\langle\zeta, x\rangle = \zeta_1 x^1 + \cdots + \zeta_n x^n.$$

Indeed,

$$P(D)e^{i\langle\zeta, x\rangle} = P(\zeta)e^{i\langle\zeta, x\rangle}$$

and it suffices to require ζ to be a *zero* of the polynomial P, i.e.,

$$P(\zeta) = 0. \tag{2.4}$$

But a polynomial on $\mathbf{C}^n$ always has infinitely many distinct zeros, provided that $n > 1$, and if $\zeta \neq \zeta'$, the functions $\exp(i\langle\zeta, x\rangle)$ and $\exp(i\langle\zeta', x\rangle)$ are linearly independent.

Let us give another example, the equation (with variable coefficients) in the plane

$$x\frac{\partial u}{\partial y} - y\frac{\partial u}{\partial x} = 0. \tag{2.5}$$

If we switch to polar coordinates, r, θ, (2.5) can be written

$$\frac{\partial u}{\partial \theta} = 0. \tag{2.6}$$

We get a solution of (2.6) by taking u to be any (once continuously differentiable) function of r alone, that is, rotation-invariant. Here again there are many linearly independent solutions.

Among the wealth of solutions of a partial differential equation, the mathematician will select a restricted number, often only one, which satisfy certain "side conditions." Let us briefly describe two of the most important examples:

Example 2.1. Let Ω be an open subset of $\mathbf{R}^3$, bounded, whose boundary is a smooth surface S. The Laplace equation $\Delta h = 0$ in Ω has infinitely many linearly independent solutions. But we may consider only those which take preassigned values on S. Given a function f defined on S (and not necessarily anywhere else), we may seek a harmonic function h in Ω which is equal to f on S. Under suitable conditions, there will be one and only one such function.

Example 2.2. Consider the wave equation

$$(\partial/\partial x)^2 u - (\partial/\partial y)^2 u = 0 \tag{2.7}$$

in the plane $\mathbf{R}^2$. Any function $u(x, y) = f(x + y) + g(x - y)$ is a solution of (2.7), but we may want to look only at those solutions such that u and $(\partial/\partial y)u$ take preassigned values when $y = 0$, say $u_0(x)$, $u_1(x)$, respectively. Under suitable assumptions of smoothness, we may write

$$f(x) + g(x) = u_0(x), \tag{2.8}$$

$$f'(x) - g'(x) = u_1(x). \tag{2.9}$$

If we denote by $U_1(x)$ any primitive of $u_1(x)$, we derive from (2.9):

$$f(x) - g(x) = U_1(x), \tag{2.10}$$

whence, by combining with (2.8),

$$f = \tfrac{1}{2}(u_0 + U_1), \qquad g = \tfrac{1}{2}(u_0 - U_1).$$

It is verified at once that the value of $u(x, y)$ thus obtained is independent of the choice of the primitive U_1 of u_1. We have therefore obtained a solution of (2.7) satisfying our side conditions. One can show, as we do eventually, that this solution is unique.

This type of problem with side conditions leads immediately to problems of existence and uniqueness: Does there exist a solution to $\Delta h = 0$ in Ω equal to f on S? And if there is at least one solution, how many are there? We have only looked at homogeneous equations, i.e., at equations where the right-hand side is equal to zero, but exactly the same considerations apply to inhomogeneous equations, where the right-hand side is a given function or distribution. But then it is only natural that before tackling the much more difficult problems with side conditions, we wish to know more about the solutions of the equation when no side conditions are imposed. For instance, there are linear partial differential equations with variable coefficients which do not have solutions at all (another difference with ordinary differential equations!). In such cases, it makes little sense to discuss problems with side conditions. It is thus somewhat more urgent that we answer the question: Does the equation

$$Pu = f$$

have solutions? Unfortunately, this is a very difficult question to answer in general, although it might not be too difficult to answer in particular cases or for particular classes of equations (once it has been more precisely formulated). One such class for which a fairly satisfactory answer can be given is that of linear PDEs with *constant* coefficients.

2.1 Existence of Solutions of Linear PDEs with Constant Coefficients in Bounded Sets

We could also discuss the problem in unbounded sets but it would then require some machinery that is not too elementary. The discussion in bounded sets, on the contrary, can be carried through with relatively elementary machinery (assuming that the student has some basic knowledge of distribution theory). We consider the equation

(2.11) $$P(D)u = f$$

in a *bounded* set $\Omega \subset \mathbf{R}^n$. As stated before, $P(\xi)$ is a polynomial in n variables with complex coefficients; we have substituted $D_j = -\sqrt{-1}\,\partial/\partial x^j$ for the variable ξ_j. We assume that there is a distribution E in $\mathbf{R}^n$ such that

(2.12) $P(D)E = \delta,$ the Dirac distribution at the origin.

First of all, there is a theorem which states that this is indeed the case, whatever the polynomial P (except in the uninteresting case where all its coefficients are equal to zero). Secondly, in all the cases we discuss, we shall effectively *construct*, i.e., give an explicit expression of E. The distribution E is called a *fundamental solution* of $P(D)$; a linear PDE in general possesses many fundamental solutions. For take arbitrarily one of them, E_0, and add to it any solution h of the homogeneous equation (2.3): $E_0 + h$ is again a fundamental solution (fundamental solutions form a linear manifold, not passing through the origin, of the space of all distributions in $\mathbf{R}^n$).

Let us now return to (2.11). We shall make the assumption that if we extend f by setting it equal to zero outside of Ω, we obtain a distribution, which we shall denote by $\tilde{f}$, in the whole space $\mathbf{R}^n$. It is not true that we can always do this. For instance, if $n = 1$ and Ω is the open interval $]0, 1[$, take

$$f = \exp(+1/x).$$

Then the extension of f which we defined above does not yield a distribution. If our assumption about f were to be eliminated, it would not be possible to handle arbitrary bounded open sets, although it would still be possible to solve our problem for some of them (but by different methods). How inconvenient is the assumption? In practice, not much: Any f belonging to a space $L^p(\Omega)$, $1 \leqq p \leqq +\infty$, is "extendable" in the way described—as a matter of fact, any distribution in Ω which does not grow at the boundary $\partial\Omega$ of Ω ("grow" in a suitable sense, adapted to distributions) faster than some power of the inverse of the distance to $\partial\Omega$ is extendable.

We use both hypotheses—the boundedness of Ω and the extendability of f—via the property that $\tilde{f}$ has *compact support* in $\mathbf{R}^n$ (the support of $\tilde{f}$ is contained in the closure $\bar{\Omega}$ of Ω). Now we may convolve two distributions, one of which, at least, has compact support—such as E and $\tilde{f}$. This yields a distribution $E * \tilde{f}$ in $\mathbf{R}^n$ which is not, in general, compactly supported. Now, in order to differentiate a convolution like $E * \tilde{f}$, we are free to differentiate any one of the factors. Thus,

$$P(D)(E * \tilde{f}) = [P(D)E] * \tilde{f} = \delta * \tilde{f} = \tilde{f}.$$

In particular, if we call u the restriction to Ω of the distribution in $\mathbf{R}^n$, $E * \tilde{f}$, we see that (2.11) is satisfied.

If we can content ourselves with solving (2.11) in a smaller set than Ω, more precisely in a relatively compact open subset Ω' of Ω, we can dispense with the extendability assumption about the right-hand side f. We arbitrarily select a C^∞ function g with compact support in Ω, equal to one in Ω' and take $u = E * (gf)$. It is easily checked that (2.11) is satisfied in Ω'.

Once fundamental solutions have been introduced, for example in the process of solving (2.11), it is soon discovered that they can be put to several uses. One of these is related to the *smoothness* of the solutions of (2.11). Let us take another look at ordinary differential equations, this time with constant coefficients. Suppose, in fact, that (2.11) represents such an equation. Suppose moreover that the right-hand side f is a C^∞ function in Ω. Then, every solution u of (2.11) is a C^∞ function. This is a striking property and we may ask whether it persists when $P(D)$ is a linear partial differential operator. The answer in general is that it does not. In order to see this, it suffices to return to Example 2.2: The right-hand side, the function zero, is certainly C^∞. But this obviously is not the case for $u(x, y) = f(x + y) + g(x - y)$ when either f or g is not a C^∞ function of one variable. Could it happen, however, for certain special equations? Now the answer is yes. It is, in fact, true in three of our basic examples: the Laplace, Cauchy–Riemann, and heat equations. Let us tackle the problem by means of the fundamental solution, but first let us introduce a definition:

Definition 2.1. The linear partial differential operator P in Ω is said to be hypoelliptic if, given any open subset U of Ω and any distribution u in U, u is a C^∞ function in U if this is true of Pu.

Suppose that P has constant coefficients, and let E be any fundamental solution of $P = P(D)$. By (2.12) we see that $P(D)E = 0$ in the complement of the origin in $\mathbf{R}^n$, $\mathbf{R}^n\backslash\{0\}$. Therefore, if $P(D)$ is hypoelliptic, E must be a C^∞ function in $\mathbf{R}^n\backslash\{0\}$. But a classical theorem of L. Schwartz states that the converse is true:

THEOREM 2.1. *If there is one fundamental solution E of $P(D)$ which is a C^∞ function in $\mathbf{R}^n\backslash\{0\}$, $P(D)$ is hypoelliptic (in $\mathbf{R}^n$).*

Theorem 2.1 provides us with a useful criterion to test whether a PDE with constant coefficients is hypoelliptic. We shall exploit it in subsequent sections.

Proof of Theorem 2.1. Let U be an arbitrary open subset of $\mathbf{R}^n$, u a distribution in U such that $f = P(D)u$ is a C^∞ function in U. Let x_0 be an arbitrary point of U. It will suffice to show that, under our hypothesis, u is a C^∞ function in some open neighborhood of x_0. Let U' be a relatively compact open subset of U containing x_0 and let g be a cutoff function of the kind already encountered, that is, $g \in C_c^\infty(U)$, $g = 1$ in U'. We have

$$P(D)(gu) = gP(D)u + v = gf + v.$$

By applying the standard Leibniz formula for differentiation of a product, we see that v is a linear combination of derivatives of g of order $\neq 0$, hence $v = 0$ where the derivatives of g vanish, in particular in U' and outside the support of g. Using the fundamental solution E in the statement of Theorem 2.1 we may write

$$E * P(D)(gu) = [P(D)E] * (gu) = gu,$$

hence

$$gu = E * (gf) + E * v.$$

But $gf \in C_c^\infty$ and the convolution of any distribution with any C^∞ function with compact support is a C^∞ function, hence everything is reduced to showing that $E * v$ is a C^∞ function in an open neighborhood of x_0: for then this would also be true of gu, which is equal to u in $U' \ni x_0$.

Let us select a number $\varepsilon > 0$ such that the set

$$V_\varepsilon = \{x \in \mathbf{R}^n;\ d(x, \mathbf{R}^n \backslash U') > \varepsilon\}$$

is a neighborhood of x_0. Let $\zeta_\varepsilon(x)$, another cutoff function, be equal to one for $|x| < \varepsilon/2$, to zero for $|x| > \varepsilon$; $\zeta_\varepsilon \in C^\infty(\mathbf{R}^n)$. We write

$$E * v = (\zeta_\varepsilon E) * v + [(1 - \zeta_\varepsilon)E] * v.$$

The second term $[(1 - \zeta_\varepsilon)E] * v$ is a C^∞ function everywhere, since by our hypothesis, $(1 - \zeta_\varepsilon)E$ is a C^∞ function and since the convolution of a distribution with compact support, here v, with any C^∞ function is a C^∞ function. On the other hand, by the standard properties of convolution,

$$\operatorname{supp}[(\zeta_\varepsilon E) * v] \subset \operatorname{supp}(\zeta_\varepsilon E) + \operatorname{supp} v.$$

By our choice of ζ_ε we see that the support of $(\zeta_\varepsilon E) * v$ is contained in the order ε neighborhood of supp v. We have already seen that $v = 0$ in U'. We reach the conclusion that $(\zeta_\varepsilon E) * v$ vanishes in V_ε and, consequently, that $E * v$ is a C^∞ function in V_ε. Q.E.D.

We have seen two important applications of fundamental solutions. We study a third one in the next section—it is closely related to the application in Theorem 2.1. In the sections following the next one, we explicitly compute certain remarkable fundamental solutions in the basic examples of Sect. 1.

Exercises

2.1. Consider the following equation in the plane

$$x\frac{\partial u}{\partial y} - y\frac{\partial u}{\partial x} = f(x^2 + y^2), \tag{2.13}$$

where $f(t)$ is a C^∞ function of the real variable t such that

$$f(t) = 0 \quad \text{if} \quad t < 1 \text{ or } t > 2, \qquad f(\tfrac{3}{2}) = 1.$$

Show that Eq. (2.13) has no distribution solution in $\mathbf{R}^2\backslash\{0\}$. [*Hint*: Rewrite $\iint |f(x^2 + y^2)|^2\, dx\, dy$ by introducing u, assumed to satisfy (2.13).]

2.2. Let Ω be a nonempty open subset of $\mathbf{R}^n$, $n > 1$, L a first-order linear partial differential operator with *real* C^∞ coefficients in Ω. Using Exercise 1.4, show that L cannot be hypoelliptic in Ω.

2.3. Consider the following equation:

$$(\partial/\partial x)^2 u - (\partial/\partial y)^2 u + \lambda(x, y)u = 0, \tag{2.14}$$

where λ is a C^∞ function in $\mathbf{R}^2$. Show that (2.14) is not hypoelliptic in $\mathbf{R}^2$.

2.4. Let $P(D)$ denote a linear partial differential operator with constant coefficients in $\mathbf{R}^n$, E a distribution in $\mathbf{R}^n$ such that

$$h = P(D)E - \delta$$

is a C^∞ function in $\mathbf{R}^n$. By suitably modifying the proof of Theorem 2.1 show that, *if E is a C^∞ function in the complement of the origin, $P(D)$ is hypoelliptic in $\mathbf{R}^n$.*

2.5. Extending Definition 2.1 to systems (whether determined or not), show that the gradient (§1.6) is hypoelliptic (in any open subset of $\mathbf{R}^n$) whereas the divergence (§1.7) is not, unless $n = 1$.

2.6. Construct all the fundamental solutions of d/dx, $d/dx - \lambda$, $d^2/dx^2 - \lambda$ (in $\mathbf{R}^1$, $\lambda \in \mathbf{C}$). Apply this, together with the change of variables in Exercise 1.5, to find a fundamental solution of $\partial^2/\partial x^2 - \partial^2/\partial y^2$ (in $\mathbf{R}^2$).

2.7. Show that we may obtain a set of solutions of the homogeneous Schrödinger equation (§1.5),

$$\frac{1}{i}\frac{\partial u}{\partial t} = \frac{\partial^2 u}{\partial x^2} \qquad (\text{in } \mathbf{R}^2), \tag{2.15}$$

by taking

$$u(x, t) = \int_{-\infty}^{+\infty} \exp[i(\tau^2 x - \tau t)]w(\tau)\, d\tau. \tag{2.16}$$

By choosing w appropriately, exhibit a solution of (2.15) which is not a C^∞ function in $\mathbf{R}^2$.

2.8. Show that there is an integral representation analogous to (2.16) for certain solutions of the wave equation in $\mathbf{R}^{n+1}$ ($n > 0$ arbitrary):

$$\frac{\partial^2 u}{\partial t^2} = \Delta_x u. \tag{2.17}$$

Show that, for any $n > 0$, Eq. (2.17) cannot be hypoelliptic.

2.9. Let $P(D)$, $Q(D)$ be two linear partial differential operators with constant coefficients in $\mathbf{R}^n$. Show that the product $P(D)Q(D)$ is hypoelliptic if and only if both $P(D)$ and $Q(D)$ are. What could we have stated if P and Q had variable coefficients?

2.10. Let $P(D)$ be a differential operator with constant coefficients in $\mathbf{R}^n$, having a fundamental solution E which is C^∞ in the complement of the origin. Let Ω be an open subset of $\mathbf{R}^n$ and denote by $\mathcal{N}_\Omega$ the space of distribution solutions of the homogeneous equation $P(D)h = 0$ [by Theorem 2.1 $\mathcal{N}_\Omega \subset C^\infty(\Omega)$]. Prove that the following topologies on $\mathcal{N}_\Omega$ are identical: (i) the C^∞ topology (uniform convergence of the functions and all their derivatives on every compact subset of Ω); (ii) the C^0 topology (uniform convergence of the functions on every compact subset of Ω); (iii) the topology induced by $\mathcal{D}'(\Omega)$ (the functions $f_\alpha \in \mathcal{N}_\Omega$ converge if the integrals $\int f_\alpha \phi\, dx$ converge for any test function $\phi \in C_c^\infty(\Omega)$, uniformly on bounded subsets of $C_c^\infty(\Omega)$: it suffices to consider *sequences* f_α and thus require the convergence of the integrals for each individual ϕ].

2.11. Derive from the result stated in Exercise 2.10 the following (we use the same notation as in Exercise 2.10 and form the same hypotheses): If $P(D)$ is hypoelliptic, and if $\{f_\alpha\}$ is a collection of solutions, in Ω, of the homogeneous equation $P(D)f = 0$, bounded (independently of α) on every compact subset of Ω, it contains a subsequence which converges in $C^\infty(\Omega)$.

Is the conclusion true if we assume, instead of boundedness on every compact subset of Ω, that to every test function $\phi \in C_c^\infty(\Omega)$ there is a constant $C(\phi) > 0$ such that, for all indices α,

$$\left| \int f_\alpha \phi\, dx \right| \leqq C(\phi)?$$

3

Analyticity of Solutions

Let us return, once more, to a linear *ordinary* differential equation with constant coefficients. We have already pointed out that if the right-hand side f is a smooth, i.e., a C^∞, function, this is also true of *all* the solutions. But there is more: If f is an *analytic* function, all the solutions will also be analytic. We recall that a complex-valued function u in an open set $\Omega \subset \mathbf{R}^n$ is analytic if its Taylor expansion about any point of Ω converges to the function in a full neighborhood of the point (cf. Definition 3.2 below). An equivalent definition is that u can be extended to an open neighborhood of Ω in $\mathbf{C}^n$ as an analytic function of the complex variables $z^1, \ldots, z^n$, or, as we shall often say, as a *holomorphic* function.

Obviously, this property that all the solutions are analytic when the right-hand side is an analytic function does not hold for an arbitrary linear PDE with constant coefficients: It does not hold for the wave equations (cf. Example 2.2); it does not hold for the heat equation either, as we show below (Example 3.1). But it holds for the Laplace equation and for the Cauchy–Riemann equation. As a matter of fact, the linear PDEs with *constant* coefficients which do possess it are exactly known. They are the *elliptic* equations. However, there are nonelliptic equations with *variable* coefficients which have the property. Without going into the meaning of ellipticity, the preceding property warrants a new definition:

Definition 3.1. (Cf. Definition 2.1) *A linear partial differential operator P in Ω is said to be analytic-hypoelliptic if, given any open subset U of Ω and any distribution u in U, u is an analytic function in U if this is true of Pu.*

Example 3.1. The heat equation is not analytic-hypoelliptic.

It suffices to look at the case where there is only one space variable:

$$L = \frac{\partial}{\partial t} - \frac{\partial^2}{\partial x^2}.$$

We define the following function $F(x, t)$ in the set $U = \{(x, t) \in \mathbf{R}^2 ; x \neq 0\}$:

$$F(x, t) = \begin{cases} t^{-1/2} \exp(-x^2/4t) & \text{when} \quad t > 0, \\ 0 & \text{when} \quad t \leqq 0, \end{cases}$$

It is checked at once that (i) $F(x, t)$ is a C^∞ function in the whole set U, (ii) $LF = 0$ in U, (iii) F is not an analytic function in U. Claim (iii) is obvious if we recall that an analytic function equal to zero in an open subset U_0 of its domain of definition, here the part of U where $t < 0$, vanishes in every connected component of U whose intersection with U_0 is not empty: In our case, the union of these components is U itself.

Suppose now that $P(D)$ is an analytic-hypoelliptic linear partial differential operator with constant coefficients and let E be any one of its fundamental solutions; since $P(D)E = 0$ in $\mathbf{R}^n \backslash \{0\}$, E must be an analytic function there (see Definition 3.2). In fact, the converse is true, and this is the analog of Theorem 2.1:

THEOREM 3.1. *If there is a fundamental solution E of $P(D)$ which is an analytic function in $\mathbf{R}^n \backslash \{0\}$, $P(D)$ is analytic-hypoelliptic in $\mathbf{R}^n$.*

Proof. By Theorem 2.1 we know that $P(D)$ is hypoelliptic (thus, *every linear partial differential operator with constant coefficients which is analytic-hypoelliptic is hypoelliptic*). It suffices to prove that if U is any open subset of $\mathbf{R}^n$, u any C^∞ function in U such that $P(D)u = f$ is analytic in U, then u is also analytic in U. We reason in the neighborhood of an arbitrary point x_0 of U. Although it is not really necessary, we shall simplify the argument by exploiting a consequence of the *Cauchy–Kovalevska theorem* (on the latter, see Sect. 17; the student unfamiliar with the Cauchy–Kovalevska theorem may want to accept the consequence of it as applied below), namely, that if the open neighborhood $W \subset U$ of x_0 is sufficiently small, there is an analytic function h in W satisfying there the equation $P(D)h = f$. Thus we have $P(D)(u - h) = 0$ in W and it is enough to show that $u - h$ is analytic in a neighborhood of x_0. Let U' be an open neighborhood of x_0 whose closure is compact and contained in W, and let $g \in C_c^\infty(W)$ be equal to one in U'. We set

$$v = P(D)[g(u - h)],$$

from which:

$$g(u - h) = E * v.$$

We must show that $E * v$ is analytic in the neighborhood of x_0, noting that $v = 0$ in U'. We use the same cutoff function ζ_ε as in the proof of Theorem 2.1, and write

$$E * v = (\zeta_\varepsilon E) * v + [(1 - \zeta_\varepsilon)E] * v.$$

The first term vanishes in the neighborhood of x_0, $V_\varepsilon = \{x; d(x, \mathbf{R}^n \backslash U') > \varepsilon\}$ ($\varepsilon > 0$ is very small); it will be enough to show that the second term is analytic in V_ε. We use the following characterization of analytic functions:

Definition 3.2. A C^∞ function φ in an open subset $\mathcal{O}$ of $\mathbf{R}^n$ is analytic in $\mathcal{O}$ if to every compact subset K of $\mathcal{O}$ there is a constant $r_K > 0$ such that

$$\sup \frac{1}{\alpha!} r_K^{|\alpha|} |D^\alpha \varphi(x)| < +\infty, \tag{3.1}$$

where the supremum is computed over all points x of K and all n-tuples α of nonnegative integers.

Since v is a C^∞ function with compact support, $w = [(1 - \zeta_\varepsilon)E] * v$ is a C^∞ function. We have

$$D^\alpha w = \{D^\alpha[(1 - \zeta_\varepsilon)E]\} * v = [(1 - \zeta_\varepsilon)D^\alpha E] * v + T * v.$$

By the Leibniz formula, the support of T is contained in the support of the gradient of ζ_ε, therefore $T * v = 0$ in V_ε. On the other hand, $(1 - \zeta_\varepsilon)D^\alpha E$ is a C^∞ function in the whole space $\mathbf{R}^n$. Therefore, when $x \in V_\varepsilon$,

$$D^\alpha w(x) = \int [1 - \zeta_\varepsilon(y)] D^\alpha E(y) v(x - y)\, dy.$$

The integration can be restricted to the set

$$K = \{y \in \mathbf{R}^n;\ |y| \geqq \varepsilon/2,\ y \in \overline{V}_\varepsilon - \operatorname{supp} v\}.$$

Now K is clearly a compact subset of $\mathbf{R}^n \backslash \{0\}$ where E is analytic; we may therefore apply (3.1) with E instead of φ (and y instead of x). We obtain

$$\frac{1}{\alpha!} r_K^{|\alpha|} |D^\alpha w(x)| \leqq C \sup_{y \in K} \left\{\frac{1}{\alpha!} r_K^{|\alpha|} |D^\alpha E(y)|\right\} \int |v(y)|\, dy,$$

where x is any point in V_ε and $C = 1 + \sup_{\mathbf{R}^n} |\zeta_\varepsilon|$. From this we conclude at once that w is analytic in V_ε. Q.E.D.

In the forthcoming sections we shall compute certain fundamental solutions of the ordinary differential equations with constant coefficients, and of the Laplace and of the Cauchy–Riemann equations. It will be seen that these fundamental solutions are all analytic in the complement of the origin. We shall thus reach the conclusion that these equations are all analytic-hypoelliptic. The result for the Laplace equation has been known for a long time as *Weyl's lemma.*

Exercises

3.1. By using the fact that the Cauchy–Riemann operator in $\mathbf{R}^2$ is analytic-hypoelliptic, derive that this is also true of the operator in $\mathbf{R}^2$

$$\frac{\partial}{\partial \bar{z}} + c(x, y),$$

where $c(x, y)$ is an analytic function in the plane.

3.2. Let $P(D)$ denote a linear partial differential operator with constant coefficients in $\mathbf{R}^n$, E a distribution in $\mathbf{R}^n$ such that

$$h = P(D)E - \delta$$

is an analytic function in $\mathbf{R}^n$. By suitably modifying the proof of Theorem 3.1 show that, *if E is an analytic function in the complement of the origin, $P(D)$ is analytic-hypoelliptic in $\mathbf{R}^n$.*

3.3. Prove that any C^∞ function $u(x, y)$ which satisfies the homogeneous wave equation in the plane, $u_{xx} - u_{yy} = 0$, is analytic with respect to (x, y) in $\mathbf{R}^2$ if (and only if) $u(0, y)$ and $u_y(0, y)$ are analytic functions of y in $\mathbf{R}^1$.

3.4. For each positive number d, introduce the following definition (cf. Definition 3.2):

Definition 3.3. A C^∞ function φ in an open subset $\mathcal{O}$ of $\mathbf{R}^n$ is said to belong to the d*th Gevrey class in $\mathcal{O}$, and one writes $\varphi \in G_d(\mathcal{O})$, if to every compact subset K of $\mathcal{O}$ there is $r_K > 0$ such that*

$$\sup_{x \in K, \alpha \in \mathbf{Z}_+^n} \left\{ \left(\frac{1}{\alpha!} \right)^d r_K^{|\alpha|} |D^\alpha \varphi(x)| \right\} < +\infty. \tag{3.2}$$

By adapting the proof of Theorem 3.1, prove the following statement:

THEOREM 3.2. *Suppose that $P(D)$ has a fundamental solution E which belongs to $G_d(\mathbf{R}^n \backslash \{0\})$. Then, given any open subset U of $\mathbf{R}^n$ and any distribution u in U, such that $P(D)u = 0$ in U, we must have $u \in G_d(U)$.*

4

Fundamental Solutions of Ordinary Differential Equations

This section is devoted to a review of certain facts concerning ordinary differential equations and their interpretation in the light of PDE theory. We limit ourselves to ODEs with constant coefficients. Later we shall consider ODEs with variable coefficients.

The simplest nontrivial ordinary differential operators are the first-order ones, of the kind

$$L = \frac{d}{dx} - a, \qquad a \in \mathbf{C}.$$

We seek the solutions of the ODE

(4.1) $$\frac{dF}{dx} - aF = \delta.$$

The answer is well known when $a = 0$. It is one of the first examples of differentiation of distributions that the derivative of the *Heaviside function* $H(x)$, equal to one when $x > 0$ and to zero when $x < 0$, is the Dirac distribution. In order to get *all* the solutions of $F' = \delta$, it suffices to add to H any constant function. Now, the solutions of (4.1) are transformed into the solutions of $F' = \delta$ by the mapping

$$F \mapsto e^{-ax}F.$$

By performing the inverse transformation, we see that all the solutions of (4.1) are given by

(4.2)† $$F = E + Ce^{ax},$$

† These fundamental solutions are analytic in the complement of the origin; therefore, by Theorem 3.1, the differential operator L is analytic-hypoelliptic. In particular, all the distribution solutions of the homogeneous equation $Lh = 0$ are "classical" solutions, and (4.2) gives all the distribution solutions of (4.1).

where C is an arbitrary complex constant and $E = H(x)e^{ax}$. Note that when $C = 0$, $F = E$ is the unique fundamental solution of L with support in the nonnegative half-line. When $C = -1$, F is the unique fundamental solution of L with support in the nonpositive half-line. When $C = \frac{1}{2}$, we obtain the "symmetric" fundamental solution

$$F = \tfrac{1}{2}\operatorname{sgn}(x)e^{ax},$$

where sgn(x) denotes the function *signum of* x, equal to $+1$ for $x > 0$ and to -1 for $x < 0$ (as a distribution, it need not be defined when $x = 0$).

Let us also observe that $U = e^{ax}$ is the unique solution of the initial value problem

$$U' - aU = 0, \qquad U\Big|_{x=0} = 1, \tag{4.3}$$

and that we have

$$E = HU.$$

Although this example may seem overly simple, it will only need straightforward elaboration to enable us to construct all the fundamental solutions of ODEs, as well as important fundamental solutions of many PDEs.

First of all, let us point out that the preceding argument extends directly to first-order *determined systems* of ODEs. Suppose that we deal with a system of p equations in p unknowns. This means that, instead of L, we study the matrix ordinary differential operator

$$\boldsymbol{L} = I\frac{d}{dx} - A,$$

where I is the $p \times p$ identity matrix and A any $p \times p$ matrix (with complex entries). Thus $\boldsymbol{L}$ acts now on functions valued in $\mathbf{C}^p$, the space of complex p-vectors, and transforms them into functions (or distributions) also valued in $\mathbf{C}^p$. Keep in mind, however, that all these functions and distributions are defined in intervals of the real line (where we continue to denote the variable by x). The problem (4.3) must be suitably generalized: U denotes now a *matrix*-valued function, the solution of the problem

$$\boldsymbol{L}U = 0, \qquad U\Big|_{x=0} = I, \qquad \text{the } p \times p \text{ identity matrix.} \tag{4.4}$$

The unique solution of (4.4) is found at once:

$$U = e^{xA}.$$

Now let K denote an arbitrary distribution with values in the space of $p \times p$ matrices or, equivalently, a $p \times p$ matrix whose entries are distributions. We have, by the Leibniz formula,

$$\boldsymbol{L}(UK) = (\boldsymbol{L}U)K + UK' = UK'.$$

Thus, if we wish to solve $\boldsymbol{L}(UK) = \delta I$, which corresponds to finding a *right*-fundamental solution of $\boldsymbol{L}$, we must solve

$$K' = \delta U^{-1} = \delta e^{-xA} = \delta I.$$

Thus, everything reduces to finding the solution of

$$K' = \delta I. \tag{4.5}$$

In the scalar case, i.e., when $p = 1$, the solutions of (4.5) are well known as we have pointed out: They are the distributions of the form $H(x) + C$, where H is the Heaviside function and C an arbitrary constant. This is what motivated our choice of the fundamental solution E of L and the expressions (4.2). When p is arbitrary, the same considerations apply: All the solutions K of (4.5) are of the form

$$K = H(x)I + C$$

where C is now a constant $p \times p$ matrix. Note that $H(x)I$ is the $p \times p$ diagonal matrix whose diagonal entries are all equal to the Heaviside function. We have thus obtained all the *right*-fundamental solutions of L. They are the matrix-valued distributions of the form

$$\boldsymbol{F} = \boldsymbol{E} + e^{xA}C, \tag{4.6}$$

where

$$\boldsymbol{E} = H(x)e^{xA} \tag{4.7}$$

is the unique fundamental solution of $\boldsymbol{L}$ with support in the nonnegative half-line.

We have used the names fundamental solution and *right*-fundamental solution. The reason is that systems of PDEs with constant coefficients have both *right*- and *left*-fundamental solutions. For instance, a *left*-fundamental solution of $\boldsymbol{L}$ will be a distribution $\boldsymbol{G}$ such that

$$\boldsymbol{G}' - \boldsymbol{G}A = \delta, \tag{4.8}$$

whereas the distributions $\boldsymbol{F}$ given by (4.6) satisfy

$$\boldsymbol{F}' - A\boldsymbol{F} = \delta. \tag{4.9}$$

If $\boldsymbol{F}$ is a right-fundamental solution and T any distribution with compact support, we have

$$\boldsymbol{L}(\boldsymbol{F} * T) = T,$$

whereas if $\boldsymbol{G}$ is a left-fundamental solution of $\boldsymbol{L}$, we have

$$\boldsymbol{G} * \boldsymbol{L}T = T.$$

All left-fundamental solutions of $\boldsymbol{L}$ are given by

$$\boldsymbol{G} = \boldsymbol{E} + Ce^{xA}, \tag{4.10}$$

where $\boldsymbol{E}$ is given by (4.7) and C is an arbitrary constant $p \times p$ matrix. There are *two-sided* fundamental solutions of L, for instance $\boldsymbol{E}$ in (4.7) and, in general, many more (see Exercise 4.1).

The next step is to extend the preceding argument to higher order ODEs with constant coefficients. Consider therefore the operator

$$L = \frac{d^m}{dx^m} + a_1 \frac{d^{m-1}}{dx^{m-1}} + \cdots + a_m .$$

We shall begin by transforming the equation $Lu = f$ into a first-order system, in the customary manner, by setting

$$u_1 = u, \qquad u_k = \frac{d^{k-1}u}{dx^{k-1}}, \qquad k = 2, \ldots, m. \tag{4.11}$$

The equation $Lu = f$ is then rewritten as

$$u_m' + a_1 u_m + a_2 u_{m-1} + \cdots + a_m u_1 = f. \tag{4.12}$$

To this we adjoin the equations

$$u_k' = u_{k+1}, \qquad k = 1, \ldots, m-1, \tag{4.13}$$

which follow from (4.11). Let us then denote by M the $m \times m$ matrix

$$M = \begin{pmatrix} 0 & 1 & 0 & \cdots & 0 \\ 0 & 0 & 1 & \cdots & 0 \\ \vdots & & & & \\ -a_m & -a_{m-1} & -a_{m-2} & \cdots & -a_1 \end{pmatrix}$$

Last, we denote by $\mathbf{u}$ the vector (i.e., the one-column matrix) with components $u_1, \ldots, u_m$, and by $\mathbf{f}$ the vector with components $f_1 = 0, \ldots, f_{m-1} = 0, f_m = f$. Equations (4.12) and (4.13) together can be written in the system form:

$$\frac{d\mathbf{u}}{dx} = M\mathbf{u} + \mathbf{f}. \tag{4.14}$$

Suppose now that for our given $\mathbf{f}$ we find a solution $\mathbf{u}$ to (4.14). Explicitly stated this means that (4.12) and (4.13) hold. But substituting then (4.13), in the equivalent form (4.11) where $u = u_1$ defines u, into (4.12) yields at once $Lu = f$.

If we seek a fundamental solution E to L it is only natural to try to solve (4.14) when $f_j = 0$ for $j < m$ and $f_m = \delta$, the Dirac distribution. We proceed more or less as we did in dealing with a general system. We set

$$\mathbf{u} = e^{xM}\mathbf{v},$$

which transforms (4.14) into

$$\frac{d\mathbf{v}}{dx} = e^{-xM}\mathbf{f}. \tag{4.15}$$

In order to find out the value of the right-hand side in (4.15), it suffices to observe that $\mathbf{f} = \delta\mathbf{e}_m$, where $\mathbf{e}_m$ is the vector with all components equal to zero, except the mth one, which is equal to one. Then

$$e^{-xM}\mathbf{f} = \delta(e^{-xM}\mathbf{e}_m)\bigg|_{x=0} = \delta\mathbf{e}_m.$$

Thus (4.15) reduces to

$$\frac{d\mathbf{v}}{dx} = \delta\mathbf{e}_m. \tag{4.16}$$

But all the solutions of (4.16) are easily found; they are

$$\mathbf{v} = H(x)\mathbf{e}_m + \mathbf{c}, \tag{4.17}$$

where $\mathbf{c}$ is a constant m-vector. Therefore

$$\mathbf{u} = H(x)e^{xM}\mathbf{e}_m + e^{xM}\mathbf{c}. \tag{4.18}$$

The last term in the expression of $\mathbf{u}$ corresponds to the solution of the system of homogeneous equations

$$\frac{d\mathbf{w}}{dx} = M\mathbf{w}. \tag{4.19}$$

A particular case of such a solution $\mathbf{w}$ is $e^{xM}\mathbf{e}_m$: As a matter of fact, it is the only solution such that

$$w_j = 0 \quad \text{if } j < m; \qquad w_m = 1 \quad \text{when } x = 0. \tag{4.20}$$

Let us set $U(x) = w_1(x)$. If we recall that $w_k = (w_1)^{(k-1)}$, $k = 2, \ldots, m$, we see that $U(x)$ is the unique solution of

$$LU = 0, \tag{4.21}$$

such that

$$U^{(j)} = 0 \quad \text{if } j < m - 1; \qquad U^{(m-1)} = 1 \quad \text{when } x = 0. \tag{4.22}$$

The fundamental solutions of L are all obtained by taking the first components u_1 of the vector-valued functions (4.18). In other words, they are all of the form

$$F = E + h, \tag{4.23}$$

where $Lh = 0$ and

$$E = HU. \tag{4.24}$$

We have now come around full circle and obtained the unique fundamental solution of L with support in the nonnegative half-line as the product of the Heaviside function with a remarkable solution of the homogeneous equation, namely the solution satisfying the *initial* conditions (4.22)—just as in the case of a single first-order equation.

Notice that all the fundamental solutions we have constructed are analytic in $\mathbf{R}^1\backslash\{0\}$: All the equations considered in this section are analytic-hypoelliptic.

Notice also that in the preceding discussion, we had no need to investigate the nature of the eigenvalues of the matrices A or M. We should also stress the link we have encountered between the construction of fundamental solutions and the initial value problem (4.21)–(4.22). This relationship dominates the theory of certain linear PDEs, such as the wave equation.

Exercises

4.1. Let I denote the $p \times p$ identity matrix, A a $p \times p$ matrix with complex entries. Describe *all* the two-sided fundamental solutions of the system of ordinary differential operators $\boldsymbol{L} = I\, d/dx - A$.

4.2. Let $I, A_1, \ldots, A_m$ denote $m + 1$ $p \times p$ matrices, I denoting the identity matrix. Describe all the fundamental solutions of the operator

$$\boldsymbol{L} = I\left(\frac{d}{dx}\right)^m + A_1\left(\frac{d}{dx}\right)^{m-1} + \cdots + A_m.$$

4.3. Describe all the two-sided fundamental solutions of the differential operator $\boldsymbol{L}$ in Exercise 4.2.

4.4. Can one solve the problem in Exercise 4.2 by transforming $\boldsymbol{L}$ into a determined system of first-order ordinary differential operators? If you think this is the case, describe the procedure. Otherwise, explain why it is not possible.

4.5. Let $\boldsymbol{L}$ denote the operator in Exercise 4.1. Give necessary and sufficient conditions on the matrix A in order that the fundamental solution of $\boldsymbol{L}$ with support in the nonnegative half-line be (1) a bounded function; (2) a function rapidly decaying at infinity.

4.6. Give necessary and sufficient conditions on the matrix A in order that the operator $\boldsymbol{L}$ in Exercise 4.1 have a fundamental solution which is rapidly decaying at infinity on the real line.

4.7. Give the expression of the fundamental solution of

$$L = \frac{d^2}{dx^2} - k^2, \qquad k \in \mathbf{R},$$

whose support is the nonnegative half-line. Do the same when k is replaced by $\sqrt{-1}\,k$.

4.8. Let U be the solution of the problem (4.21)–(4.22), H the Heaviside function, and L the ordinary differential operator in (4.21). Show by direct computation that

$$L(HU) = \delta.$$

4.9. Let $\boldsymbol{L}$ denote the operator in Exercise 4.1, $\chi_A(z) = \det(zI - A)$ the characteristic polynomial of A. Show that the solutions of the homogeneous equation $\boldsymbol{L}\mathbf{u} = 0$ are vector-valued functions whose components u_i satisfy the equation $\chi_A(d/dx)u_i = 0$. Conversely, let $\mathbf{u}$ be any vector-valued function with this property; show that there is a $p \times p$ system of ordinary differential equations with constant coefficients, of order $p - 1$, M, such that $M\mathbf{u}$ is a solution of $\boldsymbol{L}\mathbf{v} = 0$. Exactly describe M in the general case.

4.10. Let $\boldsymbol{L}$ denote the operator in Exercise 4.1. Let Γ be an invertible $p \times p$ matrix. There is a natural linear isomorphism between the spaces of solutions of $\boldsymbol{L}\mathbf{h} = 0$ and $\boldsymbol{L}_\Gamma \mathbf{h} = d\mathbf{h}/dx - \Gamma A \Gamma^{-1}\mathbf{h} = 0$. Describe this isomorphism. Next write $A = S + N$ where S is a diagonalizable $p \times p$ matrix, N a nil-potent $p \times p$ matrix (i.e., such that $N^p = 0$) such that $SN = NS$. Study the equation $\boldsymbol{L}\mathbf{h} = 0$ in the two separate cases $S = 0$ and $N = 0$: (1) Show that, when $N = 0$, this equation is equivalent (in a sense to be made precise) with the p scalar first-order ODEs

$$\frac{dw}{dx} - \lambda_i w = 0, \qquad 1 \leqq i \leqq p,$$

where the λ_i are the eigenvalues of A (repeated according to their multiplicity); (2) when $S = 0$, show that all the solutions of the equation just given are of the form $\mathbf{g}(x)$, where $\mathbf{g}$ is a vector-valued polynomial of degree $<p$. Describe precisely how $\mathbf{g}$ is determined by its value at $x = 0$, $\mathbf{g}(0)$.

Use the preceding discussion to describe exactly all the solutions of $\boldsymbol{L}\mathbf{h} = 0$ in the general case (when neither S nor N is necessarily zero).

4.11. Let $\mathscr{D}'_+$ denote the space of distributions on the real line which vanish for the strictly negative values of the variable (denoted by t). Show that the convolution of any two distributions S, $T \in \mathscr{D}'_+$ always makes sense [we may define the convolution $S * T$ by the formula

$$\langle S * T, \phi \rangle = \langle S_s, \langle T_t, \phi(s + t)\rangle\rangle, \qquad \phi \in C_c^\infty(\mathbf{R}^1)]. \tag{4.25}$$

Show that $\mathscr{D}'_+$ is a commutative convolution algebra with a unit, the Dirac measure δ.

Let L be an ordinary differential operator of order m, with constant coefficients (not all of which are equal to zero). Show that $L\delta$ has an inverse in the convolution algebra $\mathscr{D}'_+$. Compute this inverse. Derive from this that the equation $LU = F$ has a unique solution $U \in \mathscr{D}'_+$ for every $F \in \mathscr{D}'_+$ and write U in terms of F.

4.12. Let $\boldsymbol{L}$ be the operator in Exercise 4.1. Define the *transpose* ${}^t\boldsymbol{L}$ of $\boldsymbol{L}$ by the formula

$$\int_{-\infty}^{+\infty} \langle \boldsymbol{L}\boldsymbol{u}, \boldsymbol{v}\rangle\, dx = \int_{-\infty}^{+\infty} \langle \boldsymbol{u}, {}^t\boldsymbol{L}\boldsymbol{v}\rangle\, dx, \qquad \boldsymbol{u}, \boldsymbol{v} \in C_c^\infty(\mathbf{R}^1; \mathbf{C}^p). \tag{4.26}$$

Give the expression of ${}^t\boldsymbol{L}$ and relate the right- and left-fundamental solutions of ${}^t\boldsymbol{L}$ to those of $\boldsymbol{L}$.

5

Fundamental Solutions of the Cauchy–Riemann Operator

Consider the equation in the (x, y) plane $\mathbf{R}^2$:

$$\frac{\partial E}{\partial x} + i\frac{\partial E}{\partial y} = \delta. \tag{5.1}$$

We shall transform it into an ordinary differential equation by means of the *Fourier transformation* with respect to the variable y. If f is a function of x, y we set

$$\tilde{f}(x, \eta) = \int e^{-iy\eta} f(x, y)\, dy. \tag{5.2}$$

This only makes sense for suitable functions f. Here we shall use the Fourier transformation in the distribution sense, that is, acting on *tempered* distributions with respect to y. We recall that these are the distributions equal to finite sums of derivatives (with respect to y) of continuous functions of y which grow at infinity slower than some power $|y|^k$. On the space $\mathscr{S}'_y$ of tempered distributions, the Fourier transformation is a linear isomorphism. Its inverse extends the transformation defined on functions by the classical *Fourier inversion formula*. On functions of the type we consider here, this reads

$$f(x, y) = \frac{1}{2\pi}\int e^{iy\eta}\tilde{f}(x, \eta)\, d\eta. \tag{5.3}$$

Note that the right-hand side of (5.1) is the Dirac distribution $\delta(x, y)$ at the origin in $\mathbf{R}^2$. Its Fourier transformation with respect to y is equal to $\delta(x)1(\eta)$, where $1(\eta)$ stands for the function of η identically equal to one. For brevity we shall write $\delta(x)$ instead of $\delta(x)1(\eta)$. If we Fourier transform both sides of (5.1), we obtain

$$\frac{\partial \tilde{E}}{\partial x} - \eta\tilde{E} = \delta(x). \tag{5.4}$$

This is a first-order differential equation of the kind (4.1). We know all its fundamental solutions—they are given by (4.2). Here, however, the constant C need not be independent of η. We thus obtain

$$\tilde{E} = (H(x) + C(\eta))e^{\eta x}. \tag{5.5}$$

At this point there arises a serious problem: We are interested in E, not in its Fourier transform with respect to y, $\tilde{E}$. We wish, therefore, to compute the inverse Fourier transform of $\tilde{E}$. But this is only possible if E is *tempered* with respect to η. For $x > 0$, $e^{\eta x}$ is not tempered when $\eta \to +\infty$; for $x < 0$, it is not tempered when $\eta \to -\infty$. Fortunately, we have the "constant" $C(\eta)$ which we can use to correct the growth of $H(x)e^{\eta x}$. This is done in the following manner. We choose

$$C(\eta) = \begin{cases} -1 & \text{when} \quad \eta > 0, \\ 0 & \text{when} \quad \eta < 0. \end{cases} \tag{5.6}$$

This means that

$$\tilde{E} = \begin{cases} -H(-x)e^{\eta x} & \text{when} \quad \eta > 0, \\ H(x)e^{\eta x} & \text{when} \quad \eta < 0. \end{cases} \tag{5.7}$$

Observe that, when $x \neq 0$, E is a very rapidly decreasing function of η at infinity. We may compute its inverse Fourier transform by the formula (5.3):

$$\begin{aligned}(2\pi)E(x, y) &= H(x) \int_{-\infty}^{0} e^{(x+iy)\eta}\, d\eta - H(-x) \int_{0}^{+\infty} e^{(x+iy)\eta}\, d\eta \\ &= \frac{1}{x + iy}.\end{aligned}$$

This is valid when $x \neq 0$. But we observe that the function z^{-1} is locally integrable in the plane ($z = x + iy$). Indeed, the only question concerns its integrability in a neighborhood of the origin. But using polar coordinates r, θ shows that we must check the (local) integrability of $(1/r)e^{-i\theta}$ with respect to $r\,dr\,d\theta$; and this is obvious. Furthermore, z^{-1} goes to zero at infinity. Therefore, it is certainly a tempered distribution whose Fourier transform with respect to y is equal to $(2\pi)\tilde{E}$. We have reached the conclusion that $1/2\pi z$ is a solution of (5.1). Recalling that the Cauchy–Riemann operator is

$$\frac{\partial}{\partial \bar{z}} = \frac{1}{2}\left(\frac{\partial}{\partial x} + i\frac{\partial}{\partial y}\right),$$

we may state

THEOREM 5.1. *The locally integrable function* $1/\pi z$ *is a fundamental solution of the Cauchy–Riemann operator* $\partial/\partial\bar{z}$.

It is now very easy to describe *all* the fundamental solutions of $\partial/\partial\bar{z}$. Indeed, $(\pi z)^{-1}$ is an analytic function of (x, y) in $\mathbf{R}^2\backslash\{0\}$; hence, by Theorem 3.1, $\partial/\partial\bar{z}$ is analytic-hypoelliptic. Consequently, all the distribution solutions of the homogeneous Cauchy–Riemann equations are "classical" solutions, in fact C^∞ functions, and therefore holomorphic functions. Thus

COROLLARY 5.1. *Every fundamental solution of* $\partial/\partial\bar{z}$ *is of the form* $1/\pi z + h$, *where h is an entire function.*

We recall that an *entire* function of $z \in \mathbf{C}$ is a function which is holomorphic in the whole plane.

We now derive the homogeneous and inhomogeneous *Cauchy formulas*. Let Ω be a bounded open subset of $\mathbf{C}$. We shall assume that the boundary of Ω is a finite disjoint union of *Jordan curves*, i.e., of closed rectifiable curves that are not self-intersecting. Moreover, we assume that, *locally*, Ω *lies on one side of its boundary*, $\partial\Omega$: Every point of $\partial\Omega$ has an open neighborhood U such that $U\backslash\partial\Omega$ consists of two open connected components, only one of which is contained in Ω. This excludes the possibility, for instance, that Ω be the union of two half-disks such as $\{z;\ |z| < 1,\ \text{Im } z > 0\}$, $\{z;\ |z| < 1,\ \text{Im } z < 0\}$. Suppose then that

$$\frac{\partial u}{\partial \bar{z}} = f, \tag{5.8}$$

in an open neighborhood of $\bar{\Omega}$, where both u and f are defined and, for instance, once continuously differentiable (or even C^∞: the formula which we shall obtain will clearly extend to a larger class of functions). Let then $\chi = \chi_\Omega(x, y)$ denote the characteristic function of Ω, equal to one in Ω and to zero everywhere else. By the Leibniz formula we derive from (5.8):

$$(\partial/\partial\bar{z})(\chi_\Omega u) = \chi_\Omega f + u\{(\partial/\partial\bar{z})\chi_\Omega\}. \tag{5.9}$$

Since both sides of (5.9) have compact support we may write

$$\chi_\Omega u = \frac{1}{\pi z} * (\chi_\Omega f) + [u(\partial/\partial\bar{z})\chi_\Omega] * \frac{1}{\pi z}\cdot \tag{5.10}$$

Let us compute $(\partial/\partial\bar{z})\chi_\Omega$. We know that it vanishes in Ω and off $\bar{\Omega}$, hence it is a distribution supported by the boundary $\partial\Omega$. For obvious reasons, it is "uniformly distributed"; i.e., given any two points of $\partial\Omega$, it must be equal in a neighborhood of the first one (in $\partial\Omega$) to what it is in a similar neighborhood of the second one. But this does not tell us what it is. In an obvious sense, χ_Ω is a measure depending continuously on the set Ω, and $\partial/\partial\bar{z}$ is a continuous linear operator when acting on distributions. We may therefore approximate Ω by open sets which are the union of finitely many, pairwise disjoint, simple sets, like triangles or disks. Since χ_Ω is an additive functional of Ω, it suffices to prove the formula when Ω is a disk. We may take the center of the disk as the origin. The equation of $\partial\Omega$ is now $r = R$, a constant greater than zero. For any $\phi \in C_c^\infty(\Omega)$, we have

$$\langle(\partial/\partial\bar{z})\chi_\Omega, \phi\rangle = -\langle\chi_\Omega, (\partial/\partial\bar{z})\phi\rangle = -\frac{1}{2}\int_0^{2\pi}\int_0^R e^{i\theta}\left(\phi_r + \frac{i}{r}\phi_\theta\right) r\, dr\, d\theta,$$

using the expression of $\partial/\partial\bar{z}$ in polar coordinates. We may compute [setting $\tilde{\phi}(r, \theta) = \phi(r\cos\theta, r\sin\theta)$]

$$\int_0^R \phi_r r\, dr = \tilde{\phi}(R, \theta)R - \int_0^R \tilde{\phi}(r, \theta)\, dr,$$

$$\int_0^{2\pi} e^{i\theta}\phi_\theta\, d\theta = -i\int_0^{2\pi}\tilde{\phi}(r, \theta)e^{i\theta}\, d\theta,$$

whence

$$\langle(\partial/\partial\bar{z})\chi_\Omega, \phi\rangle = -\tfrac{1}{2}R\int_0^{2\pi} e^{i\theta}\tilde{\phi}(R, \theta)\, d\theta.$$

This can be rewritten in the following manner:

(5.11) $$\langle(\partial/\partial\bar{z})\chi_\Omega, \phi\rangle = -\frac{1}{2i}\oint_{\partial\Omega}\phi(x, y)\, dz.$$

Formula (5.11) clearly makes sense for arbitrary open subsets Ω of $\mathbf{R}^2$ of the type we considered at the start: The boundary $\partial\Omega$ is a finite union of disjoint Jordan curves and, locally, Ω lies on one side of it. But we must state precisely what the notation $\oint$ stands for: It cannot mean counterclockwise orientation, as the example $\Omega = \{z \in \mathbf{C}; 1 < |z| < 2\}$ readily shows. Because of our assumptions, whenever the boundary is a smooth curve, we may look at its *interior normal* N: Its unit vector, $\boldsymbol{\nu}$, is orthogonal to the line T tangent to $\partial\Omega$ and points inside Ω. Then the orientation on $\partial\Omega$ is such that the angle $(\boldsymbol{\tau}, \boldsymbol{\nu})$ is $+\pi/2$ (not $-\pi/2$; $\boldsymbol{\tau}$ is the unit tangent vector to the oriented curve $\partial\Omega$ at the same point where we consider $\boldsymbol{\nu}$). For boundaries which are not smooth (i.e., C^∞) we go to the limit over smooth approximations. Roughly speaking, $\oint$ indicates that the orientation on $\partial\Omega$ is counterclockwise *when we look from inside* Ω.

Substituting (5.11) into (5.10) yields the *inhomogeneous Cauchy formula*

$$(5.12)\quad u(x, y) = \frac{1}{\pi}\iint_\Omega f(x', y')\frac{dx'\,dy'}{(x - x') + i(y - y')}$$
$$-\frac{1}{2\pi i}\oint_{\partial\Omega} u(x', y')\frac{dx' + i\,dy'}{(x - x') + i(y - y')},\qquad (x, y) \in \Omega.$$

There is a more "symmetric" way of writing (5.12) by systematically using the two complex variables z, $\bar{z}$. If we interpret the integration of a function $h(x, y)$ over a (measurable) subset of the plane as the integration of the *two-form* $h(x, y)\,dx \wedge dy$, and the integration of $h(x, y)$ with respect to $dx + i\,dy$ over a curve as the integration, over the same curve, of the one-form $h(x, y)\,dz$, noting furthermore that

$$dz \wedge d\bar{z} = -2i\,dx \wedge dy,$$

we see that (5.12) can be rewritten as

$$(5.12')\quad u(x, y) = \frac{1}{2\pi i}\iint_\Omega \frac{\partial u}{\partial \bar{z}}(x', y')\frac{dz' \wedge d\bar{z}'}{z' - z} + \frac{1}{2\pi i}\oint_{\partial\Omega} u(x', y')\frac{dz'}{z' - z}\quad (\text{in } \Omega).$$

When $f = 0$ in Ω, (5.8) implies that u is a holomorphic function in Ω; we denote it then by $u(z)$. In this case, formula (5.12′) [or (5.12)] is simplified into the standard *Cauchy formula*:

$$(5.13)\qquad u(z) = \frac{1}{2i\pi}\oint_{\partial\Omega} u(z')\frac{dz'}{z' - z},\qquad z \in \Omega.$$

Remark 5.1. Formula (5.10) can be regarded as the extension of the inhomogeneous Cauchy formula [(5.12) or (5.12′)] to arbitrary bounded open sets Ω (without the condition that their boundary be a finite union of disjoint Jordan curves).

Exercises

5.1. Show by direct computation that, for any $\varphi \in C_c^\infty(\mathbf{R}^2)$,

$$\varphi(0) = -\frac{1}{\pi}\iint \frac{\partial\varphi}{\partial\bar{z}}\frac{dx\,dy}{x + iy}.$$

[*Hint*: Switch to polar coordinates.]

5.2. Suppose that $f(x, y) \in C^\infty(\mathbf{R}^2)$ vanishes for $|x| > 1$ and is periodic with respect to y, with period 2π. By using the Fourier expansion of f with respect to y, show that the Cauchy–Riemann equation

$$\frac{1}{2}\left(\frac{\partial u}{\partial x} + i\frac{\partial u}{\partial y}\right) = f \quad \text{in } \mathbf{R}^2, \tag{5.14}$$

has a solution which is periodic with respect to y (with period 2π). Show also that there is a distribution $E(x, y)$ in $\mathbf{R}^2$, periodic with respect to y, such that the following is a solution of (5.14):

$$u(x, y) = \int_{x'=-\infty}^{x'=+\infty} \int_0^{2\pi} E(x', y')f(x - x', y - y')\, dx'\, dy'. \tag{5.15}$$

5.3. Let D_r denote the disk $|z| < r$ $(r > 0)$ in the plane. Let L be a straight line through the origin, f a continuous (complex-valued) function in the closure D_r, holomorphic in the complement of $D_r \cap L$ in D_r. By applying the formula for the distribution derivative of a function $g(t)$ of one real variable, which is smooth for $t > 0$ and $t < 0$, and has a finite jump at $t = 0$, prove that f is holomorphic in the whole of D_r.

5.4. Let D_r be as in Exercise 5.3. By using the assertion made there, prove the *Schwarz reflection principle*, namely that if h is any continuous function in $\overline{D_r^+} = \{z = x + iy \in \bar{D}_r;\ y \geqq 0\}$, holomorphic for $y > 0$, and real for $y = 0$, it has a unique holomorphic extention to the whole of D_r.

5.5. Let μ be a distribution with compact support, in the plane $\mathbf{R}^2$, u any distribution in $\mathbf{R}^2$ satisfying

$$\frac{\partial u}{\partial \bar{z}} = \mu \quad \text{in } \mathbf{R}^2. \tag{5.16}$$

Prove that if h is any entire function in $\mathbf{C}$ and γ any closed simple rectifiable curve containing the support of μ in its interior, we have

$$\oint_\gamma u(z)h(z)\, dz = 2\sqrt{-1}\langle \mu, h\rangle. \tag{5.17}$$

5.6. Let $f_1, \ldots, f_n$ be n functions belonging to $C_c^\infty(\mathbf{R}^{2n})$ and assume $n > 1$. Assume that the following *compatibility conditions* [cf. (1.16)] are satisfied:

$$\frac{\partial f_j}{\partial \bar{z}_k} = \frac{\partial f_k}{\partial \bar{z}_j}, \qquad j, k = 1, \ldots, n. \tag{5.18}$$

By exploiting the fact that $1/\pi z_1$ is a fundamental solution of $\partial/\partial \bar{z}_1$, show that there is a C^∞ function u in $\mathbf{R}^{2n}$ which satisfies the (overdetermined) system of equations

$$\frac{\partial u}{\partial \bar{z}_j} = f_j, \qquad j = 1, \ldots, n, \tag{5.19}$$

in the whole of $\mathbf{R}^{2n}$ and *which, furthermore, has compact support.* [*Hint*: Take $|z_2| + \cdots + |z_n| \to +\infty$ in the inhomogeneous Cauchy formula with respect to z_1.]

5.7. Let K be a compact subset of $\mathbf{R}^{2n}$ whose complement is connected $(n > 1)$ and let h be a holomorphic function of $z = (z_1, \ldots, z_n)$ in $\mathbf{C}^n \backslash K$ [which means that h is, say, C^1, and satisfies the Cauchy–Riemann equations $(\partial/\partial \bar{z}_j)h = 0$, $j = 1, \ldots, n$, there]. Derive from the result in Exercise 5.6 that there is an entire function $\tilde{h}$ in $\mathbf{C}^n$ which is equal to h in $\mathbf{C}^n \backslash K$. [*Hint*: Use a cutoff function $g \in C_c^\infty(\mathbf{R}^{2n})$, equal to one in a neighborhood of K, and solve (5.19) where $f_j = (\partial/\partial \bar{z}_j)\{(1 - g)h\}$, $j = 1, \ldots, n$.] Show that the same result would not be true if $n = 1$. (The preceding extension result in dimension $n > 1$ is a classical theorem due to Hartogs.)

5.8. Let Ω be an arbitrary open subset of $\mathbf{R}^2$. Let $C^\infty(\Omega)$ denote, as usual, the space of complex C^∞ functions in Ω, equipped with the topology $\mathcal{T}$ of uniform convergence of the functions and of each one of their derivatives on the compact subsets of Ω. Let $\mathcal{T}'$ denote the topology on C^∞ in which convergence of functions means that the functions only converge uniformly on compact subsets of Ω and that their first $\bar{z}$ derivatives converge in the sense of the natural topology $\mathcal{T}$. By using the inhomogeneous Cauchy formula (5.12), show that the topologies $\mathcal{T}$ and $\mathcal{T}'$ are the same.

5.9. Show that there exists a real-analytic, but no complex-analytic (i.e., holomorphic) function u in $\mathbf{C} \backslash \{0\}$ such that

$$\frac{\partial u}{\partial \bar{z}} = \frac{1}{z} \quad \text{in } \mathbf{C} \backslash \{0\}. \tag{5.20}$$

Let $f(z)$ be a holomorphic function in $\mathbf{C} \backslash \{0\}$. By using its Laurent expansion, prove that the equation

$$\frac{\partial v}{\partial \bar{z}} = f \quad \text{in } \mathbf{C} \backslash \{0\} \tag{5.21}$$

always has a solution, and that all its (distribution) solutions are real-analytic in $\mathbf{C} \backslash \{0\}$.

5.10. Let $P(z)$ be a polynomial in one variable, with complex coefficients. Describe all solutions in $\mathbf{R}^2$ of the partial differential equation

$$P\left(\frac{\partial}{\partial \bar{z}}\right)h = 0, \tag{5.22}$$

and construct a fundamental solution of $P(\partial/\partial \bar{z})$.

6

Fundamental Solutions of the Heat and of the Schrödinger Equations

If we go back to our basic examples in Sect. 1, we see that both the heat equation and Schrödinger's equation are first order with respect to the time variable t. It is only natural to perform a Fourier transformation with respect to the space variables x, according to the formula

$$\tilde{u}(\xi, t) = \int_{\mathbf{R}^n} e^{-i\langle x, \xi\rangle} u(x, t)\, dx \tag{6.1}$$

where $\langle x, \xi\rangle = x^1\xi_1 + \cdots + x^n\xi_n$. On the other hand, the wave equation is second order with respect to all variables, and this is also true of the Laplace equation. Their treatment is therefore more difficult. Constructing a fundamental solution for the heat equation and for Schrödinger's equation is similar to constructing one for the Cauchy–Riemann equations (see pp. 34–35).

6.1 Heat Equation

We must find the solutions of the equation

$$\frac{\partial E}{\partial t} - \Delta_x E = \delta \quad \text{in } \mathbf{R}^{n+1}. \tag{6.2}$$

As announced, we perform a Fourier transformation with respect to x; (6.2) is transformed into

$$\frac{\partial \tilde{E}}{\partial t} + |\xi|^2 \tilde{E} = \delta(t). \tag{6.3}$$

A remarkable fundamental solution of (6.3) is given by

$$\tilde{E}(\xi, t) = H(t) \exp(-t|\xi|^2).$$

It is obviously tempered with respect to ξ, in fact it decays rapidly at infinity in $\mathbf{R}_n$ provided that $t \neq 0$. We may perform its inverse Fourier transformation, according to the formula

$$u(x, t) = (2\pi)^{-n} \int_{\mathbf{R}_n} e^{i\langle x, \xi\rangle} \tilde{u}(\xi, t)\, d\xi. \tag{6.4}$$

Suppose $t > 0$ and take $\tilde{u}(\xi, t) = \tilde{E}(\xi, t) = \exp(-t|\xi|^2)$. We obtain

$$u(x, t) = E(x, t) = (2\pi)^{-n} \int_{\mathbf{R}_n} \exp(i\langle x, \xi\rangle - t|\xi|^2)\, d\xi$$

$$= \left\{\prod_{j=1}^{n} (2\pi)^{-1} \int_{\mathbf{R}_1} \exp\left[-t\left(\xi_j - i\frac{x^j}{2t}\right)^2\right] d\xi_j\right\} \exp\left(-\frac{|x|^2}{4t}\right).$$

Let z be the complex variable in the plane $\mathbf{C}^1$ and consider the integral

$$I = \int_\gamma \exp(-tz^2)\, dz,$$

where γ is any horizontal line $\operatorname{Im} z = \text{const} = c$. We contend that I is independent of c. To see this it suffices to apply Cauchy's integral theorem, taking as the contour of integration the rectangle boundary shown in Fig. 6.1. The contributions to the integral of $\exp(-tz^2)$ coming from the two

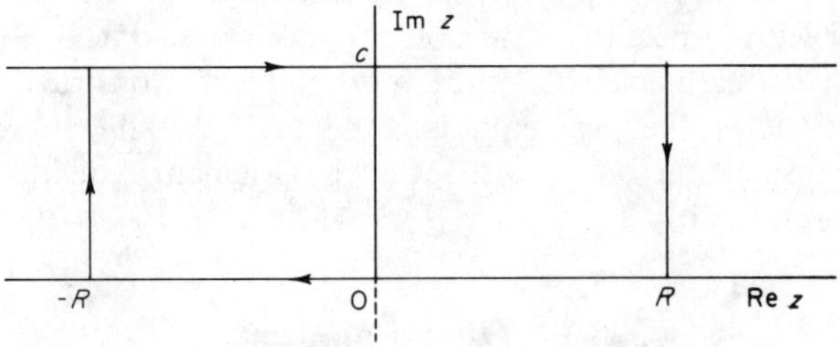

FIG. 6.1

vertical segments at $\operatorname{Re} z = \pm R$ rapidly go to zero as $R \to +\infty$. Consequently, at the limit, the sum of the integrals on the two horizontal lines, with the orientations as indicated, equals zero. This proves that

$$I = \int_{-\infty}^{+\infty} \exp[-t(\operatorname{Re} z)^2]\, d(\operatorname{Re} z) = \sqrt{\frac{\pi}{t}},$$

and, finally, that

$$E(x, t) = (2\sqrt{\pi t})^{-n} \exp\left(-\frac{|x|^2}{4t}\right).$$

Therefore, for $t \neq 0$, the inverse Fourier transform of $\tilde{E}(\xi, t)$ (with respect to x and ξ) is given by

$$E(x, t) = (2\sqrt{\pi t})^{-n} H(t) \exp\left(-\frac{|x|^2}{4t}\right). \tag{6.5}$$

As a matter of fact, if we look at things in the right way, we see easily that $E(x, t)$ is the inverse Fourier transform of $\tilde{E}(\xi, t)$ for all t. Indeed, $\tilde{E}(\xi, t)$ is a tempered distribution, in fact a bounded function, of $\xi \in \mathbf{R}_n$, with values in the space of bounded measurable functions of t [these form a Banach space, $L^\infty(\mathbf{R}_t^1)$]. As such, it has an inverse Fourier transform, which is precisely (6.5). All other fundamental solutions of the heat equation are obtained by adding to E, given by (6.5), a solution of the homogeneous heat equation. Let us point out two important properties of the fundamental solution E in (6.5):

(1) $E(x, t)$ *is invariant under the space rotations*, i.e., it depends only on the square norm $|x|^2$ as far as the space variables are concerned;

(2) $E(x, t)$ *is a* C^∞ *function of* (x, t) *in the complement of the origin*, $\mathbf{R}^{n+1}\backslash\{0\}$. In order to see this, it suffices to reason in the neighborhood of a point $x = x_0$, $t = 0$. One must show that all the partial derivatives of

$$t^{-n/2} \exp\left(-\frac{|x|^2}{4t}\right)$$

tend to zero when $t > 0$ converges to zero. This is not difficult. Notice that E is nowhere zero in the half-space $t > 0$, and everywhere zero in the half-space $t < 0$. Consequently, E cannot be analytic in $\mathbf{R}^{n+1}\backslash\{0\}$ (though it is an analytic function in the region $t \neq 0$) and the heat operator, as already pointed out, is hypoelliptic but not analytic-hypoelliptic (Definitions 2.1 and 3.1).

6.2 The Schrödinger Equation

Let us write the analogs of (6.2) and (6.3) in the present case:

$$\frac{1}{i}\frac{\partial E}{\partial t} - \Delta_x E = \delta \quad \text{in } \mathbf{R}^{n+1}, \tag{6.6}$$

$$\frac{1}{i}\frac{\partial \tilde{E}}{\partial t} + |\xi|^2 \tilde{E} = \delta(t). \tag{6.7}$$

We obtain at once a solution of (6.7) by taking

$$\tilde{E} = iH(t) \exp(-it|\xi|^2).$$

Here also $\tilde{E}$ can be looked upon as a tempered distribution, or a bounded function of $\xi \in \mathbf{R}_n$ valued in the space $L^\infty(\mathbf{R}_t^1)$. We must compute its inverse Fourier transform with respect to ξ. We observe that, when $\varepsilon > 0$ tends to zero,

$$\tilde{E}_\varepsilon = iH(t)\exp(-(\varepsilon + it)|\xi|^2)$$

converges, in the distribution sense, to $\tilde{E}$. As a consequence, the inverse Fourier transform of $\tilde{E}_\varepsilon$ converges to the one of $\tilde{E}$. The same argument applied to the function $u(x, t)$ given by (6.4) shows that

$$v(x, t) = (2\pi)^{-n} \int_{\mathbf{R}_n} \exp(i\langle x, \xi\rangle - (\varepsilon + it)|\xi|^2)\, d\xi$$

is equal to

$$(2\pi)^{-n}\left\{\int_{\mathbf{R}_n} \exp(-(\varepsilon + it)|\xi|^2)\, d\xi\right\} \exp\left(-\frac{|x|^2}{4(\varepsilon + it)}\right). \tag{6.8}$$

At this point we observe that when $t > 0$ and $\varepsilon \to 0$ ($\varepsilon > 0$), the function (6.8) has a limit, namely

$$(2\pi)^{-n} t^{-n/2} C \exp\left(-\frac{|x|^2}{4it}\right), \tag{6.9}$$

where C is equal to the nth power of the *Fresnel integral*

$$J = \int_{-\infty}^{+\infty} \exp(-i\lambda^2)\, d\lambda.$$

We have

$$J = J_0 - iJ_1, \qquad J_0 = \int_{-\infty}^{+\infty} \cos \lambda^2\, d\lambda, \qquad J_1 = \int_{-\infty}^{+\infty} \sin \lambda^2\, d\lambda.$$

There are various ways of computing J_0, J_1 (differentiation with respect to a parameter, integration by the residue methods, etc.); they show that

$$J_0 = J_1 = \sqrt{\pi/2}.$$

Since $(1 - i)/\sqrt{2} = \exp(-i\pi/4)$, we see that (6.9) equals

$$(4\pi t)^{-n/2} \exp\left(-in\frac{\pi}{4}\right) \exp\left(-\frac{|x|^2}{4it}\right)$$

and, therefore, that

$$E = H(t)\exp\left[-i(n-2)\frac{\pi}{4}\right](4\pi t)^{-n/2} \exp\left(-\frac{|x|^2}{4it}\right). \tag{6.10}$$

Now, a few remarks about the distribution E are in order:

(1) *E is rotation-invariant with respect to the space variables*;

(2) *E is a C^∞, and even an analytic function when $t \neq 0$; it is not a C^∞ function in any neighborhood of a point $x = x_0$, $t = 0$.* As a matter of fact, it is not even an integrable function in such a neighborhood, unless $n = 1$. For instance, when $n = 2$, E is a pseudofunction with respect to t, of the type

$$\text{Pf}\, H(t)t^{-1}.$$

The singularities of E on the hyperplane $t = 0$ get worse as n increases. At any rate, the Schrödinger operator is *not* hypoelliptic.

Exercises

6.1. Suppose $n = 2$. Prove directly that E, given by (6.5), is a fundamental solution of the heat operator, i.e., show that, given any test function $\varphi \in C_c^\infty(\mathbf{R}^3)$,

$$\varphi(0) = -\int_{\mathbf{R}^3} E(x, t)\left(\frac{\partial \varphi}{\partial t} - \Delta_x \varphi\right) dx\, dt.$$

[*Hint*: Use the polar coordinates r, θ in the plane of the space variables x.]

6.2. Same question as in 6.1, but replace formula (6.5) by (6.10) and the heat operator by Schrödinger's.

6.3. Let E be the distribution defined by (6.5). Prove that it is a locally integrable function. As a matter of fact, prove that its integral in the slab

$$\{(x, t) \in \mathbf{R}^{n+1};\, x \in \mathbf{R}^n,\, 0 < t < t_1\} \qquad (t_1 < +\infty)$$

is bounded.

6.4. Let E be given by (6.5). For $t > 0$, we regard E as a distribution in the space variables, depending on the parameter t. Prove that, as $t \to +0$, $E(x, t) \to \delta(x)$, the Dirac distribution.

6.5. Let E be given by (6.5). Assuming $n \geqq 3$, compute

$$F(x) = \int_0^{+\infty} E(x, t)\, dt,$$

and show, by using the result in Exercise 6.4, that

$$-\Delta F = \delta \qquad (\text{in } \mathbf{R}^n). \tag{6.11}$$

6.6. Prove that the fundamental solution of the heat equation, E, given by (6.5), belongs to $G_2(\mathbf{R}^{n+1}\backslash\{0\})$ (Definition 3.3, Exercise 3.4) and derive that the same is true of any fundamental solution of the heat equation.

6.7. Let $\mathbf{T}^n$ denote the n-dimensional torus, $\theta = (\theta_1, \ldots, \theta_n)$ the variable in it (each θ_j is an angle and varies from 0 to 2π). Construct the analog $E^{\#}(\theta, t)$ of the fundamental solution $E(x, t)$ given in (6.5), for the "periodic heat equation"

$$\frac{\partial u}{\partial t} = \Delta_\theta u \qquad \left[\text{in } \mathbf{R}^1 \times \mathbf{T}^n; \quad \Delta_\theta = \sum_{j=1}^{n} \left(\frac{\partial}{\partial \theta_j}\right)^2\right]; \tag{6.12}$$

write the Fourier series expansion of $E^{\#}(\theta, t)$.

What happens if t is also interpreted as an angle and made to vary on the one-dimensional torus?

7

Fundamental Solutions of the Wave Equation

The choice of a time variable is unambiguously† suggested to us in the wave equation

$$\frac{\partial^2 u}{\partial t^2} - \Delta_x u = f. \tag{7.1}$$

It is not so in the case of the Laplace equation (see Sect. 9). As in the study of the heat and of Schrödinger's equations, we call n the number of space variables: $x = (x^1, \ldots, x^n)$. We now construct a remarkable fundamental solution of (7.1). The case $n = 1$ is totally elementary and we begin by describing it.

Case of a Single Space Variable

We denote by x the space variable. We wish to solve the equation

$$\frac{\partial^2 E}{\partial t^2} - \frac{\partial^2 E}{\partial x^2} = \delta(t)\,\delta(x). \tag{7.2}$$

It is convenient to change variables and set

$$s = t - x, \qquad y = t + x. \tag{7.3}$$

As we are going to see, we may find a solution E of (7.2) which is a locally integrable function, also denoted by $E(t, x)$. Let φ be any element of $C_c^\infty(\mathbf{R}^2)$

† Up to a Lorentz transformation.

and let $\varphi(t, x)$ [resp. $\varphi^\natural(s, y)$] be its expression in the coordinates t, x (resp. s, y). Of course, we have

$$\varphi^\natural(s, y) = \varphi\left(\frac{s+y}{2}, \frac{y-s}{2}\right).$$

Also

$$4\left(\frac{\partial}{\partial s}\right)\left(\frac{\partial}{\partial y}\right)\varphi^\natural(s, y) = \left[\left(\frac{\partial}{\partial t}\right)^2 \varphi - \left(\frac{\partial}{\partial x}\right)^2 \varphi\right]\left(\frac{s+y}{2}, \frac{y-s}{2}\right).$$

On the other hand,

$$\langle E, \varphi\rangle = \iint E(t, x)\varphi(t, x)\, dt\, dx = \frac{1}{2}\iint E\left(\frac{s+y}{2}, \frac{y-s}{2}\right)\varphi^\natural(s, y)\, ds\, dy,$$

Thus the distribution E is defined, in the coordinates s, y, by the function

$$E^\natural(s, y) = \frac{1}{2}\, E\left(\frac{s+y}{2}, \frac{y-s}{2}\right).$$

Take now $\varphi = \{(\partial/\partial t)^2 - (\partial/\partial x)^2\}\psi$ (in the coordinates t, x). We must have, according to (7.2),

$$\psi(0, 0) = \langle E, \varphi\rangle = \iint E^\natural(s, y)(\psi_{tt} - \psi_{xx})^\natural(s, y)\, ds\, dy$$

$$= 4\iint E^\natural(s, y)\psi^\natural_{sy}(s, y)\, ds\, dy.$$

We must therefore find a locally integrable function $E^\natural$, solution of

(7.4) $$4\,\frac{\partial^2 E^\natural}{\partial s\, \partial y} = \delta(s)\, \delta(y),$$

and set, once this is done,

(7.5) $$E(t, x) = 2E^\natural(t - x, t + x).$$

We immediately find many solutions of (7.4), suitable linear combinations of

(7.6) $$E_1^\natural = \tfrac{1}{4}H(s)H(y), \quad E_2^\natural = -\tfrac{1}{4}H(s)H(-y), \quad E_3^\natural = -\tfrac{1}{4}H(-s)H(y),$$
$$E_4^\natural = \tfrac{1}{4}H(-s)H(-y).$$

By (7.5) the first of these functions leads to the fundamental solution E_1 of the wave operator in $\mathbf{R}^2$ which, in the coordinates t, x, is defined by the locally integrable function

(7.7) $$E_1(t, x) = \tfrac{1}{2}H(t - x)H(t + x).$$

Note that

$$E_1(t, x) = \tfrac{1}{2}H(t)H(t^2 - x^2).$$

The support of E_1 is equal to the following sector (which should be called the *forward light-cone* in the plane):

$$(7.8) \qquad t + x \geqq 0, \qquad t - x \geqq 0.$$

Moreover, the function $E_1(t, x)$ is constant (equal to $\frac{1}{2}$) in the interior of the set (7.8). Consequently, its *singular support*, i.e., the smallest closed set outside of which it is a C^∞ function, is exactly equal to the boundary of the set (7.8), that is, to the union of the two rays

$$(7.9) \qquad x + t = 0, \quad t \geqq 0; \qquad x - t = 0, \quad t \geqq 0.$$

We shall eventually prove that E_1 is the *unique* fundamental solution of the wave equation with support in (7.8).

Similar considerations apply to the transforms E_2, E_3, E_4 of $E_2^\natural$, $E_3^\natural$, $E_4^\natural$, respectively. The sector (7.8) must be replaced by different sectors. For instance, in the case of E_4, the role of (7.8) is played by its symmetric image with respect to the origin:

$$(7.10) \qquad x + t \leqq 0, \qquad t - x \leqq 0.$$

Analogous phenomena will appear in the case of $n > 1$ space variables, with important differences.

The General Case

After a Fourier transformation with respect to the space variables x, the equation we have to solve reads

$$(7.11) \qquad \frac{\partial^2 \tilde{E}}{\partial t^2} + |\xi|^2 \tilde{E} = \delta(t).$$

We readily obtain the solution of (7.11) with support in the half-line $t \geqq 0$ by means of formula (4.24). Indeed, it is not difficult to find what U is, in our case:

$$U(t, \xi) = \frac{\sin(|\xi| t)}{|\xi|},$$

hence

$$(7.12) \qquad \tilde{E}_+(t, \xi) = H(t)\frac{\sin(|\xi| t)}{|\xi|}.$$

The solution of (7.11) with support in the half-line $t \leqq 0$ is likewise given by

$$\tilde{E}_-(t, \xi) = -H(-t)\frac{\sin(|\xi| t)}{|\xi|}. \tag{7.13}$$

We obtain further solutions of (7.11) by taking expressions of the form

$$\alpha(\xi)\tilde{E}_+ + \beta(\xi)\tilde{E}_-, \qquad \alpha + \beta \equiv 1, \tag{7.14}$$

with α, β, say, bounded measurable functions of $\xi \in \mathbf{R}_n$ or even tempered distributions in $\mathbf{R}_n$. The first fact to be noted is that all the distributions (7.14) are then tempered with respect to ξ; their inverse Fourier transforms in these variables ξ are fundamental solutions of the wave equation. In the sequel, we limit ourselves to studying $\tilde{E}_+(t, \xi)$ and its inverse Fourier transform with respect to ξ, $E_+(t, x)$.

Let $\varphi = \varphi(x)$ denote an arbitrary test function in $\mathbf{R}^n$. We observe that $\tilde{E}_+(t, \xi)$ is a function of t with values in the space of bounded measurable functions of ξ. By definition of the Fourier transformation of tempered distributions, we have

$$\begin{aligned}\langle E_+(t, x), \varphi(x)\rangle &= \langle E_+(t, x), (\mathscr{F}\mathscr{F}^{-1}\varphi)(x)\rangle \\ &= \langle \tilde{E}_+(t, \xi), (\mathscr{F}^{-1}\varphi)(\xi)\rangle,\end{aligned}$$

where the first two brackets are those of the duality between distributions and test functions in the variable x, whereas the third one applies to the variable ξ. We have also denoted by $\mathscr{F}$ and $\mathscr{F}^{-1}$ the Fourier transformation and its inverse, respectively; the first one transforms distributions in ξ into distributions in x; $\mathscr{F}^{-1}$ transforms distributions in x into distributions in ξ. In summary, by using formula (6.4),

$$\begin{aligned}\langle E_+(t, x), \varphi(x)\rangle &= \int \tilde{E}_+(t, \xi)(\mathscr{F}^{-1}\varphi)(\xi)\, d\xi \\ &= (2\pi)^{-n}\int \tilde{E}_+(t, \xi)\left\{\int e^{i\langle x, \xi\rangle}\varphi(x)\, dx\right\} d\xi.\end{aligned}$$

We cannot interchange the two integrations and write

$$E_+(t, x) = (2\pi)^{-n}\int e^{i\langle x, \xi\rangle}\tilde{E}_+(t, \xi)\, d\xi, \tag{7.15}$$

because E is not integrable with respect to ξ (therefore, we cannot apply Fubini's theorem). But we can introduce a convergence factor, like $\exp(-\varepsilon|\xi|)$, and interpret (7.15) as meaning

$$E_+(t, x) = (2\pi)^{-n}\lim_{\varepsilon \to +0}\int \exp(i\langle x, \xi\rangle - \varepsilon|\xi|)\tilde{E}_+(t, \xi)\, d\xi. \tag{7.16}$$

We may rewrite (7.15) in the following manner:

$$E_+(t, x) = H(t)\mathcal{U}(t, x),$$

where

(7.17)

$$\begin{aligned}\mathcal{U}(t, x) &= (2\pi)^{-n} \int e^{i\langle x, \xi\rangle} \frac{\sin(|\xi| t)}{|\xi|}\, d\xi \\ &= (2\pi)^{-n} \int \left\{ \exp\left(i\left\langle \xi, x + \frac{\xi}{|\xi|} t\right\rangle\right) - \exp\left(i\left\langle \xi, x - \frac{\xi}{|\xi|} t\right\rangle\right)\right\} \frac{d\xi}{2i|\xi|},\end{aligned}$$

where the integrals should be understood in the same sense as the one in (7.15), with a hidden convergence factor. The distribution $\mathcal{U}(t, x)$ will play an important role in the theory of the Cauchy problem for the wave equation.

Taking the Fourier transform of $E_+(t, x)$ with respect to the variable x has one great defect: It breaks the Lorentz invariance, which is one of the most important features of the wave equation. In order to recover this invariance, we must also perform a Fourier transformation with respect to t. The Fourier transform of $E_+(t, x)$ with respect to (t, x) is, of course, equal to the Fourier transform of $\tilde{E}_+(t, \xi)$ with respect to t:

$$\hat{E}_+(\tau, \xi) = \int_0^{+\infty} e^{-i\tau t} \frac{\sin(|\xi| t)}{|\xi|}\, dt.$$

This is a divergent integral. We must introduce a convergence factor:

$$\hat{E}_+(\tau, \xi) = \lim_{\varepsilon \to +0} \int_0^{+\infty} e^{-it(\tau - i\varepsilon)} \sin(|\xi| t) \frac{dt}{|\xi|}.$$

The integral under the limit sign is equal to

$$\begin{aligned}&\frac{1}{2|\xi| i} \int_0^{+\infty} \{\exp[-it(\tau - i\varepsilon - |\xi|)] - \exp[-it(\tau - i\varepsilon + |\xi|)]\}\, dt \\ &\quad = -\frac{1}{2|\xi|} \{(\tau - i\varepsilon - |\xi|)^{-1} - (\tau - i\varepsilon + |\xi|)^{-1}\} \\ &\quad = -((\tau - i\varepsilon)^2 - |\xi|^2)^{-1}.\end{aligned}$$

Thus the Fourier transform we seek is given by

$$\hat{E}_+(\tau, \xi) = \lim_{\varepsilon \to +0} \frac{-1}{(\tau - i\varepsilon)^2 - |\xi|^2}. \tag{7.18}$$

Let $\varphi(t, x)$ be an arbitrary test function in $\mathbf{R}^{n+1}$ and set $\check{\varphi}(t, x) = \varphi(-t, -x)$. By definition of the Fourier transformation on tempered distributions, we have

$$\langle E_+, \check{\varphi}\rangle = -(2\pi)^{-n-1} \lim_{\varepsilon\to+0} \iint \frac{\hat{\varphi}(\tau, \xi)\, d\tau\, d\xi}{(\tau - i\varepsilon)^2 - |\xi|^2}. \tag{7.19}$$

Let us first look at the integration with respect to τ on the right-hand side of (7.19). We observe that the integrand is a holomorphic function of τ regarded as a complex variable, for $\operatorname{Im}\tau < \varepsilon$. On every horizontal line $\operatorname{Im}\tau = \text{const} < \varepsilon$, the integrand decays faster than any power of $|\operatorname{Re}\tau|^{-1}$ when the latter tends to zero. We may therefore apply Cauchy's integral theorem as we have done in §6.1. We find that the integral under consideration is equal to

$$\iint \frac{\hat{\varphi}(\tau - ia, \xi)\, d\tau\, d\xi}{(\tau - ia - i\varepsilon)^2 - |\xi|^2}, \tag{7.20}$$

where a is any number greater than zero. But in the integral (7.20) we can safely go to the limit as $\varepsilon > 0$ goes to zero. We obtain

$$\langle E_+, \check{\varphi}\rangle = -(2\pi)^{-n-1} \iint \frac{\hat{\varphi}(\tau - ia, \xi)}{(\tau - ia)^2 - |\xi|^2}\, d\tau\, d\xi. \tag{7.21}$$

The integral on the right-hand side of (7.21) may be viewed as an integral over an n-dimensional real submanifold of $\mathbf{C}^n = \mathbf{R}^{2n}$. But notice that we have gone into the nonreal space only along the τ direction; this still gives a privileged role to the time variable, as we have chosen it: this is not Lorentz-invariant. We are now going to deform the contour of integration with respect to the variables $\xi_1, \ldots, \xi_n$ (each time, by applying the Cauchy integral theorem) so as to reestablish this invariance.

We begin with the integral

$$\int_{-\infty}^{+\infty} \frac{\hat{\varphi}(\tau - ia, \xi)\, d\xi_1}{(\tau - ia)^2 - |\xi|^2}. \tag{7.22}$$

We can move the integration to a horizontal line $\{\zeta_1 \in \mathbf{C}^1;\ \zeta_1 = \xi_1 - ib_1,\ \xi_1 \in \mathbf{R}^1\}$ (b_1 real) provided that the denominator does not vanish in a slab $-b_1 - c < \operatorname{Im}\zeta_1 < +c$, $c > 0$. We have

$$(\tau - ia)^2 - (\xi_1 + i\eta_1)^2 - |\xi'|^2 = \tau^2 - |\xi|^2 - a^2 + \eta_1^2 - 2i(a\tau + \eta_1\xi_1).$$

We have used the notation $\xi' = (\xi_2, \ldots, \xi_n)$. If the quantity just given were to vanish, we should have

$$\tau^2 - |\xi|^2 = a^2 - \eta_1^2, \qquad a\tau + \eta_1\xi_1 = 0.$$

In other words, the vectors (τ, ξ) and $(a, \eta_1, 0, \ldots, 0)$ should be orthogonal while lying on the same level surface of the Lorentz quadratic form. This is impossible if we require that $a^2 - \eta_1^2 > 0$: In the interior of the *forward light-cone*, there are not two vectors which are orthogonal. Furthermore, if we require

$$a^2 - b_1^2 > 0,$$

we will also have $a^2 - \eta_1^2 > 0$ for all η_1 such that $-b_1 - c < \eta_1 < c$ for some $c > 0$ (depending on a). We conclude that the integral (7.22) is equal to

$$\int_{-\infty}^{+\infty} \frac{\hat{\varphi}(\tau - ia, \xi_1 - ib_1, \xi')\, d\xi_1}{(\tau - ia)^2 - (\xi_1 - ib_1)^2 - |\xi'|^2}.$$

Next we deform into the complex plane the contour of integration with respect to $\xi_2, \ldots, \xi_n$ successively, by using exactly the same reasoning. We move these contours to horizontal lines

$$\{\zeta_j \in C^1 ; \zeta_j = \xi_j - ib_j, \xi_j \in \mathbf{R}^1\}, \qquad b_j \text{ real}, \qquad 2 \leqq j \leqq n.$$

We may do this by the Cauchy integral theorem provided that $b = (b_1, \ldots, b_n)$ satisfies the condition $a^2 - |b|^2 > 0$. For in this case, there is no real n-vector η such that

$$-b_j - c < \eta_j < c, \qquad j = 1, \ldots, n, \qquad \text{for some} \quad c > 0,$$

and such that $\tau^2 - |\xi|^2 = a^2 - |\eta|^2$, $\langle(\tau, \xi), (a, \eta)\rangle = 0$. Finally we have proved

PROPOSITION 7.1. *Let $(a, b) \in \mathbf{R}^{n+1}$ be such that $|b|^2 < a^2$, $a > 0$. Then, given any $\varphi \in C_c^\infty(\mathbf{R}^{n+1})$,*

$$\langle E_+, \check{\varphi}\rangle = -(2\pi)^{-n-1} \iint \frac{\hat{\varphi}(\tau - ia, \xi - ib)}{(\tau - ia)^2 - (\xi - ib)^2}\, d\tau\, d\xi. \tag{7.23}$$

We have used the following notation: If ζ is a complex n-vector, $\zeta = (\zeta_1, \ldots, \zeta_n)$,

$$(\zeta)^2 = (\zeta_1)^2 + \cdots + (\zeta_n)^2.$$

We shall derive from Proposition 7.1 that E_+ is invariant under a remarkable subgroup of the Lorentz group. But we must first specify what we mean by a distribution being invariant under a linear transformation of the Euclidean space. Let the space be $\mathbf{R}^N$, y the variable in it. If the distribution under study, u, is a function, also denoted by u, and if T is any linear transformation of $\mathbf{R}^N$, the *transform of u under T* is the function

$$u^T(y) = u(T^{-1}y).$$

Viewing u and u^T as distributions this means that, for every $\varphi \in C_c^\infty(\mathbf{R}^N)$,

$$\langle u^T, \varphi\rangle = \int u(T^{-1}y)\varphi(y)\,dy = \int u(y)\varphi(Ty)|\det T|\,dy$$

$$= |\det T|\langle u, \varphi^{T^{-1}}\rangle.$$

This suggests the definition of u^T when u is an arbitrary distribution:

$$\text{(7.24)} \qquad \langle u^T, \varphi\rangle = |\det T|\langle u, \varphi^{T^{-1}}\rangle, \qquad \varphi \in C_c^\infty(\mathbf{R}^N).$$

We now ask, what is the Fourier transform of u^T? Here again we restrict ourselves to the case where u is a function, in fact a C^∞ function rapidly decaying at infinity. The formula we obtain will extend at once to *tempered* distributions, either by using the density of $\mathscr{S}$ in $\mathscr{S}'$ or by the definition of the Fourier transforms of such distributions. We have

$$(u^T)(\eta) = \int e^{i\langle y,\eta\rangle}u(T^{-1}y)\,dy = |\det T| \int e^{i\langle Ty,\eta\rangle}u(y)\,dy$$

$$= |\det T| \int \exp(i\langle y, {}^tT\eta\rangle)u(y)\,dy = |\det T|\,u({}^tTy),$$

where ${}^tT$ is the *transpose* of T. In other words

$$\text{(7.25)} \qquad \widehat{u^T} = |\det T|\hat{u}^{{}^tT^{-1}}.$$

The *Lorentz group* in $\mathbf{R}^{n+1}$ is the group of linear transformations of $\mathbf{R}^{n+1}$ which preserves the quadratic form $t^2 - |x|^2$. Among the Lorentz transformations we isolate those which preserve the forward light-cone

$$\Gamma_+ = \{(x, t) \in \mathbf{R}^{n+1};\ |x|^2 \leqq t^2,\ t \geqq 0\}.$$

These form a subgroup $\mathscr{L}_+$ of the Lorentz group. A Lorentz transformation belongs to $\mathscr{L}_+$ if, given an arbitrary point (x, t) such that $x^2 < t^2$ and $t > 0$, its transform (x', t') satisfies $t' > 0$. (The subgroup $\mathscr{L}_+$ is the connected component of the identity in the Lorentz group.) All Lorentz transformations have determinants equal to ± 1; the ones in $\mathscr{L}_+$ have determinants equal to $+1$. Note also that $T \mapsto {}^tT^{-1}$ is an isomorphism (for the topological group structure) of $\mathscr{L}_+$ onto itself (Exercise 7.10). To say that u is *invariant under* $\mathscr{L}_+$ is to say that $u^T = u$ for all $T \in \mathscr{L}_+$.

PROPOSITION 7.2. *The distribution E_+ is invariant under $\mathscr{L}_+$.*

Proof. We must prove that, for any $T \in \mathscr{L}_+$, $\langle E_+, \varphi\rangle = \langle E_+, \varphi^{T^{-1}}\rangle$. We might in fact replace $\varphi(x)$ by $\check{\varphi}(x) = \varphi(-x)$ and note that $(\check{\varphi})^{T^{-1}} = (\varphi^{T^{-1}})^\vee$. We use (7.23) and (7.25), the latter with φ in the place of u. We see that

$$\langle E_+, (\varphi^{T^{-1}})^\vee\rangle = -(2\pi)^{-n-1} \iint \frac{\hat{\varphi}({}^tT[(\tau, \xi) - i(a, b)])}{(\tau - ia)^2 - (\xi - ib)^2}\,d\tau\,d\xi.$$

We change variables in the integral, setting $(\tau', \xi') = {}^tT(\tau, \xi)$. We write $(a', b') = {}^tT(a, b)$. Then

$$(\tau - ia)^2 - (\xi - ib)^2 = (\tau^2 - |\xi|^2) - (a^2 - |b|^2) - 2i(a\tau - \langle b, \xi\rangle)$$
$$= (\tau'^2 - |\xi'|^2) - (a'^2 - |b'|^2) - 2i(a'\tau' - \langle b', \xi'\rangle),$$

for any linear transformation which preserves the quadratic form $\tau^2 - |\xi|^2$ must also preserve the associated bilinear form $(a\tau - \langle b, \xi\rangle)$. We see therefore that

$$\langle E_+, (\varphi^{T^{-1}})^\vee\rangle = -(2\pi)^{-n-1} \iint \frac{\hat{\varphi}(\tau' - ia', \xi' - ib')\, d\tau'\, d\xi'}{(\tau' - ia')^2 - (\xi' - ib')^2}.$$

But (a', b') belongs to the interior of Γ_+ since ${}^tT \in \mathscr{L}_+$. Proposition 7.2 follows therefore from Proposition 7.1. Q.E.D.

We conclude this section by deriving an important consequence of Proposition 7.2:

PROPOSITION 7.3. *$E_+(t, x)$ vanishes identically in the complement of the forward light-cone*

$$\Gamma_+ = \{(t, x) \in \mathbf{R}^{n+1};\ |x|^2 \leqq t^2, t \geqq 0\}.$$

Proof. We know that $E_+(t, x)$ vanishes for $t < 0$. Let T be some transformation belonging to $\mathscr{L}_+$; E_+^T, being equal to E_+, vanishes for $t < 0$, and therefore E_+ must vanish on the image under T^{-1} of the half-space $t < 0$. Let then $\mathscr{P}$ be an n-dimensional plane in $\mathbf{R}^{n+1}$ whose intersection with Γ_+ is reduced to the vertex of Γ_+, the origin. It suffices to show that there is a transformation belonging to $\mathscr{L}_+$ transforming $\mathscr{P}$ into the hyperplane $t = 0$. The line L, orthogonal to $\mathscr{P}$ (in the sense of the Euclidean norm on $\mathbf{R}^{n+1}$), is a straight line, half of which is in the interior of Γ_+. It suffices to show that there is a transformation in $\mathscr{L}_+$ which maps this half-line on the positive t-axis. In order to show this, one performs a space rotation, that is, a rotation affecting only the variables x, which transforms the line L into a line lying in the two-dimensional plane (t, x^1). This is clearly possible. We are thus reduced to a two-dimensional problem: We must find a Lorentz transformation involving only the two variables t, x^1, therefore preserving the quadratic form $t^2 - (x^1)^2$, and preserving the forward light-cone, mapping a straight line with equation

(7.26) $$x^1 = \lambda t, \qquad |\lambda| < 1,$$

onto the t-axis.

It is easily seen (Exercise 7.10) that all Lorentz transformations in $\mathbf{R}^2$ are of the form

(7.27) $$\begin{pmatrix} \varepsilon \cosh\theta & \varepsilon \sinh\theta \\ \eta \sinh\theta & \eta \cosh\theta \end{pmatrix},$$

where θ, ε, η are real numbers and $\varepsilon^2 = \eta^2 = 1$. The ones which preserve the forward light-cone correspond to the values $\varepsilon = \eta = +1$. If we choose the argument θ so as to have

$$\lambda + \tanh \theta = 0,$$

which is possible since $|\lambda| < 1$, the matrix (7.27) transforms the line (7.26) into the t-axis. Q.E.D.

Let us return to the distribution $\mathscr{U}(t, x)$ defined in (7.17). If we look at its Fourier transform with respect to x, $\sin(|\xi|t)/|\xi|$, we see at once that it is continuous, and in fact a C^∞ function of t, with values in the space $\mathscr{S}'_\xi$ of tempered distributions in the ξ variables. Moreover, this function vanishes at the origin. In order to see the latter, it suffices to show that, given any function $\varphi \in \mathscr{C}^\infty_c(\mathbf{R}_n)$,

$$\int \frac{\sin(|\xi|t)}{|\xi|} \varphi(\xi)\, d\xi \to 0 \qquad \text{as} \quad |t| \to 0,$$

which is obvious. We see therefore that $E_+(t, x)$ is a *continuous* function of real t, valued in the space of distributions in x,

$$E_+(t, x) = \begin{cases} \mathscr{U}(t, x) & \text{when} \quad t > 0, \\ 0 & \text{when} \quad t \leqq 0. \end{cases}$$

From this and the fact that $\mathscr{U}(-t, x) = -\mathscr{U}(t, x)$, we see that Proposition 7.3 implies

COROLLARY 7.1. *The distribution $\mathscr{U}(t, x)$ vanishes identically outside the light-cone $\Gamma = \{(x, t) \in \mathbf{R}^{n+1};\ |x|^2 \leqq t^2\}$.*

Exercises

7.1. Set $u = t^2 - r^2$ and consider a twice continuously differentiable function on $\mathbf{R}^{n+1}$, $f(t, x) = F(t^2 - r^2)$ $(r = |x|)$.

(i) Show that there is a differential operator L in the single variable u such that

$$\left\{\left(\frac{\partial}{\partial t}\right)^2 - \Delta_x\right\} f(t, x) = LF(u).$$

Give the expression of L.

(ii) Let T be the distribution on $\mathbf{R}^{n+1}$ defined by the function $f(t, x)$. Show that there is a differential operator M in u such that the distribution

$$\left\{\left(\frac{\partial}{\partial t}\right)^2 - \Delta_x\right\}T$$

is defined by the function $MF(u)$. Give the expression of M and compare with L in (i).

7.2. If you were to solve (7.2) as if E were a function, not a distribution, and if you were to make the change of variables (7.3), solve the transformed equation, and revert to the coordinates t, x, what is the value of E that you would find?

7.3. What is the inverse Fourier transform, with respect to (τ, ξ), of

$$\hat{\varphi}(\tau - ia, \xi),$$

and that of

$$(R^2 + (\tau - ia)^2 + |\xi|^2)^k \hat{\varphi}(\tau - ia, \xi),$$

where $\hat{\varphi}$ is the Fourier transform of $\varphi \in C_c^\infty(\mathbf{R}^{n+1})$, a any real number and k any nonnegative integer?

7.4. Suppose $a > 1$.

(i) Show that, for all $(\tau, \xi) \in \mathbf{R}^{n+1}$,

$$|(\tau - ia)^2 - |\xi|^2| \geqq a(\tau^2 + |\xi|^2 + a^2)^{1/2}.$$

(ii) Show that if $k > n/2$,

$$\iint \{(\tau - ia)^2 - |\xi|^2\}^{-1}(1 + a^2 + (\tau - ia)^2 + |\xi|^2)^{-k}\, d\tau\, d\xi$$

is bounded by a constant independent of $a > 1$.

7.5. Using the results in Exercises 7.3 and 7.4 show that if k is an integer greater than $n/2$, there is a constant C independent of $a > 1$ such that

(7.28)
$$\left|\iint \frac{\hat{\varphi}(\tau - ia, \xi)\, d\tau\, d\xi}{(\tau - ia)^2 - |\xi|^2}\right|$$
$$\leqq C \iint e^{-2at}\left|\left(1 + a^2 - \frac{\partial^2}{\partial t^2} - \Delta_x\right)^k \varphi(t, x)\right|^2 dt\, dx.$$

Derive from this inequality that the linear functional $\phi \to \langle E_+, \phi\rangle$, defined by (7.21), is indeed a distribution on $\mathbf{R}^{n+1}$ (ignore the earlier definitions of E_+).

7.6. Derive from Exercise 7.5 that, for all $\phi \in C_c^\infty(\mathbf{R}^{n+1})$,

$$|\langle E_+, \phi\rangle| \leqq \text{const} \int e^{2at} |(1 + a^2 - \Delta_{t,x})^k \phi(t, x)|^2 \, dt \, dx, \tag{7.29}$$

with a constant independent of $a > 1$ [$\Delta_{t,x}$ stands for the Laplacian in the $(n + 1)$ variables (t, x)]. Deduce from inequality (7.29) that, if the support of ϕ lies in the space $t < 0$, we have $\langle E_+, \phi\rangle = 0$; in other words, deduce that the support of E_+ is contained in the half-space $t \geqq 0$.

7.7. Construct explicitely a fundamental solution of the differential operator in $\mathbf{R}^2$, $(\partial/\partial t)^2 - (\partial/\partial x)^2 + \lambda$, where λ is an arbitrary complex number.

7.8. Prove the analog of Proposition 7.3 for the differential operator (sometimes called the *Klein–Gordon operator*)

$$K_m = \frac{\partial^2}{\partial t^2} - \Delta_x + m^2. \tag{7.30}$$

7.9. Let V be a nonzero real number. Prove the analog of Proposition 7.3 for the operator

$$\frac{1}{V^2}\frac{\partial^2}{\partial t^2} - \Delta_x. \tag{7.31}$$

What is the equation of the light-cone in this case?

7.10. Prove that every Lorentz transformation in $\mathbf{R}^2$ is of the form (7.27). Identify any 2×2 real matrix with a point in $\mathbf{R}^4$ by regarding its four entries as coordinates. Describe the subset of $\mathbf{R}^4$ corresponding to the Lorentz group. Show that it consists of four connected components.

7.11. Let $\mathscr{L}$ denote the Lorentz group in $\mathbf{R}^{n+1}$, $n \geqq 1$.

(i) Prove that $T \mapsto {}^tT^{-1}$ is an isomorphism (for the group structure) of $\mathscr{L}$ onto itself, and of $\mathscr{L}_+$ onto itself.

(ii) Prove that, if $T \in \mathscr{L}$, and if we write

$$B(y, y') = y_{n+1}y'_{n+1} - y_1y'_1 - \cdots - y_n y'_n, \qquad y, y' \in \mathbf{R}^{n+1},$$

then $B(Ty, Ty') = B(y, y')$.

(iii) Prove that, for any $T \in \mathscr{L}$, $\det T = \pm 1$.

7.12. Using the result in Exercise 7.10 and the properties of the group of rotations in $\mathbf{R}^n$, prove that the Lorentz group $\mathscr{L}$ in $\mathbf{R}^{n+1}$ consists exactly of four connected components.

7.13. Let us call an *orbit* of $\mathscr{L}$ (resp., of $\mathscr{L}_+$) any subset of $\mathbf{R}^{n+1}$ which is the image of $\mathscr{L}$ (resp., of $\mathscr{L}_+$) under a mapping $T \mapsto Tx_0$, for some $x_0 \in \mathbf{R}^{n+1}$. Describe all the orbits of $\mathscr{L}$ (resp., of $\mathscr{L}_+$) and point out any difference there might be between the case $n = 1$ and the cases $n > 1$.

8

More on the Supports and Singular Supports of the Fundamental Solutions of the Wave Equation

It is not easy to obtain an explicit formula for $\mathscr{U}$ and E_+ (see Sect. 7) in the coordinates (t, x), out of their Fourier integral representation (7.17), (7.19), or (7.21). Explicit formulas in the case of two and three space variables are derived in the Appendix to this section, but much valuable information can easily be extracted from the Fourier integrals, as we show here. We are going to prove two important properties of E_+, one concerning its *support* (the smallest closed set outside which it vanishes) and the other concerning its *singular support* (the smallest closed set outside which it is a C^∞ function). We begin with the property of its support (classically known as the *Huyghens principle*).

THEOREM 8.1. *If n is odd and $n > 1$, the support of E_+ is exactly equal to the boundary of the forward light-cone, i.e., to the set $\{(t, x);\ t^2 - |x|^2 = 0, t \geqq 0\}$. For all other values of n, it is exactly equal to the forward light-cone itself.*

Proof. (1) *Case $n = 1$.* In this case, we show that E_+ is equal to the distribution E_1 in the plane defined by (7.7). We have, by virtue of (7.12) and (7.15),

$$E_+ = (2\pi)^{-1} H(t) \int_0^{+\infty} \frac{\sin(\xi(x + t)) - \sin(\xi(x - t))}{\xi} d\xi.$$

Direct computation of the integral would prove our contention but it suffices to observe that $E_+ - E_1$ is a solution of the homogeneous wave equation in the plane. All the solutions of this equation are known: They are of the form $S(t - x) + T(t + x)$, where S and T are distributions in one variable. Unless they are identically zero, their support cannot be contained in the half-plane $t \geqq 0$.

(2) *Case $n > 1$, odd.* It suffices to prove that $E_+ \equiv 0$ in the interior $\mathring{\Gamma}_+$ of the forward light-cone Γ_+: For the support of E_+, which we know (by Proposition 7.2) consists of orbits of $\mathscr{L}_+$ and (by Proposition 7.3) is contained in Γ_+, then must be equal to the boundary of Γ_+ (it cannot just be $\{0\}$!). In particular, t will always be assumed to be greater than zero. In $\mathring{\Gamma}_+$, E_+ is the limit, as $\varepsilon \to +0$, of the distributions.

$$(8.1) \quad E_+^\varepsilon = (2\pi)^{-n} \int_{S_{n-1}} \int_0^{+\infty} \exp[-(\varepsilon - ir\langle\theta, \omega\rangle)\rho] \sin(\rho t)\rho^{n-2}\, d\rho\, d\omega,$$

where ρ, ω (resp. r, θ) denote the spherical coordinates in the space $\mathbf{R}_n$ (resp. $\mathbf{R}^n$); ω (resp. θ) is the variable on the unit sphere S_{n-1} (resp. S^{n-1}) [see (7.16)]. Now, clearly

$$(8.2) \qquad (2\pi)^n E_+^\varepsilon = (-1)^{n-2}\left(\frac{\partial}{\partial\varepsilon}\right)^{n-2} \int_{S_{n-1}} I_\varepsilon(s, t)\, d\omega,$$

where $s = r\langle\theta, \omega\rangle$ and

$$I_\varepsilon(s, t) = \int_0^{+\infty} e^{-(\varepsilon - is)\rho} \sin(\rho t)\, d\rho = \frac{i}{2}\left\{\frac{1}{\varepsilon - i(s-t)} - \frac{1}{\varepsilon - i(s+t)}\right\}.$$

We note that $(-1)^{n-2}(\partial/\partial\varepsilon)^{n-2}(\varepsilon - a)^{-1} = (n-2)!(\varepsilon - a)^{1-n}$, from which

$$(8.3) \quad (2\pi i)^n E_+^\varepsilon = -\frac{(n-2)!}{2} \int_{S_{n-1}} \left\{\frac{1}{(t - s - i\varepsilon)^{n-1}} + \frac{(-1)^n}{(t + s + i\varepsilon)^{n-1}}\right\} d\omega.$$

Let now $\varphi = \varphi(t) \in C_c^\infty(\mathbf{R}^1)$ with support in the positive half-line $t > 0$. Later φ will be made to depend also on x. For all $k > 0$,

$$(k-1)! \int_0^{+\infty} (t-a)^{-k}\varphi(t)\, dt = (k-2)! \int_0^{+\infty} (t-a)^{-k+1}\varphi'(t)\, dt.$$

We see therefore that

$$(8.4) \quad (2\pi i)^n \int E_+^\varepsilon(t, x)\varphi(t)\, dt$$

$$= -\frac{1}{2}\int_{S_{n-1}} \int_0^\infty \left\{\frac{1}{t - s - i\varepsilon} + \frac{(-1)^n}{t + s + i\varepsilon}\right\}\varphi^{(n-2)}(t)\, dt\, d\omega.$$

This formula is valid whatever the *parity* of n. Let us assume now that n is odd ($n = 3, 5, \ldots$). We notice that the integral with respect to ω over S_{n-1} of the function

$$g_\varepsilon(s, t) = (t - s - i\varepsilon)^{-1} - (t + s + i\varepsilon)^{-1}$$

is equal to the integral of its *even* part, $g_\varepsilon^+(s, t) = \frac{1}{2}[g_\varepsilon(s, t) + g_\varepsilon(-s, t)]$. We have

$$g_\varepsilon^+(s, t) = h_\varepsilon(t - s) + h_\varepsilon(t + s),$$

where

$$h_\varepsilon(\tau) = \tfrac{1}{2}[(\tau - i\varepsilon)^{-1} - (\tau + i\varepsilon)^{-1}] = \frac{i\varepsilon}{\tau^2 + \varepsilon^2}.$$

Consequently,

$$\int_0^{+\infty} h_\varepsilon(t - s)\varphi^{(n-2)}(t)\,dt = i\int_{-s/\varepsilon}^{+\infty} \phi^{(n-2)}(s + \varepsilon\tau)\frac{d\tau}{1 + \tau^2},$$

$$\int_0^{+\infty} h_\varepsilon(t + s)\varphi^{(n-2)}(t)\,dt = i\int_{s/\varepsilon}^{+\infty} \varphi^{(n-2)}(-s + \varepsilon\tau)\frac{d\tau}{1 + \tau^2},$$

and, as $\varepsilon \to +0$,

$$\int_0^{+\infty} g_\varepsilon^+(s, t)\varphi^{(n-2)}(t)\,dt \to i\pi\varphi^{(n-2)}(|s|).$$

At this point, we take $\varphi = \varphi(t, x) \in C_c^\infty(\mathbf{R}^{n+1})$ with support in the open cone

(8.5) $$t > |x|.$$

From (8.4) and from the preceding computation we reach the conclusion that

$$\iint E_+(t, x)\varphi(t, x)\,dt\,dx = -\tfrac{1}{4}(2\pi i)^{1-n}\int_{\mathbf{R}^n}\int_{S^{n-1}} \varphi^{(n-2)}(|s|, x)\,d\omega\,dx.$$

But $|s| \leqq r = |x|$, therefore the point $(|s|, x)$ does not belong to the set (8.5) and $\varphi(|s|, x) = 0$.

(3) *Case n even.* We return to (8.4), taking $\varphi = \varphi(t, x)$ as above and noting, as before, that $|s| < t$ on the support of φ. We can go to the limit right away in (8.4) as $\varepsilon \to +0$ and conclude that

(8.6) $$\iint E_+(t, x)\varphi(t, x)\,dt\,dx$$

$$= -(2\pi i)^{-n}\int_{\mathbf{R}^n}\int_0^{+\infty}\int_{S_{n-1}} \frac{t}{t^2 - s^2}\left(\frac{\partial}{\partial t}\right)^{n-2}\varphi(t, x)\,dt\,dx\,d\omega.$$

This means that, in the interior of the forward light-cone, E_+ is defined by the function

(8.7) $$-(2\pi i)^{-n}\left(\frac{\partial}{\partial t}\right)^{n-2}\int_{S_{n-1}} \frac{t\,d\omega}{t^2 - r^2\langle\omega, \theta\rangle^2}.$$

It is clear that (8.7) is an analytic function of (t, r^2), hence of (t, x) in the set (8.5). If it were to vanish in any open subset of (8.5), it would vanish in the whole of (8.5), since this set is connected. If this were true, the function

$$\int_{S_{n-1}} \frac{t\,d\omega}{t^2 - r^2\langle\omega, \theta\rangle^2} \tag{8.8}$$

should be a polynomial with respect to t (of degree $< n-2$). But in every cone $r < \lambda t$, $\lambda < 1$, it tends to zero as $t \to +\infty$; therefore, if it were a polynomial in t, it should be identically zero. It is not, since it is $\geqq t^{-1}|S_{n-1}|$, with $|S_{n=1}| = $ Area of S_{n-1}. Q.E.D.

We have also proved the second result we had in mind, that concerning the singular support of E_+:

THEOREM 8.2. *Off the boundary of the forward light-cone, $E_+(t, x)$ is an analytic function of* (t, x).

The assertion is evident when $n = 1$, for in this case E_+ is a constant off the boundary of the forward light-cone. When n is odd and $n > 1$, $E_+ = 0$ off that boundary. When n is even, we have shown that E_+ is defined in the interior of the forward light-cone by the function (8.7) which, as we have already said, is an analytic function of (t, r^2).

Appendix Explicit Formulas for E_+ in Space Dimensions Two and Three

Our starting point will be (8.3).

E_+ in two space dimensions

In this case, (8.3) reads

$$E_+^\varepsilon = \frac{H(t)}{8\pi^2} \int_{S_1} \left\{ \frac{1}{t - s - i\varepsilon} + \frac{1}{t + s + i\varepsilon} \right\} d\omega. \tag{8.9}$$

We recall that $s = r\langle\theta, \omega\rangle$, where θ is the variable point in S^1 and $r = |x|$. But any integral of the form $\int_{S_{n-1}} f(\langle\theta, \omega\rangle)\,d\omega$ is independent of $\theta \in S^{n-1}$; we are therefore free to choose θ as we like. We choose it to be the unit vector along the first coordinate axis. Therefore $\langle\theta, \omega\rangle = \cos\varphi$ if we agree to write $\omega = (\cos\varphi, \sin\varphi)$. We have

$$E_+^\varepsilon = \frac{H(t)}{8\pi^2} \int_0^{2\pi} \left\{ \frac{1}{t - r\cos\varphi - i\varepsilon} + \frac{1}{t + r\cos\varphi + i\varepsilon} \right\} d\varphi.$$

Let us look at the integral

$$J_+^\varepsilon = \frac{1}{2\pi}\int_0^{2\pi} \frac{d\varphi}{t + r\cos\varphi + i\varepsilon} = \oint_{|z|=1} \frac{1}{t + \frac{1}{2}r(z + z^{-1}) + i\varepsilon}\frac{dz}{2\pi i z}.$$

Let us set $a = r/(t + i\varepsilon)$. Then

$$\tfrac{1}{2}(t + i\varepsilon)J_+^\varepsilon = (2\pi i)^{-1}\int_{|z|=1} \frac{dz}{az^2 + 2z + a}.$$

Similarly,

$$J_-^\varepsilon = \frac{1}{2\pi}\int_0^{2\pi} \frac{d\varphi}{t - r\cos\varphi - i\varepsilon} = \oint_{|z|=1} \frac{1}{t - \frac{1}{2}r(z + z^{-1}) - i\varepsilon}\frac{dz}{2\pi i z},$$

from which

$$\tfrac{1}{2}(t - i\varepsilon)J_-^\varepsilon = -(2\pi i)^{-1}\int_{|z|=1} \frac{dz}{\bar{a}z^2 - 2z + \bar{a}}.$$

Let us first consider the case where $r > t \geqq 0$. If $\varepsilon > 0$ is sufficiently small, we have $|a| > 1$. In fact, let us set

$$a^{-1} = \frac{t + i\varepsilon}{r} = b + i\eta, \qquad b = \frac{t}{r} < 1,$$

$\eta \to +0$ when $\varepsilon \to +0$. The polynomial $f_+(z) = z^2 + 2(b + i\eta)z + 1$ has two roots:

$$\lambda_\alpha = -b - i\eta + \alpha i(1 - (b + i\eta)^2)^{1/2}, \qquad \alpha = \pm 1.$$

Modulo η^2, we have

$$\begin{aligned}
\lambda_\alpha &= -b - i\eta + \alpha i(1 - b^2)^{1/2}\left(1 - \frac{2ib\eta}{1 - b^2}\right)^{1/2} \\
&= -b - i\eta + \alpha i(1 - b^2)^{1/2}\left(1 - \frac{ib\eta}{1 - b^2}\right) \\
&= -b - i\eta + \alpha i(1 - b^2)^{1/2} + \alpha b\eta(1 - b^2)^{-1/2} \\
&= -b\{1 - \alpha\eta(1 - b^2)^{-1/2}\} + \alpha i\{(1 - b^2)^{1/2} - \alpha\eta\}, \\
&= [-b + \alpha i(1 - b^2)^{1/2}]\{1 - \alpha\eta(1 - b^2)^{-1/2}\}.
\end{aligned}$$

We derive from this that, for $\eta > 0$ sufficiently small,

$$|\lambda_\alpha| \;\#\; 1 - \alpha\eta(1 - b^2)^{-1/2}. \tag{8.10}$$

Similarly, the polynomial $f_-(z) = z^2 - 2(\bar{a})^{-1}z + 1 = z^2 - 2(b - i\eta)z + 1$ has two roots:

$$\mu_\alpha = b - i\eta + \alpha i(1 - (b - i\eta)^2)^{1/2}$$

such that, for small $\eta > 0$,

$$|\mu_\alpha| \;\#\; 1 - \alpha\eta(1 - b^2)^{-1/2} \tag{8.11}$$

(notice that one goes from f_+ to f_- by substituting $-b$ for b). The roots λ_α

and μ_α belong to the disk $|z| < 1$ when $\alpha = +1$ and lie outside of the closure of this disk when $\alpha = -1$. The residue of f_+^{-1} at $z = \lambda_{+1}$ is equal to

$$R_+ = \frac{1}{2i}(1 - (b + i\eta)^2)^{-1/2},$$

the one of f_-^{-1} at $z = \lambda_{+1}$ is equal to $R_- = (1/2i)(1 - (b - i\eta)^2)^{-1/2}$. Thus

$$J_+^\varepsilon + J_-^\varepsilon = \frac{1}{2ir}\left\{(1 - (b + i\eta)^2)^{-1/2} - (1 - (b - i\eta)^2)^{-1/2}\right\}$$

converges to zero when $\eta \to +0$, i.e., when $\varepsilon \to +0$. We have thus proved that $E_+ = 0$ when $r > t$, which we knew already by Proposition 7.3.

Consider now the case where $r \leqq t$; for every $\varepsilon > 0$, we have then $|a| < 1$. The polynomial f_+ given above now has the roots

$$\lambda_\alpha = -\frac{1}{a} + \alpha\frac{1}{a}(1 - a^2)^{1/2} \# -\frac{1}{a} + \alpha\frac{1}{a}\left(1 - \frac{a^2}{2}\right)$$

$$\# -\frac{1}{a}(1 - \alpha) - \alpha\frac{a}{2}.$$

If $\alpha = -1$, $\lambda_\alpha \# -(2/a)(1 - a^2/4)$ lies outside of the disk $|z| \leqq 1$, whereas if $\alpha = +1$, λ_α belongs to the open disk $|z| < 1$ (for $\varepsilon > 0$ sufficiently small). Since one passes from f_+ to f_- by substituting $-\bar{a}$ for a, a similar conclusion applies to the root μ_α of f_-: It belongs to the disk $|z| < 1$ when $\alpha = +1$, and to the exterior $|z| > 1$ when $\alpha = -1$. The residue of f_+^{-1} at λ_{+1} equals

$$R_+ = \frac{a}{2}(1 - a^2)^{-1/2},$$

the one of f_-^{-1} at μ_{+1} equals $R_- = -(\bar{a}/2)(1 - \bar{a}^2)^{-1/2}$, whence

$$\begin{aligned} J_+^\varepsilon + J_-^\varepsilon &= (t + i\varepsilon)^{-1}(1 - a^2)^{-1/2} + (t - i\varepsilon)^{-1}(1 - \bar{a}^2)^{-1/2} \\ &= 2\,\mathrm{Re}[(t + i\varepsilon)^{-1}(1 - a^2)^{-1/2}] = 2\,\mathrm{Re}[(t + i\varepsilon)^2 - r^2]^{-1/2} \end{aligned}$$

(the square root always stands for the branch that is greater than zero on the positive real numbers).

Finally we observe that the function $H(t^2 - r^2)(t^2 - r^2)^{-1/2}$ is locally integrable in $\mathbf{R}^3$ (where the measure is $r\,dr\,d\theta\,dt$). Indeed,

$$\int_{t=t_0}^{t=t_1}\int_{r=0}^{r=|t|}(t^2 - r^2)^{-1/2}r\,dr = \int_{t_0}^{t_1}|t|\,dt.$$

We see thus that $J_+^\varepsilon + J_-^\varepsilon$ converges to $2H(t^2 - r^2)(t^2 - r^2)^{-1/2}$. In other words,

$$E_+ = \begin{cases} \dfrac{1}{2\pi}(t^2 - r^2)^{-1/2} & \text{if } r < t, \\ 0 & \text{otherwise.} \end{cases} \tag{8.12}$$

E_+ in three space dimensions

When $n = 3$, (8.3) reads

$$(8.13)\quad -E_+^\varepsilon = \tfrac{1}{2}(2\pi i)^{-3}H(t)\int_{S_2}\left\{\frac{1}{(t-s-i\varepsilon)^2} - \frac{1}{(t+s+i\varepsilon)^2}\right\}d\omega$$

$$= \tfrac{1}{2}(2\pi i)^{-3}H(t)\left(-\frac{\partial}{\partial t}\right)\int_{S_2}\left\{\frac{1}{t-s-i\varepsilon} - \frac{1}{t+s+i\varepsilon}\right\}d\omega.$$

We use spherical coordinates in the ξ variables:

$$\xi_1 = \rho\cos\varphi_1\cos\varphi_2, \qquad \xi_2 = \rho\sin\varphi_1\cos\varphi_2, \qquad \xi_3 = \rho\sin\varphi_2,$$

and we choose θ, in $s = r\langle\theta, \omega\rangle$, to be the unit vector in the direction of the coordinate x^3, i.e., $\langle\theta, \omega\rangle = \sin\varphi_2$. We use the fact that

$$d\omega = \tfrac{1}{2}|\cos\varphi_2|\, d\varphi_1\, d\varphi_2,$$

if we allow both φ_1 and φ_2 to vary between 0 and 2π, and observe that

$$\int_{S_2} f(\sin\varphi_2)\, d\omega = \frac{1}{2}\int_0^{2\pi}\int_0^{2\pi} f(\sin\varphi_2)|\cos\varphi_2|\, d\varphi_1\, d\varphi_2$$

$$= 2\pi\int_0^{\pi/2}[f(\sin\varphi_2) + f(-\sin\varphi_2)]\cos\varphi_2\, d\varphi_2$$

$$= 2\pi\int_0^1 [f(u) + f(-u)]\, du,$$

which combines with (8.13) to yield

$$(8.14)\quad -E_+^\varepsilon = \frac{i}{2}(2\pi)^{-2}H(t)\left(-\frac{\partial}{\partial t}\right)\int_0^1\left\{\frac{1}{t-ru-i\varepsilon} + \frac{1}{t+ru-i\varepsilon} - \frac{1}{t+ru+i\varepsilon} - \frac{1}{t-ru+i\varepsilon}\right\}du$$

$$= (2\pi)^{-2}H(t)\frac{\partial}{\partial t}\int_0^1\left\{\frac{1}{(t+ru)^2+\varepsilon^2} + \frac{1}{(t-ru)^2+\varepsilon^2}\right\}\varepsilon\, du$$

$$= (2\pi)^{-2}H(t)\frac{1}{r}\frac{\partial}{\partial t}\int_{t-r}^{t+r}\frac{\varepsilon\, dv}{v^2+\varepsilon^2}$$

$$= (2\pi)^{-2}H(t)\frac{\varepsilon}{r}\left\{\frac{1}{(t+r)^2+\varepsilon^2} - \frac{1}{(t-r)^2+\varepsilon^2}\right\}.$$

Let now φ be an arbitrary element of $C_c^\infty(\mathbf{R}^4)$ and set (cf. Sect. 9)

$$\varphi_\natural(t, r) = (4\pi)^{-1} \int_{S^2} \varphi(t, r, \theta)\, d\theta.$$

By virtue of (8.14) we have

$$\langle E_+^\varepsilon, \varphi\rangle = 4\pi \int_0^{+\infty}\int_0^{+\infty} E_+^\varepsilon(t, r)\varphi_\natural(t, r) r^2\, dr\, dt$$

$$= -\varepsilon\pi^{-1}\int_0^{+\infty}\int_0^{+\infty} \varphi_\natural(t, r)\left\{\frac{1}{(t+r)^2+\varepsilon^2} - \frac{1}{(t-r)^2+\varepsilon^2}\right\} r\, dr\, dt.$$

In the last expression, the integral with respect to t is equal to

$$J^\varepsilon = \int_r^{+\infty} \varphi_\natural(s-r, r)\frac{ds}{s^2+\varepsilon^2} - \int_{-r}^{+\infty}\varphi_\natural(s+r, r)\frac{dt}{s^2+\varepsilon^2}$$

$$= \varepsilon^{-1}\left\{\int_{r/\varepsilon}^{+\infty}\varphi_\natural(-r+\varepsilon t, r)\frac{dt}{t^2+1} - \int_{-r/\varepsilon}^{+\infty}\varphi_\natural(r+\varepsilon t, r)\frac{dt}{t^2+1}\right\}.$$

For $r > 0$, $\varepsilon J^\varepsilon$ converges to $-\pi\varphi_\natural(r, r)$. We reach the conclusion that

$$\text{(8.15)} \qquad \langle E_+, \varphi\rangle = \int_0^{+\infty}\varphi_\natural(r, r) r\, dr.$$

In distributions notation, this can be rewritten as

$$\text{(8.16)} \qquad E_+ = (4\pi)^{-1}\frac{1}{r}\,\delta(t-r).$$

Remarks. 8.1. When $n = 2$, the support of E_+, as seen in (8.12), is identical with the set $r \leqq t$; when $n = 3$, (8.16) shows that it is the set $r = t$. Of course, this agrees with Theorem 8.1.

8.2. In both cases $n = 2$ and $n = 3$, we find that E_+ is a Radon measure; as a matter of fact, when $n = 2$, E_+ is absolutely continuous with respect to the Lebesgue measure: It is of the form $h(x, t)\, dx\, dt$, where h is locally integrable (but not bounded; compare with the case $n = 1$). When $n = 3$, E_+ is a measure carried by the surface $r = t$ of the forward light-cone, as we see in (8.15). As $n \nearrow +\infty$, E_+ becomes less and less "regular": For $n = 4$, it ceases to be a measure; it is the distribution derivative of a measure carried by the forward light-cone (the increasing singularity of E_+ with the number of variables is already visible when $n = 1, 2, 3$).

Exercises

8.1. Let m be a real number not equal to zero, $E_+(m)$ the fundamental solution of the Klein–Gordon operator (7.30) with support in the forward light-cone. State and prove the analogs of Theorems 8.1 and 8.2; in particular, show what becomes of the Huyghens principle.

8.2. Compute the fundamental solution E_+ of the wave operator in any odd space dimension $n \geqq 5$ (cf. the second part of the Appendix to Sect. 8).

8.3. Set $s = (t^2 - |x|^2)^{1/2}$ when $t \geqq |x|$ (i.e., in the forward light-cone) and $s = 0$ otherwise. Show that

$$\langle F(\alpha), \phi \rangle = \iint s^\alpha \phi(t, x)\, dt\, dx, \qquad \phi \in C_c^\infty(\mathbf{R}^{n+1}), \tag{8.17}$$

defines a holomorphic function of $\alpha \in \mathbf{C}$, Re $\alpha > -2$, valued in the space of distributions in $\mathbf{R}^{n+1}$, $\mathscr{D}'(\mathbf{R}^{n+1})$.

Prove that, for all complex numbers α such that Re $\alpha > 0$, we have

$$\Box F(\alpha) = \alpha(\alpha - 1 + n)F(\alpha - 2), \tag{8.18}$$

where $\Box$ is the d'Alembertian, $(\partial/\partial t)^2 - \Delta_x$. [*Hint*: Use spherical coordinates in space, and cf. Proposition 9.3.]

8.4. Let $F(\alpha)$ be the distribution defined in (8.17) and set

$$Z_\alpha = \pi^{-(n-1)/2} \frac{F(\alpha - n - 1)}{2^{\alpha-1}\Gamma(\alpha/2)\Gamma((\alpha - n + 1)/2)}, \tag{8.19}$$

where Γ is the Euler gamma function (the distributions Z_α have been introduced by Marcel Riesz and are called *Riesz potentials*).

Prove that $\Box Z_\alpha = Z_{\alpha-2}$ ($\Box$ is the wave operator on $\mathbf{R}^{n+1}$) and that Z_α can be extended as an *entire* function of α in $\mathbf{C}$, valued in $\mathscr{D}'(\mathbf{R}^{n+1})$. Show that the support of Z_α, for all complex α, is contained in the forward light-cone.

8.5. Let Z_α be the Riesz potential (8.19). Compute Z_2 when $n = 2$ and derive from this the value of Z_0.

Compute Z_4 for $n = 3$ and derive from this the values of Z_2 and Z_0.

8.6. Suppose n odd >1. Admitting the fact that $Z_0 = \delta$, derive an explicit formula for the fundamental solution E_+ of the wave operator in $\mathbf{R}^{n+1}$ (the one having its support in the forward light-cone) making it obvious that its support is supported by the *surface* of the forward light-cone.

8.7. Let Z_α be the Riesz potential (8.19). Prove that

$$W_\lambda = \sum_{p=0}^{+\infty} \lambda^p Z_{2p+2} \tag{8.20}$$

defines an *entire* function of the complex variable λ, valued in $\mathscr{D}'(\mathbf{R}^{n+1})$, which is a fundamental solution of $\Box - \lambda$. (The student may admit that Z_0 is the Dirac measure.)

9

Fundamental Solutions of the Laplace Equation

In this section we are concerned with solving the equation in $\mathbf{R}^n$,

$$\Delta E = \delta. \tag{9.1}$$

We recall that

$$\Delta = \left(\frac{\partial}{\partial x^1}\right)^2 + \cdots + \left(\frac{\partial}{\partial x^n}\right)^2. \tag{9.2}$$

We could proceed as in the preceding sections and reduce the problem to solving an ordinary differential equation with respect to one of the variables x, say x^n, after having performed a Fourier transformation with respect to the remaining ones. Equation (9.1) would be transformed into

$$(\partial/\partial x^n)^2 \tilde{E} - (\xi_1^2 + \cdots + \xi_{n-1}^2)\tilde{E} = \delta(x^n). \tag{9.3}$$

But under the present circumstances this procedure is strikingly unnatural. By selecting one of the variables, we violate the basic symmetry of the situation. The Laplace operator enters in many partial differential equations of physics which purport to describe phenomena taking place in *isotropic* media. It is therefore more natural, and more convenient, to use a different method—a method that does not disregard but instead fully exploits the fact that the Laplace operator Δ is *rotation-invariant*. Actually, the new method, like the one used in Sects. 5 to 7, reduces the problem to solving a linear ordinary differential equation. But the variable in the latter will be the *radial* variable,

$$r = |x| = [(x^1)^2 + \cdots + (x^n)^2]^{1/2},$$

and the equation will *not* have constant coefficients. Nevertheless we shall be able to solve it without much difficulty. [When $n \geqq 3$ a different, and perhaps quicker, method of solving (9.1) is indicated in Exercise 6.5.]

We must begin with some general considerations following from the invariance under rotations of the Laplace operator. As a matter of fact, we shall exploit the invariance of Δ, not only under rotations but under the full *orthogonal group* $O(n)$. This is the group of linear transformations of $\mathbf{R}^n$ which leave the quadratic form $|x|^2$ invariant.

PROPOSITION 9.1. *Let T be any orthogonal transformation of $\mathbf{R}^n$, u any C^∞ function with compact support in $\mathbf{R}^n$. We have*

$$(9.4) \qquad (\Delta u)^T = \Delta(u^T).$$

Proof. It suffices to apply the considerations about the action on functions of linear transformations, and their "reflection" on the Fourier transforms (see pp. 53–54). We observe that, if $T \in O(n)$, det $T = \pm 1$, hence

$$\widehat{f^T} = \hat{f}^{{}^tT^{-1}}.$$

But if $T \in O(n)$, its contragredient ${}^tT^{-1}$ belongs also to $O(n)$. By Fourier transformation, Eq. (9.4) is equivalent to

$$(9.5) \qquad (\widehat{\Delta u})^{{}^tT^{-1}} = \widehat{\Delta(u^T)}.$$

This in turn reads

$$(9.6) \qquad (|\xi|^2 \hat{u})^{{}^tT^{-1}} = |\xi|^2 \hat{u}^{{}^tT^{-1}}.$$

Recalling that $v^{{}^tT^{-1}}(\xi) = v({}^tT\xi)$, (9.6) follows from

$$|{}^tT\xi|^2 = |\xi|^2,$$

which is evident. Q.E.D.

As a matter of fact, we have also proved that if a *linear transformation T leaves Δ invariant, it is an orthogonal transformation.*

Formula (9.4) remains valid for an arbitrary distribution u, as we see readily by using the definition (7.24) of the transform of a distribution under a linear transformation of $\mathbf{R}^n$.

We introduce the *average* of a function f on a sphere of radius $r = |x|$ centered at the origin:

$$f_\natural(x) = |S^{n-1}|^{-1} \int_{S^{n-1}} f(r\dot{x})\, d\dot{x}.$$

Here $\dot{x}$ denotes the variable on the unit sphere S^{n-1}; $|S^{n-1}|$ stands for the *area* of S^{n-1} (for a computation of the area of the unit sphere, see the

Appendix to this section). Note that $f_\natural(x)$, although a function of r alone, is not a function on $\mathbf{R}^1_+$; it is a function on $\mathbf{R}^n$. Nevertheless we shall denote it sometimes by $f_\natural(r)$.

PROPOSITION 9.2. *Given any C^∞ function f in $\mathbf{R}^n$, we have*

$$\Delta(f_\natural) = (\Delta f)_\natural . \tag{9.7}$$

Proof. The proof we give here is *not* elementary; the student who is not acquainted with the theory of the *Haar measure* on locally compact groups may skip it and try to do Exercise 9.2.

We denote by dT the Haar measure on the compact group $O(n)$ normalized so as to have $\int_{O(n)} dT = 1$, and we set

$$f^\natural(x) = \int_{O(n)} f(Tx)\, dT. \tag{9.8}$$

The average $f^\natural$ is $O(n)$-invariant:

$$f^\natural(T_0^{-1}x) = \int_{O(n)} f(TT_0^{-1}x)\, dT = \int_{O(n)} f(Tx)\, dT$$

[for dT is left- and right-invariant: $O(n)$, being compact, is unimodular]. Consequently, the average of $f^\natural$ on the sphere of radius $|x|$ is equal to $f^\natural(x)$:

$$(f^\natural)_\natural = f^\natural. \tag{9.9}$$

On the other hand, $f_\natural(Tx) = f_\natural(x)$, hence

$$(f_\natural)^\natural(x) = \int_{O(n)} f_\natural(Tx)\, dT = f_\natural(x) \int_{O(n)} dT = f_\natural(x),$$

i.e.,

$$(f_\natural)^\natural = f_\natural . \tag{9.10}$$

But by virtue of Fubini's theorem, we have the right to interchange the integrations over S^{n-1} and over $O(n)$. We derive

$$(f^\natural)_\natural = (f_\natural)^\natural,$$

hence, by (9.9) and (9.10),

$$f^\natural = f_\natural . \tag{9.11}$$

Finally, by the rule of differentiation under the integral sign, we have

$$\Delta(f^\natural)(x) = \int_{O(n)} \Delta[f(Tx)]\, dT$$

which is equal, in view of Proposition 9.1, to

$$\int_{O(n)} (\Delta f)(Tx)\, dT = (\Delta f)^\natural(x),$$

i.e.,

$$(9.12) \qquad \Delta(f^\natural) = (\Delta f)^\natural.$$

Combination of (9.11) and (9.12) yields (9.7). Q.E.D.

PROPOSITION 9.3. *When acting on a function $\varphi \in C_c^\infty(\mathbf{R}^n)$ which depends only on $r = |x|$,*

$$(9.13) \qquad \Delta = \left(\frac{\partial}{\partial r}\right)^2 + \frac{n-1}{r}\frac{\partial}{\partial r}.$$

Proof. Let $f(r)$ be a function of r alone. Then

$$\left(\frac{\partial}{\partial x^j}\right) f(r) = f'(r)\frac{\partial r}{\partial x^j} = f'(r)\frac{x^j}{r},$$

hence

$$\begin{aligned}(\partial/\partial x^j)^2 f(r) &= \frac{1}{r}f'(r) + x^j(\partial/\partial x^j)\left(\frac{1}{r}f'(r)\right)\\ &= \frac{1}{r}f'(r) + x^j\left(\frac{1}{r}f'(r)\right)'\frac{x^j}{r}\\ &= \frac{1}{r}f'(r) + \frac{(x^j)^2}{r^2}f''(r) - \frac{(x^j)^2}{r^3}f'(r).\end{aligned}$$

By summing with respect to $j = 1, \ldots, n$, we obtain

$$\Delta f(r) = \left(\frac{n}{r} - \frac{1}{r^3}\sum_{j=1}^{n}(x^j)^2\right)f'(r) + f''(r). \qquad \text{Q.E.D.}$$

Now let u be a distribution in $\mathbf{R}^n$ which is defined by a locally integrable function depending on r alone, which we denote by $f(r)$. We have, given any $\varphi \in C_c^\infty(\mathbf{R}^n)$,

$$\begin{aligned}(9.14) \qquad \langle u, \varphi\rangle &= \int f(r)\varphi(x)\, dx = \iint f(r)\varphi(r\dot{x})r^{n-1}\, dr\, d\dot{x}.\\ &= |S^{n-1}|\int_0^{+\infty} f(r)\varphi_\natural(r)r^{n-1}\, dr,\end{aligned}$$

where (somewhat improperly) $\varphi_\natural(r) = \varphi_\natural(x)$ with $r = |x|$ [i.e., here $\varphi_\natural(r)$ is a function of the single real variable $r \in \mathbf{R}_+$, whereas before $\varphi_\natural(x)$ denoted a function on $\mathbf{R}^n$, constant on the spheres centered at the origin]. Let us replace φ by $\Delta\varphi$ in (9.14):

$$\begin{aligned}
\langle u, \Delta\varphi\rangle &= |S^{n-1}| \int_0^{+\infty} f(r)(\Delta\varphi)_\natural(r) r^{n-1}\, dr \\
&= |S^{n-1}| \int_0^{+\infty} f(r)\, \Delta(\varphi_\natural)(r) r^{n-1}\, dr \qquad \text{(by Proposition 9.2)} \\
&= |S^{n-1}| \int_0^{+\infty} f(r)\left(\frac{d}{dr} + \frac{n-1}{r}\right)\frac{d\varphi_\natural}{dr} r^{n-1}\, dr \qquad \text{(by Proposition 9.3)} \\
&= |S^{n-1}| \int_0^{+\infty} f(r)\frac{d}{dr}\{r^{n-1}\varphi_\natural'(r)\}\, dr.
\end{aligned}$$

We assume that the derivative of f with respect to r is also locally integrable in $\mathbf{R}^n$; we may integrate by parts:

$$\langle u, \Delta\varphi\rangle = |S^{n-1}| \left\{\left[r^{n-1} f(r)\varphi_\natural'(r)\right]_{r=0}^{+\infty} - \int_0^{+\infty} f'(r)\varphi_\natural'(r) r^{n-1}\, dr\right\}.$$

Recalling that $n > 1$ and that φ has compact support, we see that the integrated term vanishes and therefore that we have

$$\text{(9.15)} \qquad \langle u, \Delta\varphi\rangle = -|S^{n-1}| \int_0^{+\infty} \frac{df}{dr}\frac{d\varphi_\natural}{dr} r^{n-1}\, dr.$$

At this point, we note that

$$-\int_0^{+\infty} \frac{d\varphi_\natural}{dr}\, dr = \varphi_\natural(0) = \varphi(0).$$

Therefore, if we could determine the locally integrable function f, depending only on r, so as to have

$$\text{(9.16)} \qquad r^{n-1}\frac{df}{dr} = |S^{n-1}|^{-1}, \qquad r > 0,$$

we would have

$$\text{(9.17)} \qquad \langle u, \Delta\varphi\rangle = \langle \Delta u, \varphi\rangle = \varphi(0);$$

in other words u would be a fundamental solution of the Laplace operator. Moreover, u would be a *rotation-invariant fundamental solution* of Δ.

Solving (9.16) is immediate. A particular solution is obtained by taking $f = F$ with

$$(9.18) \qquad F(r) = \begin{cases} \dfrac{1}{2\pi} \log r & \text{when} \quad n = 2, \\ -\dfrac{1}{(n-2)|S^{n-1}|} \dfrac{1}{r^{n-2}} & \text{when} \quad n > 2. \end{cases}$$

Since the element of volume in $\mathbf{R}^n$ is $r^{n-1}\,dr\,d\theta$, we see at once that $f(r)$ given by (9.18) is indeed locally integrable and that its derivative with respect to r is also locally integrable [the latter is in fact evident by (9.16)].

We may apply Theorem 3.1 and obtain Weyl's lemma:

THEOREM 9.1. *The Laplace operator is analytic-hypoelliptic.*

This in turn enables us to determine *all* the rotation-invariant fundamental solutions of Δ. Each one of them differs from the one defined by the function F in (9.18) by a solution u of the homogeneous Laplace equation $\Delta u = 0$. In view of Theorem 9.1, u is defined by an analytic function in $\mathbf{R}^n$; it is necessarily rotation-invariant, and so must be the function defining u [denoted as before $f(r)$]. Note, first, that

$$r\frac{df}{dr} = \sum_{j=1}^{n} x^j \frac{\partial}{\partial x^j} f(r)$$

is an analytic function of x, and that if $k \geqq 1$, $(d/dr)(r^k\,df/dr)$ is a bounded function of $r \geqq 0$ on any finite interval $[0, R]$, analytic for $r > 0$. According to (9.15) we must have, whatever the test function φ in $\mathbf{R}^n$,

$$\int_0^{+\infty} \frac{df}{dr}\frac{d\varphi_\natural}{dr} r^{n-1}\,dr = -\int_0^{+\infty} \frac{d}{dr}\left(r^{n-1}\frac{df}{dr}\right)\varphi_\natural\,dr = 0;$$

hence, by our previous considerations,

$$\frac{d}{dr}\left(r^{n-1}\frac{df}{dr}\right) = 0.$$

This means that $df/dr = Cr^{1-n}$, hence $f = [C/(2-n)]r^{2-n} + C_1$ when $n > 2$, $f = C\log r + C_1$ when $n = 2$. In all cases, f can be bounded in intervals $[0, R]$, as it should be, only if $C = 0$. In all cases, we find $f = C_1 = \text{const.}$

Remark 9.1. We shall eventually prove that the value of a harmonic function at the center of a sphere equals its average on the sphere. This implies at once that the only harmonic functions which are rotation-invariant are the constant functions.

Let us summarize what we have obtained so far.

THEOREM 9.2. *The function $F(r)$ in* (9.18) *defines a rotation-invariant fundamental solution of the Laplace operator Δ. Every rotation-invariant fundamental solution of Δ is defined by a locally integrable function of the form $F(r)$* + const.

Let us denote by E the fundamental solution of Δ defined by $F(r)$ of (9.18). If we apply (9.14) to $u = E$, we obtain

$$\langle E, \varphi\rangle = \begin{cases} \displaystyle\int_0^{+\infty} \varphi_\natural(r)\, r \log r\, dr & \text{if } n = 2, \\ \displaystyle -\frac{1}{n-2}\int_0^{+\infty} \varphi_\natural(r)\, r\, dr & \text{if } n > 2. \end{cases} \tag{9.19}$$

Appendix Computation of the Area of the Unit Sphere

Assume $n \geqq 1$ and set $I_n = \int_{\mathbf{R}^n} \exp(-|x|^2)\, dx$. Clearly $I_n = I_1^n$ and, by going to polar coordinates, $I_n = |S^{n-1}| \int_0^{+\infty} \exp(-r^2)\, r^{n-1}\, dr$ (if $n = 1$, we take $|S^{n-1}| = 2$, assigning "area" 1 to each one of the two points $x = +1$ and $x = -1$ which make up S^0). Thus $I_2 = 2\pi \int_0^{+\infty} \exp(-r^2)\, r\, dr = \pi$, and consequently, $I_1 = \sqrt{\pi}$. Assuming now $n \geqq 2$ and setting $r^2 = s$, we obtain

$$2I_n = |S^{n-1}| \int_0^{+\infty} e^{-s} s^{-1+n/2}\, ds = |S^{n-1}|\,\Gamma(n/2),$$

where $\Gamma(z) = \int_0^{+\infty} e^{-s} s^{z-1}\, ds$ (Re $z > 0$) is the *Euler gamma function*. Thus we obtain, for all $n = 1, 2, \ldots,$

$$|S^{n-1}| = 2\pi^{n/2}/\Gamma(n/2).$$

It is easy to derive more explicit values from this. If $n = 2p$, we have

$$|S^{2p-1}| = 2\pi^p/(p-1)!.$$

If $n = 2p + 1$, $\Gamma(n/2) = \Gamma(p + \frac{1}{2}) = (p - \frac{1}{2}) \cdots \frac{1}{2}\Gamma(\frac{1}{2})$, and we have seen that $\Gamma(\frac{1}{2}) = 2I_1/|S^0| = I_1 = \sqrt{\pi}$, from which

$$|S^{2p}| = 2^{p+1}\pi^p/[1 \cdot 3 \cdots (2p-1)].$$

Exercises

[In the exercises below Δ always stands for the Laplace operator in $\mathbf{R}^n$, $\sum_{j=1}^n (\partial/\partial x^j)^2$, and r for the Euclidean norm, $|x|$. We assume $n \geqq 2$.]

9.1. For each complex number α, let V_α denote the two-dimensional linear space, over the complex numbers, having the basis formed by the two functions (in $\mathbf{R}^n\backslash\{0\}$), r^α, $r^\alpha \log r$. Show that Δ induces a linear map $V_\alpha \to V_{\alpha-2}$ and write down the 2×2 matrix representing it. Describe completely the injectivity properties of this mapping, for the various values of α.

Now viewing V_α as a linear space of distributions on $\mathbf{R}^n$, describe ΔV_{2-n}.

9.2. Let m be an arbitrary integer, $m \geqq 1$. Using the notation and the results of Exercise 9.1, construct a sequence of distributions in $\mathbf{R}^n$, E_k, $k = 1, \ldots, m$, having the following properties:

$$\Delta E_{k+1} = E_k, \quad k = 1, \ldots, m-1, \qquad \Delta E_1 = \delta; \tag{9.20}$$

$$E_k \in V_{2k-n}, \qquad k = 1, \ldots, m. \tag{9.21}$$

Show that, for a suitable choice of the constant $C_{m,n}$, the following is true:

(9.22) *if* $2m - n$ *is not an even integer* $\geqq 0$, $E_m = C_{m,n} r^{2m-n}$ *is a fundamental solution of* Δ^m;

(9.23) *if* $2m - n$ *is an even integer* $\geqq 0$, $E_m = C_{m,n} r^{2m-n} \log r$ *is a fundamental solution of* Δ^m.

9.3. Derive from (9.22) that, if n is odd (and $\lambda > 0$), a fundamental solution of $\lambda - \Delta$ is given by

$$F_\lambda = -r^{2-n} \sum_{m=0}^{+\infty} C_{m+1,n} (\sqrt{\lambda}\, r)^{2m}, \tag{9.24}$$

whereas if n is even, we may take

$$F_\lambda = -r^{2-n} \sum_{m=0}^{p-2} C_{m+1,n} (\sqrt{\lambda}\, r)^{2m} - r^{2-n} \log r \sum_{m=p-1}^{+\infty} C_{m+1,n} (\sqrt{\lambda}\, r)^{2m}, \tag{9.25}$$

where $p = n/2$ (the student must show that the series above converge).

Compute the coefficients in (9.24) when $n = 3$.

9.4. Show that, when $n = 3$ (and $\lambda \geqq 0$),

$$F = \frac{1}{4\pi} \frac{1}{r} e^{-\sqrt{\lambda}\, r}$$

is a fundamental solution of $\lambda - \Delta$. How do you reconcile this fact with the result in Exercise 9.3?

9.5. Let λ be a number greater than zero. Denote by G_λ the inverse Fourier transform of $(|\xi|^2 + \lambda)^{-1}$. Prove that G_λ is a rotation-invariant fundamental solution of $\lambda - \Delta$.

Let $p = (p_1, \ldots, p_n)$ be an arbitrary n-tuple of integers $\geqq 0$, $\hat{G}_{\lambda,p}$ the Fourier transform of $x^p G_\lambda$. Prove that there is a constant $C_p > 0$ such that

$$|\hat{G}_{\lambda,p}(\xi)| \leqq C_p (1 + |\xi|)^{-|p|-2}, \qquad \forall \xi \in \mathbf{R}_n. \tag{9.26}$$

Derive from (9.26) that the function $G_\lambda(x)$ in $\mathbf{R}^n\backslash 0$ goes to zero faster than any power of $1/r$ when $r \to +\infty$.

9.6. Let G_λ be the distribution so denoted in Exercise 9.5. Derive from (9.26) that $r^{2M}G$ is continuously differentiable as many times as we wish, provided that we choose the integer M large enough. Derive from this that G_λ is C^∞ in the complement of the origin, and consequently, that $\lambda - \Delta$ is hypoelliptic.

9.7. Let G_λ be the distribution so denoted in Exercises 9.5 and 9.6. Prove that, if n is odd, $n \geqq 3$, or if n is even, $n \geqq 4$, we have the asymptotic equivalence, as $r \sim 0$,

$$G_\lambda(x) \sim \frac{(n-2)}{|S^{n-1}|}\frac{1}{r^{n-2}}, \tag{9.27}$$

whereas, when $n = 2$, we have

$$G_\lambda(x) \sim -\frac{1}{2\pi}\log r \tag{9.28}$$

(the student must state precisely the meaning of these equivalences, by estimating the size of the difference between the two members in a neighborhood of the origin).

9.8. Let E denote the fundamental solution of Δ in $\mathbf{R}^n$ given by (9.19). Prove that $u \mapsto E * u$ is a continuous linear map $C_c^0(\mathbf{R}^n) \to C^1(\mathbf{R}^n)$.

[*Hint*: Take advantage of the fact that

$$\left(\frac{\partial}{\partial x^j}\right)(E * u)(x) = C_n \int_{\mathbf{R}^n} y^j u(x-y)|y|^{-n}\, dy.] \tag{9.29}$$

9.9. Let λ be a number greater than or equal to 0. Prove that, if f is any continuous function with compact support in $\mathbf{R}^n$, every distribution solution of the equation

$$(\lambda - \Delta)u = f \tag{9.30}$$

belongs to $C^1(\mathbf{R}^n)$.

9.10. Let E denote the fundamental solution of Δ given by (9.19). Let R be any number greater than zero, y any point in $\mathbf{R}^n$ such that $|y| < R$. Prove that the average of $E(x-y)$ over the sphere $|x| = R$ is equal to $F(R)$, given by (9.18). [*Hint*: Prove that the average of $E(x-y)$ over the sphere $|x| = R$ is a solution of the Laplace equation, in the variable y, in the interior of that sphere, and that it is rotation-invariant; then apply the assertions in Remark 9.1.]

10

Green's Formula. The Mean Value Theorem and the Maximum Principle for Harmonic Functions. The Poisson Formula. Harnack's Inequalities

We are now going to derive, for the Laplace operator, the analog of the inhomogeneous Cauchy formula (5.12) for the Cauchy–Riemann operator. It is useful to put things in the proper perspective. We are dealing with a finite-dimensional real vector space, equipped with a positive-definite quadratic form; the operator we are studying is the one canonically associated with this quadratic form (via transfer of the quadratic form to the dual by the Riesz representation theorem and via Fourier transformation): the space is $\mathbf{R}^n$, the form is the square of the norm, $|x|^2$, the operator is Laplace's. Rephrasing this slightly we may say that, associated with the operator under study, we are given a Riemannian structure on $\mathbf{R}^n$. Now let Ω be a *bounded* connected open set in $\mathbf{R}^n$ whose boundary $\partial\Omega$ is a smooth, say C^2, hypersurface. We make the hypothesis that Ω *lies on one side* of its boundary. Because of the Riemannian structure on $\mathbf{R}^n$ we may talk about the *normal* to this hypersurface. On the other hand, $\partial\Omega$ divides $\mathbf{R}^n$ into two regions: an inner region, Ω, and an outer one, the interior of $\mathbf{R}^n\backslash\Omega$. The former is relatively compact; but the latter is not. The normal line to $\partial\Omega$ at some point consists of two well-defined half-lines: the exterior normal, and the interior one (in other words, the normal to $\partial\Omega$ is naturally orientated and so is the tangent hyperplane to $\partial\Omega$). We may also define the *unit vector field* v at each point of $\partial\Omega$, normal to $\partial\Omega$ and pointing into $\mathbf{R}^n\backslash\overline{\Omega}$.

Let x_0 be an arbitrary point of $\partial\Omega$. If U is a sufficiently small open neighborhood of x_0 in $\partial\Omega$, we may find $(n-1)$ functions $y^1, \ldots, y^{n-1}$, defined and C^2 in U, such that, at every point of U, the vectors grad y^j $(j = 1, \ldots, n-1)$ define a frame in the tangent hyperplane to $\partial\Omega$, having the "correct" orientation. Then U is the image of an open subset V of $\mathbf{R}^{n-1}$ under the C^2 diffeomorphism $y' = (y^1, \ldots, y^{n-1}) \mapsto x = x(y')$. It is standard to set

$$g_{jk}(y') = \left\langle \frac{\partial x}{\partial y^j}, \frac{\partial x}{\partial y^k} \right\rangle (1 \leqq j, k < n), \; g(y') = \det[(g_{jk}(y')],$$

where $\langle\, , \rangle$ stands for the inner product in $\mathbf{R}^n$. We may define the *area element* in U as the transfer from V of the (oriented) measure $\sqrt{g(y')}\, dy'$. It is easy to check that this definition is independent of the choice of the local coordinates y' (having the right orientation). It coincides with the usual definition when U is a piece of hyperplane. By covering $\partial\Omega$ with coordinate patches such as U we define a positive (oriented) measure on $\partial\Omega$, canonically associated with the Riemannian structure and with the orientation of $\partial\Omega$ (both induced by those of $\mathbf{R}^n$), which we shall denote by $d\sigma$.

LEMMA 10.1. *Let $\varphi^1, \ldots, \varphi^n$ be n functions, defined and C^1 in an open neighborhood $\mathcal{O}$ of $\bar{\Omega}$ and valued in* $\mathbf{C}$; *set $\boldsymbol{\varphi}(x) = (\varphi^1(x), \ldots, \varphi^n(x))$. We have*

$$\int_\Omega \operatorname{div} \boldsymbol{\varphi}\, dx = \int_{\partial\Omega} \langle \boldsymbol{\varphi}, \mathbf{v} \rangle \, d\sigma. \tag{10.1}$$

Proof. By using a partition of unity, we see that it suffices to prove (10.1) when the support of $\boldsymbol{\varphi}$ is compact and contained in a small open neighborhood $\mathcal{N}$ (in $\mathbf{R}^n$) of an arbitrary point x_0 of $\partial\Omega$. We may assume that $\mathcal{N} \cap \partial\Omega = U$ is of the kind considered above, and that we have a system of coordinates $(y^1, \ldots, y^{n-1})$ in U, exactly as before; we add to this system a coordinate y^n along the axis parallel to the normal to $\partial\Omega$ at x_0. There is a C^2 function $f(y')$ (in V) such that $y^n < f(y')$ defines $\Omega \cap \mathcal{N}$, and therefore $y^n = f(y')$ defines U in $\mathcal{N}$. Finally, we may and shall suppose that the coordinate system $(y^1, \ldots, y^n)$ is orthonormal in $\mathcal{N}$, for the Riemannian structure of $\mathbf{R}^n$. Under such circumstances it is readily checked that, in V,

$$g(y') = 1 + |\operatorname{grad}_{y'} f(y')|^2. \tag{10.2}$$

On the other hand, the vector

$$\left(-\frac{\partial f}{\partial y^1}, \ldots, -\frac{\partial f}{\partial y^{n-1}}, 1 \right)$$

is normal to U and points into $\mathbf{R}^n \backslash \Omega$. Consequently the right-hand side of (10.1) reads

$$\int_{\mathbf{R}^{n-1}} \left(-\sum_{j=1}^{n-1} \varphi^j_\natural \frac{\partial f}{\partial y^j} + \varphi^n_\natural \right) dy', \tag{10.3}$$

where $\varphi_{\natural}^{j}(y') = \varphi^j\,[y^1, \ldots, y^{n-1}, f(y')]$, $j = 1, \ldots, n$. The left-hand side is

$$(10.4)\qquad \sum_{j=1}^{n} \int_{y^n < f(y')} \frac{\partial \varphi^j}{\partial y^j}\, dy.$$

In (10.4) we set $z^j = y^j$ if $j < n$, $z^n = y^n - f(y')$, $\psi^j(z) = \varphi^j\,(z^1, \ldots, z^{n-1}, z^n + f(y'))$. This turns (10.4) into

$$(10.5)\qquad \sum_{j=1}^{n-1} \int_{z^n < 0} \frac{\partial \psi^j}{\partial z^j}\, dz + \int_{z^n < 0} \left(- \sum_{j=1}^{n-1} \frac{\partial \psi^j}{\partial z^n} \frac{\partial f}{\partial y^j} + \frac{\partial \psi^n}{\partial z^n} \right) dz,$$

which, because of the nature of supp $\boldsymbol{\varphi}$, is equal to (10.3). Q.E.D.

Corollary 10.1. *Let $u \in C^2(\mathcal{O})$. We have*

$$(10.6)\qquad \int_{\Omega} \Delta u\, dx = \int_{\partial\Omega} \frac{\partial u}{\partial \nu}\, d\sigma.$$

Proof. Apply (10.1) with $\boldsymbol{\varphi} = \operatorname{grad} u$. Q.E.D.

We have denoted by $\partial u/\partial \nu$ the partial differentiation in the direction of the outer normal to $\partial\Omega$:

$$\frac{\partial u}{\partial \nu}(x) = \lim_{\substack{h \to 0 \\ h \neq 0}} \frac{1}{h} [u(x + h\boldsymbol{\nu}) - u(x)].$$

Corollary 10.2. *Let $\varphi \in C^1(\mathcal{O})$, $\mathbf{v} \in C^1(\mathcal{O}; \mathbf{C}^n)$. We have*

$$(10.7)\qquad \int_{\Omega} \operatorname{div}(\varphi \mathbf{v})\, dx = \int_{\partial\Omega} \varphi \langle \mathbf{v}, \boldsymbol{\nu} \rangle\, d\sigma.$$

Corollary 10.3. *Let $f, g \in C^2(\mathcal{O})$. We have*

$$(10.8)\qquad \int_{\Omega} (f\, \Delta g - g\, \Delta f)\, dx = \int_{\partial\Omega} \left(f \frac{\partial g}{\partial \nu} - g \frac{\partial f}{\partial \nu} \right) d\sigma.$$

Equation (10.8) is *Green's formula*.

Proof of Corollary 10.3. In (10.7) choose $\mathbf{v} = \operatorname{grad} \psi$, $\psi \in C^2(\mathcal{O})$. We obtain

$$(10.9)\qquad \int_{\Omega} (\varphi\, \Delta\psi + \langle \operatorname{grad} \varphi, \operatorname{grad} \psi \rangle)\, dx = \int_{\partial\Omega} \varphi \frac{\partial \psi}{\partial \nu}\, d\sigma.$$

In this identity take first $\varphi = f$, $\psi = g$, then $\varphi = g$, $\psi = f$. By subtraction one obtains (10.8). Q.E.D.

We now derive the analog, for the Laplace equation

$$\Delta u = f \qquad \text{(in the neighborhood } \mathcal{O} \text{ of } \overline{\Omega}\text{)},$$

of the inhomogeneous Cauchy formula [see (5.10), (5.12)]. We shall assume that $u \in C^2(\mathcal{O})$ and $f \in C^0(\mathcal{O})$.

Let χ_Ω denote the characteristic function of the (bounded) open set Ω. By the Leibniz formula we have

$$\Delta(\chi_\Omega u) = \chi_\Omega f + 2 \sum_{j=1}^{n} \frac{\partial u}{\partial x^j} \frac{\partial \chi_\Omega}{\partial x^j} + u(\Delta \chi_\Omega).$$

Since both members have compact support, we may apply $E *$ to them; we obtain

$$\text{(10.10)} \qquad \chi_\Omega u = E * (\chi_\Omega f) + 2E * \left\{ \sum_{j=1}^{n} \frac{\partial u}{\partial x^j} \frac{\partial \chi_\Omega}{\partial x^j} \right\} + E * u(\Delta \chi_\Omega).$$

where E is any fundamental solution of the Laplace operator. We need to know what the following distributions are:

$$A = \sum_{j=1}^{n} \frac{\partial u}{\partial x^j} \frac{\partial \chi_\Omega}{\partial x^j}, \qquad B = u(\Delta \chi_\Omega).$$

Let φ be an arbitrary test function in $\mathbf{R}^n$, in fact having its (compact) support in $\mathcal{O}$. We have

$$\langle A, \varphi \rangle = -\sum_{j=1}^{n} \int_\Omega \frac{\partial}{\partial x^j} \left(\varphi \frac{\partial u}{\partial x^j} \right) dx = -\int_\Omega \operatorname{div}(\varphi \operatorname{grad} u)\, dx$$

$$= -\int_{\partial\Omega} \varphi \frac{\partial u}{\partial \nu} d\sigma \qquad \text{(by Corollary 10.2).}$$

In other words,

$$\text{(10.11)} \qquad A = -\chi_{\partial\Omega} \frac{\partial u}{\partial \nu} d\sigma,$$

the measure carried by $\partial\Omega$ with density (with respect to $d\sigma$) $-\, \partial u/\partial \nu$. On the other hand,

$$\text{(10.12)} \qquad \langle B, \varphi \rangle = \int_\Omega \Delta(\varphi u)\, dx = \int_{\partial\Omega} \varphi \frac{\partial u}{\partial \nu} d\sigma + \int_{\partial\Omega} u \frac{\partial \varphi}{\partial \nu} d\sigma,$$

by Corollary 10.1. Thus $B = -A + B_1$, where

$$\text{(10.13)} \qquad \langle B_1, \varphi \rangle = \int_{\partial\Omega} u \frac{\partial \varphi}{\partial \nu} d\sigma.$$

The distribution B_1 is a distribution carried by $\partial\Omega$ but it is *not* a measure carried by $\partial\Omega$. It is called a "double layer."

We take (10.11) and (10.12) into account in (10.10), but restrict the latter to Ω. Notice that, when $x \in \Omega$, $E(x - x')$ is a C^∞ function of x' in a neighborhood of $\partial\Omega$. Consequently we may evaluate A and B, as distributions in the variable x', on this function. On the other hand, E is a locally integrable function and therefore $E * (\chi_\Omega f)$ can be computed by integration over Ω. Thus, if $x_0 \in \Omega$, we have

$$(10.14)\quad u(x_0) = \int_\Omega E(x_0 - x) f(x)\, dx - \int_{\partial\Omega} E(x_0 - x) \frac{\partial u}{\partial \nu}(x)\, d\sigma + \int_{\partial\Omega} u(x) \frac{\partial}{\partial \nu} E(x_0 - x)\, d\sigma.$$

In this formula it makes no difference which fundamental solution E of Δ we choose to use. Customarily one likes to use the fundamental solution defined by (9.18). This is particularly convenient in the special case where Ω is an open ball $\{x \in \mathbf{R}^n;\ |x - x_0| < R\}$ centered at an arbitrary point x_0 and we wish to evaluate $u(x_0)$. It is natural to switch to spherical coordinates at x_0, i.e., set $x = x_0 + r\dot{x}$, where $\dot{x}$ is the variable point on the unit sphere S^{n-1}. With the notation of (9.18) we have

$$(10.15)\quad u(x_0) = \int_0^{+\infty} \int_{S^{n-1}} F(r) f(x_0 + r\dot{x})\, r^{n-1}\, dr\, d\dot{x} - R^{n-1} F(R) \int_{S^{n-1}} \frac{\partial u}{\partial r}(x_0 + R\dot{x})\, d\dot{x} + R^{n-1} F'(R) \int_{S^{n-1}} u(x_0 + R\dot{x})\, d\dot{x},$$

since, in the present situation, $d\sigma = R^{n-1}\, d\dot{x}$. We have also used the fact that $(\partial/\partial\nu) E(x_0 - x) = (d/dr) F(r)$. If we apply (10.6) once more, we see that

$$R^{n-1} \int_{S^{n-1}} \frac{\partial u}{\partial r}(x_0 + R\dot{x})\, d\dot{x} = \int_0^{+\infty} \int_{S^{n-1}} f(x_0 + r\dot{x})\, r^{n-1}\, dr\, d\dot{x},$$

whence, by (10.15),

$$(10.16)\quad u(x_0) = \int_\Omega \{F(|x|) - F(R)\} f(x)\, dx + F'(R) \int_{\partial\Omega} u(x)\, d\sigma,$$

assuming that Ω is the open ball of radius R centered at x_0. Formula (10.16) is very important. One of its first consequences is the following striking property of harmonic functions, the so-called *mean value theorem*:

THEOREM 10.1. *If u is a harmonic function in an open neighborhood of the closed ball* $\{x \in \mathbf{R}^n;\ |x - x_0| \leqq R\}$, *we have*

$$u(x_0) = \frac{1}{|S^{n-1}|} \int_{S^{n-1}} u(x_0 + R\dot{x})\, d\dot{x}. \tag{10.17}$$

In other words, the value of a harmonic function at the center of a sphere is equal to its average on the sphere.

Proof. If $\Delta u = 0$ for $|x - x_0| < R + \varepsilon$, $\varepsilon > 0$, we have, by (10.16),

$$u(x_0) = F'(R)R^{n-1} \int_{S^{n-1}} u(x_0 + R\dot{x})\, d\dot{x}.$$

We have, of course, $|S^{n-1}|R^{n-1}F'(R) = 1$ [cf. (9.16) and 9.18)].

COROLLARY 10.4. *Under the hypothesis of Theorem 10.1, we have*

$$|u(x_0)| \leqq \sup_{|x - x_0| = R} |u(x)|. \tag{10.18}$$

Proof. One gets (10.18) at once by applying the elementary mean value theorem to (10.17).

Corollary 10.4 implies the celebrated *maximum principle*:

THEOREM 10.2. *Let* Ω *be a connected open subset of* $\mathbf{R}^n$, *u a harmonic function in* Ω. *Unless u is constant in* Ω, *we have, for every* $x_0 \in \Omega$,

$$|u(x_0)| < \sup_{\Omega} |u(x)|. \tag{10.19}$$

Proof. Suppose that $|u(x_0)| = \sup_\Omega |u(x)|$ for some $x_0 \in \Omega$. Let R be any number greater than zero such that the ball $\{x;\ |x - x_0| \leqq R\}$ is contained in Ω. Suppose, for instance, that $u(x_0) \geqq 0$ [otherwise replace u by $\bar{u}(x_0)u$]. By (10.17) we have necessarily $u(x_0) = \sup_{\dot{y} \in S^{n-1}} |u(x_0 + R\dot{y})|$ and therefore

$$\int_{S^{n-1}} \left\{ \sup_{\dot{y} \in S^{n-1}} |u(x_0 + R\dot{y})| - \operatorname{Re} u(x_0 + R\dot{x}) \right\} d\dot{x} = 0.$$

The integrand is a continuous function, everywhere $\geqq 0$, on the unit sphere; hence it must be identically zero. Thus $u(x) = u(x_0)$ for any x such that $|x - x_0| = R$; and we may take any $R < d(x_0, \partial\Omega)$. This proves that $u(x) = u(x_0)$ in a full neighborhood of x_0 in Ω, therefore in the whole of Ω: Both functions are analytic in Ω or, if one prefers, the set where they are equal, which is obviously closed, has just been shown to be open. Q.E.D.

Next we derive the Poisson formula for an open ball.

First let us consider the case of a bounded open subset Ω of $\mathbf{R}^n$, with smooth boundary $\partial\Omega$ and lying on one side of this boundary—otherwise arbitrary. We apply the identity (10.10) once more, taking (10.11) and (10.12) into account. We evaluate both sides of (10.10) at a point x_1 lying in the complement of $\bar{\Omega}$. We obtain [cf. (10.14)]

$$(10.20)\quad \int_\Omega E(x_1 - x)f(x)\,dx = \int_{\partial\Omega}\left\{E(x_1 - x)\frac{\partial u}{\partial \nu}(x) - u(x)\frac{\partial}{\partial \nu}E(x_1 - x)\right\}d\sigma,$$

recalling that $f = \Delta u$ in Ω (we may assume that u is a twice continuously differentiable function in $\bar{\Omega}$). In particular, if u is harmonic,

$$(10.21)\quad \int_{\partial\Omega} E(x_1 - x)\frac{\partial u}{\partial \nu}(x)\,d\sigma = \int_{\partial\Omega} u(x)\frac{\partial}{\partial \nu}E(x_1 - x)\,\partial\sigma.$$

Let us now assume that Ω is the open ball $|x| < R$ and let x_0 be a point of Ω different from its center, i.e., $x_0 \neq 0$. We shall apply (10.21) with $E(x)$ the fundamental solution of Δ defined by (9.18) and with

$$x_1 = (R/|x_0|)^2 x_0$$

(x_1 is the point obtained from x_0 by "reflection" with respect to the sphere $|x| = R$). We have

$$(10.22)\quad |x - x_1| = (R/|x_0|)|x - x_0|, \qquad \forall x, \quad |x| = R.$$

In view of this we have, according to (9.18),

$$(10.23)\quad E(x_1 - x) = (|x_0|/R)^{n-2}E(x_0 - x), \qquad |x| = R, \qquad \text{if } n \geqq 3;$$

$$(10.24)\quad E(x_1 - x) = E(x_0 - x) + \text{const}, \qquad |x| = R, \qquad \text{if } n = 2.$$

If we substitute this into the left-hand side of (10.21) and take into account the fact that the surface integral of $\partial u/\partial\nu$ over $\partial\Omega$, when $u \in C^2(\bar{\Omega})$ is harmonic in Ω, is equal to zero (Exercise 10.6), we see that, for any $n \geqq 2$,

$$(10.25)\quad \int_{|x|=R} E(x_0 - x)\frac{\partial u}{\partial r}\,d\sigma = (R/|x_0|)^{n-2}\int_{|x|=R} u(x)\frac{\partial}{\partial r}E(x_1 - x)\,d\sigma.$$

We may take this into account in (10.14) (where $f \equiv 0$). We obtain

$$(10.26)\quad u(x_0) = \int_{|x|=R} u(x)\frac{\partial}{\partial r}\{E(x_0 - x) - (R/|x_0|)^{n-2}E(x_1 - x)\}\,d\sigma.$$

We note that if $z \neq x$ (and $r = |x|$), then

$$(10.27)\quad r\frac{\partial}{\partial r}E(z - x) = |S^{n-1}|^{-1}\frac{(r^2 - x\cdot z)}{|x - z|^n}.$$

If we use (10.27) in the right-hand side of (10.26), we see that $u(x_0)$ is equal to the integral, over $\partial\Omega$, of $u(x)$ multiplied by

$$R^{-1}|S^{n-1}|^{-1}|x - x_0|^{-n}\left\{(R^2 - x \cdot x_0) - \left(\frac{R}{|x_0|}\right)^{n-2}\left(\frac{|x - x_0|}{|x - x_1|}\right)^n (R^2 - x \cdot x_1)\right\}.$$

If we go back to the definition of x_1 and to (10.22), we obtain at once the *Poisson formula*

$$u(x_0) = \frac{1}{|S^{n-1}|R}\int_{|x|=R} u(x)\frac{R^2 - |x_0|^2}{|x - x_0|^n}\, d\sigma. \tag{10.28}$$

It will be noticed that this formula, which was proved under the assumption that $x_0 \neq 0$, is in fact also valid when $x_0 = 0$. It then reduces to the mean value formula (10.17).

In (10.28) let us assume that $u(x) \geqq 0$ everywhere and let us write $r_0 = |x_0|$. Noting that $R - r_0 < |x - x_0| < R + r_0$ if $|x| = R$, and applying the mean value theorem (Theorem 10.1) we obtain at once *Harnack's inequalities*

(10.29)

$$\frac{1 - r_0/R}{(1 + r_0/R)^{n-1}}\, u(0) \leqq u(x_0) \leqq \frac{1 + r_0/R}{(1 - r_0/R)^{n-1}}\, u(0), \qquad \forall x_0, \quad |x_0| = r_0 < R.$$

By solving the problem in Exercise 10.7 the student may check that the two constants in (10.29) are the best possible.

From (10.29) we derive the classical result, also known as Harnack's inequalities:

THEOREM 10.3. *Let Ω be a connected open subset of $\mathbf{R}^n$, K any compact subset of Ω. There is a constant $c_K > 0$ such that every nonnegative harmonic function u in Ω satisfies the inequality*:

$$c_K\, u(y) \leqq u(x) \leqq c_K^{-1} u(y), \qquad \forall x, y \in K. \tag{10.29'}$$

The proof is a simple application of (10.29) and is left to the student.

Exercises

(In the exercises below, Ω denotes an open subset of $\mathbf{R}^n$.)

10.1. Suppose that Ω is bounded. Let u be a continuous real-valued function in $\bar{\Omega}$, twice continuously differentiable in Ω, and such that $\Delta u \geqq 0$ in Ω.

Prove that for any $x_0 \in \Omega$ and $r < d(x_0, \complement\Omega)$,

$$u(x_0) \leqq |S^{n-1}| \int_{S^{n-1}} u(x_0 + r\dot{x})\, d\dot{x}. \tag{10.30}$$

Derive from (10.30) the *maximum principle* for u:

$$\forall x_0 \in \Omega, \qquad u(x_0) \leqq \sup_{x \in \partial\Omega} u(x). \tag{10.31}$$

What is the correct statement of (10.31) when Ω is not bounded?

10.2. Let $u \in C^2(\Omega)$, $x_0 \in \Omega$ arbitrary. By using the Taylor expansion of order 2 or u about x_0, prove that

$$-\Delta u(x_0) = \lim_{r \to +0} \frac{2n}{r^2} \left\{ u(x_0) - \frac{1}{|S^{n-1}|} \int_{S^{n-1}} u(x_0 + r\dot{x})\, d\dot{x} \right\}. \tag{10.32}$$

Derive from this that, if u satisfies (10.30) for any $x_0 \in \Omega$, $\Delta u \geqq 0$ everywhere in Ω.

10.3. Let λ be a number greater than or equal to zero. Prove the maximum principle (10.31) for the solutions of the (*metaharmonic*) equation

$$(-\Delta + \lambda)u = 0 \quad \text{in } \Omega. \tag{10.33}$$

10.4. Exhibit a solution of the equation

$$\frac{d^2u}{dx^2} - u = -1 \qquad \text{in the unit interval} \quad |x| < 1, \tag{10.34}$$

which does *not* obey the maximum principle.

10.5. Let λ be a number less than zero. Construct an example in which a solution of Eq. (10.33) does *not* satisfy the maximum principle.

10.6. Assume that $\mathscr{B}$ is an open ball in $\mathbf{R}^n$ and that $u \in C^1(\bar{\mathscr{B}})$ is harmonic in $\mathscr{B}$. Prove that we then have

$$\int_{\partial\mathscr{B}} \frac{\partial u}{\partial \nu}\, \partial\sigma = 0. \tag{10.35}$$

Conversely, show that if $u \in C^1(\Omega)$ satifises (10.35) for any ball $\mathscr{B}$ with closure contained in Ω, u is harmonic in Ω.

10.7. Prove that the function

$$u(x) = (1 - |x|^2)/|z_0 - z|^n,$$

where z_0 is a point on the sphere S^{n-1}, is a harmonic function in the interior $|x| < 1$ of that sphere ($n \geqq 2$). Show that there are points x_0, $|x_0| = r_0 < 1$, where one of the two inequalities in (10.29) (with $R = 1$) is in fact an equality.

10.8. Let u be a real-valued harmonic function in $\mathbf{R}^n$. Prove that either u is constant or else u maps $\mathbf{R}^n$ onto $\mathbf{R}$.

10.9. Let f denote a measurable function on the sphere $|x| = R$ $(R > 0)$ whose values are either (finite) real numbers or else $+\infty$. Let us set [cf. (10.28)]

$$I_f(x) = \frac{1}{|S^{n-1}|R} \int_{|y|=R} f(y) \frac{R^2 - |x|^2}{|y - x|^n} \, d\sigma_y \tag{10.36}$$

(the so-called *Poisson integral* of f). Suppose that f is *lower semicontinuous* on the sphere $|x| = R$, which means that, given any y_0, $|y_0| = R$, and any $\alpha < f(y_0)$, there is an open neighborhood U_α of y_0 on the sphere such that

$$f(y) > \alpha, \qquad \forall y \in U_\alpha .$$

Prove that, given any y_0, $|y_0| = R$, the *lower limit* of $I_f(x)$ as x, $|x| < R$, tends to y_0, is larger than or equal to $f(y_0)$.

Derive from this that if f is continuous on the sphere $|x| = R$, I_f is continuous on the closed ball $|x| \leq R$ and equal to f on its boundary.

10.10. Let Ω be an open subset of $\mathbf{R}^n$. Derive from the Poisson formula (10.28) that if a sequence of harmonic functions in Ω converges uniformly on every compact subset of Ω, then they also converge in $C^\infty(\Omega)$.

10.11. Let Ω be an open subset of $\mathbf{R}^n$. Derive from the Poisson formula (10.28) that if a sequence of harmonic functions in Ω converges in the distribution sense, then they also converge in $C^\infty(\Omega)$. (Cf. Exercise 2.10.)

CHAPTER II

The Cauchy Problem

II

The Cauchy Problem for Linear Ordinary Differential Equations

Consider a first-order linear differential equation in an open interval $-T < t < T$ ($T > 0$) of the real line:

$$\frac{du}{dt} - a(t)u = f. \tag{11.1}$$

Suppose that both the coefficient $a(t)$ and the right-hand side $f(t)$ are continuous functions of t, $|t| < T$. We obtain at once a solution of (11.1) by taking

$$u(t) = e^{A(t,0)}u_0 + \int_0^t e^{A(t,t')}f(t')\,dt', \tag{11.2}$$

where u_0 is an arbitrary constant and

$$A(t, t') = \int_{t'}^{t} a(s)\,ds, \qquad |t| < T, \qquad |t'| < T.$$

As a matter of fact, all the solutions of (11.1) are obtained in this manner: They are all of the form (11.2). It is clear that $u(t)$ is completely determined by the choice of u_0. Observe that

$$u(0) = u_0. \tag{11.3}$$

In other words, the problem (11.1)–(11.3) has a unique solution, given by (11.2). Notice, incidentally, that this solution is *once continuously differentiable* in $]-T, T[$.

We now go to first-order systems of linear ODEs:

$$\frac{d\mathbf{u}}{dt} - M(t)\mathbf{u} = \mathbf{f}, \tag{11.4}$$

$$\mathbf{u}(0) = \mathbf{u}_0. \tag{11.5}$$

Here $\mathbf{u}$, $\mathbf{f}$, $\mathbf{u}_0$ are valued in $\mathbf{C}^m$ and $M(t)$ is an $m \times m$ matrix with continuous functions as entries. If all these entries were constant functions, i.e., $M(t) \equiv M$, a complex $m \times m$ matrix, the theory of (11.4)–(11.5) would be a straightforward generalization of the theory of (11.1)–(11.3). The solution would be given by

$$\mathbf{u}(t) = e^{tM}\mathbf{u}_0 + \int_0^t e^{(t-t')M}\mathbf{f}(t')\,dt'. \tag{11.6}$$

But the situation is radically different when $M(t)$ is nonconstant. In the constant coefficients case, set $U(t, t') = \exp((t - t')M)$. What lies at the root of formula (11.6) is the fact that

$$\frac{dU}{dt} = MU, \qquad U\Big|_{t=t'} = I \quad \text{(the } m \times m \text{ identity matrix).} \tag{11.7}$$

But this ceases to be true, in general, when M is nonconstant. The thing to do, therefore, is to lay aside the exponential and concentrate on the solution of (11.7). It should be noted that problem (11.7) is of the same kind as (11.4)–(11.5): It is a system of first-order linear ODEs. This time there are m^2 equations in m^2 unknowns, submitted to m^2 conditions at time $t = t'$. It follows at once from the fundamental theorem on ordinary differential equations that (11.7) has a unique solution, which furthermore is a C^1 function of t, t' for $|t| < T$, $|t'| < T$.

The correct generalization of (11.2) is then

$$\mathbf{u}(t) = U(t, 0)\mathbf{u}_0 + \int_0^t U(t, t')\mathbf{f}(t')\,dt'. \tag{11.8}$$

The function $U(t, t')$ is sometimes called the *Riemann function* of the problem (11.4)–(11.5).

PROPOSITION 11.1. *If t, t', t'' are three points in the interval $]-T, T[$, we have*

$$U(t, t')U(t', t'') = U(t, t''), \tag{11.9}$$

$$U(t, t) = I, \qquad U(t, t')U(t', t) = I. \tag{11.10}$$

Proof. The first identity in (11.10) is part of the definition of $U(t, t')$. By differentiating it (with respect to t), we obtain

$$\frac{dU}{dt} + \frac{dU}{dt'} = 0 \quad \text{when} \quad t = t',$$

whence, by combining this with (11.7),

$$\frac{dU}{dt'} = -M(t') \quad \text{when} \quad t = t'. \tag{11.10a}$$

On the other hand, if we differentiate the first equation in (11.7) with respect to t', we obtain

$$\frac{d}{dt}\left(\frac{dU}{dt'}\right) = M(t)\frac{dU}{dt'}. \tag{11.10b}$$

The solution of the Cauchy problem (11.10a)–(11.10b) is given by

$$\frac{dU}{dt'} = -U(t, t')M(t'), \tag{11.10c}$$

as can be checked directly. Set now $V(t, t') = U(t, t')U(t', t)$. We have, by (11.7) and (11.10c),

$$\begin{aligned}\frac{dV}{dt'} &= \frac{dU}{dt'}(t, t')U(t', t) + U(t, t')\frac{dU}{dt}(t', t)\\ &= -U(t, t')M(t')U(t', t) + U(t, t')M(t')U(t', t) = 0,\end{aligned}$$

hence $V(t) = V(0) = I$.

Now set $W(t) = U(t, t')U(t', t'')$. We have

$$\frac{dW}{dt} = \frac{dU}{dt}(t, t')U(t', t'') = M(t)U(t, t')U(t', t'') = MW.$$

On the other hand, $W = I$ when $t = t''$. By the uniqueness of the solution of (11.7) we conclude that $W = U(t, t'')$. Q.E.D.

Consider now an mth-order differential equation

$$Lu = \frac{d^m u}{dt^m} + a_1(t)\frac{d^{m-1}u}{dt^{m-1}} + \cdots + a_m(t)u = f, \tag{11.11}$$

where the coefficients a_j and the right-hand side f are continuous functions in $]-T, T[$. By the standard procedure, already described (see p. 29) we may transform it into an equation of the kind (11.4). This means that, here,

$$M(t) = \begin{pmatrix} 0 & 1 & 0 & \cdots & 0\\ 0 & 0 & 1 & \cdots & 0\\ \vdots & & & & \\ -a_m(t) & -a_{m-1}(t) & -a_{m-2}(t) & \cdots & -a_1(t)\end{pmatrix}$$

and the components u_j of $\mathbf{u}$ are given by

$$u_j = \frac{d^{j-1}u}{dt^{j-1}}, \qquad 1 \leqq j \leqq m. \tag{11.12}$$

We see the meaning here of the initial condition (11.5): Let us denote by $u_{0,0}, u_{0,1}, \ldots, u_{0,m-1}$ the components of $\mathbf{u}_0$. Condition (11.5) is equivalent, in the present context, to

$$u^{(k)}(0) = u_{0,k}, \qquad 0 \leqq k \leqq m-1. \tag{11.13}$$

Condition (11.13) is thus what generalizes (11.3) to higher order equations. In accordance with the terminology for partial differential equations, we shall refer to the problem (11.11)–(11.13) as the *Cauchy problem* for the differential operator L.

The next question is related to what generalizes formula (11.2) in the higher order case. This can be answered by means of formula (11.8) relative to the equivalent system (11.4)–(11.5) One must recall that $\mathbf{f}(t) = f(t)\mathbf{e}_m$. We denote by $\mathbf{e}_j (1 \leqq j \leqq m)$ the vector whose components all vanish, except the jth one, which is equal to one. Observe then that

$$U(t, 0)\mathbf{u}_0 = \sum_{k=0}^{m-1} u_{0,k}\, U(t, 0)\mathbf{e}_{k+1}.$$

Let us call $U_j(t, t')$ the *first* component of $U(t, t')\mathbf{e}_{j+1}$. In view of (11.12) we derive from (11.8)

$$u(t) = \sum_{k=0}^{m-1} u_{0,k}\, U_k(t, 0) + \int_0^t U_{m-1}(t, t') f(t')\, dt'. \tag{11.14}$$

From this identity we derive that $U_k(t, t')$ is the solution of the Cauchy problem

$$LV = 0, \tag{11.15}$$

$$\text{at } t = t', \quad \frac{d^j V}{dt^j} = \begin{cases} 0 & \text{if } j \neq k \\ 1 & \text{if } j = k \end{cases} \quad (0 \leqq j \leqq m-1). \tag{11.16}$$

In general, it is not easy to compute the functions $U_j(t, t')$. It is therefore convenient to give an equivalent statement of (11.14) in which all U_j have been eliminated but one, U_{m-1}. This can be done as follows. Introduce the function

$$\tilde{u}_0(t) = \sum_{k=0}^{m-1} u_{0,k} \frac{t^k}{k!}. \tag{11.17}$$

Clearly $\tilde{u}_0(t)$ satisfies the initial conditions (11.13). Let us set

$$v(t) = u(t) - \tilde{u}_0(t).$$

We have

$$Lv = f - L\tilde{u}_0, \tag{11.18}$$

$$v^{(k)}(0) = 0, \qquad 0 \leqq k \leqq m-1. \tag{11.19}$$

We may therefore apply (11.14) with v instead of u:

$$v(t) = \int_0^t U_{m-1}(t, t')(f(t') - Lu_0(t'))\, dt',$$

whence

$$u(t) = \tilde{u}_0(t) - \int_0^t U_{m-1}(t, t')L\tilde{u}_0(t')\, dt' + \int_0^t U_{m-1}(t, t')f(t')\, dt'. \tag{11.20}$$

In the first integral, on the right-hand side of (11.20), we may integrate by parts and use the explicit expression of $L\tilde{u}_0$. This leads to a more explicit formula for $u(t)$, involving $U_{m-1}(t, t')$ and its derivatives. We leave it to the student to compute it, if he wishes to do so. In the case where the coefficients a_j are *constant*, $U_{m-1}(t, t')$ is not difficult to compute; it is the first component of

$$e^{(t-t')M}\mathbf{e}_m,$$

We have $U_{m-1}(t, t') = U_{m-1}(t - t', 0) = U(t - t')$ where $U(x)$ is the function so denoted on page 30. With the notation of Sect. 4, we have $E = HU$ where H is the Heaviside function and E is the fundamental solution of L with support in the nonnegative half-line. Formula (11.20) in this case can be rewritten as

$$u = \tilde{u}_0 + (HU) * \{H(f - L\tilde{u}_0)\} - (\check{H}U) * \{\check{H}(f - L\tilde{u}_0)\}, \tag{11.21}$$

where $\check{H}(t) = H(-t)$. The support of the first convolution on the right-hand side of (11.21) is contained in the half-line $t \geqq 0$, the one of the second convolution is contained in the half-line $t \leqq 0$.

Let us go back, for some final remarks, to the general case of variable coefficients. Concerning the system (11.4) and its solution (11.8) we may note that if the coefficients $M(t)$ and the right-hand side $\mathbf{f}(t)$ are N times continuously differentiable functions of t, $|t| < T$, the Riemann function $U(t, t')$ is an $(N + 1)$-times continuously differentiable function of t, t' in the same interval. Consequently, $\mathbf{u}(t)$ is C^{N+1} in $]-T, T[$. Let us now return to the mth-order equation (11.11) via the equivalent first-order system (11.4). Assume now that the a_j and f are C^N functions in $]-T, T[$. Because of the relations (11.12) we reach the conclusion that $u^{(m-1)}$ is a C^{N+1} function in $]-T, T[$. We may state

PROPOSITION 11.2. *If the coefficients a_j $(1 \leqq j \leqq m)$ and the right-hand side f of Eq. (11.11) are C^N functions of t, $|t| < T$, the solution u given by (11.14) is a C^{N+m} function in the interval $]-T, T[$.*

COROLLARY 11.1. *If the coefficients a_j and the right-hand side f are C^∞ functions in the interval $]-T, T[$, the same is true of the solution u.*

In this corollary one may replace C^∞ everywhere by *analytic*. This follows at once from the fact that if the coefficient $M(t)$ in (11.4) is analytic with respect to t, the Riemann function $U(t, t')$ is analytic with respect to t, t' (in the interval $]-T, T[$).

Exercises

11.1. Suppose that for any two points t, t' in the interval $]-T, T[$ the $m \times m$ matrices $M(t)$ and $M(t')$ commute. In this case, give a simple expression of the Riemann function $U(t, t')$ of the problem (11.4)–(11.5).

11.2. Suppose that, in (11.4), the matrix M is constant. Suppose furthermore that M is negative definite, i.e., $(M\mathbf{v}, \mathbf{v}) < 0$ for all $\mathbf{v} \in \mathbf{C}^m$, $\mathbf{v} \neq 0$ [we have denoted by $(\ ,\ )$ the *Hermitian* inner product in $\mathbf{C}^m$]. Let $U(t)$ denote the solution of the problem (11.7). Show that there is a constant $c > 0$ such that the matrix norm of $U(t)$ decreases at least as fast as e^{-ct} when $t \to +\infty$.

11.3 Using Exercise 4.10, give a complete description of the solution U of problem (11.7) when M is an arbitrary $m \times m$ matrix.

11.4. Give a complete description of the function $U_{m-1}(t, t')$ in (11.14) in the case where the coefficients a_j of L are constant (cf. Exercises 4.10 and 11.3).

11.5. Let L be the differential operator defined in (11.11). Suppose that the coefficients $a_j(t)$ and the right-hand side $f(t)$ are C^∞ functions in the interval $]-T, T[$. Suppose furthermore that $f(t) = 0$ if $t < 0$ and that u is a distribution in $]-T, T[$ of the form

$$u = \left(\frac{d}{dt}\right)^p g,$$

with g a continuous function of t, $t < T$, vanishing when $t < 0$. Use the fact that $Lu = f$ to prove that u is a C^∞ function in $]-T, T[$.

11.6. We assume that the coefficients a_j $(j = 1, \ldots, m)$ of the operator L and the second member f in (11.11) are (complex-valued) C^∞ functions in the real line. It follows at once from formula (11.14) that the problem (11.11)–(11.13) has a unique solution u which is also a C^∞ function in $\mathbf{R}^1$ (the uniqueness follows from the fundamental theorem about ODEs). Let then $\tilde{u}(t)$ be the function equal to $u(t)$ for $t \geqq 0$ and to zero for $t < 0$. Compute the distribution $L\tilde{u}$.

In particular, compute the distributions $L\tilde{u}$ when $u = U_k(t, 0)$, $k = 0, \ldots, m-1$ [see (11.14) to (11.16)].

11.7. Suppose that the differential operator L in (11.11) has constant coefficients and let $\tilde{u}$ denote the function defined in Exercise 11.6. Prove that if v is any distribution in the real line, vanishing for $t < 0$, i.e., $v \in \mathscr{D}'_+$

(cf. Exercise 4.11) such that $Lv = L\tilde{u}$, then necessarily we must have $v = \tilde{u}$. Is this fact still true when L has variable (but C^∞) coefficients?

11.8. Give an example of a homogeneous first-order ODE with analytic coefficients (in the whole real line) whose solutions all vanish of infinite order at the origin.

11.9. Analyze completely the possibility of solving the "initial value problem"

$$xu'' + a_1 u' + a_2 xu = 0, \tag{11.22}$$

$$u(0) = u_0. \tag{11.23}$$

Show that, in general (this is to be made precise), the Cauchy problem, obtained by adjoining the condition

$$u'(0) = u_1, \tag{11.24}$$

cannot be solved.

Is it true that, when it can be solved, its solution is unique?

11.10. Let a, f, and v be three nonnegative continuous functions on the closed interval $[0, T]$, v_0 a nonnegative constant. By comparing v to a solution of (11.1), prove that the inequality

$$v(t) \leqq v_0 + \int_0^t f(s)\, ds + \int_0^t a(s)v(s)\, ds, \qquad 0 \leqq t \leqq T, \tag{11.25}$$

implies the *Gronwall inequality*

$$v(t) \leqq \left(v_0 + \int_0^t f(s)\, ds\right) \exp\left(\int_0^t a(s)\, ds\right), \qquad 0 \leqq t \leqq T. \tag{11.26}$$

What inequality, bearing on v, would have followed from

$$|v'(t)| \leqq v_1 + \int_0^t f(s)\, ds + \int_0^t a(s)v(s)\, ds, \qquad 0 \leqq t \leqq T, \tag{11.27}$$

assuming that $a(s)$ is monotone?

12

The Cauchy Problem for Linear Partial Differential Equations. Preliminary Observations

In Example 2.2 we showed, using slightly different notation and under suitable assumptions of smoothness about the data u_0, u_1, that the problem

$$\left(\frac{\partial}{\partial t}\right)^2 u - \left(\frac{\partial}{\partial x}\right)^2 u = 0, \qquad x \in \mathbf{R}, \qquad |t| < T, \tag{12.1}$$

$$\text{when } t = 0, \qquad u = u_0(x), \qquad \frac{\partial u}{\partial t} = u_1(x), \tag{12.2}$$

has a solution, given by

$$u(t, x) = \tfrac{1}{2}[u_0(t + x) + u_0(x - t)] + \tfrac{1}{2}[U_1(t + x) - U_1(x - t)], \tag{12.3}$$

where U_1 is an arbitrarily chosen primitive of u_1. We have also stated that this is the only solution of (12.1)–(12.2). Now, clearly, this problem is the natural generalization of the Cauchy problem (11.11)–(11.13) when the right-hand side f is identically zero. In addition to the time variable t, we now have the "transversal" or space variable x; the coefficients in (12.1) are no longer functions of t, acting as multipliers, but rather differential operators with respect to x. But we know how to revert to the multipliers' case: We perform a Fourier transformation with respect to x. Equation (12.1) is transformed into

$$\left(\frac{\partial}{\partial t}\right)^2 \tilde{u} + |\xi|^2 \tilde{u} = 0, \tag{12.4}$$

which is of the kind (11.11); as a matter of fact, its coefficients are constants depending on the *parameter* ξ.

This method may even work for the inhomogeneous equation

$$\left(\frac{\partial}{\partial t}\right)^2 u - \left(\frac{\partial}{\partial x}\right)^2 u = f(t, x), \qquad x \in \mathbf{R}, \qquad |t| < T, \tag{12.5}$$

provided that the Fourier transform of f with respect to x, $\tilde{f}(t, \xi)$, makes sense, i.e., is a tempered distribution. It suggests a general way of approaching problems of the kind (12.1)–(12.5) and may give us valuable information about the possibility of solving them. There are obvious restrictions on the kind of data we are allowed to handle: They should all be tempered with respect to x. In addition to the fact that these are rather lax restrictions, we may also hope to get rid of them, once we know how to handle the problem when they are fulfilled. The crux of the matter is that the Riemann function of the ordinary differential equation, obtained by performing a Fourier transformation with respect to x in the partial differential equation under study, should be a tempered distribution with respect to ξ.

We may put our basic examples to test along these lines. Take for instance the homogeneous *Cauchy–Riemann* equation. By Fourier transformation, it leads to the Cauchy problem

$$\frac{\partial \tilde{u}}{\partial t} - \xi \tilde{u} = 0, \qquad \tilde{u}\Big|_{t=0} = \tilde{u}_0(\xi). \tag{12.6}$$

Its solution is given by

$$\tilde{u}(t, \xi) = \tilde{u}_0(\xi) e^{\xi t}. \tag{12.7}$$

When is this solution tempered with respect to ξ for all t, $|t| < T$? Roughly speaking, it is when

$$\tilde{u}_0(\xi) e^{T|\xi|} \quad \text{is tempered.} \tag{12.8}$$

If $T' < T$, (12.8) implies that $\tilde{u}_0(\xi) e^{T'|\xi|}$ is rapidly decreasing when $|\xi| \to +\infty$. Then write

$$u_0(x) = (2\pi)^{-n} \int e^{ix\xi} \tilde{u}_0(\xi)\, d\xi.$$

It is clear that we may replace x by $z = x + iy$ provided that $|y| < T$:

$$u_0(z) = (2\pi)^{-n} \int e^{ix\xi} e^{-y\xi} \tilde{u}_0(\xi)\, d\xi.$$

Moreover, if $|y| < T$, we may differentiate with respect to $\bar{z}$ under the integral sign; and the result is obviously zero: $u_0(z)$ is a holomorphic function of z in the strip $|\mathrm{Im}\, z| < T$. Consequently, $u_0(x)$ *must be an analytic function in* $\mathbf{R}$.

This must come to us as a reminder—a reminder that (12.6) is the transform of

$$\frac{\partial u}{\partial t} + i\frac{\partial u}{\partial x} = 0, \qquad u\Big|_{t=0} = u_0(x). \tag{12.9}$$

In view of the first condition (12.9), $u(t, x)$ must be a holomorphic function of $t + ix$. Therefore $u_0(x) = u(0, x)$ must be analytic. As soon as we make this observation, we notice that a similar one will apply to all analytic-hypoelliptic differential operators. Let us state it with some generality:

PROPOSITION 12.1. *Let Ω be an open subset of $\mathbf{R}^n$, P a linear partial differential operator with respect to the variables t, x, with analytic coefficients, of order $m > 0$, in the product set $\{t \in \mathbf{R}^1;\ |t| < T\} \times \Omega$. If P is analytic-hypoelliptic and if the function $f(t, x)$ is analytic in this product set, the problem*

$$Pu = f, \qquad x \in \Omega, \qquad |t| < T, \tag{12.10}$$

$$\left(\frac{\partial}{\partial t}\right)^j u = u_j(x), \qquad j = 0, \ldots, m-1, \qquad \textit{when} \quad t = 0, \quad x \in \Omega, \tag{12.11}$$

has no (distribution) *solution unless the initial data u_j are analytic functions in Ω.*

This observation settles also the case of the *Laplace* operator: The Cauchy problem (to give it its name) (12.10)–(12.11) will not have a solution for general data $u_j \in C^\infty(\Omega)$, $j = 0, 1$, when $P = \Delta$.

Let us look now at the *heat* equation. In this case, the transform equation reads

$$\frac{\partial \tilde{u}}{\partial t} + |\xi|^2 \tilde{u} = \tilde{f}(t, \xi). \tag{12.12}$$

If we take $f \equiv 0$, all solutions of (12.12) are of the form

$$\tilde{u}(t, \xi) = \tilde{u}_0(\xi) \exp(-t|\xi|^2). \tag{12.13}$$

This is not tempered with respect to ξ when $-T < t < 0$ unless for any $T' < T$, $\tilde{u}_0(\xi) \exp(T'|\xi|^2)$ is tempered. This requires that $u_0(x)$ be analytic, in fact be extendable to the complex space $\mathbf{C}^n$ as an *entire* function. On the other hand, for $t > 0$, (12.13) is tempered with respect to ξ as soon as this is true of $\tilde{u}_0$. Although this constitutes only circumstantial evidence, it seems to indicate that we may be able to solve the Cauchy problem when t is greater than zero but not when t is less than zero. We encounter here a typical situation, where solvability might hold for a "one-sided Cauchy problem," precisely the *forward* Cauchy problem, but not for the "two-sided" one. This observation applies only to the *existence* of solutions and will be confirmed by a more rigorous analysis. As for the *uniqueness* of the solution, we already have a hint that it is not going to hold. For this we refer to Example 3.1: Take Ω to be any open subset of the complement of the origin in $\mathbf{R}^1$; the function denoted $F(x, t)$ in Example 3.1 is a C^∞ function of (x, t) in $\Omega \times \mathbf{R}^1$, vanishes identically for $t < 0$, and hence satisfies

$$F = \frac{\partial F}{\partial t} = 0 \quad \text{when} \quad t = 0, \quad x \in \Omega, \tag{12.14}$$

and is a solution of the homogeneous heat equation

$$\frac{\partial F}{\partial t} = \Delta_x F \quad \text{in } \Omega \times \mathbf{R}^1. \tag{12.15}$$

If instead of the heat equation, we look at the Schrödinger equation, we face a different situation. The transform equation then reads

$$\frac{1}{i}\frac{\partial \tilde{u}}{\partial t} + |\xi|^2 \tilde{u} = \tilde{f}(t, \xi). \tag{12.16}$$

The Riemann function of (12.16) is

$$\exp(-i(t - t')|\xi|^2), \tag{12.17}$$

which is tempered for all real numbers t, t'. This suggests that solutions to the two-sided Cauchy problem will always exist—in any case, that they exist when the *Cauchy datum* $u_0(x)$, at $t = 0$, is tempered. Uniqueness, without additional restrictions, does not hold (Exercise 12.3). In connection with this, notice that the heat and the Schrödinger equations have an important feature in common, namely, the order with respect to t of both is 1, whereas their total order is 2. This lies at the heart of the matter, as the main theorem on uniqueness, Holmgren's theorem (Sect. 21), will show (see also Exercises 12.1 to 12.3).

Last but not least, there is the *wave equation.* The Riemann function of the transformed equation

$$\frac{\partial^2 \tilde{u}}{\partial t^2} + |\xi|^2 \tilde{u} = \tilde{f}(t. \xi) \tag{12.18}$$

is the function

$$\frac{\sin(|\xi|(t - t'))}{|\xi|} \tag{12.19}$$

which is tempered for any t, $t' \in \mathbf{R}^1$. Furthermore, the order of the wave equation with respect to t is 2, the same as its total order. We are in a situation where we can expect both the existence and the uniqueness of the solution to the Cauchy problem. This is borne out by what we know in the case of a single space variable, and will now be confirmed (for any number of space variables) by a rigorous analysis, based on the information already gathered in Sects. 7 and 8.

Exercises

12.1. Let θ be a number such that $\frac{1}{2} < \theta < 1$. Prove that

$$F(t, x) = \int_{-\infty}^{+\infty} \exp[-i\tau t - (i\tau)^{1/2}x - (i\tau)^{\theta}]\, d\tau \tag{12.20}$$

defines a C^∞ function in $\mathbf{R}^2$ which is a solution of the homogeneous heat equation

$$\frac{\partial F}{\partial t} = \frac{\partial^2 F}{\partial x^2} \qquad (\text{in } \mathbf{R}^2). \tag{12.21}$$

[*Note*: *The branches of the fractional powers* z^α *are always chosen in the "natural way": they are the ones that are positive on the positive half of the real axis.*]

12.2. By applying the Cauchy integral theorem, show that the function defined in (12.20) is also equal to

$$\int_{-a-i\infty}^{-a+i\infty} \exp(-zt - z^{1/2}x - z^{\theta})\, dz \tag{12.22}$$

(complex integration performed over the vertical straight line Re $z = -a$, where a is an arbitrary positive number). Derive from this that there is a constant $C > 0$, independent of $a > 0$, such that

$$|F(t, x)| \leqq C \exp(at + Ca^{\theta}) \tag{12.23}$$

(C is allowed to depend on x but remains bounded as long as x remains bounded). Derive from (12.23) that $F(t, x) \equiv 0$ if $t < 0$.

12.3. By modifying the argument indicated in Exercises 12.1 and 12.2, construct a solution $F_1(t, x)$ of the Schrödinger equation,

$$\frac{1}{i}\frac{\partial F_1}{\partial t} = \frac{\partial^2 F_1}{\partial x^2} \qquad \text{in } \mathbf{R}^2, \tag{12.24}$$

which is a C^∞ function of (t, x) in the whole plane, and vanishes identically in the half-plane $t < 0$.

12.4. Fix $t > 0$ in (12.20). What is the behavior of $F(t, x)$ as x tends to $+\infty$ or to $-\infty$?

12.5. Let $P(\xi, \tau)$ be a polynomial of degree m in two variables, with complex coefficients. Assume that $P(0, \tau)$ is of degree $\leq m - 1$ [whereas at least one partial derivative of $P(\xi, \tau)$ of order m is nonzero]. By using the Puiseux theorem and generalizing the argument in Exercises 12.1 and 12.2, show that there is a C^∞ function of (t, x) in $\mathbf{R}^2$, $F(t, x)$, vanishing identically for $t < 0$, and satisfying, in the entire plane, the homogeneous equation

$$P\left(\frac{\partial}{\partial x}, \frac{\partial}{\partial t}\right)F = 0. \tag{12.25}$$

12.6. Let A be a complex $m \times m$ matrix and consider the Cauchy problem in $\mathbf{R}^2$,

$$\frac{\partial \mathbf{u}}{\partial t} = A\frac{\partial \mathbf{u}}{\partial x}, \qquad \mathbf{u}\Big|_{t=0} = \mathbf{u}_0(x), \tag{12.26}$$

where boldface indicates complex m-vectors. Prove that, if A is self-adjoint, i.e., $A = A^* = {}^t\bar{A}$, the problem (12.26) has a unique solution which is a C^∞ function of t (in $\mathbf{R}^1$) with values in the space $\mathscr{S}'_x(\mathbf{C}^m)$ of tempered distributions with respect to x, valued in $\mathbf{C}^m$ (i.e., of m-vectors whose components are tempered distributions in x) for every $\mathbf{u}_0 \in \mathscr{S}'_x(\mathbf{C}^m)$.

12.7. Suppose that the matrix A in Exercise 12.6 has *distinct* eigenvalues. Prove that *if none of these eigenvalues is real*, the Cauchy problem (12.26) has a solution only if $\mathbf{u}_0(x)$ is an analytic function of x.

12.8. As in Exercise 12.7, suppose that the eigenvalues of the matrix A are all distinct. Prove that, *if all these eigenvalues are real*, then the Cauchy problem (12.26) always has a solution, whatever the continuous function $\mathbf{u}_0$ in $\mathbf{R}^1$. Give an explicit expression of the solution and use this expression to describe the support of the solution when we assume that $\mathbf{u}_0(x) > 0$ for $|x| < 1$ and $\mathbf{u}_0(x) = 0$ for $|x| \geqq 1$.

12.9. Consider the Cauchy problem

$$\frac{\partial u}{\partial t} = \lambda t\, \Delta_x u \quad \text{in } \mathbf{R}^{n+1} \qquad u\Big|_{t=0} = u_0(x) \quad \text{in } \mathbf{R}^n, \tag{12.27}$$

where $u_0(x) \in L^2(\mathbf{R}^n)$ (we could have taken u_0 as well to be a tempered distribution). Show that if $\lambda > 0$, problem (12.27) has a solution, which is a distribution in $\mathbf{R}^{n+1}$, and in fact an analytic function for $t \neq 0$.

13

The Global Cauchy Problem for the Wave Equation. Existence and Uniqueness of the Solutions

In this section we study the *Cauchy problem*

$$\frac{\partial^2 u}{\partial t^2} - \Delta_x u = f(t, x) \quad \text{in } \mathbf{R}^{n+1}, \tag{13.1}$$

$$\text{when } t = 0, \qquad u = u_0(x), \qquad \frac{\partial u}{\partial t} = u_1(x) \quad \text{in } \mathbf{R}^n. \tag{13.2}$$

We shall prove the existence and uniqueness of the solution $u(t, x)$ for a very general class of data f and u_0, u_1. For the moment, we make a very restrictive assumption on the data:

(13.3) *$f(t, x)$ is a C^∞ function of (t, x) in $\mathbf{R}^{n+1}$ whose support is contained in some cylinder $\{(x, t);\ |x| \leqq R\}$;*

(13.4) *u_0, u_1 are C^∞ functions of x in $\mathbf{R}^n$, having a compact support.*

Under these circumstances, we may perform a Fourier transformation with respect to x in both (13.1) and (13.2). We obtain

$$\tilde{u}_{tt} + |\xi|^2 \tilde{u} = \tilde{f}(t, \xi) \quad \text{in } \mathbf{R}^1 \times \mathbf{R}_n, \tag{13.5}$$

$$\tilde{u}(0, \xi) = \tilde{u}_0(\xi), \qquad \tilde{u}_t(0, \xi) = \tilde{u}_1(\xi) \quad \text{in } \mathbf{R}_n, \tag{13.6}$$

where the subscripts mean differentiation.

In order to express u in terms of f and u_0, u_1 we shall apply formula (11.14). We have already computed $U_1(t, t')$ in Sect. 7. It is given by

$$U_1(t, t') = \frac{\sin(|\xi|(t - t'))}{|\xi|}. \tag{13.7}$$

Since [see (11.15), (11.16)] $U_0(t, 0)$ is the solution of

$$V_{tt} + |\xi|^2 V = 0, \qquad V\Big|_{t=0} = 1, \qquad V_t\Big|_{t=0} = 0, \tag{13.8}$$

we immediately obtain

$$U_0(t, 0) = \cos(|\xi| t). \tag{13.9}$$

By (11.14) we have

$$\tilde{u}(t, \xi) = \tilde{u}_0(\xi)\cos(|\xi| t) + \tilde{u}_1(\xi)\frac{\sin(|\xi| t)}{|\xi|} + \int_0^t \tilde{f}(t', \xi)\frac{\sin(|\xi|(t - t'))}{|\xi|}\,dt'. \tag{13.10}$$

First take $t > 0$. Then we may write

$$\tilde{u}(t, \xi) = \tilde{u}_0(\xi)H(t)\frac{\partial}{\partial t}\frac{\sin(|\xi| t)}{|\xi|} + \tilde{u}_1(\xi)H(t)\frac{\sin(|\xi| t)}{|\xi|} + \int_{-\infty}^{+\infty} H(t')\tilde{f}(t', \xi)H(t - t')\frac{\sin(|\xi|(t - t'))}{|\xi|}\,dt'. \tag{13.11}$$

Observe that the first term, on the right-hand side of (13.11) is equal to

$$\tilde{u}_0(\xi)\frac{\partial}{\partial t}\left\{H(t)\frac{\sin(|\xi| t)}{|\xi|}\right\}.$$

In view of this, (13.11) can be rewritten, for $t > 0$, as

$$\tilde{u}(t, \xi) = \tilde{u}_0(\xi)\frac{\partial}{\partial t}\tilde{E}_+(t, \xi) + \tilde{u}_1(\xi)\tilde{E}_+(t, \xi) + \{H(t)\tilde{f}(t, \xi)\} \underset{(t)}{*} \tilde{E}_+(t, \xi), \tag{13.12}$$

where $\underset{(t)}{*}$ means that the convolution should only be computed with respect to t. For $t < 0$, we have

$$\tilde{u}(t, \xi) = -\tilde{u}_0(\xi)\frac{\partial}{\partial t}\tilde{E}_-(t, \xi) - \tilde{u}_1(\xi)\tilde{E}_-(t, \xi) + \{H(-t)\tilde{f}(t, \xi)\} \underset{(t)}{*} \tilde{E}_-(t, \xi). \tag{13.13}$$

In these formulas, E_+ (resp. E_-) is the fundamental solution of the wave equation with support in the forward (resp. backward) light-cone, $\tilde{E}_+$ (resp. $\tilde{E}_-$) its Fourier transform with respect to x [recall that $\tilde{E}_+ - \tilde{E}_- =$

$\sin(|\xi|t)/|\xi|]$. If we now perform an inverse Fourier transformation with respect to x, we obtain

$$(13.14)\qquad u(t, x) = u_0(x) \underset{(x)}{*} \frac{\partial}{\partial t}(E_+ - E_-) + u_1(x) \underset{(x)}{*} (E_+ - E_-)$$
$$+ \{H(t)f(t, x)\} \underset{(t,x)}{*} E_+ + \{H(-t)f(t, x)\} \underset{(t,x)}{*} E_- .$$

Once this formula is written down, it is apparent that much of our restricting conditions (13.3)–(13.4) on the data can be lifted. Indeed, the convolutions in (13.14) make sense when $u_0(x)$, $u_1(x)$ are arbitrary distributions in $\mathbf{R}^n$ and when $f(t, x)$ is a function of t, say continuous, with values in the space $\mathscr{D}'(\mathbf{R}^n_x)$ of distributions with respect to x. This is due to the fact that, for each t, the supports of E_+ and E_- as distributions in x are contained in the ball $|x| \leqq |t|$. As for the convolutions with respect to t, they make sense because both the support of E_+ and of $H(t)f(t, x)$ are contained in the half-space $t \geqq 0$, whereas those of E_- and $H(-t)f(t, x)$ are both contained in the half-space $t \leqq 0$ (we recall that, on the real line, the distributions with support in the half-line $t \geqq 0$ form a convolution algebra).

Thus formula (13.14) extends to very general data [the only restriction is that $f(t, x)$ should be a continuous function of t with values in $\mathscr{D}'_x$]. But we may ask whether $u(t, x)$ given by (13.14) satisfies (13.1) and (13.2). That it satisfies (13.1) is obvious. Let us examine (13.2). Let Ω denote a bounded open subset of $\mathbf{R}^n$. If $|t| < T$, the supports of

$$E_+, \qquad E_-, \qquad \left(\frac{\partial}{\partial t}\right)(E_+ - E_-), \qquad E_+ - E_-,$$

viewed as distributions in x, are contained in the open ball $|x| < T$. Let us denote by Ω_T the set of points $x \in \mathbf{R}^n$ whose distance to Ω does not exceed T. Let $\alpha \in C^\infty_c(\mathbf{R}^n)$, $\alpha = 1$ in Ω_T, $\alpha = 0$ off $\Omega_{T+\varepsilon}$ ($\varepsilon > 0$). By well-known properties of convolution, we see that

(13.15) if $|t| < T, \quad x \in \Omega,$

$$u(t, x) = \{\alpha(x)u_0(x)\} \underset{(x)}{*} \left\{\frac{\partial}{\partial t}(E_+ - E_-)\right\} + \{\alpha(x)u_1(x)\} \underset{(x)}{*} (E_+ - E_-)$$
$$+ \{H(t)\alpha(x)f(t, x)\} \underset{(t,x)}{*} E_+ + \{H(-t)\alpha(x)f(t, x)\} \underset{(t,x)}{*} E_- .$$

But the right-hand side in this formula is somewhat similar to the right-hand side of (13.14) under assumptions (13.3) and (13.4). It is true that now $\alpha(x)u_0(x)$, $\alpha(x)u_1(x)$, $\alpha(x)f(t, x)$ are not necessarily smooth functions. But their supports with respect to x are contained in a fixed compact set. Therefore, by the Paley–Wiener theorem, their Fourier transforms with respect to

x are analytic functions of ξ. A formula of the kind (13.10) is also valid here: One must of course replace $u_j(x)$ by $v_j(x) = \alpha(x)u_j(x)$, $j = 0, 1$, $f(t, x)$ by $g(t, x) = \alpha(x)f(t, x)$, and also $u(t, x)$ by $v(t, x)$, equal to the right-hand side of (13.15). Suppose that f is a continuous function of t valued in $\mathscr{D}'_x$. Then g is a continuous function of t with values in the space of tempered distributions in x, $\mathscr{S}'_x$, and $\tilde{g}(t, \xi)$ is a continuous function of t valued in $\mathscr{S}'_\xi$. By virtue of (13.10) we see that, when $t \to 0$, $\tilde{v}(t, \xi)$ converges to $\tilde{v}_0(\xi)$ in $\mathscr{S}'_\xi$, while

$$\tilde{v}_t(t, \xi) = -\tilde{v}_0(\xi)|\xi| \sin(|\xi|t) + \tilde{v}_1(\xi)\cos(|\xi|t) + \int_0^t \tilde{g}(t', \xi)\cos(|\xi|(t - t'))\, dt' \tag{13.16}$$

converges to $\tilde{v}_1(\xi)$. This implies that, when $t \to 0$, $u(t, x)$ converges to $u_0(x)$ and $u_t(t, x)$ converges to $u_1(x)$ in the sense of distributions in Ω. As Ω was an arbitrary bounded open subset of $\mathbf{R}^n$, this convergence is also valid in $\mathbf{R}^n$. Moreover, (13.16) shows that $u_t(t, x)$ is a C^1 function of t with values in $\mathscr{D}'(\mathbf{R}^n)$, hence (cf. Proposition 11.2) $u(t, x)$ is a C^2 function of t valued in $\mathscr{D}'(\mathbf{R}^n)$.

The reader will notice that we are here dealing with differentiable functions valued in (infinite-dimensional) locally convex spaces, such as $\mathscr{D}'(\mathbf{R}^n)$. On the subject of differentiable functions valued in Banach spaces, see Sect. 39. If E is a locally convex Hausdorff space, it is in particular a topological space, and we know therefore the meaning of a *continuous* function f, defined in another topological space X, and valued in E. If X is an open subset of the real line, or the Euclidean space $\mathbf{R}^m$, we may define the *derivative* of f, or the *gradient* of f, at a point of X. Consider for instance the case $X = \mathbf{R}^1$ and denote by t the variable in $\mathbf{R}^1$. Then $f'(t_0) = \lim h^{-1}[f(t_0 + h) - f(t_0)]$ as $h \neq 0$ converges to zero. The only novelty, compared to the standard situation, is that the convergence takes place in E and that the latter can be equipped, in a natural way, with various topologies, for instance the strong (or initial) topology or the weak, $\sigma(E, E')$, topology. Throughout this book, we shall most of the time use the strong topology—in a Banach space, this means the topology defined by the norm. In spaces such as $\mathscr{D}'$ or $\mathscr{S}'$ (the space of tempered distributions), it does not make any difference: for sequences, or for filters with countable bases (which is what is involved in the definition of the derivative), it is equivalent to the weak topology. Thus to say that a distribution $T(t)$, depending on the real variable t (say, in some interval $]t_0, t_1[$) is continuously differentiable, resp. C^k $(0 \leqq k \leqq +\infty)$, is the same as saying that, given any test function $\phi \in C_c^\infty$, the *scalar* function $\langle T(t), \phi\rangle$ is continuously differentiable, resp. C^k. This is also valid when $T(t)$ is tempered and regarded as valued in $\mathscr{S}'$. One fact that will be used, and should perhaps be recalled, is that a function f, defined and C^∞ in an open subset of $\mathbf{R}^m$, X, valued in the space of C^∞ functions in an open subset Y of another space $\mathbf{R}^n$, is the same thing as a C^∞ function in the product $X \times Y$.

Consider now a distribution $h(t, x)$ in $\mathbf{R}^{n+1}$ which is a C^2 function of t valued in $\mathscr{D}'(\mathbf{R}^n_x)$ and which satisfies

$$h_{tt} - \Delta_x h = 0 \quad \text{in } \mathbf{R}^{n+1}, \tag{13.17}$$

$$h = h_t = 0 \quad \text{in } \mathbf{R}^n \qquad \text{when} \quad t = 0. \tag{13.18}$$

Let us set $h_+(t, x) = H(t)h(t, x)$. The Leibniz formula and the initial conditions (13.18) imply at once that

$$\left(\frac{\partial}{\partial t}\right)^2 h_+ = H(t)h_{tt},$$

hence

$$\left(\frac{\partial}{\partial t}\right)^2 h_+ - \Delta_x h_+ = 0 \quad \text{in } \mathbf{R}^{n+1}. \tag{13.19}$$

Since the support of h_+ is contained in the half-space $t \geqq 0$, the preceding considerations about convolutions in formula (13.14) apply. We may now write

$$0 = E_+ * (\Box h_+) = (\Box E_+) * h_+ = h_+$$

where $\Box$ stands for the wave operator. This proves the *uniqueness* of the solution $u(t, x)$ whose existence was proved above. Summarizing:

THEOREM 13.1. *Let u_0, u_1 be any two distributions in $\mathbf{R}^n$, $f(t,x)$ any continuous function of $t \in \mathbf{R}^1$ valued in $\mathscr{D}'(\mathbf{R}^n_x)$. There is a unique solution $u(t, x)$ of the Cauchy problem (13.1)–(13.2) which is a C^2 function of t with values in $\mathscr{D}'(\mathbf{R}^n_x)$. It is given by formula (13.14).*

Formulas (13.5) and (13.16) show also that when the data are C^∞ functions, the same is true of the solution:

THEOREM 13.2. *Let u_0, u_1 be any two C^∞ functions in $\mathbf{R}^n$, f any C^∞ function in $\mathbf{R}^{n+1}$. The solution u of problem (13.1)–(13.2) given by formula (13.14) is a C^∞ function in $\mathbf{R}^{n+1}$.*

It should be understood that Theorem 13.2 is not at all a hypoellipticity result: It is not a statement of regularity about every solution of (13.1) but only about that unique solution of (13.1) which satisfies (13.2) also.

Rather than looking for C^∞ solutions to the Cauchy problem with C^∞ data, one may want to use data with limited regularity of some kind and obtain solutions with as much regularity as possible. A good method of measuring regularity with respect to x is provided by the so-called *Sobolev spaces* in $\mathbf{R}^n$. They measure it by taking *square-integrability* as the zero-level regularity. Then the degree of regularity can be any real number s: If s is a positive integer, to say that u is regular of degree s is to say that all derivatives of u of order $\leqq s$ are L^2 functions; if s is a negative integer, it is to say that u is a sum of distribution derivatives of order $\leqq s$ of L^2 functions. The great advantage of this approach lies in the fact that the Fourier transformation works so well in the space $L^2(\mathbf{R}^n)$: One can fully exploit the Plancherel theorem. This is what enables one to define the degrees of regularity when they are not integers. The formal definition of the sth Sobolev space $H^s(\mathbf{R}^n)$ is as follows:

Definition 13.1. A tempered distribution u in $\mathbf{R}^n$ is said to belong to $H^s(\mathbf{R}^n)$ if the Fourier transform $\hat{u}$ of u is a square-integrable function with respect to the measure $(1 + |\xi|^2)^s \, d\xi$.

We repeat that here s can be any real number. One may turn $H^s = H^s(\mathbf{R}^n)$ into a Hilbert space by setting the norm to be

$$\|u\|_s = \frac{1}{(2\pi)^{n/2}} \left\{ \int |\hat{u}(\xi)|^2 (1 + |\xi|^2)^s \, d\xi \right\}^{1/2}. \tag{13.20}$$

Denoting by $\mathscr{F}$ the Fourier transformation in $\mathbf{R}^n$ and by $\mathscr{F}^{-1}$ its inverse we may introduce the operator

$$T^s u = \mathscr{F}^{-1}(1 + |\xi|^2)^{s/2} \mathscr{F} u. \tag{13.21}$$

Then we see that T^s is an *isometry* of H^s onto $H^0 = L^2$: via T^s, H^s is a copy of L^2. Let $\mathscr{S}$ denote the space of C^∞ functions in $\mathbf{R}^n$, φ, such that $|D^\alpha \varphi|$ decreases at infinity faster than $|x|^{-k}$, whatever the multi-index α and the integer k. The Fourier transformation is an isomorphism of $\mathscr{S}_x$ (equipped with its natural topology) onto $\mathscr{S}_\xi$; on the other hand, it is obvious that multiplication by $(1 + |\xi|^2)^{s/2}$ is an isomorphism of $\mathscr{S}_\xi$ onto itself. We conclude that T^s is an isomorphism of $\mathscr{S}_x$ onto itself. In fact, when s varies, *the operators T^s form a one-parameter group of automorphisms of $\mathscr{S}_x$*. In particular, T^{-s} maps $\mathscr{S} \subset L^2$ onto $\mathscr{S} \subset H^s$. Since it is an isometry and $\mathscr{S}$ is dense in L^2, we reach the following conclusion, used below:

PROPOSITION 13.1. *The space $\mathscr{S}$ of C^∞ functions in $\mathbf{R}^n$, rapidly decaying at infinity, is dense in every space H^s (s real).*

Since the space of test functions C_c^∞ is dense in $\mathscr{S}$, it is also dense in every H^s (for more on the Sobolev spaces, see Sects. 24 and 25).

Let us now return to formula (13.10). Let us multiply both members by $(1 + |\xi|^2)^{s/2}$ and observe that

$$\left| \frac{\sin(|\xi| t)}{|\xi|} \right| \leqq 2(1 + |\xi|^2)^{-1/2} \sup(1, |t|);$$

hence, by virtue of (13.20),

$$\|u(t, \cdot)\|_s \leqq \|u_0\|_s + 2\sup(1, |t|)\|u_1\|_{s-1} + 2\left|\int_0^t \|f(t', \cdot)\|_{s-1} \sup(1, |t'|)\, dt'\right|. \tag{13.22}$$

Because of the density of C_c^∞ in H^s this shows that formula (13.10) remains valid when $u_0 \in H^s$, $u_1 \in H^{s-1}$, and f is a continuous function of t valued in H^{s-1}. It is *not* true, however, that u is then a C^2 function of t with values in H^s: for instance, if we differentiate (13.10) once with respect to t, a new factor $|\xi|$ shows up, as in (13.6). Thus, if the hypotheses stated above hold, we see that u_t is a continuous function of t with values in H^{s-1}. We have

$$\|u_t(t, \cdot)\|_{s-1} \leqq \text{const}\left\{\|u_0\|_s + \|u_1\|_{s-1} + \left|\int_0^t \|f(t', \cdot)\|_{s-1}\, dt'\right|\right\}. \tag{13.23}$$

One can push this argument one step further and prove

THEOREM 13.3. *Suppose that $u_0 \in H^s$, $u_1 \in H^{s-1}$ and that f is a continuous function of $t \in \mathbf{R}^1$ valued in H^{s-1}. Then the solution u of (13.1)–(13.2) is a C^k function of t valued in H^{s-k}, $k = 0, 1, 2$.*

More generally,

THEOREM 13.4. *Suppose that $u_0 \in H^s$, $u_1 \in H^{s-1}$ and that f is a C^m function of t valued in H^{s-1}. Then the solution u of (13.1)–(13.2) is a C^k function of t valued in H^{s-k}, $k = 0, \ldots, m + 2$.*

At this stage, it is traditional to discuss the *dependence* on the data of the solution u whose existence and uniqueness are asserted in the preceding theorems. This can easily be done by means of the *closed graph theorem*, after having observed that formula (13.14) has the following obvious implication:

THEOREM 13.5. *The mapping*

$$(u_0, u_1, f) \mapsto u, \tag{13.24}$$

where u is given by (13.14), is a continuous linear map of the triple product

$$\mathscr{D}'(\mathbf{R}^n) \times \mathscr{D}'(\mathbf{R}^n) \times C^0(R_t^1; \mathscr{D}'(\mathbf{R}_x^n))$$

into $C^2(\mathbf{R}_t^1; \mathscr{D}'(\mathbf{R}_x^n))$.

Proof. It suffices to apply formula (13.15), which reduces the proof to the case where u_0, u_1 have compact support while the projection on the space $\mathbf{R}^n$ of the x variables of supp f is compact. We may then make a Fourier transformation with respect to x and verify the statement on formula (13.10), where it is obvious. Q.E.D

Then one obtains at once the following results:

THEOREM 13.6. *The mapping (13.24) is a continuous linear map of*

$$C^\infty(\mathbf{R}^n) \times C^\infty(\mathbf{R}^n) \times C^\infty(\mathbf{R}^{n+1})$$

into $C^\infty(\mathbf{R}^{n+1})$.

Indeed, the graph of this mapping is easily seen to be closed (cf. Remark 15.1), and hence continuous by the closed graph theorem. Similarly, one proves the following extension of Theorem 13.3:

THEOREM 13.7. *The mapping (13.24) is a continuous linear map of*

$$H^s(\mathbf{R}^n) \times H^{s-1}(\mathbf{R}^n) \times C^0(\mathbf{R}^1_t; H^{s-1}(\mathbf{R}^n))$$

into

$$\bigcap_{k=0,1,2} C^k(\mathbf{R}^1_t; H^{s-k}(\mathbf{R}^n)).$$

We recall that if Ω is an open subset of $\mathbf{R}^m$ and X a Banach space, the space $C^k(\Omega; X)$ is equipped with the topology of uniform convergence, on every compact subset, of the functions and of all their derivatives of order $\leqq k$. We also recall that the topology of an intersection $A \cap B$ is defined as follows: A set is the neighborhood of a point in $A \cap B$ if this is true for the topology induced on $A \cap B$ by A or for the one induced by B.

Exercises

13.1. In formula (12.3) suppose that both u_0 and u_1 have compact support. Prove that (12.3) is equivalent to (13.10) in which $n = 1$ and $f \equiv 0$.

13.2. Let θ denote the angular variable on the unit circumference Γ (thus θ varies from 0 to 2π). By using Fourier series expansions, solve the Cauchy problem

$$\frac{\partial^2 u}{\partial t^2} = \frac{\partial^2 u}{\partial \theta^2}, \qquad t \in \mathbf{R}^1, \quad 0 \leqq \theta \leqq 2\pi, \tag{13.25}$$

$$u = u_0(\theta), \qquad \frac{\partial u}{\partial t} = u_1(\theta) \qquad \text{when} \quad t = 0, \quad 0 \leqq \theta \leqq 2\pi, \tag{13.26}$$

where u_0 and u_1 are two C^∞ functions of θ (on Γ). Do the same when u_0 and u_1 are any two *distributions* on Γ. Compare with (12.3) and (13.10).

13.3. Using the same notation as in Exercise 13.2, state the analog of Theorem 13.4 for the Cauchy problem (13.25)–(13.26) (in particular the student must define the Sobolev spaces H^k on the unit circumference Γ).

13.4. Using the same notation as in Exercise 13.2, solve explicitly the problem (13.25)–(13.26) when $u_0 = 0$, $u_1 = \delta(\theta)$, the Dirac distribution on the unit circumference Γ (i.e., the mass $+1$ at $\theta = 0$). Consider then the periodic distribution $\overset{\circ}{\delta}$ on the real line, equal to the Dirac measure in the interval $-2\pi < t < 2\pi$. Solve the Cauchy problem (12.1)–(12.2) when $u_0 = 0$, $u_1 = \overset{\circ}{\delta}$ by applying both formulas (12.3) and (13.10).

13.5. Consider the Cauchy problem (13.1)–(13.2) in the case $n \geqq 2$. Suppose that the data f, u_0, u_1 are rotation-invariant functions of $x \in \mathbf{R}^n$, i.e., are functions of $|x|$ alone. Prove that the solution u is then also a function of $|x|$ alone (for fixed time t). Show that formula (13.14) can be simplified to involve integrals, and more generally expressions where only the variables t and $|x|$ intervene. Write it down explicitly in the cases $n = 2$, 3, using the expression of E_+ computed in the Appendix to Sect. 8.

13.6. Consider the *Klein–Gordon operator*

$$K_m = \frac{\partial^2}{\partial t^2} - \Delta_x + m^2. \tag{13.27}$$

Give the formulas that are the analogs of (13.7) and (13.9) when m is real, $m \neq 0$. Extend Theorem 13.1 to this case by verifying that the proof can indeed be extended. Is Theorem 13.1 still true when m is any complex number?

14

Domain of Influence, Propagation of Singularities, Conservation of Energy

The solution u of the Cauchy problem (13.1)–(13.2), whose existence and uniqueness are stated in Theorem 13.1, is the sum $v + w$ of the solution v of the problem

$$\frac{\partial^2 v}{\partial t^2} - \Delta_x v = 0 \quad \text{in } \mathbf{R}^{n+1}, \tag{14.1}$$

$$\text{when} \quad t = 0, \quad v = u_0(x), \quad \frac{\partial v}{\partial t} = u_1(x) \quad \text{in } \mathbf{R}^n, \tag{14.2}$$

and of the solution w of the following problem:

$$\frac{\partial^2 w}{\partial t^2} - \Delta_x w = f(t, x) \quad \text{in } \mathbf{R}^{n+1}, \tag{14.3}$$

$$\text{when} \quad t = 0, \quad w = 0, \quad \frac{\partial w}{\partial t} = 0 \quad \text{in } \mathbf{R}^n. \tag{14.4}$$

(14.4)†

One moment of reflection easily shows the meaning of each one of these problems. In the first one, an oscillatory phenomenon is generated at time zero, with a certain initial intensity and a certain initial velocity of propagation: It then propagates, both in the future and in the past, through a medium where no other disturbance is created. The function v measures, at any given point in space and time, the amplitude and the phase of the oscillation. In problem (14.3)–(14.4) there is no "initial impulsion": but a "field" is

† Throughout this section, unless otherwise specified, u_0 and u_1 will be two distributions in $\mathbf{R}^n$ and $f(t, x)$ will be a continuous function of t in $\mathbf{R}^1$ valued in $\mathscr{D}'_x(\mathbf{R}^n)$.

generated in the medium, in certain zones (where "sources" and "sinks" are located), in a manner that varies with time; w measures the resulting oscillation.

It is convenient to consider the two problems separately. We shall look mainly at the first, (14.1)–(14.2). Its solution is given by (13.14):

$$v(t, x) = u_0(x) \underset{(x)}{*} \frac{\partial}{\partial t}(E_+ - E_-) + u_1(x) \underset{(x)}{*} (E_+ - E_-). \tag{14.5}$$

Let us look, for instance, at v in the half-space $t > 0$. There we have

$$v(t, x) = u_0(x) \underset{(x)}{*} \frac{\partial E_+}{\partial t} + u_1(x) \underset{(x)}{*} E_+ . \tag{14.6}$$

THEOREM 14.1. *Let Ω be any open subset of $\mathbf{R}^n$, t any number greater than zero. The values of $v(t, x)$ in Ω depend only on the values of $u_0(x)$ and $u_1(x)$ in the set*

$$\Omega_t = \{x \in \mathbf{R}^n;\, \text{dist}(x, \Omega) < t\}.$$

Proof. It suffices to show that if both u_0 and u_1 vanish in Ω_t, $v(t, x)$ vanishes in Ω. We have, in such a case,

$$\text{supp}_x\left(u_0 \underset{(x)}{*} \frac{\partial E_+}{\partial t}\right) \subset \text{supp}_x\, u_0 + \text{supp}_x\, E_+ \subset \complement\Omega_t + \text{supp}_x\, E_+$$

where supp_x stands for the support of distributions in the x variables (t is fixed); $A + B$ stands for the vector addition of the subsets A and B of $\mathbf{R}^n$. Recalling that the support of $E_+(x, t)$ as a distribution in x is contained in the ball $|x| \leqq t$, we see that

$$\text{supp}_x\left(u_0 \underset{(x)}{*} \frac{\partial E_+}{\partial t}\right) \subset \complement\Omega.$$

A similar conclusion holds for $u_1 \underset{(x)}{*} E_+$. Q.E.D.

The preceding statement is usually referred to as the theorem on the *domain of influence*. The open set Ω_t can be constructed in the following manner: View Ω as a subset of $\mathbf{R}^{n+1}$, contained in the hyperplane corresponding to time t, i.e., identify Ω with $\tilde{\Omega} = \{(x, t);\, x \in \Omega\}$. Let Ω_Γ denote the union of all the translates of the *backward* light-cone with vertices at the points of $\tilde{\Omega}$. Then Ω_t is the intersection of Ω_Γ with the hyperplane corresponding to time zero (see Fig. 14.1).

Suppose for a moment that u_0 and u_1 are C^∞ functions in $\mathbf{R}^n$. Then the preceding considerations about the domain of influence can be made *punctual*: For fixed t, the value of $v(t, x)$ at x_0 (v is now a C^∞ function in $\mathbf{R}^{n+1}$) depends

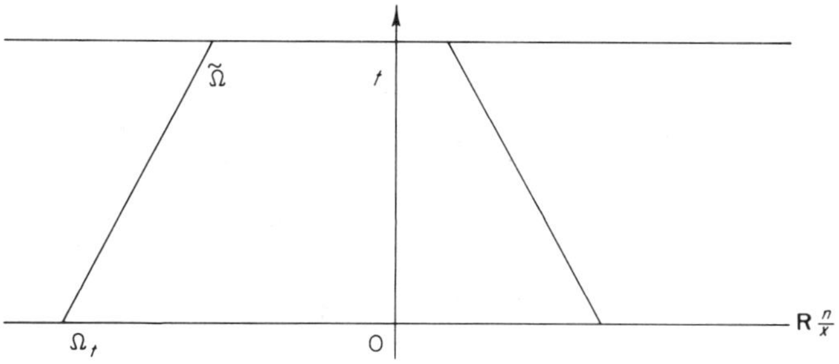

FIG. 14.1.

only on the values of u_0 and u_1 in the ball $|x - x_0| \leqq t$. This can be checked directly on the convolutions in (14.6) or else one may apply Theorem 14.1 and make Ω converge to the point x_0. As a matter of fact, for this argument to work, u_0 and u_1 need not be C^∞: They must have sufficient regularity so that v be a continuous function of (t, x).

The above considerations can be reverted in time: Rather than asking where the phenomena observed in a certain region of space, at a given time, did originate, one might ask what regions will be influenced by something happening in some set at time zero. More rigorously, this is to ask what is the support of $v(t, x)$, knowing that the supports of u_0 and u_1 are contained in K? Of course, in the half-space $t > 0$,

$$\operatorname{supp} v \subset K + \operatorname{supp} E_+ , \tag{14.7}$$

where K has been identified with the set of points $(x, 0)$, $x \in K$. Because of the Huyghens principle, there is now a significant difference between the case $n > 1$, n odd, and the other cases. This is best seen by taking K to be a point, say $K = \{0\}$ [for instance, one may then take $u_0 = \delta(x)$, $u_1 = 0$; more generally, one may take u_0, u_1 to be arbitrary linear combinations of the Dirac measure and of its derivatives, with respect to x]. This would correspond to something like flashing a light at some point in space, at time $t = 0$, with an "infinite" intensity and turning it off at time $t > 0$. If $n = 3$ or, more generally, if $n \geqq 3$ and odd, at an arbitrary time $t_0 > 0$, the oscillation, measured by $v(x, t)$, is concentrated on the *sphere* $|x - x_0| = t_0$. We may rephrase this as follows: Suppose that an observer is located at some point in space, at a distance $r > 0$ from 0: the oscillation is coming toward him as t goes from 0 to r, reaches him at time $t = r$, and leaves him behind as $t > r$ increases. The fact that it reaches the observer at time t exactly equal to r is simply due to our choice of the unities measuring time and space; we have assumed throughout that the velocity of propagation is equal to one. The general case, where the velocity is any number $V > 0$, is easy: The oscillation reaches the observer at time r/V.

When $n = 1$ or when n is even, it is also true that the observer is reached by the wave at time $t = r$ but the difference with the case $n = 3, 5, 7, \ldots$, is that now he will continue to perceive the oscillation even at times $t > r$ (if he has at his disposal instruments which are sufficiently sensitive).

The comparison with the second problem, (14.3)–(14.4), is also instructive. Its solution is given by

$$w(t, x) = \{H(t)f(t, x)\} * E_+ + \{H(-t)f(t, x)\} * E_- . \tag{14.8}$$

Here $*$ means convolution with respect to all the variables, t and x. In the half-space $t > 0$ we have

$$w = \{H(t)f\} * E_+ . \tag{14.9}$$

It is worth noticing that the only right-hand side contributions representing the effect of sources and sinks lying in the medium (and variable in time), which can influence events at time t originates between zero and t (the influence of events that occurred in negative times is suppressed by the imposition of values at zero time). This is in agreement with the *causality principle*. Of course,

$$\operatorname{supp} w \subset \operatorname{supp} H(t)f + \operatorname{supp} E_+ . \tag{14.10}$$

This is the analog of (14.7). But in (14.7) K is a compact subset of the hyperplane $t = 0$, whereas here $\operatorname{supp}(H(t)f)$ can be any subset of the closed half-space $t \geqq 0$. The difference can be best seen by looking at punctual sources. Suppose, for instance, that $f(t, x) = 1(t)\,\delta(x)$. Then $H(t)f(t, x) = H(t)\,\delta(x)$ would describe something like turning on a light, at a given point of space (taken as the origin of space), and keeping it on forever after. The support of $H(t)f(t, x)$ is the half-line $t \geqq 0$, $x = 0$. The right-hand side of (14.10) is equal to the forward light-cone Γ_+ *regardless of the parity of the space dimension n*. This agrees with the customary experience that if a source keeps broadcasting, an observer located anywhere in space will keep receiving the broadcast. Rather than sources (of light, of sound, of radio waves, etc.), one may consider sinks or obstacles, or any combinations of sources and sinks [this corresponds to the sign of the right-hand side $f(t, x)$, assumed to be a real-valued function—or, more generally, a real-valued measure]. With obstacles one would of course encounter shadows and light absorption, and so on.

We study now the so-called *propagation of singularities*. We assume that u_0 and u_1 are C^∞ functions outside of a certain closed subset S of $\mathbf{R}^n$, in other words, we assume that the *singular supports* of u_0 and u_1 are contained in S and we try to locate the singularities of the solution v of (14.1)–(14.2). The analogous question can be raised about problem (14.3)–(14.5). In both cases, the answer can be obtained by looking at a convolution

$$g = E * h \tag{14.11}$$

where E, g, h are distributions in $\mathbf{R}^{n+1}$, the support of E is contained in Γ_+, and the supports of g and h are contained in the half-space $t \geqq 0$. Observe that the two terms in the expression (14.6) of v are indeed of this kind: For instance, the second one is obtained by taking $E = E_+$ and $h(t,x) = \delta(t)u_1(x)$.

LEMMA 14.1. *Under the preceding hypotheses,*

$$\text{sing supp } g \subset \text{sing supp } E + \text{sing supp } h. \tag{14.12}$$

Proof. Let α (resp. β) be a C^∞ function in $\mathbf{R}^{n+1}$ whose support lies in the neighborhood of order $\varepsilon > 0$ of $A = \text{sing supp } E$ (resp. of $B = \text{sing supp } h$) and which is equal to one in the neighborhood of order $\varepsilon/2$ of A (resp. B). We have $E = \alpha E + (1 - \alpha)E$, $h = \beta h + (1 - \beta)h$ and both $(1 - \alpha)E$, $(1 - \beta)h$ are C^∞ functions; hence

$$g - (\alpha E) * (\beta h)$$

is a C^∞ function. But $\text{supp}[(\alpha E) * (\beta h)] \subset \text{supp } \alpha + \text{supp } \beta$ is contained in the neighborhood of order 2ε of $A + B$. By taking $\varepsilon \to 0$, we reach the conclusion that g is a C^∞ function in the complement of $A + B$. Q.E.D.

In applying this to our situation, we have $E = E_+$ or $E = \partial E_+/\partial t$, $h = \delta(t)u_j(x)$ $(j = 0, 1)$ or $h = H(t)f(t, x)$. We may therefore state

THEOREM 14.2. *Suppose that u_0 and u_1 are C^∞ functions in $\mathbf{R}^n \backslash S$ (where S is a closed subset of $\mathbf{R}^n$). Given any $t > 0$, the distribution with respect to x, $v(t, x)$, the solution of (14.1)–(14.2), is a C^∞ function in the complement with respect to $\mathbf{R}^n$ of the set*

$$S_{(t)} = \{x \in \mathbf{R}^n;\ \exists y \in S,\ |x - y| = t\}.$$

There is an analogous statement concerning the solution w of (14.3)–(14.4) which we leave to the student.

The physical interpretation of Theorem 14.2 is not difficult. Since support and singular support of E_+ coincide when $n = 3, 5, 7, \ldots$, it is best seen when $n = 1$ or n is even. Take, for instance, $n = 2$: In concrete terms, this corresponds to a two-dimensional propagation medium, like the surface of a lake or a sea. In the center of the lake, at time zero, there takes place a very brief but very intense stormlike event ("very" means of course "infinitely"); from there on, a wave, something like a tidal wave, propagates. For a fixed observer on the lake, at a distance r from the center, nothing happens until he is reached by the wave, at time $t = r$. At that moment, he is likely to find himself in a very violent situation: all the energy of the initial "explosion" or storm seems to be concentrated at the *wave front*. If our observer survives, once the wave front passes him by, he will find himself in a smooth "swell," whose amplitude, as measured from the level of the lake at rest, decreases to zero as time goes to $+\infty$.

It should be observed that the set $S_{(t)}$ has a certain thickness, determined by the thickness of the set S containing the singularities of u_0 and u_1. If S is compact, the thickness of $S_{(t)}$ will be finite: Here one could think of a star which has exploded and is at the center of an expanding shell of burning matter: $S_{(t)}$ is the shell (at time t). We may consider the *outer* boundary of $S_{(t)}$; it is the set

$$\dot{S}_{(t)} = \{x \in \mathbf{R}^n;\ \mathrm{dist}(x, S) = t\}.$$

The $\dot{S}_{(t)}$ form a one-parameter family of hypersurfaces; if $t_1 < t_2$, $\dot{S}_{(t_1)}$ is contained in the interior of $\dot{S}_{(t_2)}$. If we assume that these hypersurfaces are sufficiently smooth, say C^1, we may introduce their *orthogonal trajectories.* These are usually called *light rays*: Their interpretation in optics needs no elaboration. When suitably generalized, the concept of light ray plays an important role in the general theory of partial differential equations.

Returning once more to problem (14.1)–(14.2) we introduce the following quantity

$$E(t) = \int \left\{ \left| \frac{\partial v}{\partial t} \right|^2 + |\mathrm{grad}_x\, v|^2 \right\} dx,$$

called the *total energy* of the system. The integration is performed over the whole space $\mathbf{R}^n$ (t is regarded as a parameter). The integral makes sense when v_t is square-integrable with respect to x, and so is $\mathrm{grad}_x\, v$. We may apply Theorem 13.3, noting that here $f \equiv 0$ and taking $s = 1$. We conclude that, if $u_0 \in H^1(\mathbf{R}^n)$, $u_1 \in H^0 = L^2(\mathbf{R}^n)$, v is a C^0 function of t valued in H^1 (hence $\mathrm{grad}_x\, v$ is a C^0 function of t valued in L^2) whereas v_t is a C^0 function of t valued in L^2. In this case, $E(t)$ is well defined.

THEOREM 14.3. *Let u_0 belong to $H^1(\mathbf{R}^n)$, u_1 to $H^0 = L^2(\mathbf{R}^n)$. The total energy E is constant, i.e., for all $t \in \mathbf{R}^1$,*

$$E(t) = \int \{|u_1(x)|^2 + |\mathrm{grad}\, u_0(x)|^2\}\, dx. \tag{14.13}$$

Proof. Suppose first that both u_0 and u_1 are C^∞ functions with compact support in $\mathbf{R}^n$. Then v is a C^∞ function in $\mathbf{R}^{n+1}$ and, for each t, the support of $x \mapsto v(t, x)$ is compact. We have

$$E' = 2\,\mathrm{Re} \int (v_{tt}\, \bar{v}_t + \langle \mathrm{grad}_x\, v, \mathrm{grad}_x\, \bar{v}_t \rangle)\, dx.$$

In the second term, we may integrate by parts with respect to x:

$$E' = 2\,\mathrm{Re} \int (v_{tt} - \Delta_x v)\bar{v}_t\, dx = 0 \qquad \text{by (14.1).}$$

This proves (14.13) when $u_0, u_1 \in C_c^\infty$.

When $u_0 \in H^1$ and $u_1 \in H^0$, for each fixed t, v_t and $\text{grad}_x\, v$ belong to H_0 and, by Theorem 13.7, the mapping

$$(u_0, u_1) \mapsto (v_t, \text{grad}_x\, v)$$

from $H^1 \times H^0$ into $(L^2)^{n+1}$ is linear and continuous. On the other hand, $E(t)$ is clearly a continuous (quadratic) function on its range. Consequently, viewed as a function of (u_0, u_1) it is quadratic and continuous, and so is the right-hand side of (14.13). If the latter coincides with $E(t)$ when (u_0, u_1) belongs to the dense linear subspace $C_c^\infty \times C_c^\infty$, as it does, it must coincide with $E(t)$ for all (u_0, u_1). Q.E.D.

Exercises

14.1. Does Theorem 14.1 remain true if we replace the wave operator by the Klein–Gordon operator (13.27)?

14.2. Does Theorem 14.2 remain true if we replace the wave operator by the Klein–Gordon operator (13.27)?

14.3. Let u be a solution of the Cauchy problem

$$\frac{\partial^2 u}{\partial t^2} - \Delta_x u + \lambda u = 0 \quad \text{in } \mathbf{R}^{n+1}, \tag{14.14}$$

$$u = u_0(x), \qquad \frac{\partial u}{\partial t} = u_1(x) \qquad \text{at} \quad t = 0, \tag{14.15}$$

where u_0 and u_1 are two smooth functions with compact support in $\mathbf{R}^n$. What should be the definition of the total energy $E(t)$ of the system (at time t) if Theorem 14.3 is to be extended to all real values of λ?

14.4. Let K be a compact subset of $\mathbf{R}^n$, v the solution of (14.1)–(14.2) where the Cauchy data u_0, u_1 will be assumed to be C^∞ functions with compact support. Let us introduce the *energy of* K (at time t):

$$E_K(t) = \int_K \{|v_t|^2 + |\text{grad}_x\, v|^2\}\, dx. \tag{14.16}$$

Under the hypothesis that the number n of space variables equals *one*, prove that there is a number $T \geqq 0$ such that $E_K(t) = 0$ for all $t > T$.

14.5. Use the same notation as in Exercise 14.4. Suppose now that the number of space variables, n, is an *odd* integer $\geqq 3$. Prove that $E_K(t)$ [defined in (14.16)] vanishes for all $t > T$, $T > 0$ large enough.

14.6. Let us use the same notation as in Exercise 14.4 and also introduce the distribution $U(t, x)$ in $\mathbf{R}^{n+1}$ such that

$$U_{tt} - \Delta_x U = 0 \quad \text{in } \mathbf{R}^{n+1}, \tag{14.17}$$

$$U\Big|_{t=0} = 0, \qquad U_t\Big|_{t=0} = \delta(x), \qquad \text{the Dirac measure in } \mathbf{R}^n. \tag{14.18}$$

We know by Theorem 13.4 that there is a unique C^∞ function of t in $\mathbf{R}^1$, valued in $\mathscr{D}'(\mathbf{R}^n)$, satisfying (14.17)–(14.18). What is the Fourier transform of U?

Prove that the *energy density* $\varepsilon(x, t) = |U_t|^2 + |\text{grad}_x U|^2$ is a smooth (and, in fact, an analytic) function of (x, t) in the region $|x|^2 \neq t^2$.

14.7. Using the same notation as in Exercise 14.6, suppose that $n = 2$. Compute the energy $E_K(t)$, defined in (14.16), corresponding to $v = U$ (U defined in Exercise 14.6) and to

$$K = \{x \in \mathbf{R}^n;\ |x| \leqq R\}, \qquad R > 0,$$

for values of time $t > R$.

From the result of this calculation derive that, when v is any solution of (14.1)–(14.2) with compactly supported initial data (and when $n = 2$), $E_K(t)$ does not vanish identically for large values of t, contrary to what happens for odd values of n (Exercises 14.4 and 14.5).

14.8. Let us introduce the following

DEFINITION 14.1. *Let u be any distribution in an open subset Ω of $\mathbf{R}^n$. The smallest closed subset of Ω, in the complement of which u is an analytic function, is called the* **analytic singular support** *of u and will be denoted by* $\text{sing supp}_a\, u$.

Prove the analog of Lemma 14.1, replacing sing supp everywhere by sing supp_a (cf. Definition 3.2).

14.9. Use the result stated in Exercise 14.8 (i.e., the analog of Lemma 14.1 for analytic singular supports) to prove the following analog of Theorem 14.2:

THEOREM 14.4. *Let S be a closed subset of $\mathbf{R}^n$, u_0, u_1 two distributions in $\mathbf{R}^n$ which are analytic functions in $\mathbf{R}^n \backslash S$.*

Given any $t > 0$, the solution $v(x, t)$ of (14.1)–(14.2) is an analytic function of x in the complement of $S_{(t)} = \{y \in \mathbf{R}^n;\ \exists y' \in S,\ |y - y'| = t\}$.

15

Hyperbolic First-Order Systems with Constant Coefficients

In Sects. 13 and 14 we studied the Cauchy problem (in the large, i.e., in the whole space–time $\mathbf{R}^{n+1}$) for the wave equation, and encountered remarkable properties. One may say that the wave equation and closely related equations such as Klein–Gordon's are the best suited for such a study. In particular, a solution to the Cauchy problem always *exists*, under the weakest assumptions about the Cauchy data, is *unique*, and is a continuous linear function of the data. It is the tradition to summarize these properties by saying that *the Cauchy problem is well posed for the wave equation* (see Definition 15.1). In Sect. 12 we saw that this is not so for our other basic examples (Cauchy–Riemann, Laplace, heat, Schrödinger). However, there is a wide class of PDEs for which the Cauchy problem is well posed: the *hyperbolic* equations (the wave equation is hyperbolic). In the present section we study the hyperbolic first-order systems with constant coefficients. They are systems of linear partial differential operators of the form

$$L = \frac{\partial}{\partial t} - \sum_{j=1}^{n} A_j \frac{\partial}{\partial x^j} - A_0, \tag{15.1}$$

where the A_j's $(0 \leqq j \leqq n)$ are $m \times m$ matrices with complex coefficients. It is obviously convenient to perform a Fourier transformation with respect to x, which leads to the system of ordinary differential operators

$$\tilde{L} = \frac{\partial}{\partial t} - iA(\xi) - A_0 \tag{15.2}$$

where

$$A(\xi) = \xi_1 A_1 + \cdots + \xi_n A_n.$$

We want to study the Cauchy problem:

$$Lu = f \quad \text{in } \mathbf{R}^{n+1}, \tag{15.3}$$

$$u\Big|_{t=0} = u_0(x) \quad \text{in } \mathbf{R}^n, \tag{15.4}$$

where the values of u_0 and f are complex m-vectors.

Definition 15.1. We shall say that the Cauchy problem (15.3)–(15.4) is well posed if, given any data $u_0 \in \mathscr{D}'_x(\mathbf{R}^n)$, $f \in C_t^0(\mathscr{D}'_x(\mathbf{R}^n))$, *it has a unique solution* $u \in C_t^1(\mathscr{D}'_x(\mathbf{R}^n))$, *and this solution is a* C^∞ *function whenever this is true of the data.*

Remark 15.1. If we extend this terminology to higher order equations and systems (bringing then to the right level the required differentiability of the solution), we see that Theorems 13.1 and 13.2 state that the Cauchy problem is well posed for the wave equation.

Remark 15.2. We have posed the Cauchy problem in the most global terms: Eq. (15.3) should be verified in $\mathbf{R}^{n+1}$, condition (15.4) in $\mathbf{R}^n$. One can of course state it also in local terms. When the coefficients are variable, one cannot in general avoid it. In the local statement, we are given an open set $\Omega \subset \mathbf{R}^n$ and a number $T > 0$; f is a continuous function of t, $|t| < T$, with values in $\mathscr{D}'_x(\Omega)$ while u_0 belongs to $\mathscr{D}'_x(\Omega)$. The solution u is required to be a C^1 function of t, $|t| < T$, valued in $\mathscr{D}'_x(\Omega)$.

Remark 15.3. Of course, one may somewhat vary the regularity conditions with respect to t in Definition 15.1. But it ought to be noticed that, because of (15.4), it must be made clear what one means by the value of $u(t, x)$ when $t = 0$. For instance, if $u(t, x)$ were only L^1 with respect to t, it is not clear what we would mean by it. On the other hand, if both u and f are continuous in t (with values in $\mathscr{D}'_x$), Eq. (15.3), rewritten as

$$u_t = f + \sum_{j=1}^{n} A_j u_{x^j} + A_0 u,$$

shows that u_t would also be continuous with respect to t.

Remark 15.4. In our definition of well-posed Cauchy problems we have not included the continuous dependence of the solution on the data. This is because it automatically follows from the existence and uniqueness of the solution, by virtue of the closed graph theorem, as alluded to at the end of Sect. 13. Indeed, the closed graph theorem applies to spaces like $\mathscr{D}'_x$ and $C_t^k(\mathscr{D}'_x)$. Suppose then that nets $u_0^\nu \to u_0$ in $\mathscr{D}'_x$, $f^\nu \to f$ and $u^\nu \to u$ in $C_t^0(\mathscr{D}'_x)$ and in $C_t^1(\mathscr{D}'_x)$, respectively. Since L is a continuous linear operator on the latter space, if $Lu^\nu = f^\nu$ for every ν, we must also have $Lu = f$. On the other hand, if $u^\nu(0, x) = u_0^\nu(x)$ for every ν, it is trivially clear that $u(0, x) = u_0(x)$.

The same remark applies if we replace $\mathscr{D}'_x$ by C_x^∞ and $C_t^k(\mathscr{D}'_x)$ by $C_{t,x}^\infty$. This latter observation leads to interesting developments, as we shall now see.

If one recalls the definition of the C^∞ topology, one sees that the continuity of the mapping $(u_0, f) \mapsto u$ can be expressed as follows:

(15.5) *To every compact subset K of $\mathbf{R}^{n+1}$, to every integer $M \geqq 0$, there are another compact set $K' \subset \mathbf{R}^{n+1}$, a compact subset K_0 of $\mathbf{R}^n$, two integers M', $M_0 \geqq 0$, and a constant $C > 0$ such that, for all $u \in C^\infty(\mathbf{R}^{n+1})$,*

$$\sup_{(t,x) \in K} \sum_{k+|\alpha| \leqq M} |D_t^k D_x^\alpha u(t, x)|$$
$$\leqq C \left\{ \sup_{x \in K_0} \sum_{|\alpha| \leqq M_0} |D_x^\alpha u(0, x)| + \sup_{(t,x) \in K'} \sum_{k+|\alpha| \leqq M'} |D_t^k D_x^\alpha (Lu(t, x))| \right\}.$$

It is tempting to try to get information out of this inequality. Let us simplify the situation a little, by taking $f = Lu = 0$. We notice first that if $u_0(x) = u(0, x)$ vanishes identically in a neighborhood of K_0, $u(t, x)$ vanishes identically in K. We may rephrase this by saying that *the value of $u(t, x)$ in the neighborhood of K depends only on the $u_0(x)$ in the neighborhood of K_0*. This is the *domain of influence* phenomenon, already encountered in the study of the wave equation, where, however, the relation between K_0 and K was very precisely established (Theorem 14.1).

Next, still keeping $Lu \equiv 0$, we are going to make a special choice of u_0. We take

$$u_0(x) = e^{i\langle x, \xi \rangle} v_0,$$

where v_0 is a fixed nonzero complex m-vector and ξ any vector in $\mathbf{R}^n$. If we write $u(t, x) = e^{i\langle x, \xi\rangle} v(t, x, \xi)$, we see that v is the solution of the problem

$$\frac{\partial v}{\partial t} - \sum_{j=1}^n A_j \frac{\partial v}{\partial x^j} - \{iA(\xi) + A_0\} v = 0, \qquad v\Big|_{t=0} = v_0. \tag{15.6}$$

We may take v independent of x, hence, by (15.2), the unique solution of

$$\tilde{L} v = 0, \qquad v\Big|_{t=0} = v_0, \tag{15.7}$$

which is

$$v(t, \xi) = \exp[t\{iA(\xi) + A_0\}] v_0.$$

Let us apply (15.5), choosing K to be an interval $\{(x, t);\ x = 0,\ |t| \leqq T\}$, $T > 0$. We derive, taking $M = 0$,

$$\sup_{|t| \leqq T} |\exp[t\{iA(\xi) + A_0\}]v_0| \leqq C(1 + |\xi|)^{M_0}|v_0|. \tag{15.8}$$

Let us choose arbitrarily a unit vector $\dot{\xi}$ and set $A = A(\dot{\xi})$. Let $f(z, \varepsilon)$ denote the characteristic polynomial of $A - i\varepsilon A_0$:

$$f(z, \varepsilon) = \det(zI - A + i\varepsilon A_0).$$

If we view ε as a *real* variable, it is not difficult to see that there are m continuous functions of ε, $\lambda_j(\varepsilon)$, $j = 1, \ldots, m$, with complex values, which, at each point ε, represent the m roots of $f(z, \varepsilon)$ (the proof of this assertion is left to the student).

Suppose then that A has an eigenvalue λ_0 with *nonzero* imaginary part. There is $\varepsilon_0 > 0$ such that, for all ε, $|\varepsilon| < \varepsilon_0$, $A - i\varepsilon A_0$ has an eigenvalue λ_ε verifying $|\text{Im}\ \lambda_\varepsilon| > c > 0$. From the preceding considerations we derive that we may choose λ_ε to be a continuous function of ε. From now on let us restrict the variation of ε to the open interval $]0, \varepsilon_0[$. For each such ε we may select an eigenvector v_ε of $A - i\varepsilon A_0$ corresponding to the eigenvalue λ_ε. Let us take $|v_\varepsilon| = 1$. Let us substitute v_ε for v_0 in (15.8) and choose $\xi = \varepsilon^{-1}\dot{\xi}$. We have

$$\exp[t\{iA(\xi) + A_0\}]v_\varepsilon = \exp[it\varepsilon^{-1}(A - i\varepsilon A_0)]v_\varepsilon = \exp(it\varepsilon^{-1}\lambda_\varepsilon)v_\varepsilon,$$

from which, by (15.8),

$$\sup_{|t| \leqq T} \{\exp[-t(\text{Im}\ \lambda_\varepsilon)\varepsilon^{-1}]\} \leqq C(1 + \varepsilon^{-1})^{M_0}. \tag{15.9}$$

Because of the continuity of λ_ε with respect to ε, Im λ_ε will keep the same sign for $\varepsilon < \varepsilon_0$. In (15.9) we take

$$t = -T(\text{sign of } \lambda_\varepsilon).$$

We obtain $\exp(cT\varepsilon^{-1}) \leqq C(1 + \varepsilon^{-1})^{M_0}$, which is absurd if ε is to converge to 0.

We have thus proved

PROPOSITION 15.1. *If the Cauchy problem (15.3)–(15.4) is well posed, the eigenvalues of the matrix $A(\xi)$ are real.*

Remark 15.5. Take $n = 1$, $A_1 = 1/i$, $A_0 = 0$: $L = \partial/\partial t + i\,\partial/\partial x$. We see, by a different method than the one used in Sect. 12, that the Cauchy problem is not well posed for the Cauchy–Riemann operator.

Remark 15.6. Suppose that we had taken K to be the interval $\{(x, t);\ x = 0,\ 0 \leqq t \leqq T\}$ (rather than $|t| \leqq T$). Then the condition derived from (15.9) would have been the following one:

(15.10) *the imaginary parts of the eigenvalues of $A(\xi)$ are $\geqq 0$.*

This observation might serve as the starting point for a study of the *forward* Cauchy problem.

A more precise result than Proposition 15.1 can be proved, namely

THEOREM 15.1. *If the Cauchy problem (15.3)–(15.4) is well posed, the following is true*:

(15.11) *there is a constant* $C > 0$ *such that, given any* $\xi \in \mathbf{R}_n$ *and any eigenvalue* λ *of* $A(\xi) - iA_0$, *we have* $|\text{Im}\, \lambda| \leqq C$.

The proof of this statement is not elementary; it is based on the Seidenberg–Tarski theorem and we do not give it here. However, the underlying idea is not difficult to grasp. Indeed, one derives from (15.8) that the absolute values of the imaginary parts of the eigenvalues of $A(\xi) - iA_0$ are bounded by $C_0 \log(1 + |\xi|)$, where $C_0 > 0$ does not depend on ξ. However, these eigenvalues are the roots of the polynomial in z, $\det(zI - A(\xi) - iA_0)$, whose coefficients are polynomials with respect to ξ. If these roots or their imaginary parts must grow as $|\xi| \to +\infty$ and tend to infinity, they should not do so *logarithmically*; they do it like fractional powers of $|\xi|$. One can check this easily on simple examples and the Seidenberg–Tarski theorem enables one to prove it in general. Consequently, if the absolute values of the imaginary parts of the eigenvalues are bounded by $C_0 \log(1 + |\xi|)$, this means that they are bounded independently of ξ.

The main result of this section, Theorem 15.2, states that (15.11) implies that the Cauchy problem (15.3)–(15.4) is well posed. Thus (15.11) is both necessary and sufficient in order that this be true. This constitutes the celebrated *Gårding's theorem.* It motivates the introduction of the following terminology:

Definition 15.2. The system (15.1) is called hyperbolic if (15.11) holds.

Before proving the converse of Theorem 15.1 we give two criteria that are useful when checking if a given system is hyperbolic.

PROPOSITION 15.2. *The system (15.1) is hyperbolic whenever one of the following two conditions is fulfilled*:

(15.12) *for any* $\xi \in \mathbf{R}_n$, *the matrix* $A(\xi)$ *is self-adjoint*;

(15.13) *for any* $\xi \in \mathbf{R}_n$, $\xi \neq 0$, *the eigenvalues of* $A(\xi)$ *are real and distinct.*

Proof. Suppose that the following is true:

(15.14) *given any point* $\xi^0 \in \mathbf{R}_n$, $|\xi^0| = 1$, *there is a neighborhood* U_0 *of* ξ^0 *and an invertible* $m \times m$ *matrix* $\Gamma(\xi)$ *whose entries are continuous functions in* U_0, *such that* $S(\xi) = \Gamma(\xi)^{-1} A(\xi) \Gamma(\xi)$ *is self-adjoint, whatever* ξ *in* U_0.

We contend that (15.14) implies that the system (15.1) is hyperbolic. Let us set

$$B_0(\dot{\xi}) = \Gamma(\dot{\xi})^{-1} A_0 \Gamma(\dot{\xi}), \qquad \dot{\xi} \in U_0, \qquad |\dot{\xi}| = 1,$$

and take $\xi = \rho\dot{\xi}$. We observe that if λ is an eigenvalue of $A(\xi) - iA_0$, it is also an eigenvalue of $\rho S(\dot{\xi}) - iB_0(\dot{\xi})$. Therefore there is a unit vector v in $\mathbf{C}^m$ such that

$$\rho(S(\dot{\xi})v, v) - i(B_0(\dot{\xi})v, v) = \lambda(v, v) = \lambda.$$

We have denoted by (,) the *Hermitian* product in $\mathbf{C}^m$. By hypothesis, $S(\dot{\xi})$ is self-adjoint, therefore

$$\operatorname{Im} \lambda = -\operatorname{Re}(B_0(\dot{\xi})v, v),$$

which implies $|\operatorname{Im} \lambda| \leqq \|B_0(\dot{\xi})\|$ ($\| \ \|$ stands for the matrix norm), hence

$$|\operatorname{Im} \lambda| \leqq \|\Gamma(\dot{\xi})\| \, \|\Gamma(\dot{\xi})^{-1}\| \, \|A_0\|. \tag{15.15}$$

But $\|\Gamma(\dot{\xi})\|$ and $\|\Gamma(\dot{\xi})^{-1}\|$ are continuous functions in U_0. Therefore we may find a subneighborhood $U_0' \subset U_0$ of ξ^0 where these two functions are bounded. We may then cover the unit sphere $S_{n-1} \subset \mathbf{R}_n$ with a finite collection of such neighborhoods U_0' and we conclude easily that the numbers $|\operatorname{Im} \lambda|$ are bounded independently of ξ.

In the case (15.12), condition (15.14) is trivially satisfied. Let us show that it is also satisfied in the case (15.13). Let $\lambda_1(\dot{\xi}) < \cdots < \lambda_m(\dot{\xi})$ denote the eigenvalues of $A(\dot{\xi})$ and set $d = \inf(\lambda_{j+1}(\dot{\xi}) - \lambda_j(\dot{\xi}))$ where the infimum is taken over $j = 1, \ldots, m-1$ and $\dot{\xi} \in S_{n-1}$. Let c_j be the circle in the complex plane, centered at $\lambda_j(\dot{\xi})$ and with radius $d/2$. The circle c_j varies continuously with $\dot{\xi}$. Let us set

$$P_j(\dot{\xi}) = (2i\pi)^{-1} \oint_{c_j} (zI - A(\dot{\xi}))^{-1} \, dz, \qquad j = 1, \ldots, m.$$

The following properties are standard and, at any rate, easy to check:

$$P_j(\dot{\xi})^2 = P_j(\dot{\xi}), \qquad A(\dot{\xi})P_j(\dot{\xi}) = \lambda_j(\dot{\xi})P_j(\dot{\xi}), \qquad I = P_1(\dot{\xi}) + \cdots + P_m(\dot{\xi}).$$

For each $j = 1, \ldots, m$, let $\mathbf{v}_j^0$ be an eigenvector of $A(\xi^0)$ corresponding to $\lambda_j(\xi^0)$, say with unit length. The vectors $\mathbf{v}_1^0, \ldots, \mathbf{v}_m^0$ are linearly independent and therefore form a basis of $\mathbf{C}^m$. Let us set $\mathbf{v}_j(\dot{\xi}) = P_j(\dot{\xi})\mathbf{v}_j^0$. If $\dot{\xi}$ is close enough to ξ^0, the $\mathbf{v}_j(\dot{\xi})$ also form a basis of $\mathbf{C}^m$. Let $\mathbf{e}_k$ be the vector whose components are all zero, except the kth one, which is equal to one. We set

$$\mathbf{v}_j(\dot{\xi}) = \sum_{k=1}^{m} \gamma_j^k(\dot{\xi})\mathbf{e}_k.$$

The matrix $\Gamma(\dot{\xi}) = (\gamma_j^k(\dot{\xi}))_{1 \leqq j, k \leqq m}$ has the properties required in (15.14), as the student can easily ascertain. Q.E.D.

The following terminology is often used:

Definition 15.3. The system (15.1) is called strongly (or strictly) hyperbolic if condition (15.13) is satisfied.

We now prove the main result of this section.

THEOREM 15.2. *If the system L is hyperbolic, the Cauchy problem (15.3)–(15.4) is well posed.*

Proof. The theorem follows from the study of the matrix-valued distribution $U(t, x)$, which is the inverse Fourier transform with respect to x of the distribution

$$\tilde{U}(t, \xi) = \exp\{it[A(\xi) - iA_0]\}.$$

We recall that $A(\xi)$ is *linear* with respect to ξ. We may extend at once $\tilde{U}(t, \xi)$ to the complex values of ξ, denoted by ζ, and we immediately have that

(15.16) *there are two constants B, $C > 0$ such that, for all real t and all complex vectors $\zeta \in \mathbf{C}_n$,*

$$\|\tilde{U}(t, \zeta)\| \leqq Ce^{B|t|\,|\zeta|}.$$

Thus, $\tilde{U}(t, \zeta)$ is an *entire function of exponential* type with respect to ζ ($\|\ \|$ stands for the matrix norm).

Q.E.D.

Next we prove the following assertion:

(15.17) *There are two constants B_0, $C_0 > 0$ such that, for all real t and all real vectors $\xi \in \mathbf{R}_n$,*

$$\|\tilde{U}(t, \xi)\| \leqq C_0(1 + |\xi|)^m \exp(B_0|t|).$$

Thus $\tilde{U}(t, \xi)$ is a function of ξ in $\mathbf{R}^n$, "slowly growing" at infinity. Property (15.17) will follow from certain properties of matrices, which we now describe.

LEMMA 15.1. *Let M be any $m \times m$ matrix. There is a constant $C_m > 0$, depending only on the degree m, such that, for all complex numbers z whose distance to the spectrum of M* (i.e., the set of eigenvalues of M) *is at least one,*

(15.18) $$\|(zI - M)^{-1}\| \leqq C_m(1 + |z| + \|M\|)^{m-1}.$$

Proof. Let us write $N = zI - M$. Then by the theorem of Hamilton–Cayley, and denoting by $a_1, \ldots, a_m$ the coefficients of the characteristic polynomial of N, we have

$$N(N^{m-1} + a_1 N^{m-2} + \cdots + a_{m-1}) + a_m I = 0,$$

whereas

$$\|N^{-1}\| \leqq |a_m|^{-1}\|N\|^{m-1} + |a_1/a_m|\,\|N\|^{m-2} + \cdots + |a_{m-1}/a_m|.$$

But $\|N\| \leqq |z| + \|M\|$, and the ratios of symmetric functions a_j/a_m are bounded by universal constants and so is a_m^{-1}: indeed, these are the symmetric functions of the eigenvalues of N^{-1}, which have absolute value $\leqq 1$.

Q.E.D.

LEMMA 15.2. *Let M be any* $m \times m$ *matrix. There is a constant* $C'_m > 0$, *depending only on m, such that*

$$\|e^{iM}\| \leqq C'_m(1 + \|M\|)^m e^{I(M)}, \tag{15.19}$$

where $I(M)$ *is the largest absolute value of the imaginary parts of the eigenvalues of M.*

Proof. Let γ denote the contour of integration shown in Fig. 15.1. Since all

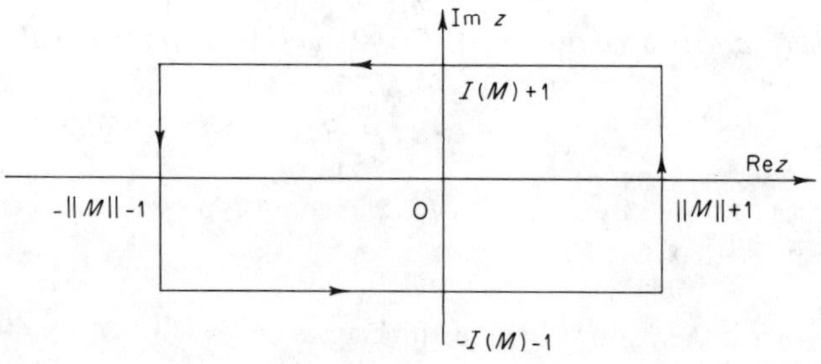

FIG. 15.1

the eigenvalues of M have absolute value $\leqq \|M\|$, the distance from γ to the spectrum of M is at least one (in fact, it is one). On the other hand, if $z \in \gamma$, $|z| \leqq 2(1 + \|M\|)$. We have

$$e^{iM} = (2\pi i)^{-1} \oint_\gamma e^{iz}(zI - M)^{-1}\, dz,$$

from which, by Lemma 15.1,

$$\|e^{iM}\| \leqq C_m \int_\gamma e^{-\mathrm{Im}\, z}(1 + |z| + \|M\|)^{m-1}\, |dz|$$

$$\leqq 3C_m(1 + \|M\|)^{m-1}\, e\, (\text{length of } \gamma)\, e^{I(M)}.$$

As the length of γ is $\leqq 4(1 + \|M\|)$, we obtain (15.19).

Proof of (15.17). It suffices to take $M = t(A(\xi) - iA_0)$ in (15.19) and use the fact that, by our hyperbolicity hypothesis, the imaginary parts of the eigenvalues of $A(\xi) - iA_0$ are bounded independently of ξ. Also keep in mind that $A(\xi)$ is linear with respect to ξ. Q.E.D.

We exploit properties (15.16) and (15.17) by applying the Paley–Wiener (–Schwartz) theorem: *If a function of $\xi \in \mathbf{R}_n$ can be extended to the complex space $\mathbf{C}_n$ as an entire function of exponential type and if, on $\mathbf{R}_n$, it grows at most like a polynomial, it is the Fourier transform of a distribution with compact support in* $\mathbf{R}^n$. We have in fact additional information, to the extent that the "exponential type" of the Fourier transform is related to the convex hull of the support of the distribution. Indeed, we may conclude from (15.16) that

(15.20) *The support of $U(t, x)$, regarded as a distribution in x* (t is fixed), *is contained in the ball* $\{x \in \mathbf{R}^n;\ |x| \leqq B|t|\}$.

Now, if in our Cauchy problem (15.3)–(15.4) the data f and u_0 were C^∞ functions with compact support, we would have the right to transform the problem by a Fourier transformation in x. The transformed problem,

$$\tilde{L}\tilde{u} = \tilde{f}(t, \xi) \qquad \text{in } \mathbf{R}^1 \times \mathbf{R}_n, \tag{15.21}$$

$$\tilde{u}(0, \xi) = \tilde{u}_0(\xi) \qquad \text{in } \mathbf{R}_n; \tag{15.22}$$

has the unique solution

$$\tilde{u}(t, \xi) = \tilde{U}(t, \xi)\tilde{u}_0(\xi) + \int_0^t \tilde{U}(t - t', \xi)\tilde{f}(t', \xi)\, dt', \tag{15.23}$$

which is the Fourier transform of

$$u(t, x) = U(t, x) \underset{(x)}{*} u_0(x) + \int_0^t U(t - t', x) \underset{(x)}{*} f(t', x)\, dt', \tag{15.24}$$

i.e.,

$$u(t, x) = U(t, x) \underset{(x)}{*} u_0(x) + \{H(t)f(t, x)\} \underset{(t,x)}{*} E_+(t, x) + \{H(-t)f(t, x)\} \underset{(t,x)}{*} E_-(t, x), \tag{15.25}$$

where

$$E_+(t, x) = H(t)U(t, x), \qquad E_-(t, x) = -H(-t)U(t, x).$$

There is no need to stress the analogy between formulas (15.25) and (13.14). From there the argument proceeds exactly in the same fashion as in Sect. 13: Formula (15.24) [or (15.25)] makes sense whatever the distributions u_0, f, since, for each fixed, the support of $U(t, x)$ is compact [by (15.20)]. Furthermore, u is a C^∞ function.

Example 15.1. We suppose $n = 3$ and consider the following 4×4 matrices:

$$K = \begin{pmatrix} I & 0 \\ 0 & -I \end{pmatrix}, \qquad A_j = \begin{pmatrix} 0 & \sigma_j \\ \sigma_j & 0 \end{pmatrix}, \qquad j = 1, 2, 3,$$

where the σ_j are the following 2×2 matrices, called *Pauli matrices* in physics texts (in K, I stands for the 2×2 identity matrix):

$$\sigma_1 = \begin{pmatrix} 0 & 1 \\ 1 & 0 \end{pmatrix}, \qquad \sigma_2 = i\begin{pmatrix} 0 & 1 \\ -1 & 0 \end{pmatrix}, \qquad \sigma_3 = \begin{pmatrix} 1 & 0 \\ 0 & -1 \end{pmatrix}.$$

We consider the 4×4 system (where m is some number at least zero):

$$L = \frac{\partial}{\partial t} - \sum_{j=1,2,3} A_j \frac{\partial}{\partial x^j} - miK. \tag{15.26}$$

We have here

$$A(\xi) = \begin{pmatrix} 0 & \sigma(\xi) \\ \sigma(\xi) & 0 \end{pmatrix}$$

where

$$\sigma(\xi) = \begin{pmatrix} \xi_3 & \xi_1 + i\xi_2 \\ \xi_1 - i\xi_2 & -\xi_3 \end{pmatrix}$$

With our previous notation, we have $A_0 = miK$. The characteristic determinant of $A(\xi) - iA_0$ is

$$\det(\tau I - A(\xi) - mK) = (\tau^2 - |\xi|^2 - m^2)^2. \tag{15.27}$$

The eigenvalues of $A(\xi) - iA_0$ are *double*; they are

$$\pm(|\xi|^2 + m^2)^{1/2}.$$

Thus the system L is hyperbolic. However, *it is not strongly hyperbolic.* The equations $Lu = f$ are one of the standard forms of *Dirac's equations*, introduced to describe the behavior of the electron in relativistic terms (notice that the characteristic polynomial, regarded as a function on $\mathbf{R}_4$, where the coordinates are τ, ξ_1, ξ_2, ξ_3, is Lorentz-invariant). The reader curious about the invariance of the Dirac operator is referred to Exercise 15.4; related questions, having to do with the Lorentz group and the Pauli matrices σ_j, are dealt with in Exercises 15.5 to 15.9.

Exercises

15.1. Consider the operator $L = I\,\partial/\partial t - A\,\partial/\partial x$ (thus $n = 1$), with I the $m \times m$ identity matrix and A another $m \times m$ matrix, with complex entries. We form the hypothesis:

(15.28) *the eigenvalues of A are all distinct* (but not necessarily real).

Then let $\mathbf{h}(z)$ be an *entire* function in $\mathbf{C}$, valued in $\mathbf{C}^m$. Consider the operator

$$\exp\left(tA\frac{\partial}{\partial x}\right)\mathbf{h} = \sum_{k=0}^{+\infty}\frac{t^k}{k!}A^k\mathbf{h}^{(k)}(x). \tag{15.29}$$

Show that there is an invertible matrix Γ ($m \times m$, with complex entries) such that, if we set

$$S_t\mathbf{w}(x) = \Gamma^{-1}\exp\left(tA\frac{\partial}{\partial x}\right)\Gamma\mathbf{w}, \tag{15.30}$$

then the components of the vector $S_t\mathbf{w}$ are related to those of $\mathbf{w}$ by the formula

$$(S_t\mathbf{w})^j(x) = \mathbf{w}^j(x + t\lambda_j), \qquad j = 1, \ldots, n. \tag{15.31}$$

Derive from this that, if all the eigenvalues of A are real, S_t can be extended as a continuous linear operator acting on the continuous functions of x in $\mathbf{R}^1$ and the Cauchy problem for L is then well posed (Definition 15.1).

15.2. Let L denote the same operator as in Exercise 15.1. Suppose that (15.28) holds but that one of the eigenvalues of A is not real. By using the matrix Γ introduced in Exercise 15.1 show that the Cauchy problem is *not* well posed (Definition 15.1) for L.

15.3. Consider an operator $L = I\,\partial/\partial t - A\,\partial/\partial x$ on $\mathbf{R}^2$, where I is the $m \times m$ identity matrix and A any $m \times m$ complex matrix. By writing A in its *Jordan canonical form*, show that the Cauchy problem for L is well posed if and only if all the eigenvalues of A are real. In this case, give the expression of a function $u(t, x) \in C^\infty(\mathbf{R}^2)$, satisfying $Lu = 0$ in $\mathbf{R}^2$, in terms of $u(0, x) \in C^\infty(\mathbf{R}^1)$.

15.4. Let A_j ($j = 1, 2, 3$) be the matrices so denoted in Example 15.1, $\mathscr{P}_4$ the real linear space formed by the 4×4 matrices

$$M(\xi) = \xi_0 I + \sum_{j=1}^{3}\xi_j A_j, \qquad \xi \in \mathbf{R}_4. \tag{15.32}$$

Consider a linear transformation T of $\mathbf{R}_4$ such that there is a 4×4 matrix S_T satisfying

$$M(T\xi) = S_T M(\xi), \qquad \forall \xi \in \mathbf{R}_4. \tag{15.33}$$

Writing $Q(\xi) = \xi_0^2 - (\xi_1^2 + \xi_2^2 + \xi_3^2)$, show that we have

$$Q(T\xi)^2 = \kappa^2 Q(\xi)^2, \qquad \forall \xi \in \mathbf{R}_4, \tag{15.34}$$

for some number κ, What is the relation between κ and S_T? Show that the set of matrices T such that the above holds true with $\kappa = 1$ forms a subgroup G of the group of 4×4 invertible matrices, with real entries, that is, of the fourth linear group $LG(4, \mathbf{R})$. Is the mapping $T \mapsto S_T$ of G into $LG(4, \mathbf{R})$ injective? What is its image?

Give an example of a linear transformation T on $\mathbf{R}_4$ such that $Q(T\xi) = -Q(\xi)$ for all $\xi \in \mathbf{R}_4$.

15.5. We denote by $M_m(\mathbf{K})$ the algebra of $m \times m$ matrices with entries in the field $\mathbf{K} = \mathbf{R}$ or $\mathbf{C}$, by $LG(m, \mathbf{K})$ the group of these matrices which are invertible, by $SL(m, \mathbf{K})$ the subgroup formed by those matrices having a determinant equal to $+1$. If M is a complex matrix, we denote by M^* its adjoint, i.e., the transpose of the complex conjugate of M. Let us denote by $\mathbf{e}_0$ the 2×2 identity matrix, and by $\mathbf{e}_j$ the Pauli matrix introduced in Example 15.1 under the notation σ_j ($j = 1, 2, 3$).

Show that

$$\zeta = (\zeta_0, \zeta_1, \zeta_2, \zeta_3) \mapsto M(\zeta) = \sum_{j=0}^{3} \zeta_j \mathbf{e}_j \tag{15.35}$$

is a linear isomorphism of $\mathbf{C}_4$ onto $M_2(\mathbf{C})$, transforming complex conjugate into adjoint, i.e., $M(\bar{\zeta}) = M(\zeta)^*$ and, consequently, mapping the real space $\mathbf{R}_4$ onto Σ_2, the space of all self-adjoint 2×2 matrices.

Establish the following formulas:

$$\det M(\zeta) = \zeta_0^2 - \sum_{j=1,2,3} \zeta_j^2, \tag{15.36}$$

$$\tfrac{1}{2} \operatorname{Tr} M(\zeta) = \zeta_0 \tag{15.37}$$

(Tr M is the *trace*, that is to say, the sum of the diagonal entries of the matrix M). Give the expression of a generic matrix corresponding, via (15.35), to a point of the hyperplane $\xi_0 = 0$ in $\mathbf{R}_4$.

Remark 15.7. The reader will check at once that the multiplication table of the $\mathbf{e}_j$ ($j = 0, 1, 2, 3$) is that shown in Table 15.1.

	$\mathbf{e}_1$	$\mathbf{e}_2$	$\mathbf{e}_3$
$\mathbf{e}_1$	$\mathbf{e}_0$	$-i\mathbf{e}_3$	$i\mathbf{e}_2$
$\mathbf{e}_2$	$i\mathbf{e}_3$	$\mathbf{e}_0$	$-i\mathbf{e}_1$
$\mathbf{e}_3$	$-i\mathbf{e}_2$	$i\mathbf{e}_1$	$\mathbf{e}_0$

TABLE 15.1

15.6. Use the same notation as in Exercise 15.5. Given any matrix $A \in M_2(\mathbf{C})$ define a transformation in $\mathbf{R}_4$, T_A, by the formula

$$AM(\xi)A^* = M(T_A \xi), \qquad \xi \in \mathbf{R}_4. \tag{15.38}$$

Prove, by using (15.36), that $A \mapsto T_A$ is a *group homomorphism* of the special linear group $SL(2, \mathbf{C})$ into the Lorentz group $\mathscr{L}$ (on $\mathbf{R}_4$). What is the kernel of this homomorphism?

15.7. Let $U(2)$ denote the 2×2 *unitary* group [$U \in U(2)$ if $UU^* = I$] and set, for U in $U(2)$,

$$UM(\xi)U^{-1} = M(T_U \xi), \qquad \xi \in \mathbf{R}_4. \tag{15.38'}$$

Prove that UMU^{-1} preserves the determinant and the trace. Use then formulas (15.36)–(15.37) to prove that, if $U \in U(2)$, T_U preserves every hyperplane $\xi_0 = \text{const}$, in $\mathbf{R}_4$, and induces an *orthogonal* transformation on $\mathbf{R}_3$ identified with the hyperplane $\xi_0 = 0$.

Show that $U \mapsto T_U$ is a group homomorphism of $U(2)$ into $O(3)$, the group of 3×3 orthogonal matrices, whose kernel is equal to the subgroup of diagonal matrices of the form $e^{i\omega}I$, $0 \leqq \omega \leqq 2\pi$.

15.8. Use the same notation as in the preceding exercises. Let $U^+(2)$ denote the subgroup of $U(2)$ consisting of the unitary matrices whose determinant is exactly equal to $+1$; $U^+(2)$ is also a subgroup of $SL(2, \mathbf{C})$ and it is obvious that the definitions (15.38) and (15.38′) of T_U are the same. We have seen, in Exercise 15.7, that $U \mapsto T_U$ is a group homomorphism of $U^+(2)$ into (not onto!) $O(3)$. What is its kernel?

Let us set

$$V(\phi) = \begin{pmatrix} e^{i\phi} & 0 \\ 0 & e^{-i\phi} \end{pmatrix}, \qquad R(\theta) = \begin{pmatrix} \cos\theta & \sin\theta \\ -\sin\theta & \cos\theta \end{pmatrix}.$$

Prove that every matrix $U \in U^+(2)$ can be written

$$U = V(\phi/2)R(\theta/2)V(\psi/2), \tag{15.39}$$

where ϕ, ψ, θ are angles (i.e., vary on the torus $\mathbf{T}^1$). Conclude from this that $U^+(2)$ is a *connected three-dimensional compact* (real analytic) manifold and that, consequently, $U \mapsto T_U$ is a group homomorphism of $U^+(2)$ onto the group of *rotations* in $\mathbf{R}_3$, $O^+(3)$ [this is the connected component of the identity in $O(3)$, also the intersection of $O(3)$ with $SL(3, \mathbf{R})$].

Compute T_U in each one of the following cases: $U = V(\phi/2)$, $U = R(\theta/2)$, and derive from this and from (15.39) the general formula for T_U.

15.9. Using the same notation as in Exercise 15.6, show that every matrix $A \in SL(2, \mathbf{C})$ such that the restriction of T_A to the plane $\xi_0 = \xi_3 = 0$ is the identity is necessarily of the form

$$A = \pm \begin{pmatrix} e^{t/2} & 0 \\ 0 & e^{-t/2} \end{pmatrix}, \qquad t \in \mathbf{R}. \tag{15.40}$$

By combining this fact with the main result in Exercise 15.8 [namely that $U \mapsto T_U$ maps $U^+(2)$ onto $O^+(3)$] derive that $A \mapsto T_A$ is a group homomorphism of $SL(2, \mathbf{C})$ *onto* the connected component of the identity in the Lorentz group, $\mathscr{L}_+$, and thus that $SL(2, \mathbf{C})$ is a *covering group of order* 2 of $\mathscr{L}_+$.

16

Strongly Hyperbolic First-Order Systems in One Space Dimension

Some of the results in the preceding sections, in particular those concerning the existence and uniqueness of the solutions to the Cauchy problem, can be extended to certain partial differential equations and systems with *variable* coefficients, generally called *hyperbolic*. Here we shall limit ourselves to first-order systems with variable coefficients, when the number of space variables is equal to *one*. These systems are of the form

$$L = \frac{\partial}{\partial t} - A(x, t)\frac{\partial}{\partial x} - B(x, t).$$

Here $A(x, t)$ and $B(x, t)$ are $m \times m$ matrices whose entries are smooth functions, say C^1 functions of (x, t) in $\Omega \times]-T, T[$ ($\Omega \subset \mathbf{R}^n$ open, $T > 0$); of course, $\partial/\partial t$ stands for $I\,\partial/\partial t$ where I is the $m \times m$ identity matrix. We shall assume that the system L is *strongly hyperbolic* (cf. Definition 15.3):

(16.1) *For every $x_0 \in \Omega$, $t_0 \in]-T, T[$, the eigenvalues of $A(x_0, t_0)$ are real and distinct.*

LEMMA 16.1. *If* (16.1) *holds, there are m C^1 functions of (x, t) in $\Omega \times]-T, T[$, $\lambda_1, \ldots, \lambda_m$, representing at each point (x, t) the eigenvalues of $A(x, t)$.*

Proof. Since the eigenvalues of $A(x, t)$ are real and distinct, we may choose their indices so as to have $\lambda_1 < \lambda_2 < \cdots < \lambda_m$ everywhere in $\Omega \times]-T, T[$. Let (x_0, t_0) be an arbitrary point in the latter set, λ_j^0 the jth eigenvalue of $A(x_0, t_0)$. Let us denote by $f(\lambda; x, t)$ the characteristic polynomial of $A(x, t) : f(\lambda; x, t) = \det(\lambda I - A(x, t))$. We have

$$f(\lambda_j^0\,; x_0, t_0) = 0, \qquad f_\lambda(\lambda_j^0; x_0, t_0) \neq 0.$$

We may therefore apply the implicit function theorem and conclude that there is a C^1 function $\lambda(x, t)$ in a neighborhood of (x_0, t_0) satisfying there

$f(\lambda(x, t); x, t) = 0$ and equal to λ_j^0 when $x = x_0$, $t = t_0$. We must necessarily have $\lambda(x, t) = \lambda_j(x, t)$. Q.E.D.

Remark 16.1. The lemma would still be true if the eigenvalues ceased to be real, provided that they remain distinct. In general, if multiple eigenvalues are allowed, the lemma is not valid. A simple example is provided by

$$A(x, t) = \begin{pmatrix} 0 & x^2 + t^2 \\ 1 & 0 \end{pmatrix}.$$

We now set

$$P_j(x, t) = (2\pi i)^{-1} \oint_{\gamma_j} (zI - A(x, t))^{-1}\, dz,$$

where γ_j is a circle in the complex plane on which no eigenvalue of $A(x, t)$ lies and whose interior contains one and only one eigenvalue of $A(x, t)$, the jth one, $\lambda_j(x, t)$. It is clear that $P_j(x, t)$ is a C^1 function of (x, t); it is the jth *spectral projector* of $A(x, t)$. The range $\mathscr{L}_j(x, t)$ of the matrix $P_j(x, t)$ acting on $\mathbf{C}^m$ is one-dimensional: It is the jth eigenspace of $A(x, t)$ (which is one-dimensional). We have

$$I = P_1 + \cdots + P_m.$$

This implies that, for each (x, t), $\mathscr{L}_1(x, t), \ldots, \mathscr{L}_m(x, t)$ span the whole space $\mathbf{C}^m$. Of course, we have $P_j^2 = P_j$ for every j.

LEMMA 16.2. *Every point (x_0, t_0) of $\Omega \times]-T, T[$ has an open neighborhood $\mathfrak{V}_0$ in this set where a matrix-valued function $\Gamma(x, t)$ is defined, having the following properties:*

(16.2) *$\Gamma(x, t)$ is C^1 in $\mathfrak{V}_0$ and invertible at every point of $\mathfrak{V}_0$;*

(16.3) *we have, for every $(x, t) \in \mathfrak{V}_0$,*

$$\Gamma^{-1}(x, t)A(x, t)\Gamma(x, t) = \begin{pmatrix} \lambda_1(x, t) & 0 & 0 \\ 0 & \cdots & 0 \\ 0 & 0 & \lambda_m(x, t) \end{pmatrix}.$$

Proof. For each j, we arbitrarily select a nonzero vector $\mathbf{v}_j^0$ in $\mathscr{L}_j(x_0, t_0)$ and we set

$$\mathbf{v}_j(x, t) = P_j(x, t)\mathbf{v}_j^0.$$

When $x = x_0$, $t = t_0$, we have $\mathbf{v}_j = \mathbf{v}_j^0$, hence the $\mathbf{v}_j$ then form a basis of $\mathbf{C}^m$ and this will remain true in a small neighborhood of (x_0, t_0), which we take to be $\mathfrak{V}_0$. Let us write

$$\mathbf{v}_j(x, t) = \sum_{k=1}^{m} \gamma_j^k(x, t)\mathbf{e}_k,$$

where $(\mathbf{e}_1, \ldots, \mathbf{e}_m)$ is the canonical basis of $\mathbf{C}^m$; i.e., for each k, all the components of $\mathbf{e}_k$ are zero, except the kth one, which is equal to one. The matrix $(\gamma_j^k(x, t))_{1 \leqq j, k \leqq m}$ will be our matrix $\Gamma(x, t)$. We leave to the student the verification that it possesses the properties (16.2) and (16.3). Q.E.D.

We are now in a position to study the following *local* Cauchy problem. In what follows, U will be a small open subset of Ω, δ a small number such that $0 < \delta < T$. The Cauchy problem is stated as follows:

$$Lu = f(x, t), \qquad x \in U, \qquad |t| < \delta, \tag{16.4}$$

$$u(x, 0) = u_0(x), \qquad x \in U. \tag{16.5}$$

About the data, we assume that $f(x, t)$ is a C^1 function of (x, t) in $\Omega \times]-T, T[$ while $u_0(x)$ is a C^1 function in Ω.

We transform the problem by setting

$$u(x, t) = \Gamma(x, t)v(x, t), \qquad f(x, t) = \Gamma(x, t)g(x, t), \tag{16.6}$$
$$u_0(x) = \Gamma(x, 0)v_0(x),$$

where $\Gamma(x, t)$ is the matrix in Lemma 16.2 when we apply it to (x_0, t_0) where x_0 is a point in U and $t_0 = 0$. From now on, until the contrary is specified, U and δ will be assumed to be sufficiently small so that $U \times]-\delta, \delta[$ is contained in $\mathfrak{V}_0$. We have

$$Lu = \Gamma v_t + \Gamma_t v - A\Gamma v_x - A\Gamma_x v - B\Gamma v$$
$$= \Gamma(v_t - \Gamma^{-1}A\Gamma v_x - B_{\#} v),$$

where we have set

$$B_{\#} = B - \Gamma_t + A\Gamma_x. \tag{16.7}$$

Problem (16.4)–(16.5) is thus transformed into

$$v_t - \Gamma^{-1}A\Gamma v_x - B_{\#} v = g(x, t), \qquad x \in U, \qquad |t| < \delta, \tag{16.8}$$

$$v(x, 0) = v_0(x), \qquad x \in U. \tag{16.9}$$

The advantage of the new system lies in the fact that $\Gamma^{-1}A\Gamma$ is diagonal [cf. (16.3)]. This advantage will be exploited fully in the solution of the Cauchy problem. It is instructive to consider first the case where $m = 1$, i.e., L is a first-order *scalar* linear partial differential operator.

The Scalar Case

In this case A and B are complex-valued functions. We shall write $\lambda(x, t)$, instead of $A(x, t)$. Hypothesis (16.1) demands that λ be *real*. We notice that the *principal part* of L,

$$L_0 = \frac{\partial}{\partial t} - \lambda(x, t)\frac{\partial}{\partial x},$$

can easily be transformed, by a change of variables, into $\partial/\partial t$. It should be emphasized, however, that this transformation is only *local*, that is, valid in a small open set (which we may identify with the set $\mathfrak{V}_0$ above). Indeed, we solve the nonlinear ordinary differential equation

$$\frac{dx}{ds} = -\lambda(x, s), \tag{16.10}$$

with initial condition

$$x\Big|_{s=0} = y, \tag{16.11}$$

where y is a point sufficiently near x_0. For small s, (16.10) has a unique solution, which we denote by $x = x(y, s)$. Consider then the change of variables

$$x = x(y, s), \qquad t = s. \tag{16.12}$$

By virtue of (16.11), the derivative of x with respect to y does not vanish if s is close to zero (for $s = 0$, it is equal to one). Consequently, (16.12) is a true C^1 change of variables. We have

$$\frac{\partial}{\partial s} = \frac{\partial}{\partial t} - \lambda(x, t)\frac{\partial}{\partial x} = L_0 .$$

The operator L is transformed into

$$L^{\#} = \frac{\partial}{\partial s} - B(x(y, s), s)$$

(rather, one should say that this is the expression of L in the new coordinates y, s). Since the lines $s = 0$ and $t = 0$ coincide and (16.11) holds, the problem (16.5)–(16.5) is transformed into

$$L^{\#}u = f(x(y, s), s), \qquad x(y, s) \in U, \qquad |s| < \delta, \tag{16.13}$$

$$u(y, 0) = u_0(y), \qquad y \in U. \tag{16.14}$$

But because of the expression of $L^{\#}$, this latter problem can be viewed as a Cauchy problem (depending on the parameter y) for a first-order linear *ordinary* differential equation, and we know how to solve it. It suffices to recall the considerations at the beginning of Sec. 11. We have

$$u(y, s) = u_0(y)e^{\beta(y, s)} + \int_0^s \exp[\beta(y, s) - \beta(y, s')]f(x(y, s'), s')\, ds'. \tag{16.15}$$

We have set

$$\beta(y, s) = \int_0^s B(x(y, s'), s')\, ds'. \tag{16.16}$$

Formula (16.15) can be interpreted in an interesting geometrical manner. Notice first that y, s are the new coordinates of the point (x, t). In particular, $s = t$. On the other hand, when s' varies from 0 to $s = t$, the point $(x(y, s'), s')$ ranges over an arc of curve joining $(y, 0)$ to (x, t). The curve in question is the integral curve of (16.10)–(16.11). If it passes through (x, t), we shall denote it by $C(x, t)$. At the point (x, t), L_0 is a partial differentiation in the direction of its tangent. This defines an orientation on $C(x, t)$, which is the same as the one defined by taking time as a parameter (we go forward if time increases). By the fundamental theorem on ordinary differential equations, applied to (16.10)–(16.11), through every point (x, t) there is one and only one arc of integral curve of (16.10)–(16.11). Of course, $C(x, t) = C(x_1, t_1)$ means that (x, t) and (x_1, t_1) lie on one and the same such integral curve. If U and δ are sufficiently small, $U \times]-\delta, \delta[$ will be "trivially fibered" by these curves $C(x, t)$, in fact by the curves $C(x, 0)$ as x ranges over U. This means that the effect of (16.12) is to "straighten up" the curves $C(x, 0)$ and transform them into segments of vertical lines $x = \text{const}$. Now, in this picture, y is completely determined by the fact that $(y, 0)$ is the intersection of $C(x, t)$ with the straight line $t = 0$. As for $\beta(y, s)$ it is equal to the integral of B along $C(x, t)$, with respect to the measure dt, from $(y, 0)$ to (x, t). We shall write $B_1(x, t)$ instead of $\beta(y, s)$. We have

$$B_1(x, t) = \int_{C(x,t)} B(x', t')\, dt',$$

keeping in mind that the bounds are 0 *and* t; (x', t') is the variable point on $C(x, t)$. As for (16.5), it reads

(16.17)

$$u(x, t) = u_0(y) \exp[B_1(x, t)] + \int_{C(x,t)} \exp[B_1(x, t) - B_1(x', t')]f(x', t')\, dt',$$

keeping in mind that y *is defined by the fact that* $(y, 0)$ *is the intersection of* $C(x, t)$ *with the axis* $t = 0$; here also the integration is performed from $t' = 0$ to $t' = t$ even when $t < 0$.

Remark 16.2. Suppose that $f \equiv 0$. Then the value of u at (x, t) depends only on the value of u_0 at y. In a sense, we may say that the "influence" of the initial data propagates along the curves $C(x, t)$.

The General Case

We return now to the general (but "diagonalized") problem (16.8)–(16.9). To solve it we shall now combine the considerations in the scalar case with Picards interation method. But first let us simplify (16.8)–(16.9) by substituting $v(x, t) - v_0(x)$ for $v(x, t)$. This amounts to the same as assuming that $v_0 \equiv 0$ (note that g must then be replaced by $g + \Gamma^{-1}A\Gamma v_{0x} + B_{\#} v_0$).

Next we define a sequence of functions $v^{(k)}$ ($k = 0, 1, 2, \ldots$) as follows:

$$(16.18) \qquad v_t^{(k)} - \Gamma^{-1}A\Gamma v_x^{(k)} = B_{\#} v^{(k-1)} + g, \qquad v^{(k)}\Big|_{t=0} = 0,$$

agreeing that $v^{(-1)} \equiv 0$. We shall show that, if U and δ are sufficiently small, the functions $v^{(k)}$ converge uniformly in $U \times \,]-\delta, \delta[$ to a C^1 function $v(x, t)$ which is a solution of (16.8)–(16.9); grad $v^{(k)}$ will converge uniformly to grad v. We set

$$w^{(k)} = v^{(k)} - v^{(k-1)}, \qquad k = 0, 1, \ldots.$$

Then we define

$$v = \sum_{k=0}^{+\infty} w^{(k)},$$

and we shall show that the sequence $\{w^{(k)}\}$ is summable. We derive from (16.18), by subtraction

$$(16.19) \qquad w_t^{(0)} - \Gamma^{-1}A\Gamma w_x^{(0)} = g, \qquad w^{(0)}(x, 0) = 0,$$

$$(16.20) \quad w_t^{(k)} - \Gamma^{-1}A\Gamma w_x^{(k)} = Bw^{(k-1)}, \qquad w^{(k)}(x, 0) = 0, \qquad k > 0.$$

Equations (16.19) and (16.20) are easy to solve. They are essentially the conjunction of m scalar equations. Indeed, let us call $w^{(k)j}$ the jth component of $w^{(k)}$, $\{B_{\#} w^{(k)}\}^j$ that of $B_{\#} w^{(k)}$, and g^j that of g. Equation (16.19) is the conjunction of the following equations (when $j = 1, \ldots, m$):

$$(16.21)_j \qquad L_j w^{(0)j} = g^j, \qquad w^{(0)j}(x, 0) = 0,$$

where $L_j = \partial/\partial t - \lambda_j(x, t)\, \partial/\partial x$, whereas (16.20) is the conjunction of the equations

$$(16.22)_j \qquad L_j w^{(k)j} = \{B_{\#} w^{(k-1)}\}^j, \qquad w^{(k)j}(x, 0) = 0, \qquad k > 0.$$

We call $C_j(x, t)$ the segment of integral curve of the vector field L_j which joins the axis $t = 0$ to the point (x, t). We apply formula (16.17):

$$w^{(k)j}(x, t) = \int_{C_j(x,t)} \{B_\# w^{(k-1)}\}^j(x', t')\, dt', \qquad k > 0, \tag{16.23}$$

whereas

$$w^{(0)j}(x, t) = \int_{C_j(x,t)} g^j(x', t')\, dt'. \tag{16.24}$$

We introduce now the following notation:

Notation 16.1. We denote by $\mathscr{D}(x, t)$ the smallest compact set containing the point (x, t) and having the following property: If $(x', t') \in \mathscr{D}(x, t)$, every curve segment $C_j(x', t')$ $(j = 1, \ldots, m)$ is contained in $\mathscr{D}(x, t)$.

We then set, for any vector-valued continuous function $\varphi(x, t)$,

$$|||\varphi(x, t)||| = \sup_{\mathscr{D}(x,t)} |\varphi(x', t')|.$$

We derive at once from (16.23),

$$|||w^{(k)}(x, t)||| \leqq C|t|\ |||w^{(k-1)}(x, t)|||, \qquad k > 0,$$

hence, by iteration,

$$|||w^{(k)}(x, t)||| \leqq (C|t|)^k |||w^{(0)}(x, t)|||, \qquad k > 0.$$

Since, on the other hand, (16.24) implies

$$|||w^{(0)}(x, t)||| \leqq C|t|\ |||g(x, t)|||,$$

we finally obtain

$$|||w^{(k)}(x, t)||| \leqq (C|t|)^{k+1} |||g(x, t)|||, \qquad k = 0, 1, \ldots. \tag{16.25}$$

Consequently, for $|t| < C^{-1}$, the series $\sum w^{(k)}$ converges uniformly. An inequality analogous to (16.25) can be derived for grad $w^{(k)}$ instead of $w^{(k)}$ (g then also has to be modified). To see this, one first estimates $w_x^{(k)}$ using formula (16.23). This is not as difficult as it may seem, if one first straightens up the curve C_j as indicated in the scalar case. To estimate $w_t^{(k)}$ one uses the estimates of $w^{(k)}$ and $w_x^{(k)}$ in combination with (16.21) and (16.22) (expressing $w_t^{(k)}$ in terms of the other quantities). In these latter estimates, one must take advantage of the fact that the eigenvalues $\lambda_j(x, t)$ are C^1 functions of (x, t) [this is then true for the curves $C_j(x, t)$] and of the assumption that the original data $f(x, t)$, $u_0(x)$ are also once continuously differentiable. We leave the details to the student.

Finally we have obtained an expression of v of the form

$$v = (I - T)^{-1} g$$

where T is an integral operator. This integral operator T is such that, as shown by formula (16.25), the value of v at the point (x, t) depends only on the values of g in the set $\mathscr{D}(x, t)$ (we are assuming here that $v_0 = 0$). We encounter once more the "domain of influence" phenomenon (cf. Theorem 14.1): In the present case, the influence propagates along the curves $C_j(x, t)$.

Example 16.1. Consider the Cauchy problem for the wave equation in one space variable:

$$(16.26) \qquad u_{tt} - u_{xx} = f, \qquad u(0, x) = u_0(x), \qquad u_t(0, x) = u_1(x).$$

Set $u^1 = u$, $u^2 = u_t - u_x$ and call $\mathbf{u}$ the vector with components u^1, u^2, $\mathbf{f}$ the one with components $0, f$, and $\mathbf{u}_0$ the one with components u_0, $u_1 - u_{0x}$. The problem (16.26) is then transformed into the following Cauchy problem:

$$(16.27) \qquad \mathbf{u}_t = \begin{pmatrix} 1 & 0 \\ 0 & -1 \end{pmatrix} \mathbf{u}_x + \begin{pmatrix} 0 & 1 \\ 0 & 0 \end{pmatrix} \mathbf{u} + \mathbf{f}, \qquad \mathbf{u}\Big|_{t=0} = \mathbf{u}_0 .$$

Here $L_1 = \partial/\partial t - \partial/\partial x$, $L_2 = \partial/\partial t + \partial/\partial x$; $C_1(x, t)$ is the curve $x' = -t' + (x + t)$, and $C_2(x, t)$ is the curve $x' = t' + (x - t)$. Supposing, for instance, that t is greater than zero, the set $\mathscr{D}(x, t)$ is the portion of the (closed) backward light-cone, with vertex at (x, t), which lies in the half-plane $t \geqq 0$.

Exercises

16.1. Here $x = (x^1, \ldots, x^n)$, $n \geqq 1$, and

$$(16.28) \qquad L = \frac{\partial}{\partial t} - \sum_{j=1}^{n} \alpha_j(x, t) \frac{\partial}{\partial x^j},$$

where the coefficients α_j are *real*-valued C^∞ functions in an open subset Ω of $\mathbf{R}^{n+1}$.

Show that every point (x_0, t_0) of Ω has an open neighborhood $\mathfrak{B}_0 \subset \Omega$ where the system of ordinary differential equations

$$(16.29) \qquad \frac{dx^j}{ds} = -\alpha_j(x, s), \qquad j = 1, \ldots, n,$$

has a unique solution $x = x(y, s)$ such that

$$(16.30) \qquad x^j\Big|_{s=t_0} = y^j, \qquad j = 1, \ldots,$$

where y is a point $\mathbf{R}^n$, close to x_0. Prove that

$$(16.31) \qquad x = x(y, s), \qquad t = s,$$

defines a C^∞ change of variables near (x_0, t_0).

Apply these facts to the solution of the Cauchy problem

$$Lu = 0, \qquad u\Big|_{t=t_0} = u_0(x) \qquad [\text{near } (x_0, t_0)], \tag{16.32}$$

where u_0 is a sufficiently regular function of x in some neighborhood of x_0 in $\mathbf{R}^n$. Solve the same problem with $Lu = f$ (instead of $Lu = 0$), f being a smooth function in an open neighborhood of (x_0, t_0) in $\mathbf{R}^{n+1}$. In particular, let u^j be the solution of (16.32) when $u_0(x) = x^j$. Relate $u = (u^1, \ldots, u^n)$ to the transformation (16.31).

16.2. Use the notation of Exercise 16.1. We assume now that $\alpha_j(x, s) = \alpha_j(x)$, independent of s ($j = 1, \ldots, n$). Denote by $y(x, t)$ the function obtained by inverting with respect to y the function $x = x(y, t)$ in (16.31). Set

$$y(x, t) = T_t x. \tag{16.33}$$

Prove that, for sufficiently small values of $|t|$, $|t'|$ and for x in a suitably small neighborhood of x_0, we have

$$T_0 = I, \qquad \text{the identity;} \qquad T_t T_{t'} = T_{t+t'}. \tag{16.34}$$

Compute T_t in the following cases:

(16.35) *the α_j are all constant*;

$$L = \frac{\partial}{\partial t} - \frac{\partial}{\partial x} \qquad (n = 1); \tag{16.36}$$

$$L = \frac{\partial}{\partial t} - \left(x^1 \frac{\partial}{\partial x^2} - x^2 \frac{\partial}{\partial x^1}\right) \qquad (n = 2). \tag{16.37}$$

16.3. Using the same notation and hypothesis as in Exercise 16.2, give an example in the case where $n = 1$ in which the operator T_t, defined in (16.33), does *not* exist for all real values of t.

16.4. Use the notation as in Exercise 16.1, with $\Omega = \mathbf{R}^{n+1}$. We suppose here that there is a constant $K > 0$ such that

$$|\text{grad}_x\, \alpha_j| \leqq K \quad \text{in the whole of } \mathbf{R}^{n+1}, \qquad j = 1, \ldots, n. \tag{16.38}$$

Prove that the solution of (16.29)–(16.30) exists for all $y \in \mathbf{R}^n$ and all $t \in \mathbf{R}$.

16.5. Consider $L = I\, \partial/\partial t - A$, where I is the identity $m \times m$ matrix and

$$A = \sum_{j=1}^{n} A_j(x) \frac{\partial}{\partial x^j}, \tag{16.39}$$

where the $A_j(x)$ are $m \times m$ matrices with entries that are analytic functions of x in some open neighborhood of the origin in $\mathbf{R}^n$. Let $\mathbf{h}$ be an arbitrary function in $\mathbf{R}^n$, valued in $\mathbf{C}^m$, which can be extended to $\mathbf{C}^n$ as an entire function. Prove that the series at the right of

$$(e^{tA}\mathbf{h})(x) = \sum_{k=0}^{+\infty} \frac{t^k}{k!} (A^k\mathbf{h})(x) \tag{16.40}$$

converges, provided that $|x|$ and $|t|$ are sufficiently small, and defines an analytic function of x in a suitably small open neighborhood of the origin.

16.6. We use the same notation as in Exercise 16.5, except we assume now that $m = 1$ (thus the coefficients A_j are *scalar* functions). Compute the operator e^{tA}, defined in (16.40), in each one of the cases (16.35), (16.36), (16.37), and also in the case

$$L = \frac{\partial}{\partial t} - x^2 \frac{\partial}{\partial x} \qquad (n = 1). \tag{16.41}$$

Compare with the operator T_t defined in (16.33).

16.7. Consider $L = \partial/\partial t - \lambda(x)\, \partial/\partial x$ (we are in the case of one space variable, x; i.e., $n = 1$) and suppose that λ is an analytic function of x in a neighborhood of zero. Give an interpretation of the operator (16.40) in this case as a "motion" in the complex plane (near the origin) and derive from it that the Cauchy problem is well posed for L if and only if λ is real-valued.

17

The Cauchy–Kovalevska Theorem. The Classical and Abstract Versions

The Cauchy problem, as we have seen, is well posed for hyperbolic equations and systems and only for these. Our definition of well-posed Cauchy problems involves the assumption that the data, the Cauchy data and the right-hand side, are arbitrary distributions (or arbitrary C^∞ functions). In practice one might be willing to settle for much less and content oneself with restricted data—restricted either from the viewpoint of smoothness or from the viewpoint of growth at infinity. In this case the classes of equations for which the solution to the Cauchy problem exists and is unique (within the chosen framework) may be considerably larger than the hyperbolic. The most classical example of this phenomenon is provided by the Cauchy–Kovalevska theorem. In its most common form it applies to the *local* Cauchy problem: The data are required to be *analytic functions* and one seeks an analytic solution. The class of equations for which it is possible to find such a solution, which is then unique, includes essentially all equations with analytic coefficients (as we shall see, there is one restrictive condition, concerning the choice of the time variable t).

Until now we have handled the PDEs under consideration by reducing them to ODEs via a Fourier transformation with respect to the space variables. This method can be modified to cover more cases, if we generalize the kind of ODE we are willing to study: We must allow the coefficients of the ODE to be not only complex- or matrix-valued functions, as before, but also *linear operators* on certain *function spaces*. For instance, we may view the heat operator $\partial/\partial t - \Delta_x$ as an ordinary differential operator (with respect to time); $-\Delta_x$ is then its zero-order coefficient. It turns out to be a linear operator acting on functions and distributions in the x variables. A similar interpretation will apply to all first-order systems considered in Sects. 15 and 16, and to many others. Generally speaking, they are of the form

$$L = \frac{\partial}{\partial t} - A(t), \tag{17.1}$$

where $A(t)$ is an operator-valued function of t. Notice that this covers also the case of the operators we have often denoted by $\tilde{L}$, resulting from a Fourier transformation with respect to x. In the operators $\tilde{L}$ the coefficients were linear operators acting on spaces of functions of the variables $\xi \in \mathbf{R}_n$ which happened to act as *multipliers*: multiplication by smooth functions of ξ. The advantage lies in the fact that multipliers are particularly easy to handle. We are now going to move in a different direction, looking toward the Cauchy–Kovalevska theorem, and deal with operators $A(t)$ of the kind

$$\sum_{j=1}^{n} A_j(x, t) \frac{\partial}{\partial x^j} + A_0(x, t) \tag{17.2}$$

acting on spaces of analytic functions of x; the coefficients $A_j(x, t)$ will be $m \times m$ matrices with entries which are *analytic* functions of x, continuous with respect to t (we will have to be more precise than this). Here x will vary in an open set Ω of $\mathbf{R}^n$ while t varies in an open interval $]-T, T[$ $(T > 0)$. We are given two functions $f(x, t)$, $u_0(x)$ analytic with respect to x in Ω; moreover, we shall assume that $f(x, t)$ is continuous with respect to t, $|t| < T$. We look then at the Cauchy problem

$$\frac{\partial u}{\partial t} = \sum_{j=1}^{n} A_j(x, t) \frac{\partial u}{\partial x^j} + A_0(x, t)u + f(x, t), \tag{17.3}$$

$$u\Big|_{t=0} = u_0(x). \tag{17.4}$$

Our version of the Cauchy–Kovalevska theorem states (roughly speaking) that if Ω' is any relatively compact open subset of Ω and if T', $0 < T' \leqq T$, is sufficiently small (depending on the choice of Ω'), there is a unique function $u(x, t)$, analytic with respect to x, once continuously differentiable with respect to t, which satisfies (17.3) when $x \in \Omega'$ and $|t| < T'$ and satisfies (17.4) when $x \in \Omega'$.

We ought to mention that, in the usual version of the Cauchy–Kovalevska theorem, the coefficients $A_j(x, t)$ and the right-hand side $f(x, t)$ are assumed to be analytic with respect to all variables, including t; the solution u is then shown to be analytic with respect to both x and t. This version, although weaker than the one we emphasize, has the advantage of treating all variables on an almost equal footing, thus preserving a certain invariance of the problem. This invariance is of interest in applications to differential geometry. At any rate, we shall also obtain the analytic version.

These various results must be stated in precise and rigorous terms. To do so we shall have to be more technical. It is advantageous to extend the variable $x = (x^1, \ldots, x^n)$ to the complex domain—where we denote it then by $z = (z^1, \ldots, z^n)$—and everywhere replace "analytic" by "holomorphic." An analytic function in $\Omega \subset \mathbf{R}^n$ is a function which can be extended holomorphically to some open neighborhood (depending on the function) of Ω in $\mathbf{C}^n$. For our purpose it is convenient to consider those functions which can be extended holomorphically to a *fixed* open neighborhood $\mathcal{O}$ of Ω in $\mathbf{C}^n$ and, as a matter of fact, which can be extended as continuous functions to the closure $\bar{\mathcal{O}}$ of $\mathcal{O}$ (assumed to be bounded). In what follows we may forget that $\mathcal{O}$ originated as a neighborhood of Ω and regard it simply as an arbitrary *bounded* open subset of $\mathbf{C}^n$.

Definition 17.1. Let $\mathcal{O}$ be a bounded open subset of $\mathbf{C}^n$. We denote by $H(\bar{\mathcal{O}})$ the space of continuous functions in $\bar{\mathcal{O}}$ which are holomorphic in the set $\mathcal{O}$. We make a normed space of $H(\bar{\mathcal{O}})$ by equipping it with the maximum norm:

$$f \mapsto |f|_{\mathcal{O}} = \sup_{x \in \mathcal{O}} |f(z)|. \tag{17.5}$$

PROPOSITION 17.1. *The normed space $H(\bar{\mathcal{O}})$ is complete, i.e., is a Banach space.*

Proof. A Cauchy sequence $\{u_j\}$ $(j = 1, 2, \ldots)$ in $H(\bar{\mathcal{O}})$ converges uniformly on $\bar{\mathcal{O}}$ to a continuous function u in $\bar{\mathcal{O}}$. In $\mathcal{O}$ we have $(\partial/\partial\bar{z}^k)u_j = 0$, $k = 1, \ldots, n$, for every $j = 1, 2, \ldots$; hence $(\partial/\partial\bar{z}^k)u = 0$, $k = 1, \ldots, n$, since the u_j converge to u in the distribution sense. Thus u is holomorphic in $\mathcal{O}$. Q.E.D.

We may also look at the continuous functions in $\bar{\mathcal{O}}$, holomorphic in $\mathcal{O}$, which are valued in $\mathbf{C}^m$. We denote the space of such functions by $H(\bar{\mathcal{O}}; \mathbf{C}^m)$ and equip it also with the maximum norm (which turns it into a Banach space).

We wish now to study the action of an operator of the kind (17.2) (with z substituted for x) on the elements of $H(\bar{\mathcal{O}}; \mathbf{C}^m)$. Clearly we might run into trouble if we look at the result of this action at the boundary of $\mathcal{O}$ but not if we stay away from the boundary. Let us take a closer look at this fact.

Let $\mathcal{O}_1$ be a second bounded open subset of $\mathbf{C}^n$ such that

$$\bar{\mathcal{O}} \subset \mathcal{O}_1.$$

Let d denote the distance from $\mathcal{O}$ to the boundary of $\mathcal{O}_1$. Observe then that the partial differentiation $\partial/\partial z^j$ is a linear operator from $H(\bar{\mathcal{O}}_1; \mathbf{C}^m)$ into $H(\bar{\mathcal{O}} : \mathbf{C}^m)$.

PROPOSITION 17.2. *The partial differentiation $\partial/\partial z^j$ $(1 \leqq j \leqq n)$ defines a bounded linear operator $H(\bar{\mathcal{O}}_1; \mathbf{C}^m) \to H(\bar{\mathcal{O}}; \mathbf{C}^m)$ with norm $\leqq d^{-1}$.*

Proof. Let z_0 be an arbitrary point of $\bar{\mathcal{O}}$, f any element of $H(\mathcal{O}_1; \mathbf{C}^m)$. We have (Cauchy's inequalities)

$$\left|\frac{\partial f}{\partial z^j}(z_0)\right| \leqq \frac{1}{d} \sup_{|z^j - z_0{}^j| < d} |f(z_0^1, \ldots, z_0^{j-1}, z^j, z_0^{j+1}, \ldots, z_0^n)|,$$

which implies at once

$$\left|\frac{\partial f}{\partial z^j}\right|_{\mathcal{O}} \leqq d^{-1}|f|_{\mathcal{O}_1},$$

from which we obtain Proposition 17.2. Q.E.D.

Let $\boldsymbol{M}_m(\mathbf{C})$ denote the Banach space of $m \times m$ matrices with complex entries, equipped with the matrix norm $\| \ \|$.

PROPOSITION 17.3. *Let $M(z)$ belong to $H(\bar{\mathcal{O}}; \boldsymbol{M}_m(\mathbf{C}))$. Then $f \mapsto Mf$ is a bounded linear operator of $H(\bar{\mathcal{O}}; \mathbf{C}^m)$ into itself, with norm*

$$\|M\|_{\mathcal{O}} = \sup_{z \in \mathcal{O}} \|M(z)\|. \tag{17.6}$$

The proof of Proposition 17.3 is very easy and we leave it to the student.

Let us now consider a matrix-valued function $M(z, t)$ which satisfies the following condition:

(17.7) *$t \mapsto M(x, t)$ is a continuous function of t, $|t| < T$ $(T > 0)$, valued in $H(\bar{\mathcal{O}}; \boldsymbol{M}_m(\mathbf{C}^m))$.*

This means that, for each t, $|t| < T$, $M(z, t)$ belongs to $H(\bar{\mathcal{O}}; \boldsymbol{M}_m(\mathbf{C}))$ [i.e., its entries belong to $H(\bar{\mathcal{O}})$], and furthermore

(17.8) *to every $\varepsilon > 0$ there is an $\eta > 0$ such that, for every t', $|t'| < T$, $|t - t'| < \eta$ implies $\|M(\cdot, t) - M(\cdot, t')\|_{\mathcal{O}} < \varepsilon$.*

(Of course, in general η depends on t.)

PROPOSITION 17.4. *If (17.7) holds, $t \mapsto (f \mapsto M(\cdot, t)f)$ is a continuous function of t, $|t| < T$, valued in the Banach space of bounded linear operators of $H(\bar{\mathcal{O}}; \mathbf{C}^m)$ into itself. For each t, $|t| < T$, the operator norm of $f \mapsto M(\cdot, t)f$ is $\leqq \|M(\cdot, t)\|_{\mathcal{O}}$.*

This follows at once from Proposition 17.3, (17.7), and (17.8).

We look now at the operator (17.2) where x has been replaced by z. Combining Propositions 17.2 and 17.4 yields

PROPOSITION 17.5. *Let $A_j(z, t)$ $(j = 0, 1, \ldots, n)$ be $n+1$ continuous functions of t, $|t| < T$, valued in $H(\bar{\mathcal{O}}; \boldsymbol{M}_m(\mathbf{C}))$, and set*

$$A(t) = \sum_{j=1}^{n} A_j(z, t) \frac{\partial}{\partial z^j} + A_0(z, t).$$

Then $A(t)$ is a continuous function of t, $|t| < T$, valued in the space of bounded linear operators of $H(\bar{\mathcal{O}}_1; \mathbf{C}^m)$ into $H(\bar{\mathcal{O}}; \mathbf{C}^m)$. For each t, $|t| < T$, the norm of $A(t)$ is not larger than

$$\frac{1}{d} \sum_{j=1}^{n} \|A_j(\cdot, t)\|_{\mathcal{O}} + \|A_0(\cdot, t)\|_{\mathcal{O}}.$$

Thus, the bound for the norm of $A(t)$ grows like $1/d$ when $d \to +0$. This observation is crucial in the proof of the Cauchy–Kovalevska theorem. In our version of this important result we shall deal with the following objects:

(17.9) *$n+1$ continuous functions of t, $|t| < T$, $A_j(z, t)$ $(j = 0, 1, \ldots, n)$, valued in $H(\bar{\mathcal{O}}_1; \boldsymbol{M}_m(\mathbf{C}))$;*

(17.10) *a continuous function of t, $|t| < T$, $f(z, t)$, valued in $H(\bar{\mathcal{O}}_1; \mathbf{C}^m)$;*

(17.11) *an arbitrary member of $H(\bar{\mathcal{O}}_1; \mathbf{C}^m)$, $u_0(z)$.*

We shall then study the following Cauchy problem:

$$\frac{\partial u}{\partial t} = \sum_{j=1}^{n} A_j(z, t) \frac{\partial u}{\partial z^j} + A_0(z, t)u + f(z, t), \tag{17.12}$$

$$u\Big|_{t=0} = u_0(z). \tag{17.13}$$

We will not be able to require that (17.12) be verified for all $z \in \mathcal{O}_1$ and $|t| < T$ and that (17.13) holds for all $z \in \mathcal{O}_1$. Rather

THEOREM 17.1. (Cauchy–Kovalevska). *Let $\mathcal{O}_0$ be a nonempty connected open subset of $\mathbf{C}^n$ whose closure is contained in $\mathcal{O}_1$. There is a number δ_0, $0 < \delta_0 \leqq T$, such that the following is true:*

There is a unique C^1 function of t, $|t| < \delta_0$, $u(z, t)$, valued in $H(\bar{\mathcal{O}}_0; \mathbf{C}^m)$, satisfying (17.12) for all $z \in \mathcal{O}_0$, t, $|t| < \delta_0$, and (17.13) for all $z \in \mathcal{O}_0$.

Let d_0 be the distance between $\mathcal{O}_0$ and the boundary of $\mathcal{O}_1$. There is no loss of generality if we decrease $\mathcal{O}_1$ and assume that

$$\mathcal{O}_1 = \{z \in \mathbf{C}^n;\ \mathrm{dist}(z, \mathcal{O}_0) < d_0\}.$$

We introduce a one-parameter family of open sets, intermediary between $\mathcal{O}_0$ and $\mathcal{O}_1$ and *all connected*:

$$\mathcal{O}_s = \{z \in \mathbf{C}^n;\ \mathrm{dist}(z, \mathcal{O}_0) < sd_0\}, \qquad 0 \leqq s \leqq 1.$$

Notice that

(17.14) $$\operatorname{dist}(\mathcal{O}_{s'}, \mathbf{C}^n \backslash \mathcal{O}_s) = (s - s')d_0, \qquad 0 \leqq s' < s \leqq 1.$$

We shall exploit (17.14) in relation to Proposition 17.5. We shall transform our Cauchy problem (17.12)–(17.13) into the abstract problem

(17.15) $$u_t = A(t)u + f(t),$$

(17.16) $$u\Big|_{t=0} = u_0.$$

We view u_0 as an element of the Banach space $E_1 = H(\bar{\mathcal{O}}_1; \mathbf{C}^m)$ and $f(t)$ as a continuous function of t, $|t| < T$, valued in E_1—which they are. More generally let us write

(17.17) $$E_s = H(\bar{\mathcal{O}}_s; \mathbf{C}^m), \qquad 0 \leqq s \leqq 1.$$

It is clear that we have

(17.18) *If $s' < s$, $E_s \subset E_{s'}$ and the norm in E_s is not smaller than the norm induced on E_s by $E_{s'}$.*

Of course, the natural injection $E_s \subsetneq E_{s'}$ is obtained by taking functions belonging to $H(\bar{\mathcal{O}}_s; \mathbf{C}^m)$ and restricting them to $\bar{\mathcal{O}}_{s'}$. We may also reinterpret Proposition 17.5 in the following manner [replacing $\mathcal{O}_1$ by $\mathcal{O}_s$ and $\mathcal{O}$ by $\mathcal{O}_{s'}$ and taking (17.14) into account]:

(17.19) *If $s' < s$, $A(t)$ is a continuous function of t, $|t| < T$, valued in the Banach space of bounded linear operators $E_s \to E_{s'}$.*

For each t, $|t| < T$, the operator norm or $A(t) : E_s \to E_{s'}$ is not greater than $C(t)/(s - s')$, where

$$C(t) = \frac{1}{d_0} \sum_{j=1}^{n} \|A_j(\cdot, t)\|_{\mathcal{O}_1} + \|A_0(\cdot, t)\|_{\mathcal{O}_1}.$$

The important thing about $C(t)$ is that it is bounded on any closed subinterval of $]-T, T[$. Thus, if we slightly decrease $T > 0$, we may assume $C(t)$ is bounded by a constant $C > 0$ and therefore that

(17.20) *For each t, $|t| < T$, the operator norm of $A(t) : E_s \to E_{s'}$ $(0 \leqq s' < s \leqq 1)$ is not greater than $C/(s - s')$ where the positive constant C does not depend on s, s' nor on t.*

Also note that, as soon as we have decreased T (however slightly), we may assume that f is a continuous function in the *closed* interval $[-T, T]$ with values in E_1. Notice also that (17.20) is not modified if we increase C. For technical reasons it is convenient to assume that

(17.21) $$(Ce)^{-1} \leqq T.$$

We are now in a position to apply the following result, which can be regarded as the *abstract version* of the Cauchy–Kovalevska theorem.

THEOREM 17.2. *Under the preceding hypotheses, given any* $u_0 \in E_1$ *and any continuous function of* t, $|t| \leqq T, f$, *valued in* E_1, *the following is true*:

(I) *There is a function* u *in the interval* $|t| < (Ce)^{-1}$, *valued in* E_0, *which, for any* s, $0 \leqq s < 1$, *is a* C^1 *function of* t, $|t| < (Ce)^{-1}(1-s)$, *valued in* E_s *and which, furthermore, satisfies (17.15) when* $|t| < (Ce)^{-1}$, *together with (17.16).*

(II) *If, for some number* T', $0 < T' \leqq T$, *and for some* s, $0 < s \leqq 1$, *there are two* C^1 *functions of* t, $|t| < T'$, *valued in* E_s *and satisfying (17.15) in this interval, together with (17.16), they must be equal.*

Proof. (I) *Existence of the solution.* We define a sequence $v_k(t)$ ($k = 0, 1, \ldots$) of continuous functions of t, $|t| \leqq T$, valued in E_s (for any value of s, $0 \leqq s < 1$), as follows:

$$v_0(t) = u_0 + \int_0^t f(t')\, dt',$$

$$v_{k+1}(t) = u_0 + \int_0^t f(t')\, dt' + \int_0^t A(t')v_k(t')\, dt', \qquad k = 0, 1, \ldots.$$

By (17.19) we know that, if $v_k(t)$ is a continuous function of t, $|t| < T$, valued in E_s, $A(t)v_k(t)$ is a continuous function in the same interval, valued in $E_{s'}$, for any $s' < s$, and the same is true of $v_{k+1}(t)$. As s is arbitrarily close to one, the same is true of s', from which arises our assertion that each v_k is a continuous function of t, $|t| < T$, valued in any E_s, $0 \leqq s < 1$. Next we set

$$w_0 = v_0, \qquad w_k = v_k - v_{k-1} \qquad \text{if} \quad k > 0.$$

We have

$$w_{k+1}(t) = \int A(t')w_k(t')\, dt', \qquad k = 0, 1, \ldots.$$

Each w_k is a continuous function of t, $|t| < T$, valued in any E_s, $s < 1$. By induction on $k = 0, 1, \ldots$, we shall prove the following inequality:

$$N_s(w_k(t)) \leqq M(t)\left(\frac{Ce|t|}{1-s}\right)^k, \qquad |t| < T, \tag{17.22}$$

where N_s denotes the norm in E_s and where

$$M(t) = N_1(u_0) + \left|\int_0^t N_1(f(t'))\, dt'\right|.$$

Observe that $M(t)$ is a nondecreasing function of $|t|$ on the half-intervals $[0, T]$ and $[-T, 0]$, respectively.

Inequality (17.22) trivially holds when $k = 0$. Suppose that it holds up to k and let us prove it for $k + 1$. We return to the definition of w_{k+1} and apply (17.20). If $0 \leqq s' < s < 1$,

$$N_{s'}(w_{k+1}(t)) \leqq \frac{C}{s - s'}\left|\int_0^t N_s(w_k(t'))\,dt'\right|$$
$$\leqq M(t)\frac{C}{s - s'}\left(\frac{Ce}{1 - s}\right)^k \frac{|t|^{k+1}}{k + 1} \qquad \text{[by (17.22)].}$$

We now choose $s = s' + (1 - s')/(k + 1)$, so that $1 - s = [k/(k + 1)](1 - s')$. We obtain

$$N_{s'}(w_{k+1}(t)) \leqq M(t)\left(\frac{C|t|}{1 - s'}\right)^{k+1} e^k\left(1 + \frac{1}{k}\right)^k,$$

and its suffices to observe that $(1 + 1/k)^k \leqq e$.

Recalling that f is continuous on the *closed* interval $|t| \leqq T$, we see that $M(t) \leqq M < +\infty$. From (17.22) we derive that the series

$$\sum_{k=0}^{+\infty} w_k(t)$$

converges absolutely in E_s, uniformly in every closed subinterval of the open interval $|t| < (Ce)^{-1}(1 - s)$: Its sum $u(t)$ will be the sought solution of (17.15)–(17.16). Of course, u is the limit of the v_k as $k \to +\infty$ [in E_s, uniformly on compact subsets of the interval $|t| < (Ce)^{-1}(1 - s)$], and from the definition of the v_k we immediately obtain

$$u(t) = u_0 + \int_0^t f(t')\,dt' + \int_0^t A(t')u(t')\,dt'. \tag{17.23}$$

Let now ε be such that $0 < \varepsilon < 1 - s$. We know that u is a continuous function in the interval $|t| < (Ce)^{-1}(1 - s - \varepsilon)$ with values in $E_{s+\varepsilon}$. Therefore, by virtue of (17.19), $A(t)u(t)$ is a continuous function in the same interval but with values in E_s. From (17.23) we derive that u is a C^1 function in the same interval with values in E_s. By taking $\varepsilon \to +0$ we derive that u is a C^1 function of t, $|t| < (Ce)^{-1}(1 - s)$, valued in E_s. That it satisfies (17.15)–(17.16) is obvious from (17.23).

(II) *Uniqueness of the solution.* We must show that if a C^1 function of t, $|t| < T'$ $(0 < s \leqq 1)$, valued in E_s, satisfies

$$h_t = A(t)h, \qquad h\Big|_{t=0} = 0, \tag{17.24}$$

it must vanish identically in that interval. The set of points where h vanishes is not empty, as it contains $t = 0$, and is obviously closed; we are going to show that it is open in $]-T, T[$, which will imply what we want. Let t_0 be any point such that $h(t_0) = 0$. Note that we have

$$h(t) = \int_{t_0}^{t} A(t')h(t')\,dt'. \tag{17.25}$$

Let s' be any number such that $0 \leqq s' < s$. We shall prove, by induction on $k = 0, 1, \ldots$, that

$$N_{s'}(h(t)) \leqq M_1(t)(s - s')^{-k}(Ce)^k |t - t_0|^k, \tag{17.26}$$

where

$$M_1(t) = \sup N_s(h(t')),$$

the supremum being taken over the segment joining t and t_0. Inequality (17.26) is trivial when $k = 0$. By (17.20) and (17.25) we have

$$N_{s'}(h(t)) \leqq M_1(t)C\varepsilon^{-1}(Ce)^k(s - s' - \varepsilon)^{-k}\frac{|t - t_0|^{k+1}}{k+1}$$

where $\varepsilon = (s - s')/(k + 1)$. This immediately yields (17.26) with $k + 1$ substituted for k.

Now (17.26) shows that if $|t - t_0| < (Ce)^{-1}(s - s')$, then $N_{s'}(h(t)) = 0$. But since E_s is injected in $E_{s'}$ this implies that $N_s(h(t)) = 0$. Q.E.D.

Let us indicate right away that a "holomorphic version" of Theorem 17.2 is valid. In this version, t is a complex variable, varying in the *disk* $|t| \leqq T$. Hypothesis (17.19) is replaced by

(17.27) *If $s' < s$, $A(t)$ is a holomorphic function of t, $|t| < T$, valued in the Banach space of bounded linear operators $E_s \to E_{s'}$.*

Hypothesis (17.20) stands unchanged, of course. But now we assume that f is a continuous function of t, $|t| \leqq T$, valued in E_1, holomorphic when $|t| < T$. The proof of Theorem 17.2 remains valid: One must simply keep in mind that t is now a complex variable; the integrations are performed on straight line segments joining the limits of integration. Since uniform limits (over open sets) of holomorphic functions are holomorphic, the statement of parts (I) and (II) in Theorem 17.2 can be kept unchanged, but then C^1 must be interpreted in the complex sense, which is identical with holomorphic.

The classical holomorphic version of the Cauchy–Kovalevska theorem follows at once from the holomorphic version of Theorem 17.2.

Exercises

17.1. Denote by $\mathfrak{A}_R$ ($R > 0$) the space of C^∞ functions $f(x)$ in $\mathbf{R}^n$ such that

$$\sup_{\alpha \in \mathbf{Z}_+{}^n} \sup_{x \in \mathbf{R}^n} \frac{R^{|\alpha|}}{\alpha!} |\partial_x^\alpha f(x)| < +\infty, \tag{17.28}$$

and equip $\mathfrak{A}_R$ with the norm defined by the left-hand side in (17.28). Prove that $\mathfrak{A}_R$ is a Banach space and that every element f of $\mathfrak{A}_R$ can be extended as a holomorphic function of $z = x + iy$ in the slab $\{x \in \mathbf{R}^n;\ |y| < R\}$.

Let $0 < R_0 < R_1 < +\infty$ and set $E_s = \mathfrak{A}_{(1-s)R_0 + sR_1}$, $0 \leqq s \leqq 1$. Prove that conditions (17.18), (17.19), and (17.20) are satisfied when

$$A(t) = \sum_{j=1}^{n} A_j(x, t) \frac{\partial}{\partial x^j} + A_0(x, t), \tag{17.29}$$

where the coefficients $A_j (j = 0, \ldots, n)$ are continuous functions of t, $|t| < T$, valued in $\mathfrak{A}_{R_1}$.

17.2. Let $g(\xi)$ be a continuous function in $\mathbf{R}_n$, real-valued, and denote by $\hat{K}^g$ the space of measurable functions f in $\mathbf{R}_n$ such that

$$e^g f \in L^2. \tag{17.30}$$

Prove that $\hat{K}^g$, equipped with the norm

$$\|f\|_{\hat{K}^g} = \left\{ \int |e^g f|^2 \frac{d\xi}{(2\pi)^n} \right\}^{1/2},$$

is a Hilbert space in which $C_c^\infty(\mathbf{R}_n)$ is dense.

Let $\tilde{\mathscr{D}}(\mathbf{R}^n)$ denote the space of functions in $\mathbf{R}^n$ whose Fourier transform belongs to $C_c^\infty(\mathbf{R}_n)$ (by the Paley–Wiener theorem, $\tilde{\mathscr{D}} = \mathscr{S} \cap \mathrm{Exp}$, the space of the C^∞ *functions rapidly decaying at infinity* which can be extended to $\mathbf{C}^n$ as *entire functions of exponential type*). Denote by K^g the completion of $\tilde{\mathscr{D}}(\mathbf{R}^n)$ for the norm

$$\|u\|_{K^g} = \|\hat{u}\|_{\hat{K}^g}. \tag{17.31}$$

Prove that K^{-g} can be canonically identified to the dual of K^g. What is the canonical antilinear isometry of the Hilbert space K^g onto its dual, K^{-g}? Prove that if, for some constant $C > 0$,

$$-g(\xi) \leqq C \log(1 + |\xi|), \qquad \forall \xi \in \mathbf{R}_n, \tag{17.32}$$

K^g can be continuously embedded in a Sobolev space H^s for some suitable real number s (see Definition 13.1).

Prove that if

$$g(\xi)/\log(1 + |\xi|) \text{ tends to } +\infty \quad \text{as} \quad |\xi| \to +\infty, \tag{17.33}$$

then K^g is continuously embedded in $C^\infty(\mathbf{R}^n)$ and all the Sobolev spaces H^s, $s \in \mathbf{R}$, are continuously embedded in its dual, K^{-g}.

17.3. Let K^g be the space defined in Exercise 17.2. Prove the following result:

PROPOSITION 17.6. *Suppose that*

$$\sum_{j=1}^{n} |\xi_j|^{1/d} \leqq g(\xi) + a, \qquad \forall \xi \in \mathbf{R}_n, \tag{17.34}$$

where d is a number strictly positive and a is a real number.

Then any element u of K^g can be identified with a C^∞ function in $\mathbf{R}^n$ such that

$$\sup_{\alpha \in \mathbf{Z}_+{}^n} \sup_{x \in \mathbf{R}^n} \left\{ \frac{R^{|\alpha|}}{(\alpha!)^d} |\partial_x^\alpha u(x)| \right\} < +\infty. \tag{17.35}$$

17.4. Let K^g be the space defined in Exercise 17.2. We shall take

$$g(\xi) = sp(\xi), \qquad s \in \mathbf{R}, \tag{17.36}$$

where p is a function in $\mathbf{R}_n$, continuous, $p > 0$. Then let $A(\xi)$ be an arbitrary continuous function in $\mathbf{R}_n$ and set

$$A(D_x)u(x) = (2\pi)^{-n} \int e^{i\langle x, \xi\rangle} A(\xi)\hat{u}(\xi)\, d\xi. \tag{17.37}$$

Prove the following result:

PROPOSITION 17.7. *Suppose that, for some constant $C_0 > 0$,*

$$|A(\xi)| \leqq C_0 p(\xi), \qquad \xi \in \mathbf{R}_n. \tag{17.38}$$

Then, for any pair of real numbers $s' < s$, the operator $A(D_x)$ [in (17.37)] *defines a bounded linear operator $K^{sg} \to K^{s'g}$ with norm $\leqq C_0 e^{-1}(s - s')^{-1}$.*

Derive from this that the Cauchy problem

$$\frac{\partial u}{\partial t} - A(D_x)u = f, \tag{17.39}$$

$$u\Big|_{t=0} = u_0, \tag{17.40}$$

where $u_0 \in K^{s_0 p}$, $f \in C^0([-T, T]; K^{s_0 p})$, has a unique solution u which is a C^1 function of t in some interval $|t| < \delta_0 \leqq T$, valued in $K^{s_1 p}$, where $s_1 < s_0$.

Can you give an upper bound for the interval radius δ_0 in terms of C_0 and $(s_0 - s_1)$?

17.5. Show that the Cauchy problem for the heat equation

$$\frac{\partial u}{\partial t} = \Delta_x u + f, \qquad u\Big|_{t=0} = u_0, \tag{17.41}$$

is well posed if we restrict ourselves to data and solutions which belong to suitable spaces K^g where g satisfies (17.34) for an appropriate value of d (and, say, $a = 0$). What is this value of d? [*Hint*: Use the result established in Exercise 17.4.]

17.6. Consider the Cauchy problem (in one space variable)

$$\frac{\partial^2 u}{\partial t^2} = \frac{\partial u}{\partial x} + f, \qquad u\Big|_{t=0} = u_0, \qquad \frac{\partial u}{\partial t}\Big|_{t=0} = u_1. \tag{17.41'}$$

Show that if the data f, u_0, u_1 are valued in suitable Gevrey classes (Definition 3.3) with respect to the variable x, problem (17.41′) has a unique solution valued also in such a class. [*Hint*: Apply the results of Exercises 17.3 and 17.4.]

17.7. Let $\phi(r)$ be a continuous function of $r > 0$, nondecreasing, $\phi(r) > 0$ for all r, B a complex Banach space. We recall that a function h in $\mathbf{C}$ is called an *entire analytic* function with values in B if it is a C^1 function $\mathbf{R}^2 \to B$ satisfying everywhere the Cauchy–Riemann equations. Denote by $\mathfrak{A}_\phi(B)$ the space of such functions which satisfy

$$\sup_{z \in C} e^{-\phi(|z|)} \|h(z)\|_B < +\infty, \tag{17.42}$$

where $\| \ \|_B$ is the norm in B. Prove that if we equip $\mathfrak{A}_\phi(B)$ with the norm defined by the left-hand side of (17.42), it becomes a Banach space.

For $0 \leqq s \leqq 1$, set $E_s = \mathfrak{A}_\phi(B)$ where $\phi(r) = \phi_0((1 - s/2)r)$ for a given ϕ_0 (continuous, greater than zero, and nondecreasing). Prove the following:

Proposition 17.8. *Given* $0 \leqq s' < s \leqq 1$, $\partial/\partial z$ *defines a bounded linear operator* $E_s \to E_{s'}$ *with norm not exceeding* $[\exp \phi_0(1)]/(s - s')$.

Apply this to the Cauchy problem

$$\frac{\partial u}{\partial t} = A\frac{\partial u}{\partial z} + f(t, z), \qquad u\Big|_{t=0} = u_0(z), \tag{17.43}$$

where A is a bounded linear operator on B and the data f and u_0 are valued in $\mathfrak{A}_\phi(B)$ for a suitable choice of ϕ (and f depends continuously on t in $[-T, T]$).

Can we modify the choice of the functions ϕ so as to obtain results about the equation

$$\frac{\partial u}{\partial t} = A\frac{\partial^2 u}{\partial z^2} + f, \tag{17.44}$$

analogous to those described in Exercise 17.5?

17.8. Consider a first-order system with *constant* coefficients:

$$L = I\frac{\partial}{\partial t} - \sum_{j=1}^{n} A_j \frac{\partial}{\partial z^j} - A_0; \tag{17.45}$$

the coefficients I, A_j $(0 \leqq j \leqq n)$ are $m \times m$ complex matrices (I is the identity). Show that we can apply Theorem 17.1 choosing, as domain $\mathcal{O}_0$, an arbitrary open ball $B_R(z_0) = \{z \in \mathbf{C}^n;\ |z - z_0| < R\}$ $(R > 0)$. Obtain an upper bound for the radius δ_0 of the interval (or disk) in the t-line (or t-plane) in which the solution of the following problem exists:

$$L\mathbf{u} = \mathbf{f}(t, z), \qquad \mathbf{u}\Big|_{t=0} = \mathbf{u}_0(z), \tag{17.46}$$

where $\mathbf{f}$ is an entire function in $\mathbf{C}^{n+1}$ and $\mathbf{u}_0$ an entire function in $\mathbf{C}^n$, both valued in $\mathbf{C}^m$. Show that δ_0 can be chosen independently of $z_0 \in \mathbf{C}^n$. Derive from this fact that the problem (17.46) has a unique solution $\mathbf{u}(t, z)$ which is an entire analytic function in $\mathbf{C}^{n+1}$, valued in $\mathbf{C}^m$.

17.9. Let $u_0(z)$ be an entire function in the complex plane (valued in $\mathbf{C}$). Show that if $u = u(t, z)$ is an entire function in $\mathbf{C}^2$ satisfying

$$\frac{\partial u}{\partial t} = z^2\frac{\partial u}{\partial z} \quad \text{in } \mathbf{C}^2, \qquad u\Big|_{t=0} = u_0(z), \tag{17.47}$$

we must have $u(t, z) = u_0(z/(1 - zt))$. Derive from this that the Cauchy problem (17.47) does *not* have, in general, an entire function as its solution (cf. the conclusion in Exercise 17.8).

17.10. Let H^0, H^1 be two complex Hilbert spaces, $H^1 \subsetneq H^0$, with continuous injection. Let A be a bounded linear operator $H^1 \to H^0$ such that, for some constant $c_0 > 0$,

$$c_0\|u\|_0^2 \leqq (Au, u)_0, \qquad \forall u \in H^1. \tag{17.48}$$

[The inner product in H^0 is denoted by $(\ ,\ )_0$, the norm by $\|\ \|_0$.] Prove that, whatever the number $t \geqq 0$, $(I + tA/m)^{-m}$ defines a bounded operator on H^0, having a limit as $m \to +\infty$ in the sense of the strong convergence of bounded linear operators on H^0. Let us denote by e^{-tA} this limit. Show that it has the semigroup properties

$$e^{-sA}e^{-tA} = e^{-(s+t)A}, \qquad e^{-tA}\Big|_{t=0} = I. \tag{17.49}$$

For $s \geqq 0$, let E_s be the image of $H^0 = E_0$ under the map e^{-sA} equipped with the norm

(17.50) $$\|e^{-sA}h\|_s = \|h\|_0 .$$

Prove that E_s is a Hilbert space for this norm, and that E_s is dense in H^0 for all $s \geqq 0$.

For $s < 0$ define E_s as the *completion* of H^0 for the norm $\|e^{sA}h\|_0$. Prove the following statements:

(17.51) *Given any two real numbers $s > s'$, E_s can be identified with a dense linear subspace of $E_{s'}$; the natural injection $E_s \subsetneq E_{s'}$ is continuous and has norm $\leqq 1$.*

(17.52) *If $s > s'$, A defines a bounded linear operator $E_s \to E_{s'}$ with norm $\leqq e^{-1}/(s - s')$.*

Exploit these facts in order to solve the Cauchy problem

(17.53) $$\frac{du}{dt} - \lambda(t)Au = f, \qquad u\Big|_{t=0} = u_0 ,$$

where λ is an arbitrary complex-valued continuous function on an interval $|t| < T$, f a continuous function in that same interval, valued in E_{s_0}, and u_0 is an element of E_{s_0} (s_0 is an arbitrary real number).

Prove also that, whatever the real number s, E_s can be "canonically" identified with the dual of E_{-s}. What is the canonical antilinear isometry of E_s onto its dual, E_{-s}?

18

Reduction of Higher Order Systems to First-Order Systems

We have proved the Cauchy–Kovalevska theorem for first-order systems of linear PDEs and this may seem to lack sufficient generality. We are going to show, however, that this is not so—precisely, that every determined system of linear PDEs of arbitrary order, in which the time variable plays a privileged role, can be transformed into a system of the kind (17.3), in such a way that any significant statement concerning the Cauchy problem for the former is equivalent to the analogous statement about the latter (for instance, as to the existence or the uniqueness of the solution). However, let us remark that it is quite easy to modify the proof of the abstract Cauchy–Kovalevska theorem (Theorem 17.2) so as to have an analogous result for abstract differential equations in t whose order is $\geqq 1$. The modification of the proof that is required is not difficult and we advise the student to try to do it.

We deal with a system of linear PDEs of order $m > 1$, of the following kind:

$$D_t^m u = \sum_{\substack{\alpha_0 + |\alpha| \leqq m \\ \alpha_0 < m}} c_{\alpha_0, \alpha}(x, t) D_t^{\alpha_0} D_x^{\alpha} u + f(x, t), \tag{18.1}$$

with Cauchy conditions

$$D_t^k u \Big|_{t=0} = v^k(x), \qquad k = 0, \dots, m-1. \tag{18.2}$$

It is not true that every system of linear PDEs can be put into the form (18.1): The restriction $|\alpha_0| < m$ in the summation on the right-hand side assigns a privileged role to the t variable. We return to this question later, in the case of a single (i.e., scalar) equation. Let us point out for the moment that the order of the system with respect to t is equal to m, no less, and that the coefficient of D_j^m in the system is the $N \times N$ identity matrix. The other coefficients $c_{\alpha_0, \alpha}(x, t)$ are $N \times N$ matrices; N is some positive integer. Their entries are functions of (x, t) in some open subset Ω of $\mathbf{R}^{n+1}$ (or $\mathbf{C}^{n+1}$, or $\mathbf{C}^n \times \mathbf{R}^1$, if we wish to consider complex variables.) For simplicity we shall always assume that they are C^∞ functions in Ω. We also assume that $f(x, t)$ is a C^∞ function of (x, t) in Ω with values in $\mathbf{C}^N$ and that the $v^k(x)$ are C^∞ functions of x in the x projection of Ω, also valued in $\mathbf{C}^N$.

We are going to perform a change of "unknown" in (18.1)–(18.2). We set

$$(18.3) \qquad u_0 = u, \qquad u_j = D_{x^j} u \quad (1 \leqq j \leqq n), \qquad u_{n+1} = D_t u,$$

and call U the vector $(u_0, u_1, \ldots, u_n, u_{n+1})$ which is valued in $\mathbf{C}^{N'}$ with $N' = N(n+2)$. We make the substitutions (18.3) in (18.1). We obtain an equation of the form

$$(18.4) \qquad D_t^{m-1} u_{n+1} = \sum_{j=0}^{n+1} \sum_{\substack{\alpha_0 + |\alpha| \leqq m-1 \\ \alpha_0 \leqq m-2}} c'^{j}_{\alpha_0, \alpha}(x, t) D_t^{\alpha_0} D_x^{\alpha} u_j + f.$$

There are several ways of achieving this, as one can easily see. We choose any one of these ways. Observe, on the other hand, that by differentiating $m-1$ times the first $n+1$ equations of (18.3) we obtain

$$(18.5) \qquad D_t^{m-1} u_0 = D_t^{m-2} u_{n+1},$$

$$(18.6) \qquad D_t^{m-1} u_j = D_t^{m-2} D_{x^j} u_{n+1}, \qquad j = 1, \ldots, n.$$

We may combine Eqs. (18.4), (18.5), and (18.6) and write

$$(18.7) \qquad D_t^{m-1} U = \sum_{\substack{\alpha_0 + |\alpha| \leqq m-1 \\ \alpha_0 \leqq m-2}} C_{\alpha_0, \alpha}(x, t) D_t^{\alpha_0} D_x^{\alpha} U + F$$

where the $C_{\alpha_0, \alpha}$ are $N' \times N'$ matrices and F is the N'-vector whose components are all zero except the last N ones, which are equal to f. The restriction $\alpha_0 < m-2$ (which is all important) was made possible by the fact that the same restriction held in the right-hand sides of (18.4), (18.5), and (18.6).

Next we must transform the Cauchy conditions (18.2). We have

$$(18.8) \qquad D_t^k u_0 \Big|_{t=0} = v^k(x), \qquad k = 0, \ldots, m-2.$$

$$(18.9) \qquad D_t^k u_j \Big|_{t=0} = D_{x^j} v^k(x), \qquad k = 0, \ldots, m-2, \quad j = 1, \ldots, n,$$

$$(18.10) \qquad D_t^k u_{n+1} \Big|_{t=0} = v^{k+1}(x), \qquad k = 0, \ldots, m-2.$$

These equations can be summarized as

$$(18.11) \qquad D_t^k U \Big|_{t=0} = V^k(x), \qquad k = 0, \ldots, m-2.$$

We have transformed the Cauchy problem (18.1)–(18.2), which is of order m, into the Cauchy problem (18.7)–(18.11), which is of order $m - 1$. We may pursue this reduction step by step until we obtain a Cauchy problem of order 1 (beyond which we cannot pursue it further).

We raise now the question of the equivalence of statements about uniqueness and/or existence of solutions to the original problem (18.1)–(18.2) and analogous statements about the transformed problem (18.7)–(18.11). The latter must be reinterpreted, however, in full generality; that is, the data F, V^k must be allowed to take all values in $\mathbf{C}^{N'}$ and not only the special kind originating in (18.1)–(18.2).

(1) Suppose first that, given any choice of F and V^k $(0 \leqq k \leqq m - 2)$, the Cauchy problem (18.7)–(18.11) possesses *at least* one solution (we do not want and need not be specific about the regularity of this solution). Then this is in particular true when

(18.12)
$$F = (0, \dots, 0, f),\ V^k = (v^k, D_{x^1}v^k, \dots, D_{x^n}v^k, v^{k+1}) \quad (0 \leqq k \leqq m - 2).$$

We obtain a solution to (18.1)–(18.2) by taking u equal to the first N components of any solution U of (18.7)–(18.11) in the case (18.12) (what we have denoted u_0 above).

(2) Suppose next that the problem (18.7)–(18.11) admits two different solutions for some choice of the data [not necessarily submitted to (18.12)]. By subtraction we see that the homogeneous problem (18.7)–(18.11) (where all data, F and the V^k, are identically zero) admits a nontrivial solution U. By the argument used in (1), the first N components u_0 of U constitute a solution of the homogeneous problem (18.1)–(18.2). Suppose that u_0 were to vanish identically. By the equations (18.5) we could conclude that $D_t^{m-2}u_{n+1} \equiv 0$, hence that u_{n+1} is a polynomial with respect to t of degree $\leqq m - 3$. But from (18.10), where $v^{k+1} \equiv 0$, we derive that all the derivatives of order $\leqq m - 2$ of u_{n+1}, with respect to t, vanish at $t = 0$. Thus $u_{n+1} \equiv 0$. Also, by (18.6) we see that $D_t^{m-1}u_j \equiv 0$ $(1 \leqq j \leqq n)$, hence u_j is a polynomial with respect to t of degree $\leqq m - 2$; the derivatives of order $\leqq m - 2$ of this polynomial at $t = 0$ are all zero, by (18.9), therefore the polynomial itself is identically zero. We have reached the conclusion that all the components of U are zero, contrary to our assumption. Thus $u = u_0$ cannot vanish identically and the homogeneous problem (18.1)–(18.2) has a nontrivial solution.

(3) If the homogeneous problem (18.1)–(18.2) (i.e., when f and the v^k vanish identically) has a nontrivial solution u, it is clear that the homogeneous problem (18.7)–(18.11) also does.

(4) We assume at last that the problem (18.1)–(18.2) has a solution for any choice of the data f, v^k and prove that the same is true of the problem (18.7)–(18.11) [without the restriction (18.12)].

Observe first that we may assume the Cauchy data V^k to be all identically zero. Indeed, it suffices to substitute

$$U(x, t) - \sum_{k=0}^{m-2} \frac{t^k}{k!} V^k(x)$$

for $U(x, t)$. Of course, this requires a redefinition of F. Once this is done, we perform a further transformation on the "unknown" U, in the following manner. Let us write $F = (F_0, F_1, \ldots, F_n, F_n, F_{n+1})$ where each F_j $(0 \leqq j \leqq n + 1)$ is valued in $\mathbf{C}^N$. For $j \leqq n$, we call w_j the (unique) solution of the problem

$$D_t^{m-1} w_j = F_j, \qquad D_t^k w_j \Big|_{t=0} = 0, \qquad k = 0, \ldots, m-2.$$

Then we substitute $\tilde{U} = U - (w_0, w_1, \ldots, w_n, 0)$ for U. We leave it to the student to check that

$$D_t^{m-1} \tilde{U} = \sum_{\substack{\alpha_0 + |\alpha| \leqq m-1 \\ \alpha_0 \leqq m-2}} C_{\alpha_0, \alpha}(x, t) D_t^{\alpha_0} D_x^{\alpha} \tilde{U} + \tilde{F}, \tag{18.13}$$

$$D_t^k \tilde{U} \Big|_{t=0} = 0, \qquad k = 0, \ldots, m-2, \tag{18.14}$$

where $\tilde{F} = (0, \ldots, 0, F_{n+1})$. But problem (18.13)–(18.14) is the transform of problem (18.1)–(18.2) where we make the choices $f = F_{n+1}$, $v^k = 0$ for all k. We know that it has a solution. We may retrace our steps and reconstruct the solution U of the problem (18.7)–(18.11).

The sought equivalence has been completely proved.

Remark 18.1. If we follow through our changes of unknowns from u to U and vice versa, we notice that certain basic features have been preserved: For instance, if the original system (18.1) has constant (resp. polynomial, analytic, C^∞, etc.) coefficients, the same is true of the "final" system (18.7) and if u is C^∞ (resp. analytic, a polynomial, etc.) with respect to (x, t), the same is true of the "end product" U; and this is also true if we go from U to u.

Exercises

18.1. Apply the reduction procedure, from second-order equations to first-order systems, to the Laplace equation in n variables, and to the wave equation in n space variables ($n \geqq 1$).

18.2 Let the coefficients $c_{\alpha_0,\alpha}$ in the system (18.1) be *constant*. We may then define the *determinant* of its *symbol*, as the determinant of the matrix

$$I\tau^m - \sum_{\substack{\alpha_0+|\alpha|\leqq m\\ \alpha_0>m}} c_{\alpha_0,\alpha}\tau^{\alpha_0}\xi^\alpha. \tag{18.15}$$

Prove that the determinant of the symbol of the transformed system (18.7) is equal to the determinant of (18.15) multiplied by a power of τ.

18.3 Let $P(\xi, \tau)$ be a polynomial with complex coefficients, of degree m, in $n+1$ variables, $\xi = (\xi_1, \ldots, \xi_n)$ and τ, of the form

$$P(\xi, \tau) = \tau^m - \sum_{\substack{\alpha_0+|\alpha|\leqq m\\ \alpha_0\leqq m}} c_{\alpha_0,\alpha}\tau^{\alpha_0}\xi^\alpha \qquad (c_{\alpha_0,\alpha} \in \mathbf{C}). \tag{18.16}$$

By using the result in Exercise 18.2, prove the following:

In order that the operator $P(D_x, D_t)$ be transformable, by the procedure described in Sect. 18, into a hyperbolic first-order system (Definition 15.2), it is necessary and sufficient that it satisfy the property

(18.17) *there is a positive constant C such that, for every $\xi \in \mathbf{R}_n$ and every $\tau \in \mathbf{C}$, if $P(\xi, \tau) = 0$, then necessarily $|\mathrm{Im}\,\tau| \leqq C$.*

(In connection with this, we mention the important

Definition 18.1. The operator $P(D_x, D_t)$ is called hyperbolic with respect to t if (18.17) holds [assuming that $P(\xi, \tau)$ has the form (18.16); cf. Definition 15.2].)

18.4 Prove that the polynomial $P(\xi, \tau)$ of Exercise 18.3 has property (18.17) if its *principal symbol* (Definition 19.1) has the following property:

(18.18) $\forall \xi \in \mathbf{R}_n$, $\xi \neq 0$, *the roots of $P_m(\xi, \cdot)$ are real and distinct.*

[If (18.18) holds, $P(D_x, D_t)$ is called *strongly*, or *strictly*, *hyperbolic*; cf. Definition 15.3.]

19

Characteristics. Invariant Form of the Cauchy–Kovalevska Theorem

As we have said, not every system of linear PDEs, not even every single equation, can be put into the form (18.1). We would like to understand what it means that it can. Obviously this is a feature of the *principal part* of the equation, that is, of its terms of order m. Let us write

$$P(x, t, D_x, D_t) = D_t^m - \sum_{\substack{\alpha_0+|\alpha|\leq m \\ \alpha_0<m}} c_{\alpha_0,\alpha}(x, t) D_t^{\alpha_0} D_x^{\alpha},$$

$$P_m(x, t, D_x, D_t) = D_t^m - \sum_{\substack{\alpha_0+|\alpha|= m \\ \alpha_0<m}} c_{\alpha_0,\alpha}(x, t) D_t^{\alpha_0} D_x^{\alpha}.$$

We shall restrict ourselves to the scalar case, i.e., where the coefficients $c_{\alpha_0,\alpha}$ are complex-valued. The general case of systems is more difficult to analyze. We observe that, given any (x, t) in the region where the equation is studied, we have

$$P_m(x, t, 0, 1) = 1.$$

More generally, suppose we were dealing with a differential operator

$$P(x, t, D_x, D_t) = \sum_{\alpha_0+|\alpha|\leq m} a_{\alpha_0,\alpha}(x, t) D_t^{\alpha_0} D_x^{\alpha}$$

with *principal part*

$$P_m(x, t, D_x, D_t) = \sum_{\alpha_0+|\alpha|= m} a_{\alpha_0,\alpha}(x, t) D_t^{\alpha_0} D_x^{\alpha}.$$

Assume that the coefficients are smooth complex-valued functions in an open set $\Omega \subset \mathbf{R}^{n+1}$. We may "reduce" the equation

$$P(x, t, D_x, D_t)u = g \tag{19.1}$$

to the form (18.1) if we know that

$$P_m(x, t, 0, 1) \neq 0, \qquad (x, t) \in \Omega. \tag{19.2}$$

Indeed, (19.2) means that the coefficient $a_{m,0}$ does not vanish at any point of Ω and it suffices to take, in (18.1),

$$c_{\alpha_0,\alpha} = -a_{\alpha_0,\alpha}/a_{m,0}, \qquad f = g/a_{m,0}.$$

One expresses the condition (19.2) by saying that *the covector* (0, 1) *or, equivalently, the t direction, is not characteristic for* $P(x, t, D_x, D_t)$ *at any point of* Ω. Let us introduce this notion of characteristic covectors in a somewhat more "invariant" setting. Assume that we deal with an open subset $\mathcal{O}$ of the Euclidean space $\mathbf{R}^N$ where the variable is denoted by y; we denote by η the variable in the dual space $\mathbf{R}_N$. We consider a linear partial differential operator

$$P(y, D_y) = \sum_{|\alpha| \le m} a_\alpha(y) D_y^\alpha, \qquad D_y = -\sqrt{-1}\,\frac{\partial}{\partial y},$$

whose coefficients are complex C^∞ functions in $\mathcal{O}$ (the requirement that they be C^∞ is obviously too much; at the present stage they could as well be merely continuous). Set

$$P_m(y, \eta) = \sum_{|\alpha| = m} a_\alpha(y)\eta^\alpha,$$

where η is the variable in $\mathbf{R}_N$. Notice that $P_m(y, \eta)$ is a homogeneous polynomial of degree m with respect to η whose coefficients are C^∞ functions of y in $\mathcal{O}$.

Definition 19.1. The function $P_m(y, \eta)$ *on* $\mathcal{O} \times \mathbf{R}_N$ *is called the principal symbol of the differential operator* $P(y, D_y)$.

Of course, the differential operator $P_m(y, D_y)$ corresponding to $P_m(y, \eta)$ is the principal part of $P(y, D_y)$.

Fixing y in $\mathcal{O}$, we consider the set of zeros of $P_m(y, \eta)$ as a function of η:

$$C_P(y) = \{\eta \in \mathbf{R}_N;\, P_m(y, \eta) = 0\}.$$

Since $P_m(y, \eta)$ is homogeneous (of degree m) with respect to η, for each y the set $C_P(y)$ is a cone with vertex at the origin.

Definition 19.2. The set $C_P(y)$ *is called the characteristic cone of* $P(y, D_y)$ *at the point* $y \in \mathcal{O}$. *Every covector* $\eta \in C_P(y)$ *different from zero is said to be characteristic with respect to* $P(y, D_y)$ *at the point* y.

In the case of the differential operator $P(x, t, D_x, D_t)$ considered above we see that for no $(x, t) \in \Omega$ did the covector (0, 1) belong to the characteristic cone at (x, t).

We consider now a C^1 *hypersurface* $S \subset \mathcal{O}$. By this we mean a subset S of $\mathcal{O}$ defined as follows: Every point y_0 of $\mathcal{O}$ has an open neighborhood U_0 such that there is a C^1 function $\varphi(y)$ in U_0 with the following properties: grad φ does not vanish anywhere in U_0; $S \cap U_0$ is exactly the set of points $y \in U_0$ such that $\varphi(y) = 0$. We may consider the *normal direction* to S at any point $y \in S \cap U_0$: It is the straight line in $\mathbf{R}_N$, passing through the origin, spanned by (grad $\varphi)(y)$. Any nonzero convector η belonging to this line will be called *a covector normal to S at the point y.*

Definition 19.3. The hypersurface $S \in \mathcal{O}$ is said to be characteristic at the point $y \in S$ with respect to $P(y, D_y)$ if any normal covector η to S at the point y is characteristic with respect to $P(y, D_y)$ at that point, i.e., belongs to $C_P(y)$.

Of course, if one covector η, normal to S at y, belongs to $C_P(y)$, the same is true of all the others. A hypersurface S is said to be *characteristic* if it is characteristic at every one of its points (this terminology is somewhat confusing but it is traditionally accepted). In our example above, the pieces of hyperplanes $\{(x, t) \in \Omega;\ t = \text{const}\}$ are nowhere characteristic with respect to $P(x, t, D_x, D_t)$. Furthermore, if a hypersurface $S \subset \mathcal{O}$ is noncharacteristic at some point y_0 [with respect to $P(y, D_y)$], we may choose new coordinates $x = (x^1, \ldots, x^n)$, t (where now $n = N - 1$) in an open neighborhood U_0 of y_0 such that $S \cap U_0$ is exactly the set of points $(x, 0)$. Furthermore, in the new coordinates, if $P(x, t, D_x, D_t)$ is the expression of the operator $P(y, D_y)$ and if U_0 is sufficiently small, we will have $P_m(x, t, 0, 1) \neq 0$ for every $(x, t) \in U_0$. The equation $P(y, D_y)u = g$ will be transformable into (18.1).

Example 19.1. The principal symbol of the Cauchy–Riemann operator $\frac{1}{2}(\partial/\partial x + i\, \partial/\partial y)$ on $\mathbf{R}^2$ is $\frac{1}{2}(\xi - i\eta)$. The one of the Laplace operator $\Delta_x = (\partial/\partial x^1) + \cdots + (\partial/\partial x^n)^2$ on $\mathbf{R}^n$ is $-|\xi|^2$. For none of these operators are there any (nonzero) characteristic covectors.

In relation to Example 19.1 we introduce the following important definition:

Definition 19.4. The differential operator $P(y, D_y)$ in $\mathcal{O}$ is said to be elliptic at the point $y_0 \in \mathcal{O}$ if $C_P(y_0) = \{0\}$, i.e., if there are no characteristic covectors at y_0. The operator $P(y, D_y)$ is said to be elliptic in a subset $\mathcal{O}_1$ of $\mathcal{O}$ if it is elliptic at every point of $\mathcal{O}_1$.

The set of points where a given operator is elliptic is obviously *open*. The Cauchy–Riemann operator on $\mathbf{R}^2$ and the Laplace's on $\mathbf{R}^n$ are everywhere elliptic.

Example 19.2. The principal symbol of the heat operator $\partial/\partial t - \Delta_x$ on $\mathbf{R}^{n+1}$ is $|\xi|^2$. Its characteristic cone (at any point) is the one-dimensional linear subspace of the (ξ, τ) space $\mathbf{R}_{n+1}$ defined by $\xi = 0$. The same holds for the Schrödinger operator.

Example 19.3. The (principal) symbol of the wave operator $\partial^2/\partial t^2 - \Delta_x$ on $\mathbf{R}^{n+1}$ is $|\xi|^2 - \tau^2$. Its characteristic cone (at any point) is the surface of the light-cone $|\xi|^2 = \tau^2$. Note, however, that if, instead of the form of the wave operator above, we consider the form

$$\frac{1}{V^2}\frac{\partial^2}{\partial t^2} - \Delta_x,$$

with $V > 0$ but not necessarily $V = 1$, the characteristic cone is defined by the equation

$$V^2|\xi|^2 = \tau^2. \tag{19.3}$$

Now, unless $V = 1$, this is *not* the equation of the surface of the *light-cone*. It is much more preferable to consider the light-cone as a subset of $\mathbf{R}^{n+1}$, not of the dual space $\mathbf{R}_{n+1}$ (after all, remember that the light-cone is meant to contain the supports of the remarkable fundamental solutions E_+ and E_- of the wave equation—and these are distributions of $\mathbf{R}^{n+1}$, not of $\mathbf{R}_{n+1}$). Then the equation of the surface of the light-cone is

$$|x|^2 = V^2t^2. \tag{19.4}$$

The relation between (19.3) and (19.4) is not difficult to find: Any normal direction to the characteristic cone [given by (19.3)], where such a direction is defined, that is, at every point of the cone which is not its vertex, can be regarded as a straight line in $\mathbf{R}^{n+1}$ passing through the origin; the union of all these straight lines is the surface of the light-cone [given by (19.4)].

Note that the covector given by $\xi = 0$, $\tau = 1$ does not belong to the characteristic cone of the wave operator. This means that the hyperplanes $t = \text{const}$ are nowhere characteristic.

Example 19.4. Consider, in $\mathbf{R}^2$, the first-order operator

$$L = \partial/\partial t - a(x, t)\,\partial/\partial x$$

where $a(x, t)$ is a real-valued C^∞ function. The characteristic cone of L at the point (x_0, t_0) is the straight line in the plane $\mathbf{R}_2$ of the variables (ξ, τ) defined by $\tau = a(x_0, t_0)\xi$. Consider now the curve $x = x(t)$ in the (x, t) space defined by the equation

$$\frac{dx}{dt} = -a(x, t), \qquad x\Big|_{t=t_0} = x_0. \tag{19.5}$$

The *tangent* direction to this curve at the point (x_0, t_0) is spanned by $(-a(x_0, t_0), 1)$. The *normal* to the curve at (x_0, t_0) is therefore characteristic with respect to L. This is of course true at every point of the curve. We reach the conclusion that *any curve defined by conditions* (19.5) (possibly with different x_0, t_0) *is characteristic with respect to L* (at every one of its points; we view here a C^1 curve in $\mathbf{R}^2$ as a hypersurface in $\mathbf{R}^2$, which it is). Conversely, let γ be any C^1 curve in $\mathbf{R}^2$ everywhere characteristic with respect to L. Let (x_0, t_0) be any one of its points and let $\varphi(x, t) = 0$ be an equation of γ in an open neighborhood U_0 of (x_0, t_0). We know that grad $\varphi = (\varphi_x, \varphi_t)$ does not vanish near (x_0, t_0), say in U_0, and furthermore, that $\varphi_t = a\varphi_x$ in $U_0 \cap \gamma$. In particular this requires that φ_x never vanishes in a neighborhood of $U_0 \cap \gamma$, which we can take to be U_0 itself. The vector field $(-\varphi_t/\varphi_x, 1)$ is tangent to $U_0 \cap \gamma$ at every point, and we have just seen that it is equal to $(-a, 1)$. This means that the conditions (19.5) define γ in U_0. Thus, *every characteristic curve of L is* (locally) *defined by conditions* (19.5). We may summarize this in the following way: *The characteristic curves of L* (viewed as a differential operator) *are exactly the integral curves of L* (viewed as a vector field).

It should be pointed out that this statement only makes sense when the number of independent variables is equal to two: Otherwise "characteristic curves" should be replaced by "characteristic hypersurfaces" which cannot be identified with "integral curves" (a meaningful concept in all dimensions). There is, however, a way to salvage the statement above, but this requires the notion of *bicharacteristic curve* (see the Appendix to the present section).

We conclude this section by stating an invariant version of the Cauchy–Kovalevska theorem. By invariant we mean here "coordinates independent": in other words, there are no more time or space variables, and hyperplanes (such as the one on which $t = 0$) should not have any privileged role. We deal with an open subset $\mathscr{U}$ of $\mathbf{R}^N$ and with a linear partial differential operator $P(y, D_y)$ of order $m > 0$ in $\mathscr{U}$, whose coefficients are analytic functions in $\mathscr{U}$. Furthermore, we are given an *analytic* hypersurface Σ in $\mathscr{U}$ which is nowhere characteristic with respect to $P(y, D_y)$ and whose *exterior* normal at every point is well defined.

THEOREM 19.1. *Let f be an analytic function in $\mathscr{U}$, u_j $(0 \leqq j \leqq m-1)$ m analytic functions in Σ. There is an open neighborhood $\mathscr{V}$ of Σ in $\mathscr{U}$ and a unique analytic function u in $\mathscr{V}$ such that*

$$P(y, D_y)u = f \quad \text{in } \mathscr{V}, \tag{19.6}$$

$$\text{for every} \quad j = 0, \ldots, m-1, \quad (\partial/\partial\nu)^j u = u_j \quad \text{in } \Sigma. \tag{19.7}$$

We have denoted by $\partial/\partial\nu$ the partial differentiation in the exterior normal direction to Σ.

Proof. It suffices to prove that every point y_0 of Σ has an open neighborhood $\mathscr{V}(y_0)$ where there is a unique analytic function $u(y; y_0)$ satisfying $P(y, D_y)u = 0$ in $\mathscr{V}(y_0)$ and $(\partial/\partial\nu)^j u = u_j$ $(0 \leqq j < m)$ in $\mathscr{V}(y_0) \cap \Sigma$. Indeed, if $\mathscr{V}(y_0) \cap \mathscr{V}(y_1) \neq \varnothing$, by the uniqueness we would have $u(y; y_0) = u(y; y_1)$ in this intersection and there is an analytic function u in the union $\mathscr{V}$ of all the $\mathscr{V}(y_0)$, $y_0 \in \Sigma$, whose restriction to each $\mathscr{V}(y_0)$ is equal to $u(y; y_0)$.

If the neighborhood $\mathscr{V}(y_0)$ is sufficiently small, we may perform a change of variables so as to "flatten" the hypersurface $\Sigma \cap \mathscr{V}(y_0)$: In the new coordinates, say $x^1, \ldots, x^n, t$ ($n = N - 1$), this piece of hypersurface becomes a piece of the hyperplane $t = 0$. As pointed out at the beginning of this section, the fact that Σ is nowhere characteristic implies that the expression of $P(y, D_y)$ in the new coordinates is of the form

$$a(x, t)D_t^m + \sum_{j=1}^{m} Q_j(x, t, D_x)D_t^{m-j},$$

where $a(x, t)$ and the coefficients of the $Q_j(x, t, D_x)$ are analytic functions in $\mathscr{V}(y_0)$, $a(x, t)$ vanishes at no point of this set, and the degree of $Q_j(x, t, \xi)$ with respect to ξ is not greater than j. After division by $a(x, t)$ we may transform our equation into a first-order system of the kind (17.3) and the Cauchy conditions into conditions of the kind (17.4). After extending the coefficients and the data to the complex domain we may apply Theorem 17.1 (or rather the version of Theorem 17.1 where the coefficients and the data are assumed to be holomorphic with respect to t, see the end of Sect. 17). We easily reach the desired conclusion, namely that our local Cauchy problem has a unique analytic solution. Q.E.D.

Remark 19.1. One could call Theorem 19.1 the "real-analytic" version of the Cauchy–Kovalevska theorem. There is a "holomorphic" version of the same, which in a sense is to be preferred to the real-analytic one. In the holomorphic version we must assume that $\mathscr{U}$ is an open subset of $\mathbf{C}^n$ and Σ is an analytic submanifold of codimension 1 of $\mathscr{U}$: By this we mean that every point z_0 of $\mathscr{U}$ has an open neighborhood U_0 where there is defined a holomorphic function $\varphi_0(z)$ such that $\Sigma \cap U_0$ is exactly the set of points z of U_0 satisfying the equation $\varphi_0(z) = 0$ and such, moreover, that grad φ does not vanish at any point of U_0. We deal then with a differential operator $P(z, \partial/\partial z)$ of order m, with holomorphic coefficients in $\mathscr{U}$. We assume that Σ is nowhere characteristic with respect to $P(z, \partial/\partial z)$ which means that, in any neighborhood U_0 such as the one above,

$$P_m(z, \operatorname{grad} \varphi_0(z)) \neq 0 \qquad \text{if} \quad z \in U_0,$$

where φ_0 is a holomorphic function defining $U_0 \cap \Sigma$ as before.

We also suppose given a holomorphic vector field $\partial/\partial v$ in an open neighborhood $\mathscr{U}' \subset \mathscr{U}$ of Σ, normal to Σ at each one of its points, that is, a linear combination

$$\frac{\partial}{\partial v} = \sum_{j=1}^{n} \gamma_j(z) \frac{\partial}{\partial z^j}$$

with coefficients γ_j which are holomorphic functions in $\mathscr{U}'$. Furthermore, on $U_0 \cap \Sigma$, where U_0 is the neighborhood of z_0 considered above, we have

$$\gamma_j(z) = \lambda(z) \frac{\partial \varphi_0}{\partial z^j}(z), \qquad j = 1, \ldots, n,$$

where λ is a holomorphic function, nowhere vanishing, in $U_0 \cap \Sigma$ (λ is independent of j). Then we may state

THEOREM 19.2. *There is an open neighborhood $\mathscr{V}$ of Σ in $\mathscr{U}$ such that, to every holomorphic function f in $\mathscr{U}$ and to every set of m holomorphic functions $u_0, \ldots, u_{m-1}$ on Σ, there is a unique holomorphic function u in $\mathscr{V}$ such that*

(19.8) $$P\left(z, \frac{\partial}{\partial z}\right) u = f \quad \text{in } \mathscr{V},$$

(19.9) $$\text{for every} \quad j = 0, \ldots, m-1, \quad (\partial/\partial z)^j u = u_j \quad \text{in } \Sigma.$$

Not only does Theorem 19.2 imply (trivially) Theorem 19.1, but it adds precision to it: Indeed, the defect of Theorem 19.1 is that the neighborhood $\mathscr{V}$ of Σ in its statement depends not only on the hypersurface Σ and on the operator $P(y, D_y)$ but also on the data f, u_j. This is not so in Theorem 19.2! As a consequence, Theorem 19.2 enables us to specify on what properties of the data f, u_j, the choice of $\mathscr{V}$ in Theorem 19.1 does depend. It depends on the "size" of the subsets of the complex space to which these data can be holomorphically continued.

Appendix Bicharacteristics and the Integration of the Characteristic Equation

In this Appendix, we consider a linear partial differential operator $P(y, D_y)$ of order m, with coefficients defined and C^∞ in an open subset of $\mathbf{R}^N$, $\mathcal{O}$. We assume that its principal symbol $P_m(y, \eta)$ is *real*. We then associate with this principal symbol the following system of $2N$ equations:

(19.10) $$\frac{dy}{dt} = \operatorname{grad}_\eta P_m(y, \eta), \qquad \frac{d\eta}{dt} = -\operatorname{grad}_y P_m(y, \eta),$$

to which we adjoin the "initial" conditions:

(19.11) $$y = y_0, \qquad \eta = \eta^0 \qquad \text{when} \quad t = 0.$$

Here y_0 is an arbitrary point of $\mathcal{O}$, η^0 an arbitrary point of $\mathbf{R}_N \backslash \{0\}$. By the fundamental theorem about ordinary differential equations, the problem

(19.10)–(19.11) has a unique solution, at least for small values of $|t|$. The solution

$$y = f(y_0, \eta^0, t), \qquad \eta = g(y_0, \eta^0, t) \tag{19.12}$$

is smooth with respect to all the arguments; for t fixed, and sufficiently close to zero, it defines a C^∞ diffeomorphism, $(y_0, \eta^0) \mapsto (y, \eta)$, in sufficiently small subsets of $\mathcal{O} \times (\mathbf{R}_N \backslash \{0\})$.

The equations (19.10)–(19.11) are known as the *Hamilton–Jacobi equations* for P_m. If the (total) gradient of P_m does not vanish in a full neighborhood of (y_0, η^0), the solution (y, η), given by (19.12), describes a true curve through (y_0, η^0). The tangential differentiation along this curve is given by the *Hamiltonian field* of P_m,

$$H_{P_m} = \sum_{j=1}^{N} \left(\frac{\partial P_m}{\partial \eta_j} \frac{\partial}{\partial y^j} - \frac{\partial P_m}{\partial y^j} \frac{\partial}{\partial \eta_j} \right). \tag{19.13}$$

This is a vector field over $\mathcal{O} \times \mathbf{R}_N$ which, it ought to be pointed out, is well-defined even when the solution of (19.10)–(19.11) is not a curve (i.e., is a single point), or even, for that matter, when P_m is complex (in which case there might not be any solution at all!). The integral curves of H_{P_m} are obviously curves on which the function P_m remains constant, since $H_{P_m} P_m = 0$. Therefore, if any one of them meets the *characteristic set* of $P(y, D_y)$, i.e., the set

$$\{(y, \eta) \in \mathcal{O} \times (\mathbf{R}_N \backslash \{0\});\ P_m(y, \eta) = 0\}, \tag{19.14}$$

it is entirely contained in this set. Such an integral curve of H_{P_m}, contained in (19.14), is called a *bicharacteristic strip* of P_m; its projection in the y space, that is, in $\mathcal{O}$, is called a *bicharacteristic curve* (or simply a bicharacteristic). Of course, it might reduce to a single point. It is a true curve, in $\mathcal{O}$, when $\mathrm{grad}_\eta P_m$ does not vanish on it.

The preceding notions are important in various aspects of PDE theory, in particular in the matter of solving the *characteristic equation*:

$$P_m(y, \mathrm{grad}\, w) = 0, \tag{19.15}$$

in some open neighborhood U of y_0. Usually, one seeks a *real*-valued solution w, defined and C^∞ in U, moreover satisfying a condition

$$\mathrm{grad}\, w \Big|_{y = y_0} = \eta^0. \tag{19.16}$$

In order that (19.15) and (19.16) be compatible we must have

$$P_m(y_0, \eta^0) = 0, \tag{19.17}$$

in other words (y_0, η^0) must belong to the characteristic set (19.14).

It is not always possible to find a (smooth) solution w to (19.15)–(19.16). We are going to show that, under certain favorable circumstances, it can be done. We are going to make the following assumption:

$$(\operatorname{grad}_\eta P_m)(y_0, \eta^0) \neq 0. \tag{19.18}$$

Then let θ be a vector (in $\mathbf{R}^n$) not orthogonal to $\operatorname{grad}_\eta P_m(y_0, \eta^0)$. We may choose the coordinates in $\mathbf{R}^n$ in such a way that θ becomes the unit vector to the y^N-axis. This implies at once that

(19.19) $\dfrac{\partial P_m}{\partial \eta_N}$ *does not vanish in some neighborhood* $\mathcal{N}$ *of* (y_0, η^0).

We may therefore apply the implicit functions theorem and write

$$P_m(y, \eta) = Q(y, \eta)[\eta_N - \lambda(y, \eta_1, \ldots, \eta_{N-1})] \quad \text{in } \mathcal{N}, \tag{19.20}$$

provided that $\mathcal{N}$ is small enough. Both Q and λ are C^∞ functions in $\mathcal{N}$, and Q does not vanish at any point of $\mathcal{N}$. Consequently, the characteristic set (19.14) is defined, in $\mathcal{N}$, by the equation

$$\eta_N = \lambda(y, \eta_1, \ldots, \eta_{N-1}). \tag{19.21}$$

On this set [which is a piece of hypersurface in $\mathcal{O} \times (\mathbf{R}_N \backslash \{0\})$], the Hamiltonian field of P_m is equal to that of the factor $\eta_N - \lambda(y, \eta_1, \ldots, \eta_{N-1})$ multiplied by Q; hence its integral curves, that is, the bicharacteristic strips of P_m, can be defined by the Hamilton–Jacobi equations for that factor; that is

$$\frac{\partial y^j}{\partial t} = -\frac{\partial \lambda}{\partial \eta_j}, \qquad j = 1, \ldots, N-1; \qquad \frac{\partial y^N}{\partial t} = 1; \tag{19.22}$$

$$\frac{\partial \eta_j}{\partial t} = \frac{\partial \lambda}{\partial y_j}, \qquad j = 1, \ldots, N. \tag{19.23}$$

Several observations can be made, on these equations: First of all, the last equation in (19.22) shows that we can take the variable y^N as the parameter along the bicharacteristic strip. Second, all the right-hand sides are independent of η_N, which must be given by (19.21), since the strip lies entirely on the hypersurface defined by (19.21)—as we have observed earlier. It is therefore convenient to change notation and to write t instead of y^N; we shall write x^j instead of y^j and ξ_j instead of η_j if $j < N$. In fact, we shall write $n = N - 1$, to further underline the analogy with a decomposition of the space $\mathbf{R}^N$ into a space hyperplane (where x varies) and a time axis, where the variable is t. The system (19.22)–(19.23) reads, in the new notation,

$$\frac{dx}{dt} = -\lambda_\xi(x, t, \xi), \qquad \frac{d\xi}{dt} = \lambda_x(x, t, \xi), \tag{19.24}$$

where the subscripts mean differentiations. We may further simplify the picture by choosing the coordinates (x, t) in $\mathbf{R}^{n+1}$ so as to have $t = 0$ at the point y_0, and those in the dual $\mathbf{R}_{n+1}$ so as to have $\tau = 0$ at the point η^0 (τ is the "covariable" corresponding to t; in the new coordinates, η^0 will be represented by $(\xi^0, 0)$; we cannot have $\xi^0 = 0$ since we do not have $\eta^0 = 0$). We shall obtain a solution of (19.15)–(19.16) if we solve

$$w_t = \lambda(x, t, w_x), \qquad w_x\Big|_{t=0} = \xi^0. \tag{19.25}$$

THEOREM 19.3. *If the principal symbol $P_m(y, \eta)$ is real and if (19.17) and (19.18) hold, the problem (19.15)–(19.16) has a smooth solution in a sufficiently small neighborhood of y_0.*

The solution cannot be unique (unless $N = 1$, and even then uniqueness can only be understood modulo constants), since every solution of (19.25) (which is the problem we are going to solve) is a solution of (19.15)–(19.16), and the choice of the time direction t could have been made otherwise. Note also that the "initial" condition in (19.25) could have been chosen differently, e.g.,

$$w_x\Big|_{t=0} = \xi^0 + O(|x - x_0|^2).$$

Proof. This proof is based on the following observation. Let

$$x = F(x_1, t, \xi^1), \qquad \xi = G(x_1, t, \xi^1), \tag{19.26}$$

represent the solution of (19.24) equal to (x_1, ξ^1) when $t = 0$. We look at the restriction of $w_x(x, t)$ to a bicharacteristic of P_m, say to the bicharacteristic obtained by projecting into x space the curve defined by (19.26). In (x, t) space (i.e., in $\mathcal{O}$) this bicharacteristic goes through the point $x = x_1$, $t = 0$. Let us write

$$\theta(t) = w_x(F(x_1, t, \xi^1), t).$$

We have

$$\frac{d\theta}{dt} = w_{xx} F_t + w_{xt} = w_{xx} F_t + \lambda_x(x, t, \theta) + \lambda_\xi(x, t, \theta) w_{xx}$$

$$= \lambda_x(x, t, \theta) + w_{xx}\{\lambda_\xi(x, t, \theta) - \lambda_\xi(x, t, G)\}.$$

We observe that, by the second equation (19.24), this is satisfied if $\theta(t) = G(x_1, t, \xi^1)$. But since $\theta(0) = \xi^0$, we must have $\xi^1 = \xi^0$. Thus

$$w_x(F(x_1, t, \xi^0), t) = G(x_1, t, \xi^0). \tag{19.27}$$

The solution of our problem follows easily from (19.27). We are interested in $w_x(x, t)$ and therefore we must express x_1 in terms of $x = F(x_1, t, \xi^0) : x_1 = x_1(x, t, \xi^0)$, and

(19.28) $$w_x(x, t) = G(x_1(x, t, \xi^0), t, \xi^0).$$

If we put this into (19.25), we obtain

(19.29) $$w(x, t) = \langle x, \xi^0 \rangle + \int_0^t \lambda(x, s, G(x_1(x, s, \xi^0), s, \xi^0))\, ds.$$

In order to check that (19.29) indeed defines a solution of (19.25), it suffices to prove that (19.28) holds, and for this it suffices to show that

$$\begin{aligned}
&\frac{d}{dt}\{w_x(x, t) - G(x_1(x, t, y^0), t, \xi^0)\} \\
&\quad = \lambda_x(x, t, G(x_1, t, \xi^0)) + \lambda_\xi(x, t, G(x_1, t, \xi^0))G_{x_1}\frac{\partial x_1}{\partial x} \\
&\qquad - G_t(x_1, t, \xi^0) - G_{x_1}\frac{\partial x_1}{\partial t} \\
&\quad = G_{x_1}\frac{\partial x_1}{\partial x}\left\{\lambda_\xi(x, t, G) - \frac{\partial x}{\partial x_1}\frac{\partial x_1}{\partial t}\right\}
\end{aligned}$$

vanishes, which follows from the fact that $F(x_1(x, t, \xi^0), t, \xi^0) = x$, that is,

$$\begin{aligned}
0 &= F_t(x_1, t, \xi^0) + F_{x_1}(x_1, t, \xi^0)\frac{\partial x_1}{\partial t} \\
&= -\lambda_\xi(F, t, G) + \frac{\partial F}{\partial x_1}\frac{\partial x_1}{\partial t}.
\end{aligned}$$

Q.E.D.

Under hypothesis (19.18), if w is a solution, in an open neighborhood U of y^0, of (19.15)–(19.16), the pair $(y, \operatorname{grad} w(y))$, regarded as a point in $U \times \mathbf{R}_N$, varies on a smooth hypersurface, the characteristic set of P_m, (19.14). This set is fibered by the integral curves of the Hamiltonian field H_{P_m}, (19.13). When y varies along a bicharacteristic of P_m contained in U, $(y, \operatorname{grad} w(y))$ varies along a bicharacteristic *strip* of P_m contained in $U \times \mathbf{R}_N$.

Exercises

19.1. Consider the operator in $\mathbf{R}^N$ ($N \geqq 2$)

(19.30) $$\left(\frac{\partial}{\partial y^N}\right)^2 - V^2\sum_{j=1}^{N-1}\left(\frac{\partial}{\partial y^j}\right)^2, \qquad V \neq 0.$$

Describe its characteristic set [see (19.14)], its bicharacteristic strips, and its bicharacteristics through the origin in $\mathbf{R}^N$ (see the Appendix to this section).

19.2. Consider a first-order operator in $\mathbf{R}^N$,

$$L = \sum_{j=1}^{N} \alpha^j(y) \frac{\partial}{\partial y^j},$$

with real coefficients α^j, defined and C^∞ in an open subset of $\mathbf{R}^N$, $\mathcal{O}$, under the hypothesis that

$$\sum_{j=1}^{N} |\alpha^j(y)| \neq 0, \qquad \forall y \in \mathcal{O}. \tag{19.31}$$

Show that there is identity between the integral curves of the vector field L and the bicharacteristics of L. Redo the proof of Theorem 19.3 in the case where $P_m(y, D_y) = -\sqrt{-1}\, L$ and indicate the analogy with Exercise 16.1.

19.3. Suppose that the principal symbol $P_m(y, \eta)$ has analytic coefficients (in some open subset $\mathcal{O}$ of $\mathbf{R}^N$) and extend it as a holomorphic function of (y, η) in $\tilde{\mathcal{O}} \times \mathbf{C}_N$, where $\tilde{\mathcal{O}}$ is an open neighborhood of $\mathcal{O}$ in $\mathbf{C}_N$. Prove that the characteristic equation (19.15) can be integrated in the complex sense, i.e., that we may find a solution w, holomorphic in a (complex) neighborhood of $y_0 \in \tilde{\mathcal{O}}$, satisfying (19.16) where now η^0 is any nonzero complex N-vector—provided that we make the hypothesis (19.18).

Describe the complex characteristic set of the Laplacian in N variables and its complex bicharacteristics. Solve (19.15)–(19.16) when $P_m(y, \eta) = -|\eta|^2$ and η^0 is an element of $\mathbf{C}_N \backslash \{0\}$ (take $y_0 = 0$).

19.4. Suppose that the principal symbol P_m has constant coefficients and write then $P_m = P_m(\eta)$. Under assumption (19.18) show that Q and λ in (19.20) can be chosen independently of y. In this case, solve problem (19.25) first directly, then by following, step by step, the proof of Theorem 19.3.

19.5. Let S be a C^∞ hypersurface in $\mathbf{R}^N$, passing through the origin, and characteristic at that point with respect to the operator (19.30) (Definition 19.3). Let Γ_0 denote the bicharacteristic strip of this operator passing through $(0, \eta^0)$ where η^0 is the covector normal to S at 0. Prove that the corresponding bicharacteristic (i.e., the projection in $\mathbf{R}^N$ of the strip Γ_0) is necessarily tangent to S at the origin.

Give an example of such a hypersurface S which is tangent at the origin to more than one bicharacteristic of (19.30).

19.6. Consider the following operator in $\mathbf{C}^2$:

$$L = \frac{1}{2}\left(\frac{\partial}{\partial z^1} + \sqrt{-1}\,\frac{\partial}{\partial z^2}\right). \tag{19.32}$$

Describe all the solutions $h = h(z_1, z_2)$ of the homogeneous equation

$$Lh = 0, \tag{19.33}$$

which are holomorphic in the polydisk

$$\{(z_1, z_2) \in \mathbf{C}^2;\ |z_1| < r_1,\ |z_2| < r_2\}, \qquad r_1, r_2 > 0. \tag{19.34}$$

If $h(z_1, z_2)$ is such a solution, prove that $h_0(z_1) = h(z_1, 0)$ can be extended as a holomorphic function in the disk

(19.35) $$\{z_1 \in \mathbf{C}^1;\ |z_1| < r_1 + r_2\}.$$

Derive from this that the Cauchy problem for (19.33) with Cauchy condition

(19.36) $$h = h_0(z_1) \quad \text{when} \quad |z_1| < r_1, \quad z_2 = 0,$$

does *not* have, in general, a holomorphic solution in (19.34).

Describe the maximum set to which one can extend holomorphically all the solutions of (19.33) defined and holomorphic in the polydisk (19.34).

19.7. Let $P(D_y)$ denote the differential operator (19.30) and consider the Cauchy problem

(19.37) $$P(D_y)h = 0,$$

(19.38) $$h(y', 0) = h_0(y'), \qquad \frac{\partial h}{\partial y^N}(y', 0) = h_1(y'), \qquad y' = (y^1, \ldots, y^{N-1}).$$

Let B' be the open unit ball in $\mathbf{R}^{N-1}$. Describe the maximum open subset S of $\mathbf{R}^N$ to which the solution h of (19.37)–(19.38) can be extended as a C^∞ function—*whatever* the Cauchy data h_0, h_1 belonging to $C^\infty(B')$. What is the relation of S to the bicharacteristics of $P(D_y)$?

19.8. Let $P(y, D_y)$ be a linear partial differential operator of order $m \geqq 0$, with coefficients defined and C^∞ in an open subset $\mathcal{O}$ of $\mathbf{R}^N$. Let $P_m(y, \eta)$ be its principal symbol. Let $y \mapsto y' = \phi(y)$ be a C^∞ mapping of $\mathcal{O}$ onto another open subset, $\mathcal{O}'$, of $\mathbf{R}^N$, which is bijective and whose Jacobian determinant does not vanish at any point of $\mathcal{O}$. Define the transform $P^\phi(y', D_{y'})$ of $P(y, D_y)$ by the formula

(19.39) $$P^\phi(y', D_{y'})u(y') = [P(y, D_y)u(\phi(y))]_{y = \overset{-1}{\phi}(y')}, \qquad u \in C^\infty(\mathcal{O}').$$

Show that $P^\phi(y', D_{y'})$ is a linear partial differential operator of order m with coefficients defined and C^∞ in $\mathcal{O}'$, whose principal symbol is related to that of $P(y, D_y)$ by the formula

(19.40) $$P_m^\phi(y', \eta') = P_m(\overset{-1}{\phi}(y'), {}^tJ_\phi(y')\eta'), \qquad y' \in \mathcal{O}', \quad \eta' \in \mathbf{R}_N,$$

where ${}^tJ_\phi(y')$ is the transpose of the Jacobian matrix of ϕ evaluated at the point $\overset{-1}{\phi}(y')$.

19.9. Apply formula (19.40) in the case where $\mathcal{O} = \mathcal{O}' = \mathbf{R}^2$, the diffeomorphism ϕ is represented by the change of coordinates

(19.41) $$y = x, \qquad s = t + \tfrac{1}{2}x^2,$$

and the operator under study is $P = \partial^2/\partial x^2 - \partial/\partial t$. Give the total expression of the transformed operator P^ϕ, and derive from this that the lower order terms are not "invariant" [whereas the principal symbol is, in the sense of transforming according to the law (19.40)].

20

The Abstract Version of the Holmgren Theorem

In this section we study certain uniqueness results which follow from the *dual form* of the abstract Cauchy–Kovalevska theorem (Theorem 17.2) and which are going to be used (in Sect. 21) to derive Holmgren's theorem.

We deal with a scale of Banach spaces $\{E_s\}$ $(0 \leqq s \leqq 1)$ as in Sect. 17; they satisfy condition (17.18). We are given an operator $A(t)$ satisfying (17.19) and (17.20), but now we add the following hypothesis:

(20.1) *If* $0 \leqq s' < s \leqq 1$, E_s *is dense in* $E_{s'}$.

It implies that the *transpose* of the natural injection of E_s into $E_{s'}$ is an *injective* continuous linear map of the dual of $E_{s'}$, $E'_{s'}$, into the one of E_s, E'_s. We shall refer to the latter map as the natural injection of $E'_{s'}$ into E'_s. Since transposition preserves the norm of the operators, this injection has nom $\leqq 1$. As a matter of fact, we set

$$F_s = (E_{1-s})', \qquad 0 \leqq s \leqq 1,$$

and we see that the Banach spaces F_s form a scale analogous to the scale $\{E_s\}$.

Let ${}^tA(t)$ denote the transpose of the operator $A(t)$; it is seen at once that properties (17.19) and (17.20) hold if we substitute F_s for E_s (for each s) and ${}^tA(t)$ for $A(t)$ (for each t). This means that Theorem 17.2 is valid with these substitutions. As we shall use it, let us stress the uniqueness part.

Lemma 20.1. *Suppose that (17.18), (17.19), and (17.20) hold. Suppose also that (20.1) holds.*

Let $v(t)$ be a C^1 function of t, $|t| < T$, valued in E'_s for some $s > 0$ and satisfying, for $|t| < T$, the homogeneous equation

$$\frac{dv}{dt} = {}^tA(t)v. \tag{20.2}$$

If $v(0) = 0$, the function v vanishes identically in the interval $]-T, T[$.

We wish to take a look at distribution solutions of (20.2) (that is, distributions in the variable t: no other variable is involved here; on the subject of distributions valued in a Banach space, we refer the reader to Sect. 39, although only the most elementary information will be needed—and provided—here). For this to make sense we must strengthen our smoothness requirements on the operator ${}^tA(t)$, that is, on $A(t)$. Incidentally we use the following lemma, whose proof we leave to the student:

LEMMA 20.2. *Suppose that the following holds*:

(20.3) *if* $0 \leqq s' < s \leqq 1$, $A(t)$ *is a* C^∞ *function of* t, $|t| < T$, *valued in the Banach space of bounded linear operator* $E_s \to E_{s'}$.

Then the analogous property holds for the transpose ${}^tA(t)$:

(20.4) *if* $0 \leqq s' < s \leqq 1$, ${}^tA(t)$ *is a* C^∞ *function of* t, $|t| < T$, *valued in the Banach space of bounded linear operators* $E'_{s'} \to E'_s$.

As a matter of fact, the same is true if we substitute C^m with $m < +\infty$ for C^∞ or if we substitute analytic (or holomorphic when t is a complex variable).

We suppose in the sequel that (20.4) holds.

Note that E'_0 is the *smallest* of the duals E'_s (it has the largest norm).

If v is a distribution in the interval $]-T, T[$ with values in E'_0, ${}^tA(t)v$ is a distribution in the same interval, with values in E'_s for any $s > 0$. We shall be interested in distributions which can be written in the form

$$v = \partial_t^k g, \qquad \partial_t = \frac{d}{dt}, \tag{20.5}$$

where g is a continuous function of t, $|t| < T$, valued in E'_0, and vanishing for $t < 0$. In this simple case, the multiplicative product ${}^tA(t)v$ is particularly easy to define (although it is very easy also in the general case; see §39.4). Indeed, we have, by the Leibniz formula,

$${}^tA(t)\partial_t^k g = \sum_{l=0}^{k} (-1)^l \partial_t^l [A^{(k-l)}(t) g]. \tag{20.6}$$

Since ${}^tA$ satisfies (20.4), the $g_l = (-1)^l [\partial_t^{k-l}({}^tA)] g$ are continuous functions of t, $|t| < T$, valued in E'_s (for an $s > 0$).

We shall exploit the following regularity result:

LEMMA 20.3. *Suppose that* (17.18), (20.3), *and* (17.20) *hold, and also* (20.1). *Let* v *be the distribution given by* (20.5).

If $v_t - {}^tA(t)v$ *is a* C^∞ *function of* t, $|t| < T$, *valued in* E'_0, *then* v *itself is a* C^∞ *function of* t, $|t| < T$, *valued in* E'_s, *for any* $s > 0$.

Proof. By (20.6) we have

$$f = v_t - {}^tA(t)v = \partial_t^{k+1} g - \sum_{l=0}^{k} \partial_t^l g_l .$$

But we may write

$$g_l = \partial_t^{k-l+1} h_l, \qquad f = \partial_t^{k+1} f_1,$$

with h_l $(0 \leqq l \leqq k)$ and f_1 vanishing for $t < 0$ (this determines them uniquely). Note furthermore that f_1 is a C^∞ function of t, $|t| < T$, valued in E_0' and that, if g is a C^μ function of t, $|t| < T$, valued in every E_s', $s > 0$, $h = \sum_{l=0}^k h_l$ is a $C^{\mu+1}$ such function (recall the definition of g_l). We have

$$\partial_t^{k+1}(g - h - f_1) = 0,$$

which means that $g - h - f_1$ is a polynomial of degree $\leqq k$ (valued in E_s'). But since it must vanish identically for $t < 0$, we have

$$g = h + f_1.$$

We know that g is C^0 (with values in E_s', $s > 0$), hence h is C^1, hence g is C^1, hence h is C^2, and so on. We reach the conclusion that g is a C^∞ function of t, $|t| < T$, valued in E_s' $(s > 0)$, and Lemma 20.3 is proved.

COROLLARY 20.1. *Using the same hypotheses as in Lemma 20.3, let v be given by (20.5). If $v_t = {}^tA(t)v$ in $]-T, T[$, we have $v \equiv 0$ in this same interval.*

Proof. By Lemma 20.3 we know that v is a C^∞ function of t, $|t| < T$, valued in E_s' for any $s > 0$. By Lemma 20.1 we may then conclude that $v \equiv 0$.

Q.E.D.

We describe now the situation to which we are going to apply the preceding considerations.

We introduce a one-parameter family of polydisks $\mathcal{O}_s$ $(0 \leqq s \leqq 1)$ in $\mathbf{C}^n$:

$$\mathcal{O}_s = \{z \in \mathbf{C}^n;\ |z^j| < r_0 + s d_0,\ j = 1, \ldots, n\} \qquad (r_0, d_0 > 0).$$

Notice that

(20.7) If $0 \leqq s' < s \leqq 1$, $\mathcal{O}_{s'} \subset \mathcal{O}_s$ and $\operatorname{dist}(\mathcal{O}_{s'}, \partial\mathcal{O}_s) = d_0(s - s')$.

We are going to choose

(20.8) $$E_s = H(\bar{\mathcal{O}}_s; \mathbf{C}^m) \qquad \text{(see Definition 17.1).}$$

LEMMA 20.4. *With the choice (20.8), condition (20.1) holds.*

Proof. If we reason componentwise, it suffices to prove that if $s' < s$, $H(\bar{\mathcal{O}}_s)$ is dense in $H(\bar{\mathcal{O}}_{s'})$. Let h be an arbitrary element of $H(\bar{\mathcal{O}}_{s'})$ and set $h_\delta(z) = h[(1 - \delta)z]$, $0 < \delta < 1$. As h is uniformly continuous in $\bar{\mathcal{O}}_{s'}$, h_δ converges to h

uniformly on $\bar{\mathcal{O}}_{s'}$ when $\delta \to +0$. But h_δ can be extended to $(1+\delta)\bar{\mathcal{O}}_{s'}$ as a continuous function, holomorphic in the interior of this polydisk. The Taylor expansion of h_δ converges uniformly to h_δ on every compact subset of this interior, in particular on $\bar{\mathcal{O}}_{s'}$. By the diagonal process we obtain a sequence of polynomials in z converging to h in $H(\bar{\mathcal{O}}_{s'})$. Q.E.D.

We shall be interested in distributions v of the kind (20.5) defined by distributions with respect to (x, t) whose support has an x projection contained in a fixed compact subset K of $\mathcal{O}_0 \cap \mathbf{R}^n$. How does this definition work?

By possibly decreasing T, we may restrict ourselves to distributions

$$u = \sum_{\alpha_0 + |\alpha| \leqq k} \partial_t^{\alpha_0} D_x^\alpha f_{\alpha_0, \alpha}(x, t),$$

where the $f_{\alpha_0, \alpha}$ are continuous functions of (x, t) valued in $\mathbf{C}^m$ (k is an integer at least zero). We assume that u vanishes identically when $x \notin K \subset\subset \mathcal{O}_0 \cap \mathbf{R}^n$. We may then choose the functions $f_{\alpha_0, \alpha}$ vanishing for $x \notin K'$, where K' is an arbitrary compact neighborhood of K contained in $\mathcal{O}_0 \cap \mathbf{R}^n$. Let us then set

$$g_{\alpha_0}(x, t) = \sum_{|\alpha| \leqq k - \alpha_0} D_x^\alpha f_{\alpha_0, \alpha}(x, t).$$

This distribution g_{α_0} defines a continuous function t, $|t| < T$, with values in the dual of $H(\bar{\mathcal{O}}_0; \mathbf{C}^m)$, by the formula

$$h \mapsto \sum_{|\alpha| \leqq k - \alpha_0} (-1)^{|\alpha|} \int f_{\alpha_0, \alpha}(x, t)\, D_x^\alpha h(x)\, dx. \tag{20.9}$$

This makes good sense, since the integration can be restricted to the compact subset K' of $\mathcal{O}_0 \cap \mathbf{R}^n$. Let us denote by $G_{\alpha_0}(t)$ the linear functional (20.9). By Cauchy's inequalities we have

$$\sup_{x \in K'} |D_x^\alpha h(x)| \leqq \alpha!\, d^{-|\alpha|} \sup_{z \in \mathcal{O}_0} |h(z)|,$$

where $d = \operatorname{dist}(K', \partial\mathcal{O}_0)$. This shows that (20.9) is continuous on $H(\bar{\mathcal{O}}_0; \mathbf{C}^m)$. Likewise,

$$|\langle G_{\alpha_0}(t) - G_{\alpha_0}(t'), h\rangle| \leqq \left\{ \sum_{|\alpha| \leqq k - \alpha_0} \alpha!\, d^{-|\alpha|} \int |f_{\alpha_0, \alpha}(x, t) - f_{\alpha_0, \alpha}(x, t')|\, dx \right\} |h|_{\mathcal{O}_0},$$

which implies at once that $G_{\alpha_0}(t)$ is a continuous function of t, $|t| < T$, valued in $H(\bar{\mathcal{O}}_0; \mathbf{C}^m)'$. We write then

$$v = \sum_{\alpha_0 \leqq k} \partial_t^{\alpha_0} G_{\alpha_0}(t).$$

Observe that if the G_{α_0} vanish for $t \leqq 0$, we may write

$$G_{\alpha_0} = \partial_t^{k - \alpha_0} G_{\alpha_0}^\natural,$$

where $G^{\natural}_{\alpha_0}$ is also continuous in $]-T, T[$ [with values in $H(\bar{\mathcal{O}}_0; \mathbf{C}^m)'$] and vanishes for $t < 0$. Setting then

$$g = \sum_{\alpha_0=0}^{k} G^{\natural}_{\alpha_0},$$

we obtain the expression (20.5). Thus with the distribution u in the (x, t) variable we have associated the distribution v with respect to t, with values in $H(\bar{\mathcal{O}}_0; \mathbf{C}^m)'$.

As we shall seek a uniqueness theorem, we want to know whether the fact that v vanishes identically implies that u also does. We may evaluate v on test functions of the form $h\varphi(t)$ where h is an arbitrary element of $H(\bar{\mathcal{O}}_0; \mathbf{C}^m)$ and φ an arbitrary C^∞ with compact support in $]-T, T[$ (and complex values). From our definition we have

$$\langle v, h\varphi(t)\rangle = \langle u, h(x)\varphi(t)\rangle,$$

where the brackets express the appropriate dualities. Thus if $v = 0$, u vanishes on all products of the form $h(x)\varphi(t)$ where $\varphi \in C_c^\infty(]-T, T[)$ and h is the restriction to $\mathcal{O}_0 \cap \mathbf{R}^n$ of an arbitrary element of $H(\bar{\mathcal{O}}_0; \mathbf{C}^m)$, in particular when h is an arbitrary *polynomial* on $\mathbf{R}^n$. But linear combinations of products of the form $h(x)\varphi(t)$ are dense in $C_c^\infty(\mathbf{R}^n \times]-T, T[)$ (this is a variant of the Stone–Weierstrass theorem), hence u must vanish identically. Let us summarize by stating that

(20.10) *the correspondence $u \mapsto v$ is injective*

Remark 20.1. We have considered distributions in $]-T, T[$ which vanish for $t < 0$; but the origin could obviously have been replaced by any other point t_0 of the open interval $]-T, T[$. Also we could have derived analogous results about distributions vanishing for $t > 0$.

Consider now a differential operator

$$\sum_{j=1}^{n} A_j(z, t)\frac{\partial}{\partial z^j} + A_0(z, t), \tag{20.11}$$

where the A_j $(0 \leqq j \leqq n)$ are $m \times m$ matrices with entries which are *holomorphic* functions of (z, t) in an open neighborhood, in $\mathbf{C}^{n+1}$, of some closed polydisk containing $\mathcal{O}_0 x]-T, T[$ (we momentarily regard t as a complex variable). It is clear that when $z = x \in \mathbf{R}^n$ and t is real, (20.11) acts on distributions like u considered above. Let us denote by $A(t)$ the action of (2.11) transferred to the distribution in t, v, associated with u. In other words, $A(t)v$ is associated, in the way described earlier, with $\mathfrak{A}(t)u$, where

$$\mathfrak{A}(t) = \sum_{j=1}^{n} A_j(x, t)\frac{\partial}{\partial x^j} + A_0(x, t).$$

If $\varphi \in C_c^\infty(]-T, T[)$ and $h \in H(\bar{\mathcal{O}}_0; \mathbf{C}^m)$ are arbitrary, we have

$$\text{(20.12)} \qquad \langle A(t)v, h\varphi(t)\rangle = \langle \mathfrak{A}(t)u, h\varphi\rangle = \langle u, {}^t\mathfrak{A}(t)(h\varphi)\rangle = \langle u, \varphi A'(t)h\rangle,$$

where

$$\text{(20.13)} \qquad A'(t) = -\sum_{j=1}^{n} {}^tA_j(z, t)\frac{\partial}{\partial z^j} + {}^tA_0(z, t) - \sum_{j=1}^{n} \frac{\partial\, {}^tA_j}{\partial z^j}(z, t)$$

(${}^tA_j$ is the transpose of the matrix A_j). This follows by integration by parts. Now, it is clear (cf. Sect. 17) that $A'(t)$ defines a bounded linear operator $H(\bar{\mathcal{O}}_s; \mathbf{C}^m) \to H(\bar{\mathcal{O}}_{s'}; \mathbf{C}^m)$. What formula (20.12) tells us is that $A(t)$ is equal to the transpose of $A'(t)$.

It is clear that (20.3) and (20.4) are satisfied if we substitute A' for A and we choose E_s according to (20.8). By virtue of (20.7) we have the analog of Proposition 17.5 with A' substituted for A. Then (17.20) holds. In view of Lemma 20.4 all the hypotheses in Lemma 20.3 are fulfilled. We may therefore apply Corollary 20.1. We reach the following conclusion, which as we shall see is one of the versions of Holmgren's theorem:

THEOREM 20.1. *Let u be a distribution in $(\mathcal{O}_0 \cap \mathbf{R}^n) \times]-T, T[$ vanishing when x does not belong to a certain compact subset K of $\mathcal{O}_0 \cap \mathbf{R}^n$.*

Suppose that

$$\text{(20.14)} \qquad \frac{du}{dt} = \mathfrak{A}(t)u$$

and that $u = 0$ for $t < 0$. Then $u = 0$ in $(\mathcal{O}_0 \cap \mathbf{R}^n) \times]-T, T[$.

The reader might object that u is not necessarily a finite sum of derivatives of continuous functions in $(\mathcal{O}_0 \cap \mathbf{R}^n) \times]-T, T[$ as we have supposed in the preceding argument. But u equals such a sum if we replace $]-T, T[$ by any subinterval $]-T', T'[$, $T' < T$. We have seen that $u = 0$ if $|t| < T'$, hence if $|t| < T$ as one sees by taking $T' \to T$.

Exercises

20.1. By modifying (slightly) the proof of the "uniqueness" in Theorem 17.2, prove the following stronger version of part (II) of Theorem 17.2:

THEOREM 20.2. *Let $\{E_s\}$ $(0 \leqq s \leqq 1)$ be a scale of Banach spaces satisfying the condition*

(20.15) *if $0 \leqq s' \leqq s \leqq 1$, $E_s \subsetneq E_{s'}$ and the natural injection of E_s into $E_{s'}$ has norm $\leqq 1$.*

Let then $u(t)$ be any C^1 function of t, $|t| < T$, valued in E_1, satisfying, for every pair of numbers $0 \leqq s' < s \leqq 1$, the inequality

$$\|u'(t)\|_{s'} \leqq \frac{C}{s - s'} \|u(t)\|_s, \qquad \forall t \in]-T, T[, \tag{20.16}$$

where C is a positive constant, independent of s, s', t.

If u vanishes at some point of the interval $]-T, T[$, it vanishes throughout this interval.

20.2. We use the notation and the concepts introduced in Exercise 17.10. Let $u(t)$ be a continuous function valued in H^1, once continuously differentiable when regarded as being valued in H^0, such that

$$\|u'(t)\|_0 \leqq \text{const } \|Au(t)\|_0, \qquad |t| < T. \tag{20.17}$$

Prove that if $u(0) = 0$, then $u(t) = 0$ for all t, $|t| < T$.

20.3. Consider the partial differential operator in $\mathbf{R}^n$,

$$P(t, D_x) = \sum_{|p| \leq m} c_p(t) D_x^p, \tag{20.18}$$

where the coefficients $c_p(t)$ are continuous functions of t, $|t| < T$, with complex values.

Prove that, if u is a C^1 function of $t, |t| < T$, valued in some space $H^s(\mathbf{R}^n)$ ($s \in \mathbf{R}$), satisfying

$$\frac{\partial u}{\partial t} = P(t, D_x)u \quad \text{in } \mathbf{R}^n \times]-T, T[, \tag{20.19}$$

$$u\Big|_{t=0} = 0, \tag{20.20}$$

then u vanishes identically in $\mathbf{R}^n \times]-T, T[$.

21

The Holmgren Theorem

Holmgren's theorem is a stronger version of Theorem 20.1. Let Ω be an open subset of $\mathbf{R}^n$. Consider the first-order system of linear partial differential equations

$$\frac{\partial u}{\partial t} = \sum_{j=1}^{n} A_j(x, t) \frac{\partial u}{\partial x^j} + A_0(x, t)u, \tag{21.1}$$

about which we assume that

(21.2) *the entries of the* $m \times m$ *matrices* $A_j(x, t)$ $(0 \leqq j \leqq n)$ *are analytic functions of* (x, t) *in* $\Omega \times]-T, T[$.

THEOREM 21.1. *Suppose that (21.2) holds. There is an open neighborhood* $\mathscr{U}$ *of* $\Omega \times \{0\}$ *in* $\Omega \times]-T, T[$ *such that every distribution* u *in* $\Omega \times]-T, T[$, *satisfying (21.1) in this set and vanishing when* $-T < t < 0$, *must also vanish in* $\mathscr{U}$.

The main difference between this statement and Theorem 20.1 is that, in Theorem 21.1, u is not required to have a compact support with respect to x. Before proving Theorem 21.1 let us show how it implies the following "classical" version of Holmgren's theorem:

COROLLARY 21.1. *Let* u *be a* C^1 *function satisfying (21.1) in* $\Omega \times]-T, T[$. *If* $u(x, 0) = 0$ *for all* $x \in \Omega$, *then* $u = 0$ *in the neighborhood* $\mathscr{U}$ *of* $\Omega \times \{0\}$.

Proof. Consider the function $\tilde{u}(x, t) = H(t)u(x, t)$, where H is Heaviside's function, equal to one on the positive half-line and to zero on the negative half-line. Since $u(x, 0) = 0$, we have $(\partial/\partial t)\tilde{u} = H(t)(\partial/\partial t)u$, hence $\tilde{u}$ also satisfies (21.1) and vanishes, of course, for $t < 0$. It suffices then to apply Theorem 21.1.

Q.E.D.

Proof of Theorem 21.1. It suffices to prove that $u = 0$ in an open neighborhood, which does not depend on u, of an arbitrary point $(x_0, 0)$, $x_0 \in \Omega$. For simplicity, let us take $x_0 = 0$. We may thus assume that Ω is some neighborhood of the origin in $\mathbf{R}^n$ which we have the right to shrink a finite number of times, as we wish; we shall also decrease T when needed. We make the following change of variables

$$x = y, \qquad t = s - |y|^2,$$

i.e.,

$$y = x, \qquad s = t + |x|^2.$$

We have

$$\frac{\partial}{\partial x^j} = \frac{\partial}{\partial y^j} + 2y^j \frac{\partial}{\partial s} \quad (1 \leqq j \leqq n), \qquad \frac{\partial}{\partial t} = \frac{\partial}{\partial s}.$$

The system (21.1) is transformed into

(21.3) $$M(y, s) \frac{\partial u}{\partial s} = \sum_{j=1}^{n} B_j(y, s) \frac{\partial u}{\partial y^j} + B_0(y, s),$$

where $B_j(y, s) = A_j(y, s - |y|^2)$ $(0 \leqq j \leqq n)$ and

$$M(y, s) = I - 2 \sum_{j=1}^{n} y^j B_j(y, s).$$

If Ω is small enough, and $y \in \Omega$, $M(y, s)$ is invertible; of course, its inverse is also analytic with respect to (y, s). Equation (21.3) is equivalent to

(21.4) $$\frac{\partial u}{\partial s} = \sum_{j=1}^{n} C_j(y, s) \frac{\partial u}{\partial y^j} + C_0(y, s),$$

where $C_j = M^{-1} B_j$. The system is of the same kind as (21.1). The only difference in our assumptions is that the solution u now vanishes when $s < |y|^2$. We could have made this assumption to start with: Reverting to (21.1) and to the coordinates (x, t), we may assume that $u = 0$ when $t < |x|^2$. If T is sufficiently small, the projection into the x space of the region $\{(x, t);\, x \in \Omega, |t| < T, |x|^2 \leqq t\}$, which contains supp u, is contained in a compact subset K of Ω. Finally we contract the whole configuration so that Ω is contained in a polydisk $\mathcal{O}_0$ in $\mathbf{C}^n$, as the polydisk in Theorem 20.1, and so as to be able to assume that the coefficients A_j can be extended as holomorphic functions to an open neighborhood of the polydisk

$$\{(z, t) \in \mathbf{C}^{n+1};\, z \in \overline{\mathcal{O}}_0,\, |t| \leqq T\}.$$

It suffices to apply Theorem 20.1: We conclude that $u = 0$ in $\Omega \times]-T, T[$. Since the choice of Ω and $T > 0$ was made independently of u, this is the desired conclusion. Q.E.D.

We conclude this section by stating an invariant form of Holmgren's theorem. As we have done for the Cauchy–Kovalevska theorem, we do so only in the case of a single equation, that is, in the case of a scalar differential operator of order $m > 0$, $P(y, D_y)$ [$y = (y^1, \ldots, y^N)$ is the variable in an open subset $\mathscr{U}$ of $\mathbf{R}^N$]. The principal symbol of $P(y, D_y)$ will be denoted, as usual, by $P_m(y, \eta)$.

THEOREM 21.2. *Suppose that the coefficients of the differential operator $P(y, D_y)$ are analytic in the open set $\mathscr{U}$. Let Σ be a C^1 hypersurface in $\mathscr{U}$, subdividing $\mathscr{U}$ into two parts, $\mathscr{U}_+$ and $\mathscr{U}_-$, and nowhere characteristic with respect to $P(y, D_y)$.*

There is an open neighborhood of Σ in $\mathscr{U}$, $\mathscr{N}$, such that every distribution u in $\mathscr{U}$, satisfying there $P(y, D_y)u = 0$ and vanishing on one side of Σ (that is, in $\mathscr{U}_+$ or in $\mathscr{U}_-$), also vanishes in $\mathscr{N}$.

That Σ is a C^1 hypersurface in $\mathscr{U}$ means that every point of $\mathscr{U}$ has an open neighborhood $\mathscr{O}$ where there is defined a real-valued C^1 function $\varphi(y)$, with nowhere vanishing gradient (in $\mathscr{O}$), such that $\Sigma \cap \mathscr{O}$ is exactly the set of points in $\mathscr{O}$ where $\varphi = 0$.

That Σ subdivides $\mathscr{U}$ into two parts $\mathscr{U}_+$, $\mathscr{U}_-$, means that $\mathscr{U} = \Sigma \cup \mathscr{U}_+ \cup \mathscr{U}_-$ where $\mathscr{U}_+$ and $\mathscr{U}_-$ are disjoint connected open sets which do not intersect Σ. For the meaning of characteristic, see Definition 19.3.

Proof of Theorem 21.2. It suffices to reason in the neighborhood of an arbitrary point of Σ, which we take to be the origin in $\mathbf{R}^N$. Let φ be a C^1 function in an open neighborhood $\mathscr{V}$ of 0 such that $\Sigma \cap \mathscr{V}$ is the set of $y \in \mathscr{V}$ such that $\varphi(y) = 0$; we assume, of course, that grad φ does not vanish at any point of $\mathscr{V}$. Suppose that we have chosen the coordinates y^j so as to have $\varphi(y) = y^N + o(|y|)$ in $\mathscr{V}$. Setting $y' = (y^1, \ldots, y^{N-1})$, there is a number $\delta > 0$ such that $\mathscr{V}_+ = \{y \in \mathscr{V};\ \varphi(y) > 0\}$ is contained in the set

$$y \in \mathscr{V}, \qquad y^N > -\delta|y'|. \tag{21.5}$$

Note also that, as we shrink $\mathscr{V}$ about the origin, we may decrease $\delta > 0$. Let ε be a number greater than zero and call $\mathscr{W}_\varepsilon$ the open subset of $\mathbf{R}^N$ defined by

$$|y'| < 2\varepsilon\delta, \qquad |y^N| < 4\delta^2\varepsilon^2.$$

For ε sufficiently small, $\mathscr{W}_\varepsilon \subset \mathscr{V}$. Set then

$$\psi(y) = y^N + \frac{1}{\varepsilon}|y'|^2.$$

Let $\mathcal{T}_\varepsilon$ denote the intersection of the paraboloid $\psi = 0$ with $\mathcal{W}_\varepsilon$. We make the following three assertions (assuming that δ is sufficiently small):

(21.6) *$P_m(y, \text{grad } \Psi)$ does not vanish anywhere in $\mathcal{W}_\varepsilon$; in particular $\mathcal{T}_\varepsilon$ is nowhere characteristic;*

(21.7) *if y belongs to the boundary of $\mathcal{T}_\varepsilon$, we have $y^N < -\delta|y'|$;*

(21.8) *$x = y', t = \Psi(y)$ defines a diffeomorphism of $\mathcal{W}_\varepsilon$ onto a neighborhood of the origin in $\mathbf{R}^N$ [as usual, we have set $n = N - 1$, $x = (x^1, \ldots, x^n)$].*

The first assertion, (21.6), follows from the fact that

$$|\text{grad } \Psi - (0, \ldots, 0, 1)| < 2\delta,$$

and that $P_m[0, (0, \ldots, 0, 1)] \neq 0$. Assertion (21.7) follows from the fact that if y belongs to the boundary of $\mathcal{T}_\varepsilon$, we have $|y'| = 2\varepsilon\delta$ and

$$y^N = -4\varepsilon\delta^2 = -2\delta|y'|.$$

Assertion (21.8) is quite obvious. We are in the situation described in Fig. 21.1.

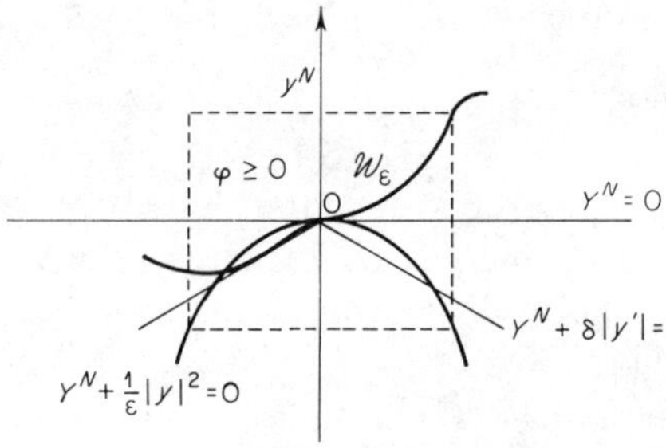

FIG. 21.1

If we make the change of variables contemplated in (21.8), the expression of the operator becomes

$$P(x, t, D_x, D_t) = a(x, t)D_t^m + \sum_{j=1}^{m} Q_j(x, t, D_x)D_t^{m-j},$$

where $a(x, t)$ and the coefficients of the Q_j are analytic functions of (x, t) in a neighborhood of the origin; moreover, $a(x, t)$ does not vanish anywhere near the origin and the $Q_j(x, t, \xi)$ are polynomials with respect to ξ of degree $\leq j$, respectively.

Let u be a distribution in $\mathscr{W}_\varepsilon$, satisfying there the homogeneous equation $P(y, D_y)u = 0$ and vanishing for $\varphi(y) < 0$. In the new coordinates (x, t) this implies the following: We have $P(x, t, D_x, D_t)u = 0$ in a set $\mathscr{W} = \{(x, t) \in \mathbf{R}^{n+1}; |x| < r, -t_0 < t < t_1\}$ (where t_0 and t_1 are greater than zero); there is r', $0 < r' < r$, such that $u = 0$ if $|x| > r'$, $-t_0 < t < t_1$; and finally, there is t_2, $0 < t_2 < t_0$, such that $u = 0$ if $|x| < r$, $-t_0 < t < -t_2$. After transformation of the equation $P(x, t, D_x, D_t)u = 0$ into a first-order system, according to the method described in Sect. 18, we find ourselves in a position to apply Theorem 20.1. We reach the conclusion that $u = 0$ in the whole set $\mathscr{W}$.

Exercises

21.1. Let $P(x, D)$ be an *elliptic* differential operator of order m, with analytic coefficients in a *connected* open subset Ω of $\mathbf{R}^n$ (Definition 19.4). Let S be a piece of C^∞ hypersurface contained in Ω, h an analytic function in Ω, satisfying $P(x, D)h = 0$ there. Prove that if all derivatives of order $<m$ of h vanish on S, then necessarily $h = 0$ identically in Ω.

21.2. Let $P(D)$ be a differential operator with constant coefficients on $\mathbf{R}^n$, μ an arbitrary distribution with compact support in $\mathbf{R}^n$. By applying the Holmgren theorem (Theorem 21.2) prove that the support of μ is necessarily contained in the *convex hull* of (i.e., the smallest convex set containing) the support of $P(D)\mu$.

21.3. By using the Holmgren theorem, prove that the fundamental solution of the wave operator vanishing for $t < 0$ must have its support in the forward light-cone.

21.4. Let $P(\xi)$ be a *homogeneous* polynomial in n variables (with complex coefficients), of degree $m \geqq 1$. Suppose that, for some $\xi^0 \in \mathbf{R}_n$, $\xi^0 \neq 0$, $P(\xi^0) = 0$. Construct a C^∞ function $h(x)$ in $\mathbf{R}^n$ with $h(x) = 0$ if $\langle x, \xi^0 \rangle \leqq 0$ and $h(x) \neq 0$ if $\langle x, \xi^0 \rangle > 0$, such that $P(D_x)h = 0$ in $\mathbf{R}^n$.

21.5. Let $P(\xi)$ be a polynomial of degree $m \geqq 1$ on $\mathbf{R}_n$, $P_m(\xi)$ its homogeneous part of degree m. By using the result in Exercise 12.5 show that if $\xi^0 \neq 0$ satisfies $P_m(\xi^0) = 0$, there is a C^∞ function h in $\mathbf{R}^n$ such that $P(D_x)h = 0$ and, furthermore, such that $h(x) = 0$ if $\langle x, \xi^0 \rangle \leqq 0$, $h(x) \neq 0$ if $\langle x, \xi^0 \rangle > 0$.

21.6. Let $P(D)$ be a differential operator with constant coefficients in $\mathbf{R}^n$. Prove that every distribution u such that $P(D)u = 0$ in $\mathbf{R}^n$, whose support is contained in a half-space whose boundary (which is a hyperplane in $\mathbf{R}^n$) is noncharacteristic with respect to $P(D)$, must vanish identically in $\mathbf{R}^n$.

21.7. Let Γ be an open convex cone with vertex x_0 in $\mathbf{R}^n$, $P(D)$ a differential operator with constant coefficients in $\mathbf{R}^n$ having the following property: Every characteristic hyperplane through the origin intersects Γ. Prove then that every distribution u in Γ which vanishes outside a bounded subset of Γ, and satisfies $P(D)u = 0$ in Γ, vanishes identically in Γ.

21.8. Consider the operator $L = \partial/\partial t + t\,\partial/\partial x$ in $\mathbf{R}^2$. Show that the $x = 0$ is characteristic (with respect to L) at the origin and that, nevertheless, every distribution u satisfying $Lu = 0$ in some open disk Ω centered at the origin, whose support lies in the half-disk $\{(x, t) \in \Omega;\ x \leqq 0\}$, vanishes identically in Ω.

21.9. Let

$$L = \sum_{j=1}^{n} \alpha^j(x) \frac{\partial}{\partial x^j} + \alpha^0(x)$$

be a first-order differential operator with C^∞ coefficients in an open subset Ω of $\mathbf{R}^n$. Assume that the coffiecients of the principal part of L, $\alpha^j, j = 1, \ldots, n$, are real and that they never vanish all together at any point of Ω. Let S be a C^∞ hypersurface contained in Ω, characteristic with respect to L at each one of its points. Let x_0 be an arbitrary point of S. Prove that there is an open neighborhood U of x_0 in Ω, exactly subdivided into two parts by S, and a C^∞ function h in U, satisfying $Lh = 0$ in U, vanishing identically on one side of S but at no point on the other side of S (and of course vanishing on $S \cap U$).

CHAPTER III

Boundary Value Problems

22

The Dirichlet Problem. The Variational Form

We begin by describing, in rather loose terms, the kind of problems with which we shall be concerned in most of this chapter. They apply to a large class of *elliptic* (Definition 19.4) linear partial differential equations, a class whose prototype is the Laplace equation or (even better, in a sense) the *metaharmonic* equation $(-\lambda + \Delta)u = f$, $\lambda > 0$. Most of the time we deal with the latter equation, but we frequently explore the possibility of extending the results obtained to more general second-order (elliptic) operators

$$L = \sum_{j,k=1}^{n} a^{jk}(x) \frac{\partial^2}{\partial x^j \, \partial x^k} + \sum_{j=1}^{n} b^j(x) \frac{\partial}{\partial x^j} + c(x).\dagger \tag{22.1}$$

The coefficients are complex functions, defined in an open subset Ω of $\mathbf{R}^n$, and usually endowed with some degree of regularity (depending on the purpose of the study: in some theories the coefficients are merely assumed to be measurable and bounded). Part of the problem will be to solve the equation

$$Lu = f \quad \text{in } \Omega. \tag{22.2}$$

But we shall not be content with any solution; we shall demand that the solution satisfy certain *boundary conditions*. In view of this, the boundary of Ω will play an important role. We denote it by $\partial\Omega$. In the most important problem of the kind studied in this chapter, the boundary condition reads

$$u = g \quad \text{on } \partial\Omega. \tag{22.3}$$

This is what we shall call the *Dirichlet problem* (it is often called the *inhomogeneous Dirichlet problem* and some people like to reserve the name of Dirichlet problem to the same, but where the right-hand side in (22.2) is

† At the end of this Chapter, in Sects. 36 and 38, we shall examine boundary value problems for higher order elliptic equations.

zero). Admittedly it is a very vague statement of the problem. In particular, the meaning of condition (22.3) is unclear. It bears on the values of the solution u at the boundary $\partial\Omega$, and thus presumes that u can be extended to the closure $\overline{\Omega}$ of Ω. The possibility of extending u, in a natural way, up to the boundary, depends very much on the nature of the boundary, and also, of course, on the regularity properties of the data, f and g, as well as on those of the coefficients of L, a^{jk}, b^j $(1 \leqq j, k \leqq n)$, and c.

The purpose of this section is almost solely introductory. We shall therefore make a number of simplifying assumptions, so as to get some hold on the problem. Later on we are going to show how to relax many of the restrictions imposed here.

First of all, we limit ourselves (for the time being) to the operator

$$L = -\Delta + \lambda, \qquad \lambda \geqq 0, \qquad \Delta = \sum_{j=1}^{n} \left(\frac{\partial}{\partial x^j}\right)^2. \tag{22.4}$$

We are going to assume, throughout the section, that

(22.5) *the open set Ω is bounded.*

It will be seen, in the following sections, that condition (22.5) is to a large extent superfluous when the zero-order coefficient λ is greater than zero, but not when $\lambda = 0$, in the all important case of the Laplace equation. In many applications Ω is *not* a bounded set: for instance, Ω may happen to be the complement of a bounded set (e.g , when studying the distribution of the electrostatic potential in the region surrounding a charged body). In this case, extra conditions at infinity are needed. We do not concern ourselves with such a problem here.

We shall also assume, momentarily, that

(22.6) *the boundary $\partial\Omega$ of Ω is a* (compact) *connected C^1 hypersurface.*

To this we adjoin the requirement that $\partial\Omega$ subdivide the whole space $\mathbf{R}^n$ into two regions (an extension of *Jordan's curve theorem*) and that Ω is the bounded connected component of the complement of $\partial\Omega$. Thus we may talk about the *normal* to $\partial\Omega$ at each one of its points and orient this normal, say, positively toward the exterior.

What kind of right-hand sides f should we consider? Our choice should be strict enough to imply the regularity of the solution needed for a reasonable interpretation of the boundary condition (22.3), yet not so restrictive as to exclude the cases of interest in the applications. It turns out that, from both these viewpoints,

$$f \in L^2(\Omega) \tag{22.7}$$

is an excellent requirement.

We shall be a bit more restrictive in what concerns the boundary value g. We shall suppose that

$$g \in C^1(\partial\Omega). \tag{22.8}$$

The fact that g is defined only on $\partial\Omega$ is somewhat of a nuisance. We shall avail ourselves of properties (22.6) and (22.8) to extend g off $\partial\Omega$. This can be done in a number of ways. One simple procedure is the following. Cover $\partial\Omega$ with a finite number of open subsets $\mathcal{O}_1, \ldots, \mathcal{O}_r$ of $\mathbf{R}^n$ such that, for each j, there is a C^1 diffeomorphism φ_j of $\mathcal{O}_j$ onto the open unit ball B of $\mathbf{R}^n$ mapping $\mathcal{O}_j \cap \Omega$ onto the open half-ball $B^+ = \{x \in B;\ x^n > 0\}$. This is possible by virtue of (22.6). For each $j = 1, \ldots, r$, let us set

$$h_j(y) = g(\overset{-1}{\varphi_j}(y^1, \ldots, y^{n-1}, 0)), \qquad y \in B,$$

and

$$g_j(x) = h_j(\varphi_j(x)), \qquad x \in \mathcal{O}_j.$$

Clearly $g_j \in C^1(\mathcal{O}_j)$, $g_j = g$ on $\mathcal{O}_j \cap \partial\Omega$. Let then $\zeta_1, \ldots, \zeta_r$ be r functions such that $\zeta_j \in C_c^\infty(\mathcal{O}_j)$ for each j and

$$\sum_{j=1}^{r} \zeta_j \equiv 1 \qquad \text{in a neighborhood of} \quad \partial\Omega.$$

At last we set $g^\# = \sum_{j=1}^r \zeta_j g_j$. We see that $g^\# \in C_c^1(\mathbf{R}^n)$ and that $g^\# = g$ on $\partial\Omega$. As a matter of fact we shall only make use of the restriction of $g^\#$ to $\overline{\Omega}$, which we denote by $\tilde{g}$. Condition (22.3) can be rewritten as

$$u - \tilde{g} = 0 \quad \text{on } \partial\Omega. \tag{22.9}$$

This suggests that we change the unknown function (or distribution) and consider

$$U = u - \tilde{g}.$$

Our problem has now become that of finding U such that

$$(-\Delta + \lambda)U = F \quad \text{in } \Omega, \tag{22.10}$$

$$U = 0 \quad \text{on } \partial\Omega. \tag{22.11}$$

We have used the notation

$$F = f + \Delta\tilde{g} - \lambda\tilde{g}.$$

Observe that if we only form the hypothesis (22.8), we cannot assert that $\Delta\tilde{g}$ is a function (in Ω). We merely know that it is a sum of first-order derivatives of continuous functions, as a matter of fact of functions which are continuous on $\overline{\Omega}$:

$$\Delta\tilde{g} = \sum_{j=1}^{n} \frac{\partial}{\partial x^j}\left(\frac{\partial \tilde{g}}{\partial x^j}\right).$$

As a consequence, we cannot expect that the solution U, if it exists, will be very smooth. But we may still hope that U will be smooth enough to enable us to assign a reasonable meaning to (22.11).

Let us go back to (22.10). As the right-hand side F is now a distribution, we shall want to solve (22.10) in the sense of distributions. This means that, given any $\varphi \in C_c^\infty(\Omega)$, we require

$$\langle U, (-\Delta + \lambda)\varphi\rangle = \langle F, \varphi\rangle. \tag{22.12}$$

But of course F is not just any kind of distribution: It is continuous on $C_c^\infty(\Omega)$ for some special seminorms (or norms). Indeed,

$$\langle F, \varphi\rangle = \int (f - \lambda\tilde{g})\varphi \, dx + \int (\text{grad } \tilde{g}) \cdot (\text{grad } \varphi) \, dx.$$

In view of our hypothesis (22.7) on f and the properties of $\tilde{g}$ (namely, that $\tilde{g}$ and grad $\tilde{g}$ are continuous in $\overline{\Omega}$), we obtain at once

$$|\langle F, \varphi\rangle| \leqq \text{const } \|\varphi\|_1, \tag{22.13}$$

where

$$\|\varphi\|_1 = \left\{\int_\Omega |\varphi|^2 \, dx + \int_\Omega |\text{grad } \varphi|^2 \, dx\right\}^{1/2}.$$

This defines a norm on $C_c^\infty(\Omega)$ which is going to play a very important role in the forthcoming [notice that, by virtue of the *Plancherel theorem*, it is equal to the norm (13.20) when $s = 1$]. In order to get some idea of this role, it is helpful to study first the case $\lambda = 1$. We begin by assuming that there exists a solution U to Eq. (22.12) which belongs to $L^2(\Omega)$ and whose first-order derivatives (in the distribution sense) also belong to $L^2(\Omega)$. In this case, we may integrate by parts *once* in the left-hand side of (22.12). After exchange of φ and $\bar{\varphi}$, the result can be written as

$$(U, \varphi)_1 = \langle F, \bar{\varphi}\rangle, \qquad \varphi \in C_c^\infty(\Omega), \tag{22.14}$$

where $(\, , \,)_1$ is the Hermitian product associated with the norm $\| \; \|_1$:

$$(U, V)_1 = \int_\Omega U\overline{V} \, dx + \sum_{j=1}^n \int_\Omega \frac{\partial U}{\partial x^j} \frac{\partial \overline{V}}{\partial x^j} \, dx.$$

The new equation (22.14) strongly suggests that we try to take advantage of the *Riesz representation theorem*. Indeed, the right-hand side defines a continuous antilinear functional on $C_c^\infty(\Omega)$ for the norm $\| \; \|_1$. The latter is obviously a Hilbert space norm. For the moment, let us call $\mathscr{H}$ the completion of $C_c^\infty(\Omega)$ for this norm. By the canonical isomorphism of a Hilbert space onto its antidual, we know that there is a unique element U in $\mathscr{H}$ which satisfies (22.14). In a sense we have solved this equation but the defect of our method is that the solution we have found lies in an abstract space, the completion of $C_c^\infty(\Omega)$ for a certain norm. It would be helpful if we could "concretize" the solution U as a function—in other words, if we could identify the Hilbert space $\mathscr{H}$ with a functions space. We have derived (22.14) from (22.12) by virtue of the hypothesis that U and grad U are square-integrable in Ω. This suggests exactly what must be done, and leads to the following very important definition:

Definition 22.1. The space of functions $w \in L^2(\Omega)$ whose first derivatives (in the sense of distributions) belong to $L^2(\Omega)$ is denoted by $H^1(\Omega)$.

$H^1(\Omega)$ is the first, and probably most important, example of *Sobolev space*. It is closely related to the space H^s encountered in the study of the Cauchy problem (Definition 13.1). It is equipped with the Hermitian product $(\ ,\ )_1$ and the norm $\|\ \|_1$.

PROPOSITION 22.1. *$H^1(\Omega)$ is a Hilbert space.*

Proof. Let $\{w^\nu\}$ be a Cauchy sequence in $H^1(\Omega)$. It is, a fortiori, a Cauchy sequence in $L^2(\Omega)$, and the same is true of $\{\partial w^\nu/\partial x^j\}$ $(j = 1, \ldots, n)$. These Cauchy sequences have limits in $L^2(\Omega)$—w and w_j, respectively. But $(\partial/\partial x^j)$ is a continuous linear operator on $\mathscr{D}'(\Omega)$, the space of distributions in Ω, therefore $(\partial/\partial x^j)w^\nu$ converges to $(\partial/\partial x^j)w$ in $\mathscr{D}'(\Omega)$; but it also converges there to w_j, hence its two limits must be equal. Q.E.D.

Now the completion $\mathscr{H}$ of $C_c^\infty(\Omega)$ for the norm $\|\ \|_1$ can be identified with the closure of $C_c^\infty(\Omega)$ in $H^1(\Omega)$.

Definition 22.2. The closure of $C_c^\infty(\Omega)$ in $H^1(\Omega)$ is denoted by $H_0^1(\Omega)$.

We have seen that Eq. (22.14) has a unique solution $U \in \mathscr{H}$. We may now interpret U as an element of $H_0^1(\Omega)$. Observe that this means that U is the limit, for the norm $\|\ \|_1$, of C^∞ functions vanishing identically in a neighborhood of the boundary of Ω. This fact can be interpreted as meaning that, in a sense, U vanishes on $\partial\Omega$. Of course, that sense is not quite the same as that applied to continuous functions. Yet the two are related, as shown by the next result.

PROPOSITION 22.2. *Assume that (22.5) and (22.6) hold. If $U \in C^0(\bar{\Omega}) \cap H^1(\Omega)$, the following two properties are equivalent:*

(a) $U = 0$ *on* $\partial\Omega$;
(b) $U \in H_0^1(\Omega)$.

Proof. We shall sketch it, leaving the details to the student, as an exercise about Sobolev spaces.

We begin by extending U to the whole space $\mathbf{R}^n$ by setting it equal to zero in the complement of Ω. We use the covering $\{\mathcal{O}_j\}_{1 \leqq j \leqq r}$ of $\partial\Omega$ and the functions ζ_j $(1 \leqq j \leqq r)$ introduced on page 191. By slightly shrinking each set $\mathcal{O}_j$ we may assume that the Jacobian matrix of the diffeomorphism φ_j and that of its inverse, $\overset{-1}{\varphi_j}$, are both bounded uniformly in their domains of definition. This ensures that $w \mapsto w \circ \varphi_j$ is an isomorphism (for the structure of topological vector space) of $H^1(B^+)$ onto $H^1(\mathcal{O}_j \cap \Omega)$, whose inverse is $v \mapsto v \circ \overset{-1}{\varphi_j}$.

Clearly, $U = 0$ on $\partial\Omega$ if and only if each $\zeta_j U$ vanishes on $\mathcal{O}_j \cap \partial\Omega$. On the other hand, if $U \in H_0^1(\Omega)$, each $\zeta_j U$ belongs to $H_0^1(\mathcal{O}_j \cap \Omega)$: Indeed, it suffices to consider a sequence $f_\nu \in C_c^\infty(\Omega)$ converging to U in the sense of the norm $\| \ \|_1$; then the $\zeta_j f_\nu \in C_c^\infty(\mathcal{O}_j \cap \Omega)$ and converge to $\zeta_j U$ in $H^1(\mathcal{O}_j \cap \Omega)$. Conversely, suppose that every $\zeta_j U$ belongs to $H_0^1(\Omega)$. Then so does U. Indeed, it is very easy to check that $U_0 = (1 - \zeta_1 - \cdots - \zeta_r)U$ belongs in any case to $H_0^1(\Omega)$, because $U \in H^1(\Omega)$ and the support of U_0 is a compact subset of Ω. It suffices to make use of the standard mollifiers ρ_ε:

$$\rho_\varepsilon(x) = \varepsilon^{-n}\rho(x/\varepsilon), \qquad \rho \in C_c^\infty(\mathbf{R}^n), \qquad \rho \geqq 0, \qquad \int_{\mathbf{R}^n} \rho(x)\, dx = 1.$$

Then, for $\varepsilon > 0$ sufficiently small, $\rho_\varepsilon * U_0 \in C_c^\infty(\Omega)$. That it converges to U_0 in $H^1(\Omega)$ or, which is exactly the same, in the space $H^1(\mathbf{R}^n)$ (see Definition 13.1), is easy to check.

To summarize, it suffices to prove the equivalence when $\mathcal{O}_j \cap \Omega$ is substituted for Ω and $\zeta_j U$ for U $(j = 1, \ldots, r)$. Thus let us fix j once and for all. Set

$$V(y) = (\zeta_j U)(\overset{-1}{\varphi_j}(y)), \qquad y \in B, \quad \text{the open unit ball in } \mathbf{R}^n.$$

We know that $V \in C^0(\overline{B^+}) \cap H^1(B^+)$ and, furthermore, that the support of V is a compact subset of B. We must prove the equivalence of the following two properties:

(a′) $V = 0$ when $y^n = 0$,
(b′) $V \in H_0^1(B^+)$.

(I) (a′) $\Rightarrow$ (b′). It suffices to prove that, if (a′) holds, V not only belongs to $H^1(B^+)$ but even belongs to $H^1(\mathbf{R}^n)$ (we recall that $V = 0$ in the complement of B). For then the translates $V^\varepsilon(y) = V(y^1, \ldots, y^{n-1}, y^n - \varepsilon)$ also belong to $H^1(\mathbf{R}^n)$. But for $\varepsilon > 0$ sufficiently small, they have their support compact and contained in B^+, hence they belong to $H_0^1(B^+)$ (as one sees by regularization;

cf. the argument about U_0 above). We know that V belongs to $L^2(\mathbf{R}^n)$. We must show that the gradient of V, in the sense of distributions on $\mathbf{R}^n$, also belongs to L^2. The assertion is trivial for each $(\partial/\partial x^j)V$ if $j < n$, and needs only to be proved when $j = n$. Let $\varphi \in C_c^\infty(\mathbf{R}^n)$ be arbitrary. We have

$$\left\langle \frac{\partial V}{\partial y^n}, \varphi \right\rangle = -\int V(y) \frac{\partial \varphi}{\partial y^n}(y)\, dy$$

$$= -\lim_{\varepsilon \to +0} \int_{y^n > \varepsilon} V(y) \frac{\partial \varphi}{\partial y^n}(y)\, dy$$

$$= \lim_{\varepsilon \to +0} \left\{ \int (V\varphi)(y^1, \ldots, y^{n-1}, \varepsilon)\, dy^1 \cdots dy^{n-1} \right\}$$

$$+ \lim_{\varepsilon \to +0} \left\{ \int_{y^n > \varepsilon} \frac{\partial V}{\partial y^n}(y)\varphi(y)\, dy \right\}.$$

In the last member, the first limit is equal to zero since $V(y^1, \ldots, y^{n-1}, \varepsilon)$ converges to zero uniformly with respect to $y^1, \ldots, y^{n-1}$ as $\varepsilon \to +0$. The second limit is equal to

$$\int_{y^n > 0} \frac{\partial V}{\partial y^n} \varphi\, dy,$$

which proves what we wanted.

(II) (b′) ⇒ (a′). Under our various hypotheses on V we have, for almost all $y' = (y^1, \ldots, y^{n-1})$ and every ε, $0 < \varepsilon < 1$,

$$V(y', \varepsilon) = -\int_\varepsilon^1 \frac{\partial V}{\partial y^n}(y)\, dy^n.$$

Let $\{w^\nu\}$ be a sequence in $C_c^\infty(B^+)$ converging to V in $H^1(B^+)$. We have

$$\left\{ \int | V(y', \varepsilon) - w^\nu(y', \varepsilon)|^2\, dy' \right\}^{1/2}$$

$$\leqq \int_\varepsilon^1 \left\{ \int \left| \frac{\partial V}{\partial y^n}(y', y^n) - \frac{\partial w^\nu}{\partial y^n}(y', y^n) \right|^2 dy' \right\}^{1/2} dy^n$$

since the norm of the integral does not exceed the integral of the norm [here, this is the norm in $L^2(\mathbf{R}^{n-1})$]. If we apply the Cauchy–Schwarz inequality to the last member, we see that it is at most equal to $\| V - w^\nu \|_1$. We may choose ν so large that this last quantity will not exceed an arbitrarily preassigned number $\eta > 0$. We keep ν fixed at this value and choose $\varepsilon > 0$ so small that $w^\nu(y', \varepsilon) = 0$ for all y'. We thus obtain

$$\int | V(y', \varepsilon)|^2\, dy' \leqq \eta^2.$$

At the limit we see that $\int |V(y', +0)|^2\, dy' = 0$. Since $V(y', +0)$ is a continuous function of y', this means that it must vanish identically. Q.E.D.

Proposition 22.2 strongly suggests that we replace condition (22.11) by

$$U \in H_0^1(\Omega), \tag{22.15}$$

and try to find such a function U which satisfies (22.10) (until now we have done it only in the case $\lambda = 1$). In the next section, we describe the solution of this problem completely. Let us say, right now, that this solution does not answer all our questions. For instance, the restrictions on the kind of open set Ω we can handle, i.e., conditions (22.5) and (22.6), are quite narrow: squares in the plane are excluded. In our solution of problem (22.10)–(22.15) the second condition, (22.6), will be quite superfluous. But it should be emphasized that, once we drop it, the relationship between our original boundary condition (22.11) and our new condition (22.15) becomes unclear. For example, what is the relation when Ω is a square in the plane, and $U \in C^0(\bar{\Omega})$? How do we obtain a solution of the Dirichlet problem (22.2)–(22.3) when the data f and g are continuous—a solution, moreover, which is acceptable from a classical viewpoint? We discuss these and other related questions in the forthcoming sections.

As a conclusion to this section, let us recall that the reinterpretation of the Dirichlet problem (22.2)–(22.3) in the context of the Sobolev spaces, precisely in the form (22.10)–(22.15), is often referred to as the *variational form* of the Dirichlet problem. The reason for this is not hard to find. Let us introduce the following notation, which is used throughout the sequel:

$$a_\lambda(U, V) = \int_\Omega (\text{grad } U) \cdot (\text{grad } \bar{V})\, dx + \lambda \int_\Omega U\bar{V}\, dx, \quad U, V \in H^1(\Omega). \tag{22.16}$$

Observe that, if $\varphi \in C_c^\infty(\Omega)$,

$$\langle F, \bar{\varphi} \rangle = \int_\Omega f\bar{\varphi}\, dx - a_\lambda(\tilde{g}, \varphi).$$

Thus our weak problem (22.10)–(22.15) can be restated as follows:

Problem 22.1. Find $U \in H_0^1(\Omega)$ such that the antilinear functional on $C_c^\infty(\Omega)$,

$$\varphi \mapsto \left\{ a_\lambda(U + \tilde{g}, \varphi) - \int_\Omega f\bar{\varphi}\, dx \right\}, \tag{22.17}$$

vanishes identically.

Because of the properties of $\tilde{g}$ the form (22.17) is continuous for the norm $\|\ \|_1$, as already pointed out, and can therefore be extended, in a unique manner, to $H_0^1(\Omega)$.

Let us now restrict ourselves, for the sake of simplicity, to real-valued functions and distributions (the complex case can be settled by handling the real and imaginary parts separately). Then the functional (22.17) is the *Fréchet derivative*, at the point U of $H_0^1(\Omega)$, of the (nonlinear) functional

$$Q(V) = \tfrac{1}{2}a_\lambda(V, V) + a_\lambda(\tilde{g}, V) - \int_\Omega fV\,dx. \tag{22.18}$$

Indeed, the value of (22.17) at φ is equal to

$$\frac{d}{dt} Q(U + t\varphi)\bigg|_{t=0}. \tag{22.19}$$

The vanishing of (22.19) is a necessary condition for Q to have an *extremum* at U_0. Let t be any real number, w any element of $H_0^1(\Omega)$. We have

$$Q(U + tw) = \tfrac{1}{2}a_\lambda(U, U) + a_\lambda(\tilde{g}, U) - \int_\Omega fU\,dx$$

$$+ t\left\{a_\lambda(U, w) + a_\lambda(\tilde{g}, w) - \int_\Omega fw\,dx\right\} + \frac{t^2}{2}\,a_\lambda(w, w),$$

i.e., by virtue of the vanishing of (22.17),

$$Q(U + tw) - Q(U) = \frac{t^2}{2}\,a_\lambda(w, w), \tag{22.20}$$

which, I claim, is strictly positive if $tw \neq 0$. The claim is evident when $\lambda > 0$. When $\lambda = 0$, $a_\lambda(w, w) = 0$ is equivalent to grad $w = 0$, i.e., $w = \text{const}$, which, by Proposition 22.2, is incompatible with $w \in H_0^1(\Omega)$ unless, of course, $w \equiv 0$. Thus, if the solution $U \in H_0^1(\Omega)$ to $(\lambda - \Delta)U = F$ exists, it is a *strict minimum* of the functional $Q(V)$. Our weak problem (22.10)–(22.15) is thus seen to be a problem in the *calculus of variations*.

Exercises

22.1. Let Ω be a bounded open interval in the real line $\mathbf{R}^1$. Prove that every function $u \in H_0^1(\Omega)$ is an absolutely continuous function vanishing at the boundary points of Ω.

22.2. Let Ω be as in Exercise 22.1. Prove that $H^1(\Omega)$ is equal to the direct sum of $H_0^1(\Omega)$ and the space of affine functions $h(x) = Ax + B$ (A, B complex constants) restricted to Ω.

22.3. Let Ω be an open disk in the plane, centered at the origin, and with radius $R > 0$. Prove that $H^1(\Omega)$ is the space of functions

$$u(x, y) = \sum_{m=-\infty}^{+\infty} c_m(r)e^{im\theta} \qquad (x + iy = re^{i\theta}) \tag{22.21}$$

whose Fourier coefficients have the following properties:

(22.22) *for each* $m \in \mathbf{Z}$, $c_m(r)$ *is a locally integrable function of r in the open interval* $0 < r < R$, *and so is its distribution derivative* $c_m'(r)$;

$$\sum_{-\infty}^{+\infty} \int_0^R |c_m(r)|^2 r\, dr < +\infty; \tag{22.23}$$

$$\sum_{-\infty}^{+\infty} \int_0^R \left(r|c_m'(r)|^2 + \frac{m^2}{r} |c_m(r)|^2 \right) dr < +\infty. \tag{22.24}$$

22.4. Using the same notation as in Exercise 22.3, the Fourier representation (22.21), and the properties (22.22) to (22.24), show that the space $C^\infty(\overline{\Omega})$ (of C^∞ functions in $\mathbf{R}^2$ restricted to Ω) is dense in $H^1(\Omega)$ and that $H^1(\Omega)$ is the image of $H^1(\mathbf{R}^2)$ under the restriction mapping to Ω.

22.5. Use the same notation as in Exercise 22.3. Let $u \in C_c^\infty(\Omega)$ be given by (22.21). Derive from the conditions (22.23) and (22.24) that we have, if $0 < r_0 \leqq R$,

$$r_0 |c_m(r_0)|^2 \leqq 2 \left| \operatorname{Re} \int_{r_0}^R \overline{c_m(r) c_m'(r)} r\, dr \right| + \frac{1}{r_0} \int_{r_0}^R |c_m(r)|^2 r\, dr, \tag{22.25}$$

and conclude from this that, given any element u of $H_0^1(\Omega)$, the coefficients $c_m(r) (m \in \mathbf{Z})$ are continuous functions of r in the semiclosed interval $]0, R]$, converging to zero as $r \to R$.

22.6. Use the results in Exercises 22.5 and 22.6 to show that, if u is an arbitrary function in $H^1(\Omega)$, its Fourier coefficients $c_m(r)$ are continuous functions in the semiopen interval $]0, R]$, and that $u \in H_0^1(\Omega)$ if and only if $c_m(R) = 0$ for all $m \in \mathbf{Z}$.

22.7. Use the same notation as in Exercise 22.3. Let $u \in C_c^\infty(\Omega)$ be given by (22.21). Derive from (22.24) that, if $0 \leqq r_0 \leqq R$,

$$m |c_m(r_0)|^2 \leqq \int_{r_0}^R \left(r|c_m'(r)|^2 + \frac{m^2}{r} |c_m(r)|^2 \right) dr. \tag{22.26}$$

Conclude from this that, for every $u \in H_0^1(\Omega)$, if $m \neq 0$, $c_m(r)$ is a continuous function in the closed interval $[0, R]$ and (cf. Exercise 22.6) that this is also true when $u \in H^1(\Omega)$.

Suppose $R < 1$ and take $u(x) = [\log(1/r)]^\alpha$. Prove that, if $0 < \alpha < \frac{1}{2}$, u belongs to $H^1(\Omega)$.

22.8. Derive from Exercise 22.7 that, for some constant $C > 0$, for all r, $R/2 \leqq r \leqq R$, and all $u \in H^1(\Omega)$,

$$\sum_{m=-\infty}^{+\infty} (1 + m^2)^{1/2} |c_m(r)|^2 \leqq C \|u\|_{H^1(\Omega)}. \tag{22.27}$$

22.9. Use the same notation as in Exercise 22.3. Let $u \in H^1(\Omega)$ be given by (22.21). Prove that

$$\sum_{m=-\infty}^{+\infty} (1 + m^2)^{1/2} |c_m(R) - c_m(r)|^2 \to 0 \quad \text{as} \quad r \to R \quad (0 < r < R) \tag{22.28}$$

and derive from this that, if we write $u^{\#}(r, \theta) = u(x, y)$,

$$\int_0^{2\pi} |u^{\#}(R, \theta) - u^{\#}(r, \theta)|^2 \, d\theta \to 0 \quad \text{as} \quad r \to R. \tag{22.29}$$

[*Hint*: Prove that (22.28) is true when $u \in C^\infty(\overline{\Omega})$ and then exploit the density of $C^\infty(\overline{\Omega})$ in $H^1(\Omega)$, and the inequality (22.27).]

22.10. Let S^1 denote the unit circumference, θ the variable in S^1, $H^{1/2}(S^1)$ the space of functions

$$g(\theta) = \sum_{m=-\infty}^{+\infty} c_m e^{im\theta} \tag{22.30}$$

such that

$$\|g\|_{1/2}^2 = \sum_{m=-\infty}^{+\infty} (1 + m^2)^{1/2} |c_m|^2 < +\infty. \tag{22.31}$$

Prove that if $H^{1/2}(S^1)$ were contained in the space $C^0(S^1)$ of continuous functions on S^1 (equipped with the maximum norm), then the space of *Radon measures* on S^1, $\mathcal{M}(S^1)$, would be contained in the space of distributions on S^1,

$$v = \sum_{m=-\infty}^{+\infty} v_m e^{im\theta}, \tag{22.32}$$

such that

$$\|v\|_{-1/2}^2 = \sum_{m=-\infty}^{+\infty} (1 + m^2)^{-1/2} |v_m|^2 < +\infty \tag{22.33}$$

[space denoted by $H^{-1/2}(S^1)$]. Prove that the Dirac measure δ at the origin, in S^1, does not satisfy (22.33).

22.11. Let $g \in H^{1/2}(S^1)$ (we are using the notions and notation of Exercise 22.10) be given by (22.30). Set

$$u = c_0 + \sum_{m=1}^{+\infty} (c_m z^m + c_{-m} \bar{z}^m), \qquad z = x + iy, \quad \bar{z} = x - iy. \tag{22.34}$$

By using the criterion established in Exercise 22.3 [and condition (22.31)] prove that $u \in H^1(\Omega)$ and satisfies $\Delta u = 0$ in Ω (taking here $R = 1$, i.e., taking Ω to be the unit disk in the plane). Derive from this, and from the preceding exercises, that $H^{1/2}(S^1)$ is exactly the space of "boundary values" of functions belonging to $H^1(\Omega)$, and that not all of these boundary values are continuous functions on S^1.

22.12. Let $u \in H_0^1(\Omega)$ be *real*-valued (Ω is an arbitrary bounded open subset of $\mathbf{R}^n$), θ an arbitrary number greater than zero. Show that the function $x \mapsto \sup(u(x) - \theta, 0)$ belongs to $H_0^1(\Omega)$. Derive from this fact that the function $|u|$ belongs to $H_0^1(\Omega)$.

23

Solution of the Weak Problem. Coercive Forms. Uniform Ellipticity

Let Ω denote an *arbitrary* open subset of $\mathbf{R}^n$. The natural injection

$$C_c^\infty(\Omega) \hookrightarrow H_0^1(\Omega)$$

can be transposed; since it has dense image, its transpose is a continous linear injection

$$(H_0^1(\Omega))' \hookrightarrow \mathscr{D}'(\Omega).$$

We may therefore identify the dual of $H_0^1(\Omega)$ with a certain space of distributions in Ω. More precisely:

PROPOSITION 23.1. *The dual of $H_0^1(\Omega)$ is the space $H^{-1}(\Omega)$ of the distributions in Ω which are equal to a (finite) sum of derivatives of order ≤ 1 of functions belonging to $L^2(\Omega)$.*

Proof. A distribution T in Ω defines a continuous linear functional on $H_0^1(\Omega)$ if and only if the linear form $\varphi \mapsto \langle T, \varphi \rangle$ is continuous on $C_c^\infty(\Omega)$ for the norm $\| \ \|_1$. If

$$T = f_0 + \sum_{j=1}^{n} \left(\frac{\partial}{\partial x^j} \right) f_j, \qquad f_j \in L^2(\Omega) \quad (0 \leq j \leq n),$$

this is clearly so. Conversely, suppose that the linear functional $\varphi \mapsto \langle T, \varphi \rangle$ can be extended continuously to $H_0^1(\Omega)$. By the Riesz representation theorem we know that there is a unique element V of $H_0^1(\Omega)$ such that

$$(\varphi, V)_1 = \langle T, \varphi \rangle, \qquad \varphi \in C_c^\infty(\Omega).$$

The left-hand side can be rewritten as $\langle(1-\Delta)\bar{V}, \varphi\rangle$ from which we conclude that

$$T = (1-\Delta)\bar{V}. \tag{23.1}$$

It suffices then to observe that $\Delta\bar{V} = \sum_{j=1}^{n} \partial/\partial x^j(\partial\bar{V}/\partial x^j)$ is a sum of first-order derivatives of L^2 functions (the $\partial\bar{V}/\partial x^j$). Q.E.D.

PROPOSITION 23.2. *The canonical antilinear isometry of $H_0^1(\Omega)$ onto its dual, $H^{-1}(\Omega)$, is the mapping $V \mapsto (1-\Delta)\bar{V}$.*

This is essentially the meaning of (23.1). We leave the details of the proof to the student. Proposition 23.2 has the following consequence:

COROLLARY 23.1. *$C_c^\infty(\Omega)$ is dense in $H^{-1}(\Omega)$.*

Indeed, $C_c^\infty(\Omega)$ is dense in $H_0^1(\Omega)$ and is mapped into itself by $(1-\Delta)$. Corollary 23.1 shows that $H^{-1}(\Omega)$ is the closure of $C_c^\infty(\Omega)$ in $H^{-1}(\mathbf{R}^n)$.

Remark 23.1. If we go back, momentarily, to the hypotheses (22.5) through (22.8), we see that the right-hand side

$$F = f - (\lambda - \Delta)\tilde{g}$$

in Eq. (22.10) belongs to $H^{-1}(\Omega)$.

The problem we wish to solve, in the present stage, is the following:

Problem 23.1. Given any $F \in H^{-1}(\Omega)$ find $U \in H_0^1(\Omega)$ such that

$$(-\Delta + \lambda)U = F. \tag{23.2}$$

Equation (23.2) is equivalent to [cf. (22.16)]

$$a_\lambda(U, V) = \langle F, \bar{V}\rangle, \qquad V \in H_0^1(\Omega). \tag{23.3}$$

In order to solve Problem 23.1 we shall combine the Riesz representation theorem with the following results:

PROPOSITION 23.3. *If $\lambda > 0$, $a_\lambda(U, V)$ is a positive-definite Hermitian form on $H^1(\Omega)$. The topological vector space structure defined on $H^1(\Omega)$ by $a_\lambda(U, V)$ is independent of λ. In particular, it is the same as that defined by $a_1(U, V) = (U, V)_1$.*

PROPOSITION 23.4. *Suppose Ω is bounded. Then*

$$a_0(U, V) = \int_\Omega (\text{grad } U) \cdot (\text{grad } \bar{V})\, dx$$

is a positive-definite Hermitian form on $H_0^1(\Omega)$, *defining a topological vector space structure on* $H_0^1(\Omega)$ *equivalent to that defined by* $(U, V)_1$.

Proof of Proposition 23.3. If $\lambda > \mu > 0$,

$$a_\mu(U, U) = a_\lambda(U, U) \leqq (\mu^{-1}\lambda)a_\mu(U, U),$$

from which we arrive at our assertion.

Proof of Proposition 23.4. Obviously $a_0(U, U) \leqq (U, U)_1$. Let then $\varphi \in C_c^\infty(\Omega)$ be arbitrarily. We have

$$\varphi(x) = \int_{-\infty}^{x^n} \frac{\partial \varphi}{\partial x^n}(x', t)\, dt, \qquad x' = (x^1, \ldots, x^{n-1}),$$

whence

$$\left\{\int |\varphi(x)|^2\, dx'\right\}^{1/2} \leqq \int_{-\infty}^{x^n} \left\{\int \left|\frac{\partial \varphi}{\partial x^n}(x', t)\right|^2 dx'\right\}^{1/2} dt,$$

$$\leqq (x^n - a)^{1/2}\left\{\int \left|\frac{\partial \varphi}{\partial x^n}\right|^2 dx\right\}^{1/2}$$

by applying the Cauchy–Schwarz inequality and assuming that the x^n projection of Ω is contained in the (compact) interval $[a, b]$. Finally by integrating with respect to x^n from a to b the squares of the extreme members, we get

$$\left\{\int |\varphi|^2\, dx\right\}^{1/2} \leqq \frac{(b - a)}{\sqrt{2}}\left\{\int \left|\frac{\partial \varphi}{\partial x^n}\right|^2 dx\right\}^{1/2}. \tag{23.4}$$

We derive from this

$$\|\varphi\|_0 \leqq C(\Omega)\|\text{grad}\, \varphi\|_0 \tag{23.5}$$

where $\| \ \|_0$ denotes the norm in $L^2(\mathbf{R}^n)$; (23.5) implies, in turn,

$$\|\varphi\|_1^2 \leqq [1 + C^2(\Omega)]a_0(\varphi, \varphi).$$

This can be clearly extended to all $\varphi \in H_0^1(\Omega)$, and the proposition is proved.

Remark 23.2. In the proof of Proposition 23.4 we have only used the fact that Ω was bounded in the x^n direction—that is, in view of the rotation invariance of the whole situation, in *some* direction.

Remark 23.3. Proposition 23.3 implies that, when λ is strictly positive, $a_\lambda(U, V)$ defines on $H_0^1(\Omega)$ a topological linear space structure identical to that defined by $(U, V)_1$.

From Proposition 23.3 and Remark 23.3, on one hand, and from Proposition 23.4 on the other, we derive that the dual of $H_0^1(\Omega)$ when this space is equipped with the structure defined by $a_\lambda(U, V)$ (assuming that Ω is bounded when $\lambda = 0$) is identical to $H^{-1}(\Omega)$. It is checked at once that the canonical antilinear isometry of the Hilbert space $(H_0^1(\Omega), a_\lambda)$ is nothing else but the mapping $V \mapsto (\lambda - \Delta)\bar{V}$. We thus obtain

THEOREM 23.1. *Let λ be strictly positive. The differential operator*

$$\lambda - \Delta$$

defines an isomorphism of $H_0^1(\Omega)$ onto $H^{-1}(\Omega)$.

If Ω is bounded, this is also true of the Laplace operator Δ.

By virtue of Remark 23.1 we see that we have proved that Problem 22.1, that is, the problem (22.10)–(22.15), has a unique solution.

It is easy to see that Theorem 23.1 extends to a large class of operators L of the form (22.1). Indeed, whether it can be extended depends on the properties of the sesquilinear form which ought to play the role of a_λ in the proof of Theorem 23.1. In order to define the form in question, let us rewrite the operator L in the following manner:

$$L = \sum_{j,k=1}^{n} \frac{\partial}{\partial x^k} a^{jk}(x) \frac{\partial}{\partial x^j} + \sum_{j=1}^{n} \tilde{b}^j(x) \frac{\partial}{\partial x^j} + c(x), \tag{23.6}$$

where we have set

$$\tilde{b}^j(x) = b^j(x) - \sum_{k=1}^{n} \frac{\partial a^{jk}}{\partial x^k}(x).$$

It is then natural to define the form associated with L by the formula

$$\begin{aligned} a(u, v) = &- \sum_{j,k=1}^{n} \int_\Omega a^{jk}(x) \frac{\partial u}{\partial x^j}(x) \frac{\partial \bar{v}}{\partial x^k}(x)\, dx \\ &+ \sum_{j=1}^{n} \int_\Omega \tilde{b}^j(x) \frac{\partial u}{\partial x^j}(x)\bar{v}(x)\, dx + \int_\Omega c(x)u(x)\bar{v}(x)\, dx. \end{aligned} \tag{23.7}$$

Clearly, if we suppose that a^{jk}, $\tilde{b}^j$, c all belong to $L^\infty(\Omega)$, $a(u, v)$ will be a continuous sesquilinear form on $H^1(\Omega) \times H^1(\Omega)$; i.e., it will satisfy, for some $C > 0$,

$$|a(u, v)| \leqq C\|u\|_1\|v\|_1 \qquad \text{for all} \quad u, v \in H^1(\Omega). \tag{23.8}$$

The crucial fact, in the proof of Theorem 23.1, was that, for some $c_1 > 0$, we had

$$a_\lambda(u, u) \geqq c_1\|u\|_1^2 \qquad \text{for all} \quad u \in H_0^1(\Omega).$$

We need the analog for our more general form $a(u, v)$. As a matter of fact it suffices to have

$$(23.9) \qquad |a(u, u)| \geqq c_1 \|u\|_1^2 \qquad \text{for all} \quad u \in H_0^1(\Omega) \quad (c_1 > 0).$$

Because of the far-reaching consequences of the argument, it is worthwhile to make it a bit more abstract. Let us consider a Hilbert space $\mathbf{E}$, with inner product $(\ ,\ )_{\mathbf{E}}$ and norm $\|\ \|_{\mathbf{E}}$, and a continuous sesquilinear form $a(\mathbf{u}, \mathbf{v})$ on $\mathbf{E} \times \mathbf{E}$. The map

$$\mathbf{u} \mapsto a(\mathbf{u}, \cdot)$$

defines a *bounded* linear operator A of $\mathbf{E}$ onto its *antidual* $\overline{\mathbf{E}}'$ (the space of continuous antilinear functionals on $\mathbf{E}$). The importance of condition (23.9) justifies the following

Definition 23.1. The form $a(\mathbf{u}, \mathbf{v})$ *is called coercive if there is a constant* $c_1 > 0$ *such that*

$$(23.10)\dagger \qquad |a(\mathbf{u}, \mathbf{u})| \geqq c_1 \|\mathbf{u}\|_{\mathbf{E}}^2 \qquad \textit{for all} \quad \mathbf{u} \in \mathbf{E}.$$

We have then the following simple result, known as the *Lax–Milgram theorem*:

LEMMA 23.1. *If the form* $a(\mathbf{u}, \mathbf{v})$ *is coercive, the operator* A *is an isomorphism of* $\mathbf{E}$ *onto* $\overline{\mathbf{E}}'$.

We shall denote by $\langle \mathbf{u}', \mathbf{v} \rangle^-$ the antiduality bracket between $\mathbf{E}$ and $\mathbf{E}'$. By definition of A we have $a(\mathbf{u}, \mathbf{v}) = \langle A\mathbf{u}, \mathbf{v} \rangle^-$, $\mathbf{u}, \mathbf{v} \in \mathbf{E}$. Let A^* denote the *antitranspose of* A: $\langle A\mathbf{u}, \mathbf{v} \rangle^- = \langle A^*\mathbf{v}, \mathbf{u} \rangle^-$.

Proof of Lemma 23.1. From (23.10) we derive that, for every $\mathbf{u} \in \mathbf{E}$,

$$(23.11) \qquad \|\mathbf{u}\|_{\mathbf{E}} \leqq c_1^{-1} \|A\mathbf{u}\|_{\mathbf{E}'}, \qquad \|\mathbf{u}\|_{\mathbf{E}} \leqq c_1^{-1} \|A^*\mathbf{u}\|_{\mathbf{E}'},$$

which implies at once the assertion (the first estimate shows that A is injective and has a closed image, the second one that this image is dense). Q.E.D.

Let us return to the special case of the form $a(u, v)$ given by (23.7). We may identify $H^{-1}(\Omega)$, the dual of $H_0^1(\Omega)$, with its antidual, by the formula $\langle u', v \rangle^- = \langle u', \bar{v} \rangle$, $u' \in H^{-1}(\Omega)$, $v \in H_0^1(\Omega)$, $\bar{v}$ being the complex conjugate of v. Observe that, under our hypothesis that a^{jk}, $\tilde{b}^j$, $c \in L^\infty(\Omega)$, L acts continuously from $H_0^1(\Omega)$ into $H^{-1}(\Omega)$: It suffices to look at its expression (23.6)

† Many authors call *coercive* the forms $a(\mathbf{u}, \mathbf{v})$ such that $\operatorname{Re} a(\mathbf{u}, \mathbf{u}) + \lambda_0 \|\mathbf{u}\|_{\mathbf{H}}^2 \geqq c_1 \|\mathbf{u}\|_{\mathbf{E}}^2$, where λ_0 is a nonnegative constant and $\|\mathbf{u}\|_{\mathbf{H}}$ a norm on $\mathbf{E}$ which is weaker than $\|\mathbf{u}\|_{\mathbf{E}}$ [for instance, the L^2 norm on $H^1(\Omega) = \mathbf{E}$]. Definition 23.1 is simpler and perfectly adequate for all our needs.

to see this. If we restrict $a(u, v)$ to $u, v \in C_c^\infty(\Omega)$, we check at once that the operator $A : H_0^1(\Omega) \to H^{-1}(\Omega)$ associated with $a(u, v)$ is identical with L. We have thus obtained

THEOREM 23.2. *Suppose that the coefficients a^{jk}, $\tilde{b}^j$, c of L in the expression (23.6) all belong to $L^\infty(\Omega)$ and that the form (23.7) is coercive on $H_0^1(\Omega)$.*
Then L is an isomorphism of $H_0^1(\Omega)$ onto $H^{-1}(\Omega)$.

Often one refers to the expression (23.6) of L as the *variational form* of L; the justification for this lies, of course, in a remark like that at the end of Sect. 22.

We conclude this section by showing that the coercivity hypothesis on the form $a(u, v)$ has an important consequence concerning the leading terms of the differential operator L. Indeed, let us take

$$u(x) = \varphi(x)e^{ix\cdot\xi}, \qquad \xi \in \mathbf{R}_n, \quad \varphi \in C_c^\infty(\Omega).$$

If we go back to (23.7), we derive for large $|\xi|$,

$$a(u, u) = -\sum_{j,k=1}^{n} \int a^{jk}(x)\xi_j \xi_k |\varphi(x)|^2 \, dx + O(|\xi|).$$

On the other hand,

$$\|u\|_1^2 = |\xi|^2 \int |\varphi|^2 \, dx + O(|\xi|).$$

If we apply (23.9) and make $|\xi|$ go to $+\infty$, we have (setting $\dot{\xi} = \xi/|\xi|$)

$$c_1 \int |\varphi|^2 \, dx \leqq \left| \int \sum_{j,k=1}^{n} a^{jk}(x)\dot{\xi}_j \dot{\xi}_k |\varphi(x)|^2 \, dx \right|. \tag{23.12}$$

It is not difficult to see that this, which has to be true for any $\varphi \in C_c^\infty(\Omega)$, implies

$$c_1 \leqq \left| \sum_{j,k=1}^{n} a^{jk}(x)\dot{\xi}_j \dot{\xi}_k \right| \quad \text{almost everywhere in } \Omega,$$

that is, by homogeneity:

$$c_1 |\xi|^2 \leqq \left| \sum_{j,k=1}^{n} a^{jk}(x)\xi_j \xi_k \right| \qquad \text{for all } \xi \in \mathbf{R}_n \text{ and almost all } x \in \Omega. \tag{23.13}$$

Indeed, set

$$\alpha(x, \xi) = \sum_{j,k=1}^{n} a^{jk}(x)\dot{\xi}_j \dot{\xi}_k .$$

Fixing ξ arbitrarily on the unit sphere of $\mathbf{R}_n$, we can find a sequence of functions in $L^\infty(\Omega)$ which converge to $\alpha(\cdot, \xi)$ in this space, and which have the following property: For each ν, α_ν is continuous in the complement of a closed subset of measure zero of Ω, N_ν. From (23.12) we derive that, to every $\varepsilon > 0$, there is $\nu(\varepsilon)$ such that $\nu > \nu(\varepsilon)$ implies, for all $\varphi \in C_c^\infty(\Omega)$,

$$(c_1 - \varepsilon) \int |\varphi|^2 \, dx \leqq \int |\alpha_\nu| |\varphi|^2 \, dx. \tag{23.14}$$

Fix $\nu > \nu(\varepsilon)$; let x_0 be an arbitrary point of $\Omega \backslash N_\nu$, which is open; take φ with support in $\Omega \backslash N_\nu$ and let $|\varphi|^2$ converge to the Dirac measure $\delta(x_0)$ at x_0. We derive at once from (23.14):

$$(c_1 - \varepsilon) \leqq |\alpha_\nu(x_0)|, \qquad \forall x_0 \in \Omega \backslash N_\nu. \tag{23.15}$$

In the complement of the union N of all the N_ν's, the α_ν converge (for the essential supremum norm) to $\alpha(\cdot, \xi)$. Since $\mathrm{meas}(N) = 0$, we easily obtain what we want.

Condition (23.13) is very important and thus qualifies for a definition:

Definition 23.2. The differential operator L is said to be uniformly elliptic in Ω if (23.13) holds.

Remark 23.4. Condition (23.13) is *not* sufficient to ensure that the sesquilinear form $a(u, v)$ be coercive. Take $L = -\Delta + \lambda$, with $\lambda < 0$ and $|\lambda|$ very large. The corresponding $a_\lambda(u, v)$ will not be coercive on $H_0^1(\Omega)$. However, we have:

Remark 23.5. When the a^{jk} make up a matrix which is *Hermitian*, i.e., when $a^{kj} = \overline{a^{jk}}$, the function we have denoted by $\alpha(x, \xi)$ will always be real (for $\xi \in \mathbf{R}_n$). If it does not vanish in $\mathbf{R}_n \backslash \{0\}$, unless $n = 1$ it must keep the same sign in $\mathbf{R}_n$ (indeed, unless $n = 1$, $\mathbf{R}_n \backslash \{0\}$ is connected). It follows from this that $\alpha(x, \xi)$, as a quadratic form on $\mathbf{R}_n$, will be either positive or negative definite.

PROPOSITION 23.5. *Let $a^{jk} = \overline{a^{kj}}$ $(1 \leqq j, k \leqq n)$ belong to $L^\infty(\Omega)$; let $c \in L^\infty(\Omega)$ be real-valued. Suppose furthermore that*

$$L = \sum_{j,k=1}^{n} \frac{\partial}{\partial x^k} a^{jk}(x) \frac{\partial}{\partial x^j} + c(x) \tag{23.16}$$

is uniformly elliptic in Ω and the matrix $\{a^{jk}\}$ is negative definite a.e. in Ω.

Then if $c(x) > c_0 > 0$ for almost all $x \in \Omega$, the form (23.7) is coercive on $H_0^1(\Omega)$. The same is true when Ω is bounded and $c(x) \geqq 0$.

Proof. From (23.13) we derive, for $u \in H_0^1(\Omega)$ *real-valued*,

$$c_1 |\text{grad } u|^2 \leqq - \sum_{j,k=1}^{n} a^{jk}(x) \frac{\partial u}{\partial x^j} \frac{\partial u}{\partial x^k},$$

whence, by integration over Ω,

$$c_1 \int_\Omega |\text{grad } u|^2 \, dx + c_0 \int_\Omega |u|^2 \, dx \leqq a(u, u). \tag{23.17}$$

But since the matrix $\{a^{jk}\}$ is Hermitian, $a(u + iv, u + iv) = a(u, u) + a(v, v)$ when $u, v \in H_0^1(\Omega)$ are real-valued. Thus (23.17) holds also for complex-valued $u \in H_0^1(\Omega)$. In order to obtain Proposition 23.5, it suffices to apply Propositions 23.3 and 23.4.

Exercises

(Unless otherwise specified, Ω will be a *bounded* open subset of $\mathbf{R}^n$.)

23.1. Prove that there is a number $\lambda_0 < 0$ such that the sesquilinear form a_λ, defined in (22.16), is coercive on $H_0^1(\Omega)$ for all $\lambda > \lambda_0$. Compute the value of λ_0 when $n = 1$ and Ω is an open interval, of finite length l, in the real line.

23.2. Let λ_0 be the number so denoted in Exercise 23.1. Prove that there is a function u belonging to $H_0^1(\Omega)$ such that $\Delta u = \lambda_0 u$ in Ω. Compute λ_0 when Ω is a product of bounded intervals,

$$\Omega = \{x \in \mathbf{R}^n; a_j < x^j < b_j, j = 1, \ldots, n\}.$$

23.3. Let Ω be the open unit disk in $\mathbf{R}^2$, and let λ_0 be the number so denoted in Exercise 23.1, relative to Ω. Prove that $\lambda_0 = -s_0$, where s_0 is the smallest zero of the functions $J_0(\sqrt{s})$ ($s \geqq 0$) where we have used the standard notation for the functions

$$J_0(z) = \frac{1}{2\pi} \int_0^{2\pi} \cos(z \sin\theta) \, d\theta. \tag{23.18}$$

Prove that there is a positive number x such that $J_0(x) = 0$.

23.4. Prove that, if $n \geqq 2$, the restrictions to Ω of the harmonic functions in $\mathbf{R}^n$ form an infinite-dimensional linear subspace of $H^1(\Omega)$. Derive from this that $H_0^1(\Omega)$ has a topological supplementary in $H^1(\Omega)$ whose dimension is infinite. If $n = 1$ and Ω is a finite interval, what is the analogous statement?

23.5. We consider a complex Hilbert space E (E could as well be real: the argument would be essentially the same) and a linear subspace Φ of E on which is defined a norm, $|||\ |||$, larger than the one induced by E. We are given a sesquilinear form on the product $E \times \Phi$, having the property that

(23.19) *for every* $\phi \in \Phi$, $u \mapsto a(u, \phi)$ *is a continuous linear form on* E.

Prove the following (useful) generalization of the Lax–Milgram lemma (Lemma 23.1):

LEMMA 23.2. *Suppose that there is a constant* $c_0 > 0$ *such that*

$$c_0 |||\phi|||^2 \leqq |a(\phi, \phi)| \qquad \text{for all} \quad \phi \in \Phi. \tag{23.20}$$

Then, given any antilinear form on Φ, μ, *continuous for the norm* $||| \; |||$, *there is an element* u *of* E *such that*

$$a(u, \phi) = \mu(\phi) \qquad \text{for all} \quad \phi \in \Phi. \tag{23.21}$$

23.6. Let Ω be a bounded open subset of $\mathbf{R}^n$. Prove that $\lambda - \Delta$ ($\lambda \geqq 0$) maps $H^{-m}(\Omega)$ *onto* $H^{-m-2}(\Omega)$ for every $m = 0, 1, \ldots$. Describe the image and the cokernel of the transpose mapping.

23.7. Let $\mathbf{R}^n_+$ denote the open half-space $x^n > 0$ in $\mathbf{R}^n$ ($n \geqq 2$). Prove that the sesquilinear form $\int_{\mathbf{R}^n} (\text{grad } u) \cdot (\text{grad } \bar{v}) \, dx$ is *not* coercive on $H^1_0(\mathbf{R}^n_+)$.

23.8. Let $\mathcal{O}$ be the open subset of $\mathbf{R}^3$ defined by

$$z > (1 + \tfrac{1}{2}(x^2 + y^2))^{1/2}$$

[the variable in $\mathbf{R}^3$ is denoted by (x, y, z)]. Prove the following facts:

(i) the sesquilinear form $\int_{\mathcal{O}} (\text{grad } u) \cdot (\text{grad } \bar{v}) \, dx$ is *not* coercive on $H^1_0(\mathcal{O})$;

(ii) there are no harmonic functions in $\mathcal{O}$ belonging to $H^1_0(\mathcal{O})$;

(iii) there is a harmonic function in $\mathcal{O}$, continuous in $\overline{\mathcal{O}}$, vanishing on the boundary of $\mathcal{O}$ (we are not including the points at infinity in this boundary).

24

A More Systematic Study of the Sobolev Spaces

If we wish to study the *regularity* of the solutions to the weak Dirichlet problem (as we strengthen the hypotheses on the regularity of the data), it is convenient to introduce the higher order Sobolev spaces, and also the Sobolev spaces constructed on L^p for any p, $1 \leqq p \leqq +\infty$.

Definition 24.1. Let Ω be any open subset of $\mathbf{R}^n$. We denote by $H^{m,p}(\Omega)$ the space of functions $u \in L^p(\Omega)$ such that $(\partial/\partial x)^\alpha u \in L^p(\Omega)$ for all $\alpha = (\alpha_1, \ldots, \alpha_n) \in \mathbf{Z}^n_+$, $|\alpha| = \alpha_1 + \cdots + \alpha_n \leqq m$.

The partial differentiations $(\partial/\partial x)^\alpha = (\partial/\partial x^1)^{\alpha_1} \cdots (\partial/\partial x^n)^{\alpha_n}$ must be understood in the distribution sense. One usually equips $H^{m,p}(\Omega)$ with the norm

$$\|u\|_{m,p} = \left\{ \sum_{|\alpha| \leqq m} \|(\partial/\partial x)^\alpha u\|^p_{L^p(\Omega)} \right\}^{1/p}, \qquad 1 \leqq p < +\infty,$$

$$\|u\|_{m,\infty} = \sup_{|\alpha| \leqq m} \|(\partial/\partial x)^\alpha u\|_{L^\infty(\Omega)}.$$

PROPOSITION 24.0. *$H^{m,p}(\Omega)$ is a Banach space. If $1 < p < +\infty$, it is a reflexive Banach space. If $p = 2$, it is a Hilbert space.*

Proof. The first assertion, i.e., about the completeness of $H^{m,p}(\Omega)$, is proved like Proposition 22.1. We leave it to the student. Let $N(m, n)$ be the number of n-tuples α such that $|\alpha| \leqq m$. There is a natural injection of $H^{m,p}(\Omega)$ onto a closed linear subspace of $(L^p(\Omega))^{N(m,n)}$, namely

$$u \mapsto ((\partial/\partial x)^\alpha u)_{|\alpha| \leqq m}. \tag{24.1}$$

We know that $L^p(\Omega)$ is reflexive for $1 < p < +\infty$, that a finite product of reflexive Banach spaces and a closed linear subspace of a reflexive Banach space are reflexive Banach spaces, from which we obtain the second assertion.

The third assertion is evident. Let us point out that, for $p = 2$, one writes, customarily,

$$H^m(\Omega) \quad \text{instead of} \quad H^{m,2}(\Omega).$$

In $H^m(\Omega)$, the norm $\| \ \|_{m,2} = \| \ \|_m$ is associated with the Hermitian product

$$(u, v)_m = \sum_{|\alpha| \leqq m} \int_\Omega (\partial/\partial x)^\alpha u (\partial/\partial x)^\alpha \bar{v} \, dx,$$

which, when $m = 1$, coincides with the product $(u, v)_1$. Q.E.D.

PROPOSITION 24.1. *If* $1 \leqq p < +\infty$, $C^\infty(\Omega) \cap H^{m,p}(\Omega)$ *is dense in* $H^{m,p}(\Omega)$.

Proof. Let $\{\Omega_\nu\}$ ($\nu = 0, 1, \ldots$) be a sequence of relatively compact open subsets of Ω, whose union is equal to Ω and such that, for each $\nu = 0, 1, \ldots$, $\bar{\Omega}_\nu \subset \Omega_{\nu+1}$. Set $\Omega_1' = \Omega_1$ and for $\nu > 1$, $\Omega_\nu' = \Omega_\nu - \bar{\Omega}_{\nu-2}$. Clearly the Ω_ν' ($\nu = 1, 2, \ldots$) form an open covering of Ω. Note that at most two Ω_ν' ever have a nonempty intersection. Let now $\{\zeta_\nu\}$ ($\nu = 1, 2, \ldots$) be a C^∞ partition of unity subordinate to the covering $\{\Omega_\nu'\}$ (the supports of the ζ_ν are compact). Let $\rho \in C_c^\infty(\mathbf{R}^n)$, $\rho \geqq 0$, $\int_{\mathbf{R}^n} \rho \, dx = +1$ and set, as usual, for $\varepsilon > 0$, $\rho_\varepsilon(x) = \varepsilon^{-n}\rho(x/\varepsilon)$. For each ν we select a number $\varepsilon_\nu > 0$ such that the neighborhood of order ε_ν of supp ζ_ν has its closure contained in Ω_ν'. We write then $v_\nu = \rho_{\varepsilon_\nu} * (\zeta_\nu u)$; of course, $v_\nu \in C_c^\infty(\Omega_\nu')$. We shall now apply the following lemma:

LEMMA 24.1. *Let* $1 \leqq p < +\infty$. *If* $f \in L^p(\mathbf{R}^n)$, $\rho_\varepsilon * f$ *converges to* f *in* $L^p(\mathbf{R}^n)$ *as* $\varepsilon \to +0$.

Proof. By Hölder's inequalities for convolution (TVS, D&*K*, Theorem 26.1) we have

$$\|\rho_\varepsilon * f\|_{L^p} \leqq \|\rho_\varepsilon\|_{L^1} \|f\|_{L^p}.$$

But $\|\rho_\varepsilon\|_{L^1} = \int \rho_\varepsilon(x) \, dx = \int \rho(x) \, dx = 1$. This shows that, as ε varies from 1 to 0 (excluding 0), the convolutions $\rho_\varepsilon *$ form an equicontinuous (i.e., bounded) set of linear operators on L^p. Now, for such a set, the pointwise convergence on L^p and that on any dense subset of L^p are identical. Thus it suffices to prove that $\rho_\varepsilon * f$ converges to f in L^p for each f in a suitable dense subset of L^p. But the assertion is evident if we take this subset to be, e.g., $C_c^\infty(\mathbf{R}^n)$ [or $C_c^0(\mathbf{R}^n)$]. Q.E.D.

In view of Lemma 24.1, given any $\varepsilon > 0$ we may choose ε_ν so small as to have, for every n-tuple, α, $|\alpha| \leqq m$,

$$\|(\partial/\partial x)^\alpha\{(\zeta_\nu u) - v_\nu\}\|_{L^p} = \|(\partial/\partial x)^\alpha(\zeta_\nu u) - \rho_{\varepsilon_\nu} * (\partial/\partial x)^\alpha(\zeta_\nu u)\|_{L^p} \leqq \varepsilon 2^{-\nu},$$

from which it follows that we may choose ε_ν so small as to have

$$\|\zeta_\nu u - v_\nu\|_{m,p} \leqq \varepsilon 2^{-\nu-1}.$$

Let us set $v = \sum_{\nu=1}^{+\infty} v_\nu$. Since, for each $\nu > 1$, the support of v_ν meets at most only those of $v_{\nu-1}$ and $v_{\nu+1}$, v clearly belongs to $C^\infty(\Omega)$. Moreover, if K is any compact subset of Ω, there is $\nu(K)$ such that $\zeta_\nu u$ and v_ν vanish identically in a neighborhood of K for $\nu \geqq \nu(K)$. Therefore

$$\left\{\sum_{|\alpha|\leqq m} \int_K |(\partial/\partial x)^\alpha (u - v)|^p \, dx\right\}^{1/p} \leqq \sum_{\nu<\nu(K)} \left\{\sum_{|\alpha|\leqq m} \int_K |(\partial/\partial x)^\alpha (\zeta_\nu u - v_\nu)|^p \, dx\right\}^{1/p}$$

$$\leqq \sum_{\nu<\nu(K)} \|\zeta_\nu u - v_\nu\|_{m,p} \leqq \varepsilon.$$

This proves everything we wanted: first of all that $u - v$, hence v belongs to $H^{m,p}(\Omega)$—and then, that u is the limit in $H^{m,p}(\Omega)$ of a sequence of functions such as v. Q.E.D.

Remark 24.0. Proposition 24.1 shows that $H^{m,p}(\Omega)$ is a "concrete" realization of the completion of $C^\infty(\Omega) \cap H^{m,p}(\Omega)$ for the norm $\| \ \|_{m,p}$ (when p is finite).

Remark 24.1. If we inspect the proof of Proposition 24.1 and make now the assumption that the support of the element u of $H^{m,p}(\Omega)$ is *compact*, we see that this is true also of the support of the approximating element,

$$v \in C_c^\infty(\Omega).$$

Finally let us point out that Proposition 24.1 is evidently false for $p = +\infty$: It is then false when $m = 0$.

Definition 24.2. The closure of $C_c^\infty(\Omega)$ in $H^{m,p}(\Omega)$ is denoted by $H_0^{m,p}(\Omega)$ [by $H_0^m(\Omega)$ when $p = 2$].

Let us point out the obvious inclusion relations (which give rise to the natural injections—bounded linear, with norm $\leqq 1$):

$$H^{m,p}(\Omega) \subset H^{m',p}(\Omega), \qquad H_0^{m,p}(\Omega) \subset H_0^{m',p}(\Omega) \qquad \text{if} \quad m > m'.$$

When Ω is *bounded*, we have $L^p(\Omega) \subsetneq L^q(\Omega)$ if $p > q$. Hence, in this case, $H^{m,p}(\Omega) \subsetneq H^{m,q}(\Omega)$ and $H_0^{m,p}(\Omega) \subsetneq H_0^{m,q}(\Omega)$ (however, one should be careful and not think that the norms of these injections are $\leqq 1$).

The case $p = +\infty$ is not often used—this is because $L^\infty(\Omega)$ has some unpleasant properties, as is well known: Not only is it not reflexive [this is also true of $L^1(\Omega)$], but it is not *separable* [which $L^1(\Omega)$ is]. Let us also point out that $H_0^{m,\infty}(\Omega)$ consists of C^m functions in $\bar{\Omega}$ which vanish at the boundary of Ω (or at infinity, if $\Omega = \mathbf{R}^n$), together with all their derivatives of order $\leqq m$. In any case, the injection $C_c^\infty(\Omega) \to H_0^{m,p}(\Omega)$ has a dense image, therefore its transpose

$$(H_0^{m,p}(\Omega))' \to \mathscr{D}'(\Omega)$$

is injective and we may identify the dual of $H_0^{m,p}(\Omega)$ with the image of this transpose, which is a space of distributions in Ω. When $p < +\infty$ the description of the latter is very simple:

PROPOSITION 24.2. *If $1 \leqq p < +\infty$, the dual of $H_0^{m,p}(\Omega)$ is the space $H^{-m,p'}(\Omega)$ of distributions in Ω of the form*

$$T = \sum_{|\alpha| \leqq m} (\partial/\partial x)^\alpha f_\alpha, \qquad f_\alpha \in L^{p'}(\Omega) \quad (|\alpha| \leqq m), \tag{24.2}$$

where $p' = p/(p-1)$.

Proof. That every distribution T of the form (24.2) defines a continuous linear functional on $H_0^{m,p}(\Omega)$ is evident. Conversely, consider the mapping (24.1): It is an isomorphism of $H_0^{m,p}(\Omega)$ onto a closed linear subspace of $(L^p(\Omega))^{N(m,n)}$. By virtue of the Hahn–Banach theorem, its transpose is a continuous linear surjection of the dual of the latter onto that of $H_0^{m,p}(\Omega)$. Of course, canonically,

$$\{(L^p(\Omega))^{N(m,n)}\}' = (L^{p'}(\Omega))^{N(m,n)},$$

i.e., a continuous linear functional on $(L^p(\Omega))^{N(m,n)}$ can be identified with a collection $(f_\alpha)_{|\alpha| \leqq m}$ of functions belonging to $L^{p'}(\Omega)$, and the linear functional it induces on $H_0^{m,p}(\Omega)$ can be defined as

$$\langle T, u\rangle = \sum_{|\alpha| \leqq m} \int f_\alpha (\partial/\partial x)^\alpha u \, dx.$$

If we want to know precisely what the distribution T in Ω is equal to, we restrict this formula to $u \in C_c^\infty(\Omega)$. We obtain

$$T = \sum_{|\alpha| \leqq m} (-1)^{|\alpha|} (\partial/\partial x)^\alpha f_\alpha. \qquad \text{Q.E.D.}$$

There is a natural norm on $H^{-m,p'}(\Omega)$. If $p' < +\infty$, it is

$$T \mapsto \|T\|_{-m,p'} = \inf\left(\sum_{|\alpha| \leqq m} \|f_\alpha\|_{L^{p'}(\Omega)}^{p'}\right)^{1/p'},$$

where the *infimum* is taken over all families $(f_\alpha)_{|\alpha| \leqq m}$ in $(L^{p'}(\Omega))^{N(m,n)}$ such that (24.2) is true. When $p' = +\infty$, it is

$$T \mapsto \|T\|_{-m,\infty} = \inf\left(\sup_{|\alpha| \leqq m} \|f_\alpha\|_{L^\infty(\Omega)}\right),$$

where inf has the same meaning as in the case $p' < +\infty$.

If we identify the dual of $H_0^{m,p}(\Omega)$ with $H^{-m,p'}(\Omega)$, we have the right to ask whether the norm $\| \;\|_{-m,p'}$ is equal to the dual norm on $H^{-m,p'}(\Omega)$. That it is so can easily be seen by inspection of the proof of Proposition 24.2. We leave this verification to the student.

We shall need the following result, which is an immediate consequence of Remark 24.1 (and which is obviously false when $p = +\infty$):

PROPOSITION 24.3. *If $1 \leqq p < +\infty$, every element u of $H^{m,p}(\Omega)$ with compact support in Ω belongs to $H_0^{m,p}(\Omega)$.*

PROPOSITION 24.4. *Let $1 \leqq p \leqq +\infty$ be arbitrary. Then the multiplication*

$$(f, u) \mapsto fu \tag{24.3}$$

is a continuous bilinear map of $H^{m,\infty}(\Omega) \times H^{m,p}(\Omega)$ into $H^{m,p}(\Omega)$.

The proof is left to the student.

PROPOSITION 24.5. *If $p < +\infty$, the mapping (24.3) is a continuous bilinear map of $H^{m,\infty}(\Omega) \times H_0^{m,p}(\Omega)$ into $H_0^{m,p}(\Omega)$.*

Proof. Suppose that $u \in H^{m,p}(\Omega)$ has a compact support contained in Ω; then the same is true of fu, whatever $f \in H^{m,\infty}(\Omega)$. It suffices therefore to combine Propositions 24.3 and 24.4. Q.E.D.

Remark 24.2. Proposition 24.5 is false when $p = +\infty$.

Let us now introduce a second open subset Ω' of $\mathbf{R}^n$ and a mapping $\psi : \Omega \to \Omega'$ with the following properties:

(24.4) *ψ is bijective;*

(24.5) *$\psi = (\psi^1, \ldots, \psi^n)$ and $\psi^{-1} = (\chi^1, \ldots, \chi^n)$ are m times continuously differentiable* (that is, their components are) *in Ω and Ω', respectively;*

(24.6) *for every $j = 1, \ldots,$ every n-tuple α, $0 < |\alpha| \leqq m$, $(\partial/\partial x)^\alpha \psi^j$ is bounded in Ω, $(\partial/\partial x)^\alpha \chi^j$ is bounded in Ω'.*

We are supposing $m > 0$. Then

PROPOSITION 24.6. *Under the preceding hypotheses, whatever $1 \leqq p \leqq +\infty$, $u \mapsto u \circ \psi$ is an isomorphism of $H^{m,p}(\Omega')$ onto $H^{m,p}(\Omega)$. Its inverse is $v \mapsto v \circ \psi^{-1}$.*

The proof, a simple application of the chain rule and of the mean value theorem, is left to the student.

PROPOSITION 24.7. *The hypotheses are the same as in Proposition 24.6 except that p must be finite. Then $u \mapsto u \circ \psi$ is an isomorphism of $H_0^{m,p}(\Omega')$ onto $H_0^{m,p}(\Omega)$.*

Proof. In view of Propositions 24.3 and 24.6, it suffices to show that if the support of $u \in H^{m,p}(\Omega')$ is a compact subset of Ω', that of $u \circ \psi$ is a compact subset of Ω, which is obvious, since the latter is the image of the former under ψ^{-1}. Q.E.D.

We consider now the case where $\Omega = \mathbf{R}^n$ and $p = 2$. We recall the definition of the spaces H^s for s real arbitrary (Definition 13.1): H^s is the space of tempered distributions u in $\mathbf{R}^n$ whose Fourier transform is square-integrable with respect to the measure

$$(2\pi)^{-n}(1 + |\xi|^2)^s \, d\xi.$$

The norm of $\hat{u}$ in the space L^2 with respect to this measure is, by definition, the norm of u in H^s, which we have denoted by $\|u\|_s$.

Now, if we go back to Definition 24.1 and perform a Fourier transformation, we see that $H^m(\mathbf{R}^n) = H^{m,2}(\mathbf{R}^n)$ is the space of L^2 functions u in $\mathbf{R}^n$ whose Fourier transform $\hat{u}$ is such that

$$\left(\sum_{|\alpha| \leqq m} |\xi^\alpha|^2 \right)^{1/2} \hat{u}(\xi) \tag{24.7}$$

belongs to $L^2(\mathbf{R}_n)$ and, again by virtue of the Plancherel theorem, $\|u\|_{m,2}$ is equal to the L^2 norm of (24.7) divided by $(2\pi)^{n/2}$. We conclude easily that the space $H^m(\mathbf{R}^n)$ defined according to Definition 24.1 is identical to the one defined according to Definition 13.1. However, the norms $\| \ \|_m$ and $\| \ \|_{m,2}$ are not equal, unless $m = 0$ or 1; they are merely equivalent.

We are now going to consider H^s for arbitrary, possibly nonintegral s. It is clear that we have $H^s \subset H^{s'}$ for $s' < s$ and that the injection is continuous and has norm $\leqq 1$. On the other hand, Proposition 13.1 shows that, when $\Omega = \mathbf{R}^n$, $H^m(\Omega) = H_0^m(\Omega)$. Let in fact s be any real number and introduce the isometry of H^s onto $H^0 = L^2$ defined in (13.21):

$$T^s u = \mathscr{F}^{-1}\{(1 + |\xi|^2)^{s/2}\hat{u}\}, \qquad u \in H^s.$$

By transposing the sequence

$$C_c^\infty \xrightarrow{\text{natural injection}} H^s \xrightarrow{T^s} L^2,$$

and by identifying, as usual, L^2 with its own dual, we obtain

$$L^2 \xrightarrow{{}^tT^s} (H^s)' \xrightarrow{\text{natural injection}} \mathscr{D}'(\mathbf{R}^n),$$

where the second arrow is injective. If we identify $(H^s)'$ with its image in $\mathscr{D}'$, by restricting ${}^t(T^s)$ to functions belonging to C_c^∞ (or to $\mathscr{S}$) we see at once that it is equal to T^s itself. But the transpose of a surjective isometry is a surjective isometry, and T^s is such an isometry from L^2 onto H^{-s}. If the student thinks a little about all this, he will see that we have proved

PROPOSITION 24.8. *The dual of H^s is canonically isomorphic to H^{-s}. The canonical antilinear isometry of the Hilbert space H^s onto its dual is transformed, by this isomorphism, into the mapping*

$$T^{2s} : u \mapsto \mathscr{F}^{-1}\{(1 + |\xi|^2)^s \mathscr{F}\bar{u}\},$$

where $\mathscr{F}$ denotes the Fourier transformation and $\mathscr{F}^{-1}$ its inverse.

It is checked at once, again by using Fourier transformation, that Proposition 24.8 does not contradict Proposition 24.2 (when $s = m$ and $p = p' = 2$): $H^{-m,2}(\Omega) = H^{-m}(\mathbf{R}^n)$ when $\Omega = \mathbf{R}^n$. However, it should be pointed out that the norm $\| \ \|_{-m,2}$ we have defined on page 213 is not equal to the norm $\| \ \|_s$ defined in (13.20) when $s = -m$, except in the case where $m = 0$ or 1. Otherwise they are merely equivalent.

The spaces H^s for nonintegral s intervene in a natural way in several problems of PDE theory—we have encountered them in the study of the Cauchy problem. They will intervene once more, in an unavoidable manner, when we study the *traces*, or boundary values, of functions belonging to $H^m(\Omega)$ (Sect. 26). We shall then take a closer look at their properties.

We conclude this section by extending the property that $H^m(\mathbf{R}^n) = H_0^m(\mathbf{R}^n)$ to the $H^{m,p}$ spaces for $p \neq 2$:

PROPOSITION 24.9. *Whatever the integer $m \geqq 0$, whatever p, $1 \leqq p < +\infty$, $C_c^\infty(\mathbf{R}^n)$ is dense in $H^{m,p}(\mathbf{R}^n)$.*

Proof. Let $\zeta \in C_c^\infty(\mathbf{R}^n)$, $\zeta(x) = 1$ for $|x| < 1$, $\zeta(x) = 0$ for $|x| > 2$; set $\zeta_\nu(x) = \zeta(\nu^{-1}x)$, $\nu = 1, 2, \ldots$. It is checked at once that, if $u \in H^{m,p}(\mathbf{R}^n)$, $\zeta_\nu u$ converges to u in $H^{m,p}(\mathbf{R}^n)$ when $\nu \to +\infty$. But the support of each $\zeta_\nu u$ is a compact subset of $\mathbf{R}^n$, hence, by Proposition 24.3, $\zeta_\nu u \in H_0^{m,p}(\mathbf{R}^n)$ for every ν. This proves our assertion.

Remark 24.3. One should not think that $\mathbf{R}^n$ is the only open set such that, for some $m > 0$ and some p, $1 \leqq p < +\infty$, $H^{m,p}(\Omega) = H_0^{m,p}(\Omega)$. An interesting case is that of $p = 2$. One has the following result (for a proof, see [1, Exposé 15]):

THEOREM. *Let Ω be a nonempty open subset of $\mathbf{R}^n$, m an integer greater than zero. Then $H^m(\Omega) = H_0^m(\Omega)$ if and only if $\mathbf{R}^n \backslash \Omega$ is m-polar, i.e., has the following property*:

(24.8) *a distribution u in $\mathbf{R}^n$ whose compact support is contained in $\mathbf{R}^n \backslash \Omega$ cannot belong to $H^{-m}(\mathbf{R}^n)$ unless $u = 0$.*

In relation with this, let us note that we have

PROPOSITION 24.10. *Let Ω be a bounded subset of $\mathbf{R}^n$, m an integer greater than zero. Then $H_0^{m,p}(\Omega) \neq H^{m,p}(\Omega)$.*

Proof. If Ω is bounded, the constant functions belong to $H^{m,p}(\Omega)$; it suffices to show that the function identically equal to one, $\mathbf{1}$, does not belong to $H_0^{m,p}(\Omega)$. It suffices to show that $\mathbf{1}$ does not belong to $H_0^{1,1}(\Omega)$, the largest of all those spaces. Let $\varphi \in C_c^\infty(\Omega)$ be arbitrary. We have

$$\|\mathbf{1} - \varphi\|_{1,1} = \int_\Omega |\mathbf{1} - \varphi| \, dx + \int |\text{grad } \varphi| \, dx.$$

On the other hand, there is a constant $C > 0$, depending only on Ω, such that

$$\int |\varphi| \, dx \leqq C \int |\text{grad } \varphi| \, dx,$$

from which

$$\|\mathbf{1} - \varphi\|_{1,1} \geqq \text{meas}(\Omega) - (C - 1) \int |\text{grad } \varphi| \, dx$$

$$\geqq \text{meas}(\Omega) - (C - 1)\|\mathbf{1} - \varphi\|_{1,1},$$

that is,

$$\text{meas}(\Omega) \leqq C\|\mathbf{1} - \varphi\|_{1,1},$$

which shows that the distance in $H^{1,1}(\Omega)$ between the function $\mathbf{1}$ and the subspace $C_c^\infty(\Omega)$ is bounded away from zero. Q.E.D.

Appendix The Sobolev Inequalities

In this section we have looked at the regularity of functions and distributions as measured by the scale of spaces $H^{m,p}$ or H^s and, actually, throughout this text, we have been and shall be using, almost exclusively, the notion of regularity defined and measured by the spaces H^s (i.e., with "base" L^2). But in numerous problems one needs to have a more precise definition, or at any rate a different measure of the regularity. For instance, an often-used notion of regularity is that of Hölder continuity. It so happens that there is a link between Hölder continuity and regularity in the sense of the Sobolev spaces, and although we shall not explore this aspect, we shall give in this Appendix the simplest and most celebrated results in such a direction, linking Sobolev regularity to continuity. These are the *Sobolev inequalities*. The reason for stating and proving them here is not that they will be used again in this book, but only to have them at the disposal of the reader who needs them, or is curious regarding their statement (or proof). For further facts, and the relationship with Hölder continuity, we refer the reader to [F, Part I].

We begin by proving the following global (that is, in $\mathbf{R}^n$) result:

LEMMA 24.2. *Suppose* $1 \leqq p < n$ *and define* p^* *by*

$$\frac{1}{p^*} = \frac{1}{p} - \frac{1}{n}. \tag{24.9}$$

Then there is a constant $C = C(n, p) > 0$ *such that*

$$\|u\|_{L^{p*}} \leqq C\left\{\sum_{j=1}^{n} \int \left|\frac{\partial u}{\partial x^j}\right|^p dx\right\}^{1/p}, \qquad \forall u \in C_c^\infty(\mathbf{R}^n). \tag{24.10}$$

Proof. If $u \in C_c^\infty(\mathbf{R}^n)$, we have

$$|u(x)| \leqq \int_{-\infty}^{x^j} \left|\frac{\partial u}{\partial x^j}(x^1, \ldots, x^{j-1}, t, x^{j+1}, \ldots, x^n)\right| dt,$$

and a similar inequality with integration performed from x^j to $+\infty$, from which follows with obvious notation,

$$2|u(x)| \leqq \int_{-\infty}^{+\infty} \left|\frac{\partial u}{\partial x^j}\right| dx^j,$$

and therefore

$$2^n|u(x)|^n \leqq \prod_{j=1}^{n} \int \left|\frac{\partial u}{\partial x^j}\right| dx^j. \tag{24.11}$$

We integrate both sides of (24.11) with respect to dx^n and apply the standard Hölder's inequalities

$$\int |u_1 \cdots u_r| \, d\mu \leqq \prod_{k=1}^{r} \left(\int |u_k|^{\alpha_k} \, d\mu'\right)^{1/\alpha_k}, \tag{24.12}$$

where $1 \leqq \alpha_k \leqq +\infty$ and $\alpha_1^{-1} + \cdots + \alpha_k^{-1} = 1$. Here we take $r = n - 1$ (we are assuming $n \geqq 2$), $\alpha_1 = \cdots = \alpha_r = n - 1$, and

$$u_k = \left(\int \left|\frac{\partial u}{\partial x^k}\right| dx^k\right)^{1/(n-1)}, \qquad k = 1, \ldots, n-1.$$

We obtain

$$\int |2u(x)|^{n/(n-1)} \, dx^n \leqq \left\{\prod_{j=1}^{n-1} \iint \left|\frac{\partial u}{\partial x^j}\right| dx^j \, dx^n\right\}^{(n-1)^{-1}} \left\{\int \left|\frac{\partial u}{\partial x^n}\right| dx^n\right\}^{(n-1)^{-1}}.$$

Next we integrate with respect to dx^{n-1}, applying once more (24.12) with $r = n - 1$, but choosing this time

$$u_k = \left\{\iint \left|\frac{\partial u}{\partial x^k}\right| dx^k \, dx^n\right\}^{(n-1)^{-1}}, \qquad k < n-1; \qquad u_n = \left\{\int \left|\frac{\partial u}{\partial x^n}\right| dx^n\right\}^{(n-1)^{-1}}.$$

We repeat this operation until we obtain

$$\|u\|_{L^{n/(n-1)}} \leqq \frac{1}{2}\left\{\prod_{j=1}^{n}\left\|\frac{\partial u}{\partial x^j}\right\|_{L^1}\right\}^{1/n}. \tag{24.13}$$

In order to obtain the estimate (24.10) when $p = 1$, it suffices to apply the elementary inequality

$$a_1 \cdots a_n \leqq \frac{1}{n!}(a_1 + \cdots + a_n)^n, \qquad a_j \geqq 0 \quad (1 \leqq j \leqq n). \tag{24.14}$$

Assume now $p > 1$. We observe that, if $u \in C_\infty^c(\mathbf{R}^n)$, the function $|u|^{[(n-1)/n]p^*}$ belongs to $C_c^1(\mathbf{R}^n)$, since $[(n-1)/n]p^* = [(n-1)/(n-p)]p > 1$. We also observe that the estimate (24.13) extends to functions belonging to $C_c^1(\mathbf{R}^n)$. We substitute in it $|u|^{[(n-1)/n]p^*}$ for u, and use the fact that

$$\int |u|^{[(n-1)/n]p^*-1}\left|\frac{\partial u}{\partial x^j}\right| dx \leqq \left\{\int |u|^{q\{[(n-1)/n]p^*-1\}}\, dx\right\}^{1/q}\left\|\frac{du}{dx^j}\right\|_{L^p}, \qquad q = \frac{p}{p-1}.$$

We observe that $q\{[(n-1)/n]p^* - 1\} = p^*$ and that $(n-1)/n - 1/q = 1/p^*$. We obtain at once

$$\|u\|_{L^{p*}} \leqq \frac{p}{2}\frac{n-1}{n-p}\prod_{j=1}^{n}\left\|\frac{\partial u}{\partial x^j}\right\|_{L^p}^{1/n}, \tag{24.15}$$

from which we obtain (24.10) for $p > 1$ by applying (24.14) once again.

Q.E.D.

We come now to the Sobolev inequalities valid for functions in a bounded open subset Ω of $\mathbf{R}^n$. We shall subdivide their study into two cases, and make different assumptions about Ω in each of the cases.

Our first assumption, which we make for values of the integration exponent p close to *one*, will be that elements of $H^{1,p}(\Omega)$ can be extended into elements of $H^{1,p}(\mathbf{R}^n)$ by a continuous linear operator:

(24.16) *The mapping "restriction to* Ω*",* r_Ω*, maps* $H^{1,p}(\mathbf{R}^n)$ *onto* $H^{1,p}(\Omega)$*, and it has a continuous right inverse,* $\varepsilon_\Omega : H^{1,p}(\Omega) \to H^{1,p}(\mathbf{R}^n)$ [which means that $r_\Omega \varepsilon_\Omega$ is the identity on $H^{1,p}(\Omega)$].

It will be shown in the Appendix to Sect. 26 (Theorem 26.A.3), that (24.16) is satisfied whenever the boundary $\partial\Omega$ of Ω is a C^1 hypersurface, with Ω lying on one side of it.

THEOREM 24.1. *Suppose that (24.16) holds and that*

$$1 \leqq p < n/m \qquad (m \text{ is an integer } \geqq 0). \tag{24.17}$$

If we write

$$\text{(24.18)} \qquad \frac{1}{p'} = \frac{1}{p} - \frac{m}{n},$$

then $H^{m,p}(\Omega) \subset L^{p'}(\Omega)$, *and there is a constant* $C > 0$ *such that*

$$\text{(24.19)} \qquad \|u\|_{L^{p'}(\Omega)} \leqq C\|u\|_{H^{m,p}(\Omega)}, \qquad \forall u \in H^{m,p}(\Omega).$$

In the second case, corresponding to "large" values of p, we shall assume that Ω has the *cone property*. Let A be a subset of the unit sphere in $\mathbf{R}^n$; let us denote by $\Gamma(A, h)$ the set of vectors u in $\mathbf{R}^n$ such that $0 < |u| < h$ and such that $u/|u| \in A$. We shall assume that there is a number $h > 0$ such that the following holds:

(24.20) *Given any point* $x \in \Omega$, *there is an open subset* $\mathcal{O}_x$ *of* S^{n-1}, *whose (surface) measure is no less than* h, *and such that*

$$x + \Gamma(\mathcal{O}_x, h) \subset \Omega.$$

If the boundary of Ω is a C^1 hypersurface, and Ω lies on one side of it, or if Ω is convex, then Ω has the cone property (Exercise 24.5); this also holds in other cases (Exercise 24.6).

THEOREM 24.2. *Suppose that (24.20) holds and that*

$$\text{(24.21)} \qquad \frac{n}{m} < p \leqq +\infty. \qquad m \geqq 1.$$

Then $H^{m,p}(\Omega) \subset C^0(\Omega) \cap L^\infty(\Omega)$ *and there is a constant* $C > 0$ *such that*

$$\text{(24.22)} \qquad \sup_\Omega |u(x)| \leqq C\|u\|_{H^{m,p}(\Omega)}, \qquad \forall u \in H^{m,p}(\Omega).$$

Proof of Theorem 24.1. We consider first the case $m = 1$. Then p', defined in (24.18), is equal to p^*, defined in (24.9). If we combine Lemma 24.2 with Proposition 24.9, we see that $H^{1,p}(\mathbf{R}^n) \subset L^{p^*}(\mathbf{R}^n)$ and that we have, for a suitable constant $C > 0$,

$$\text{(24.23)} \qquad \|v\|_{L^{p^*}(\mathbf{R}^n)} \leqq C\|v\|_{H^{1,p}(\mathbf{R}^n)}, \qquad \forall v \in H^{1,p}(\mathbf{R}^n).$$

We then use the extension mapping ε_Ω, whose existence (and continuity) is postulated in (24.16). We get, for some $C_1 > 0$ and all $u \in H^{1,p}(\Omega)$,

$$\|\varepsilon_\Omega u\|_{H^{1,p}(\mathbf{R}^n)} \leqq C_1\|u\|_{H^{1,p}(\Omega)}.$$

Since obviously $\|u\|_{L^{p^*}(\Omega)} \leqq \|\varepsilon_\Omega u\|_{L^{p^*}(\mathbf{R}^n)}$, we get (24.19) in this case.

Let now m be arbitrary and greater than one. To say that $u \in H^{m,p}(\Omega)$ is equivalent to saying that $D^\alpha u \in H^{1,p}(\Omega)$ for all α, $|\alpha| \leqq m - 1$. By applying (24.19), where $m = 1$, to $D^\alpha u$, $|\alpha| \leqq m - 1$, instead of u, we obtain

$$\|u\|_{H^{m-1,p^*}(\Omega)} \leqq C_2\|u\|_{H^{m,p}(\Omega)}, \qquad \forall u \in H^{m,p}(\Omega).$$

We may iterate this inequality m times, observing that the operation $p \mapsto p^*$, applied m times to p, yields exactly p' given by (24.18). Q.E.D.

Proof of Theorem 24.2. Let x be an arbitrary point of Ω. We may assume that it is the origin and use polar coordinates r, θ centered at this point. Let us call Γ the truncated cone $x + \Gamma(\mathcal{O}_x, h)$ in condition (24.20). Let g denote a nonnegative C^∞ function on the real line, $g(t) = 1$ for $t < h/2$, $g(t) = 0$ for $t \geqq 3h/4$.

Let $u \in C^\infty(\Omega) \cap H^{m,p}(\Omega)$; for $\theta \in \mathcal{O}_0$,

$$u(0) = -\int_0^h (\partial/\partial r)[g(r)u_1(r, \theta)]\, dr$$

$$= \frac{(-1)^m}{(m-1)!}\int_0^h r^{m-n}\{(\partial/\partial r)^m[g(r)u_1(r, \theta)]\}r^{n-1}\, dr,$$

after $m - 1$ integrations by parts, and writing $u_1(r, \theta)$ for the expression of u in the coordinates r, θ. We integrate now with respect to θ over $\mathcal{O}_0$ whose measure is $\geqq h$. We obtain

$$|u(0)| \leqq \frac{h^{-1}}{(m-1)!}\int_\Gamma r^{m-n}|(\partial/\partial r)^m[g(r)u]|\, dx$$

$$\leqq \frac{h^{-1}}{(m-1)!}\|r^{m-n}\|_{L^q(\Gamma)}\|(\partial/\partial r)^m[g(r)u]\|_{L^p(\Gamma)},$$

by Hölder's inequalities $[q = p/(p-1)]$. If $p = 1$, in which case $q = +\infty$, we have $m > n$. If $p > 1$, $q(m - n) + n - 1 > -1$ because $m > n/p$.

In order to conclude the proof, it suffices to observe that $(\partial/\partial r)^m$ is a linear combination of the $(\partial/\partial x)^\beta$, $|\beta| = m$, with coefficients belonging to $L^\infty(\Gamma)$. The fact that $\Gamma \subset \Omega$ implies at once (24.22) for $u \in C^\infty(\Omega) \cap H^{m,p}(\Omega)$ and Proposition 24.1 implies the full conclusion of Theorem 24.2. Q.E.D.

Exercises

24.1. Let E be the fundamental solution of the Laplace operator Δ in $\mathbf{R}^n$ defined in (9.19). Prove that, whatever the bounded open set Ω, the restriction of E to Ω belongs to $H^{1,p}(\Omega)$ for all $p < n/(n-1)$ (we suppose $n \geqq 2$). Derive from this fact that the Dirac measure δ_{x_0} at some point x_0 of Ω belongs to $H^{-1,p}(\Omega)$ for those same values of p.

24.2. Let Ω be a bounded open subset of $\mathbf{R}^n$. We suppose $n \geqq 2$. Let μ be an arbitrary Radon measure in $\mathbf{R}^n$, positive and carried by Ω (this means that the complement of Ω is of μ-measure zero). Let E be the fundamental solution of the Laplace operator considered in Exercise 24.1. Prove that, for all $p < n/(n-1)$, the restriction to Ω of the convolution $E * \mu$ belongs to $H^{1,p}(\Omega)$. [*Hint*: Write

$$(E * \mu)(x) = \int E(x - y)\, d\mu(y); \tag{24.24}$$

limiting the variation of x to Ω, regard $E(x - y)$ as a function of y in Ω, valued in the space $H^{1,p}(\Omega)$ with respect to x, by using the result in Exercise 24.1 applied to translates of E. Use then the fact that the norm, here in $H^{1,p}(\Omega)$, of the integral, is not greater than the integral of the norm, together with the fact that

$$\sup_{y \in \Omega} \| E(x - y) \|_{H^{1,p}(\Omega_x)} < +\infty.]$$

Derive from the above that all the Radon measures in Ω, whose total mass is finite, belong to $H^{-1,p}(\Omega)$, $p < n/(n-1)$. Derive the same result from Theorem 24.2, when Ω has the cone property.

24.3. Let m be an integer, $0 < m < n$, E_m the distribution defined in Exercise 9.2. Prove that if $p < n/(n-m)$, and if Ω is an arbitrary bounded open subset of $\mathbf{R}^n$, $E_m \in H^{m,p}(\Omega)$.

24.4. Let Ω be a bounded open subset of $\mathbf{R}^n$. Let m be an integer, $0 < m < n$, p a positive number, $1 \leqq p < n/m$. Let R be a positive number, so large that $\bar{\Omega} + \bar{\Omega}$ is contained in the open ball $|x| < R$ (we shall denote by χ_R the characteristic function of this ball). Let E_m be the distribution so denoted in Exercise 24.3. Prove that

$$\| D^{\alpha+\beta}(E_m * u) \|_{L^r(\Omega)} \leqq \| D^\alpha u \|_{L^p} \| \chi_R D^\beta E_m \|_{L^q}, \qquad \forall u \in H_0^{m,p}(\Omega), \tag{24.25}$$

if $|\alpha| \leqq m$, $|\beta| \leqq m$, $q < n/(n-m)$, and $1/r = 1/p + 1/q - 1$. Derive from this fact that $u \in L^r$ for all r such that $1 \leqq r < np/(n-m)$. Compare this with the conclusion of Theorem 24.1.

24.5. Let Ω be a bounded open subset of $\mathbf{R}^n$. Prove that Ω has the cone property (24.20) in the following cases: (1) Ω is convex; (2) Ω is a C^1 hypersurface and Ω lies on one side of it.

24.6. Give an example in $\mathbf{R}^2$ of a bounded open set which does not have the cone property. Using the fact that any finite union of sets having the cone property has the cone property, give an example of an open subset of $\mathbf{R}^2$ which is not convex, whose boundary has "corners," and which does have the cone property.

24.7. Let Ω be a bounded open subset of $\mathbf{R}^n$, having the cone property (24.20). Given any integer m, $0 \leqq m < +\infty$, denote by $\mathscr{B}^m(\Omega)$ the space of complex C^m functions in Ω whose derivatives, of order $\leqq m$, all belong to $L^\infty(\Omega)$. Suppose that we have

$$1 \leqq p \leqq +\infty, \qquad m > k + n/p. \tag{24.26}$$

Prove then that $H^{m,p}(\Omega)$ is continuously embedded in $\mathscr{B}^k(\Omega)$.

24.8. Prove that if $p > n$, there is a constant $C > 0$ such that

$$\frac{|u(x) - u(y)|}{|x - y|^{1-n/p}} \leqq C \sum_{j=1}^{n} \left\| \frac{\partial u}{\partial x^j} \right\|_{L^p}, \qquad \forall u \in C_c^\infty(\mathbf{R}^n). \tag{24.27}$$

[*Hint*: Write

$$\text{meas}(\Omega)|u(x) - u(y)| \leqq \int_\Omega |u(x) - u(z)| \, dz + \int_\Omega |u(y) - u(z)| \, dz,$$

where Ω is the set of points z such that $|z - x| \leqq |x - y|$, $|z - y| \leqq |x - y|$. Estimate then each integral on the right by changing to polar coordinates, centered either at x or at y, and by applying Hölder's inequalities.]

24.9. Let Ω be any open subset of $\mathbf{R}^n$, α a number such that $0 \leqq \alpha \leqq 1$. We shall denote by $\mathscr{B}^\alpha(\Omega)$ the space of bounded continuous functions u in Ω such that

$$\sup_{x, y \in \Omega} \frac{|u(x) - u(y)|}{|x - y|^\alpha} < +\infty. \tag{24.28}$$

[Functions having the property (24.28) are called uniformly *Hölder continuous*, with *Hölder exponent* α if $0 < \alpha \leqq 1$; when $\alpha = 0$, they are simply the functions which are continuous and bounded; when $\alpha = 1$, they are called uniformly *Lipschitz continuous*.] Suppose Ω is bounded and show that the left-hand side of (24.28) defines a norm on $\mathscr{B}^\alpha(\Omega)$ which turns it into a Banach space.

Prove that $\mathscr{B}^1(\Omega)$ is exactly equal to the space of continuous functions in Ω whose distribution derivatives belong to $L^\infty(\Omega)$.

24.10. Let Ω be a bounded open subset of $\mathbf{R}^n$ having the extension property (24.16). Let $0 < \alpha < 1$ and denote by $\mathscr{B}^{m+\alpha}(\Omega)$ the space of C^m functions in Ω whose derivatives of order m belong to $\mathscr{B}^\alpha(\Omega)$ (see Exercise 24.9). Derive from the result in Exercise 24.8 that $H^{m,p}(\Omega)$ is continuously embedded in $\mathscr{B}^{m-1+\alpha}(\Omega)$ if $p = n/(1 - \alpha)$.

Suppose now that $m > n/p$ $(1 \leqq p < +\infty)$, n/p *not* an integer. Prove then that $H^{m,p}(\Omega)$ is continuously embedded in $\mathscr{B}^{m-n/p}(\Omega)$.

25

Further Properties of the Spaces H^s

The study of boundary values, or traces on $\partial\Omega$, of the functions belonging to $H^m(\Omega)$ requires a deeper study of the global Sobolev spaces $H^s = H^s(\mathbf{R}^n)$ for nonintegral values of s.† This section is devoted to such a study.

We begin by looking at some elementary operations on H^s, and first of all, at multiplication by smooth functions. We recall that $\mathscr{S} = \mathscr{S}(\mathbf{R}^n)$ is the space of C^∞ functions φ in $\mathbf{R}^n$ whose derivatives of all order decay at infinity faster than any power of $1/|x|$.

PROPOSITION 25.1. *The multiplication* $(\varphi, u) \mapsto \varphi u$ *is a bilinear map of* $\mathscr{S} \times H^s$ *into* H^s, *such that*

$$\|\varphi u\|_s \leqq \|u\|_s \int (1 + |\eta|)^{|s|} |\hat{\varphi}(\eta)| \, d\eta. \tag{25.1}$$

We have denoted by $\hat{\varphi}$ the Fourier transform of φ:

$$\hat{\varphi}(\xi) = \int e^{-i\langle x, \xi\rangle} \varphi(x) \, dx.$$

It is well known that the Fourier transformation is an isomorphism of $\mathscr{S}$ onto itself.

Proof. It suffices to prove the assertion when $s \geqq 0$. Indeed, in H^{-s}, the multiplication by $\varphi \in \mathscr{S}$ is the transpose of the same operation in H^s. We have

$$\|\varphi u\|_s^2 = \int (1 + |\xi|^2)^s \left| \int \hat{\varphi}(\eta) \hat{u}(\xi - \eta) \, d\eta \right|^2 d\xi.$$

But, as easily seen,

$$(1 + |\xi|^2)^{s/2} \leqq (1 + |\eta|)^s (1 + |\xi - \eta|^2)^{s/2}, \tag{25.2}$$

† In the subsequent discussions, unless otherwise specified, s will be an *arbitrary* real number.

and therefore

$$\|\varphi u\|_s \leqq \left\| \int (1 + |\eta|)^s |\hat{\varphi}(\eta)| (1 + |\xi - \eta|^2)^{s/2} |\hat{u}(\xi - \eta)| \, d\eta \right\|_{L^2}.$$

We obtain (25.1) by applying the classical Hölder inequality:

$$\|f * g\|_{L^p} \leqq \|f\|_{L^1} \|g\|_{L^p}, \qquad f \in L^1, \quad g \in L^p \quad (1 \leqq p \leqq +\infty). \quad \text{Q.E.D.}$$

Remark 25.1. The estimate (25.1), although sufficient in many applications, is not the best possible. For instance, when $s = m$, a nonnegative integer, we have the better estimate:

$$\|\varphi u\|_m \leqq \text{const } \|u\|_m \sup_{|\alpha| \leqq m} \|D^\alpha \varphi\|_{L^\infty}. \tag{25.3}$$

Generalization of (25.3) to H^s for $|s| \leqq m$ is possible but less easy to prove. Note also that, when $s = m$ (or $s = -m$), the multiplication φu, $u \in H^s$, is defined for far more functions φ than just those belonging to $\mathscr{S}$. In particular, it is defined for $\varphi \in \mathscr{B}^\infty$, the space of C^∞ functions whose derivatives of all order are bounded in $\mathbf{R}^n$. For such φ the estimate (25.3) is valid, and can be extended to $|s| \leqq m$.

We go on, next, to look at convolution. In a sense, this is a neater operation than multiplication on the Sobolev spaces H^s. Let us denote by A^s the space of tempered distributions u in $\mathbf{R}^n$ whose Fourier transform $\hat{u}$ is a (measurable) function with the property that

$$(1 + |\xi|^2)^{s/2} \hat{u}(\xi) \in L^\infty.$$

PROPOSITION 25.2. *The convolution $(u, v) \mapsto u * v$ is a bilinear mapping of $H^s \times A^t$ into H^{s+t}, such that*

$$\|u * v\|_{s+t} \leqq \|u\|_s \|(1 + |\xi|^2)^{t/2} \hat{v}\|_{L^\infty}. \tag{25.4}$$

The assertion is practically evident.

An important case is that where $\hat{v}$ is a polynomial $P(\xi)$ of degree $\leqq m$; note that, then, $v \in H^{-m,\infty}$. As for the convolution with v, it is the (linear) partial differential operator with constant coefficients usually denoted by $P(D)$, $D = (-i\,\partial/\partial x^1, \ldots, -i\,\partial/\partial x^n)$. We thus see

COROLLARY 25.1. *Let $P(\xi)$ be a polynomial of degree $\leqq m$ on $\mathbf{R}_n$. Then $u \mapsto P(D)u$ is a continuous linear map of H^s into H^{s-m}.*

We may take advantage of Propositions 25.1 and 25.2 to add some precision to the statement that $\mathscr{S}$ is dense in H^s (Proposition 13.1). First of all, let $\zeta \in C_c^\infty(\mathbf{R}^n)$ be equal to *one* in the open unit ball, and set $\zeta_\nu(x) = \zeta(x/\nu)$, $\nu = 1, 2, \ldots$. We have $\hat{\zeta}_\nu(\xi) = \nu^n \hat{\zeta}(\nu\xi)$, and it follows at once from (25.1) that the mappings $u \mapsto \zeta_\nu u$ form an equicontinuous set of linear mappings $H^s \to H^s$. For such a set, pointwise convergence in H^s and that on a dense subset of H^s are one and the same thing. Since when $u \in \mathscr{S}$, $\zeta_\nu u$ converges to u in H^s (as a matter of fact, in $\mathscr{S}$), we reach the following conclusion:

PROPOSITION 25.1′. *If the functions ζ_ν are as described, for each $u \in H^s$, the cutoffs $\zeta_\nu u$ converge to u in H^s as $\nu \to +\infty$.*

Let now ρ belong to $C_c^\infty(\mathbf{R}^n)$, $\int \rho(x)\,dx = 1$ and set $\rho_\varepsilon(x) = \varepsilon^{-n}\rho(x/\varepsilon)$ for $\varepsilon > 0$. Observe that $\hat{\rho}_\varepsilon(\xi) = \hat{\rho}(\varepsilon\xi)$.

PROPOSITION 25.2′. *If $u \in H^s$, $\rho_\varepsilon * u$ converges to u in H^s as $\varepsilon \to +0$.*

Proof. We see that $\rho_\varepsilon \in A^0$ and, by (25.4), $u \mapsto \rho_\varepsilon * u$ form an equicontinuous set of linear mappings of H^s into itself. Since the assertion is well known when $u \in C_c^\infty$, it is true for any $u \in H^s$. Q.E.D.

By combining Propositions 25.1′ and 25.2′ we see that every $u \in H^s$ is the limit of the $\rho_\varepsilon * (\zeta_\nu u)$ when $\varepsilon + \nu^{-1} \to +0$. In particular, each u is the limit of functions u_k ($k = 1, 2, \ldots$) belonging to $C_c^\infty(\mathbf{R}^n)$ whose supports converge to that of u, in the sense that, given any compact set $K \subset \mathbf{R}^n$, $K \cap (\operatorname{supp} u_k)$ converges to $K \cap (\operatorname{supp} u)$.

We denote by C_0^m the space of m-times continuously differentiable functions in $\mathbf{R}^n$ whose derivatives of order $\leqq m$ all tend to zero at infinity.

PROPOSITION 25.3. *If $s > m + n/2$, $H^s \subset C_0^m$.*

Proof. In view of Corollary 25.1, it suffices to prove the assertion when $m = 0$. Let $u \in H^s$. There is a function $v \in L^2$ such that

$$\hat{u} = (1 + |\xi|^2)^{-s}\hat{v}.$$

If $s > n/2$, $(1 + |\xi|^2)^{-s}$ belongs also to L^2, hence $\hat{u} \in L^1$. It suffices then to apply a classical theorem of Lebesgue. Q.E.D.

PROPOSITION 25.4. *The Dirac measure δ belongs to $H^{-n/2-\varepsilon}$ for any $\varepsilon > 0$. No distribution with support contained in a finite set of points belongs to $H^{-n/2}$ (unless it is identical to zero).*

Proof. The Fourier transform of δ is one; and $(1 + |\xi|^2)^{s/2} \in L^2$ if and only if $s < -n/2$. This proves the first assertion.

If $u \in H^s$ has its support contained in a finite set of points, there is a function $\varphi \in C_c^\infty$ such that φu is not identical to zero (unless u is!) and $\operatorname{supp}(\varphi u)$ is one point. Since the spaces H^s are easily seen to be invariant under translations, we may suppose that this set is the origin. Now, all distributions with support at the origin are of the form $P(D)\delta$, where $P(\xi)$ is a polynomial. The Fourier transform of $P(D)\delta$ is precisely $P(\xi)$. Suppose that the degree of P is m. Then $(1 + |\xi|^2)^{s/2}P(\xi)$ belongs to L^2 if and only if $s + m < -n/2$, from which we obtain the last part of the statement. Q.E.D.

Let K now denote a *compact* subset of $\mathbf{R}^n$. We denote by $H^s(K)$ the subspace of H^s consisting of those u which vanish in $\mathbf{R}^n \backslash K$, equipped with the norm $\| \ \|_s$. It is a closed linear subspace of H^s, hence a Hilbert space for the induced structure. We shall now compare the norms $\| \ \|_s$ and $\| \ \|_{s'}$ when restricted to $H^s(K)$, for $s' < s$.

PROPOSITION 25.5. *Let s, s' be any two real numbers such that $s' < s$ and let K be any compact subset of $\mathbf{R}^n$. The natural injection of $H^s(K)$ into $H^{s'}$ is compact.*

Proof. We must show that, if a sequence $\{u_\nu\}$ converges weakly to zero in $H^s(K)$, it converges strongly to zero in $H^{s'}$. For distributions, strong and weak convergence of sequences are one and the same thing. Form

$$\hat{u}_\nu(\zeta) = \langle u_\nu, e^{-i\langle x, \zeta\rangle}\rangle, \qquad \zeta \in \mathbf{C}^n.$$

From the easy part of the Paley–Wiener theorem we know that the $\hat{u}_\nu(\zeta)$ are entire functions of exponential type in $\mathbf{C}^n$. When ζ remains in a bounded subset of $\mathbf{C}^n$, the exponentials $\exp(-i\langle x, \zeta\rangle)$ form a bounded subset of $C^\infty(\mathbf{R}^n)$. We know that, since they have their supports in a fixed compact set, the u_ν converge to zero uniformly on such subsets of $C^\infty(\mathbf{R}^n)$ and therefore the $\hat{u}_\nu$ converge to zero uniformly on every bounded subset of $\mathbf{C}^n$. On the other hand, we know by the "principle of uniform boundedness," that the norms $\|u_\nu\|_s$ are bounded. We have

$$\begin{aligned}\|u_\nu\|_{s'}^2 &= \int_{|\xi|>R} (1 + |\xi|^2)^{s'-s}(1 + |\xi|^2)^s |\hat{u}_\nu(\xi)|^2 \, d\xi \\ &\quad + \int_{|\xi|<R} (1 + |\xi|^2)^{s'} |\hat{u}_\nu(\xi)|^2 \, d\xi \\ &\leqq (1 + R^2)^{s'-s}\|u_\nu\|_s^2 + \int_{|\xi|<R} (1 + |\xi|^2)^{s'} |\hat{u}_\nu(\xi)|^2 \, d\xi.\end{aligned}$$

We may choose R independently of ν, and so big that the first term in the last member is $\leqq \varepsilon/2$, then ν so large as to make this true also for the second term. Q.E.D.

PROPOSITION 25.6. *Let s, s' be any two real numbers such that $s' < s$ and $s \geqq -n/2$. To every $\varepsilon > 0$ there is $\delta > 0$ such that, if the diameter of K is $\leqq \delta$,*

$$\|u\|_{s'} \leqq \varepsilon \|u\|_s \qquad \textit{for every} \quad u \in H^s(K). \tag{25.5}$$

Proof. Suppose that the assertion in Proposition 25.6 were not true. After a translation, we could find a sequence of elements $\{u_\nu\}$ in H^s with supp $u_\nu \subset \{x \in \mathbf{R}^n;\ |x| < \nu^{-1}\}$, such that $\|u_\nu\|_s \leqq 1$, $\|u_\nu\|_{s'} > c > 0$ for all ν. From the first inequality we conclude that a subsequence of the u_ν converges weakly in H^s, hence, by Proposition 25.5, it converges strongly to some element $u \in H^s$ in $H^{s'}$, and from the second inequality we conclude that $u \neq 0$. But the support of u must be $\{0\}$, and this contradicts Proposition 25.4. Q.E.D.

If Ω is any open subset of $\mathbf{R}^n$, we write

$$H_c^s(\Omega) = \bigcup_K H^s(K),$$

where K ranges over the collection of all compact subsets of Ω. When $\Omega = \mathbf{R}^n$, one often writes $H_c^s = H_c^s(\mathbf{R}^n)$.

We denote by $H_{\mathrm{loc}}^s(\Omega)$ the space distributions u in Ω such that $\varphi u \in H^s$ for all $\varphi \in C_c^\infty(\Omega)$. One usually equips $H_{\mathrm{loc}}^s(\Omega)$ with the topology defined by the seminorms $u \mapsto \|\varphi u\|_s$ as φ ranges over $\mathbf{C}_c^\infty(\Omega)$. It is easy to find a sequence $\{\varphi_\nu\}$ $(\nu = 1, 2, \ldots)$ in $C_c^\infty(\Omega)$ which converges to the function identically equal to one in a strong manner: If K_ν is the (compact) set of points $x \in \Omega$ such that $\varphi(x) = 1$, K_ν is contained, for each ν, in the interior of $K_{\nu+1}$, and Ω is equal to the union of the K_ν's. Then the seminorms $u \mapsto \|\varphi_\nu u\|_s$ form a *basis* of continuous seminorms on $H_{\mathrm{loc}}^s(\Omega)$, which shows that this space is metrizable. By using the fact that H^s is complete, one checks at once that $H_{\mathrm{loc}}^s(\Omega)$ is also complete. Thus the latter is a Fréchet space. By using the reflexivity of H^s (which is a Hilbert space!), one can easily check that $H_{\mathrm{loc}}^s(\Omega)$ is also reflexive. We have $C_c^\infty(\Omega) \subset H_{\mathrm{loc}}^s(\Omega)$ and the natural injection is continuous and has a dense image; its transpose is a continuous linear injection of the dual of $H_{\mathrm{loc}}^s(\Omega)$ into $\mathscr{D}'(\Omega)$, and enables us to identify the dual of $H_{\mathrm{loc}}^s(\Omega)$ with the image of this transpose. By using Proposition 24.9, one sees at once that this image is nothing else but $H_c^{-s}(\Omega)$.

We wish now to prove the *invariance of the spaces* $H_{\mathrm{loc}}^s(\Omega)$ *under a diffeomorphism*. For that, we shall need a new norm on H^s, equivalent to the norm $\|\ \|_s$.

LEMMA 25.1. *If* $0 \leqq s \leqq 1$,

$$\|u\|_s^2 \leqq (2\pi)^{-n} \int |\hat{u}(\xi)|^2 (1 + |\xi|^{2s})\, d\xi \leqq 2\|u\|_s^2, \qquad u \in H^s. \tag{25.6}$$

Proof. The proof follows at once from the obvious inequality (for $0 \leqq s \leqq 1$)

$$(1 + |\xi|^2)^s \leqq 1 + |\xi|^{2s} \leqq 2(1 + |\xi|^2)^s, \qquad \xi \in \mathbf{R}_n.$$

LEMMA 25.2. *If* $0 < s < 1$, *we have, for all* $u \in C_c^\infty(\mathbf{R}^n)$,

$$(25.7) \quad (2\pi)^{-n} \int |\hat{u}(\xi)|^2 (1 + |\xi|^{2s})\, d\xi$$

$$= \int |u|^2\, dx + C_s \int |u(x) - u(y)|^2 |x - y|^{-n-2s}\, dx\, dy,$$

where C_s *depends only on n and s.*

Proof. By Parseval's formula we have

$$\iint |u(x) - u(y)|^2 |x - y|^{-n-2s}\, dx\, dy$$

$$= \iint |u(x + y) - u(y)|^2 |x|^{-n-2s}\, dx\, dy$$

$$= (2\pi)^{-n} \iint |e^{i\langle x, \xi\rangle} - 1|^2 |x|^{-n-2s} |\hat{u}(\xi)|^2\, dx\, d\xi.$$

Indeed, the Fourier transform of $u(x + y)$ with respect to y is equal to $\hat{u}(\xi) \exp(i\langle x, \xi\rangle)$. Note that, since s is smaller than one, the integral

$$I(\xi) = \int |e^{i\langle x, \xi\rangle} - 1|^2 |x|^{-n-2s}\, dx$$

is convergent. Let R be a rotation in $\mathbf{R}_n$. We have

$$I(R\xi) = \int |\exp(i\langle {}^t R x, \xi\rangle) - 1|^2 |x|^{-n-2s}\, dx$$

$$= \int |e^{i\langle x', \xi\rangle} - 1|^2 |x'|^{-n-2s}\, dx' \qquad \text{setting} \quad x' = {}^t R^{-1} x.$$

Thus $I(R\xi) = I(\xi)$, which shows that $I(\xi)$ depends only on the norm $|\xi|$. On the other hand,

$$I(t\xi) = \int |e^{i\langle tx, \xi\rangle} - 1|^2 |x|^{-n-2s}\, dx = t^{2s} I(\xi), \qquad t > 0.$$

This shows that $I(\xi)$ is a positively homogeneous function of ξ of degree $2s$, hence $I(\xi) = C_s |\xi|^{2s}$ for a suitable constant $C_s > 0$. This easily implies what we wanted. Q.E.D.

Remark 25.2. By combining Lemmas 25.1 and 25.2 we see that the square root of the right-hand side of (25.7) defines on H^s a norm equivalent to $\| \ \|_s$. When s is an arbitrary positive number, possibly greater than one, but in any case different from an integer, the following is the square of a norm on H^s equivalent to $\| \|_s$:

$$(25.8) \qquad \sum_{|\alpha| \leqq [s]} \int |D^\alpha \varphi|^2 \, dx + C_s \sum_{|\alpha| = [s]} \iint \frac{|D^\alpha u(x) - D^\alpha u(y)|^2}{|x - y|^{n + 2(s - [s])}} \, dx \, dy,$$

where $[s]$ denotes the largest integer $\leqq s$ and C_s is a suitable constant depending only on s and n $(C > 0)$.

Remark 25.3. Let us go back to the case where $0 < s < 1$. We may then define H^s as the space of functions $u \in L^2(\mathbf{R}^n)$ such that

$$|x - y|^{-s - n/2}[u(x) - u(y)] \in L^2(\mathbf{R}^{2n}).$$

Consider now a C^∞ diffeomorphism of Ω onto another open subset Ω' of $\mathbf{R}^n$, χ. If $v \in C_c^\infty(\Omega')$, we set $(v \circ \chi)(x) = v(\chi(x))$, $x \in \Omega$; of course, $v \circ \chi \in C_c^\infty(\Omega)$. If now $v \in \mathscr{D}'(\Omega')$, we set

$$(25.9) \qquad \langle v \circ \chi, \varphi \rangle = \left\langle v, \left| \frac{d\chi}{dx} \right|^{-1} (\varphi \circ \chi^{-1}) \right\rangle, \qquad \varphi \in C_c^\infty(\Omega).$$

We have denoted by $d\chi/dx$ the *Jacobian determinant* of χ.

PROPOSITION 25.7. *Let K' be an arbitrary compact subset of Ω', s any real number. There is a constant $C_s(K')$ such that*

$$(25.10) \qquad \|u \circ \chi\|_s \leqq C_s(K') \|u\|_s \qquad \textit{for every} \qquad u \in H^s(K').$$

Proof. It suffices to prove (25.10) when $u \in C_c^\infty(K')$. Indeed, let K_1' be a compact neighborhood of K' in Ω'. We can then assert that $u \mapsto u \circ \chi$ is a continuous linear map of $C_c^\infty(K_1')$, equipped with the norm $\| \|_s$, into H^s and hence extends to the closure of $C_c^\infty(K_1')$ in H^s. This closure contains $H^s(K')$ as we see by regularization (Proposition 25.2′).

Case I: $0 < s < 1$. We apply Lemmas 25.1 and 25.2. We write $x' = \chi(x)$, $y' = \chi(y)$. We have

$$\iint |u(x') - u(y')|^2 |x - y|^{-n - 2s} \, dx \, dy$$
$$= \iint |u(x') - u(y')|^2 |x' - y'|^{-n - 2s} \omega(x', y') \, dx' \, dy',$$

where

$$\omega(x', y') = \left| \frac{x' - y'}{x - y} \right|^{-n - 2s} \left| \frac{dx}{dx'} \frac{dy}{dy'} \right|$$

and where dx/dx' and dy/dy' stand for the Jacobian determinants. Clearly ω is bounded on $K' \times K'$.

Case II: $s = 0, 1, \ldots$. This is a particular case of Proposition 24.6.

Case III: $s > 0$ arbitrary. Let $[s]$ denote the largest integer $\leqq s$, and set $\theta = s - [s]$. Now, that u belongs to H^s means that $D^\alpha u \in H^\theta$ for all α, $|\alpha| \leqq [s]$. And $\|u\|_s$ is equivalent to

$$\sum_{|\alpha| \leqq s} \| D^\alpha u\|_\theta .$$

By combining this remark with the result relative to θ, we easily obtain the result for s.

Case IV: $s < 0$. We have

$$\|u \circ \chi\|_s = \sup \left| \int u(\chi(x))v(x)\, dx \right|$$

where the supremum is taken over $v \in C_c^\infty$ such that $\|v\|_{-s} = 1$. As a matter of fact, we may limit ourselves to those v whose support lies in $\bar{\chi}^1(K_1')$, K_1' a compact neighborhood of K' in Ω'. Now, if we set $x' = \chi(x)$, we see that

$$\|u \circ \chi\|_s = \sup \left| \int u(x')(v \circ \bar{\chi}^1)(x') \left| \frac{dx}{dx'} \right| dx' \right|$$

$$\leqq \left\| \left| \frac{dx}{dx'} \right| (v \circ \bar{\chi}^1) \right\|_{-s} \|u\|_s .$$

The absolute value of the Jacobian determinant $|dx/dx'|$ is a C^∞ function in an open neighborhood of K' (in fact, in Ω'), therefore multiplication by this function defines a bounded linear operator on $H^{-s}(K')$. On the other hand, we may apply (25.10) with v substituted for u and $\bar{\chi}^1$ for χ, when s is replaced by $-s$. We obtain

$$\|u \circ \chi\|_s \leqq \text{const } \|v\|_{-s}\|u\|_s = \text{const } \|u\|_s . \qquad \text{Q.E.D.}$$

We are now in a position to prove the announced invariance:

PROPOSITION 25.8. *Let χ be a diffeomorphism of Ω onto Ω'. Then $u \mapsto u \circ \chi$ is an isomorphism of $H^s_{\text{loc}}(\Omega')$ onto $H^s_{\text{loc}}(\Omega)$. Its inverse is the mapping $v \mapsto v \circ \bar{\chi}^1$.*

Proof. Let $\varphi \in C_c^\infty(\Omega)$ be arbitrary. Set $\varphi' = \varphi \circ \bar{\chi}^1$. We have

$$\varphi(u \circ \chi) = (\varphi' u) \circ \chi,$$

hence $\|\varphi(u \circ \chi)\|_s \leqq C(K)\|\varphi' u\|_s$ by Proposition 25.7 [$C(K) > 0$ depends on the compact set $K = \text{supp } \varphi$]. This shows that the mapping $u \mapsto u \circ \chi$ is continuous from $H^s_{\text{loc}}(\Omega')$ into $H^s_{\text{loc}}(\Omega)$, which, of course, is enough. Q.E.D.

We are now able to define $H^s(\mathfrak{M})$ when $\mathfrak{M}$ is a compact C^∞ *manifold* (without boundary). This means that $\mathfrak{M}$ is a compact Hausdorff topological space, carrying a C^∞ structure. The latter can be described as follows: We are given an open covering $\{U_\alpha\}$ $(\alpha \in A)$ of $\mathfrak{M}$ and, for each $\alpha \in A$, a homeomorphism φ_α of U_α onto an open subset of $\mathbf{R}^n$ (the dimension n is independent of α and is called the dimension of the manifold $\mathfrak{M}$). These homeomorphisms φ_α are submitted to the following condition. Suppose that $U_\alpha \cap U_\beta = U_{\alpha\beta} \neq \varnothing$; the restrictions of φ_α and φ_β to $U_{\alpha\beta}$ must have the property that

(25.11) $\quad \varphi_\beta \bar{\varphi}_\alpha^1$ *is a* C^∞ *diffeomorphism of* $\varphi_\alpha(U_{\alpha\beta})$ *onto* $\varphi_\beta(U_{\alpha\beta})$.

This gives a meaning to the statement that a function f is C^∞ in $\mathfrak{M}$: It means that, given any $\alpha \in A$, $f \circ \bar{\varphi}_\alpha^1$ is a C^∞ function in $\varphi_\alpha(U_\alpha)$. The dual of $C^\infty(\mathfrak{M})$ is the space $\mathscr{D}'(\mathfrak{M})$ of distributions in $\mathfrak{M}$. Clearly $u \mapsto u \circ \varphi_\alpha$ is an isomorphism of $C_c^\infty(\varphi_\alpha(U_\alpha))$ onto $C_c^\infty(U_\alpha)$; its transpose is an isomorphism of $\mathscr{D}'(U_\alpha)$ onto $\mathscr{D}'(\varphi_\alpha(U_\alpha))$ and we may denote by $H^s_{\text{loc}}(U_\alpha)$ the preimage of $H^s_{\text{loc}}(\varphi_\alpha(U_\alpha))$ under this transpose.

Definition 25.1. We denote by $H^s(\mathfrak{M})$ *the space of distributions u in* $\mathfrak{M}$ *such that, for every* α, *the restriction of u to* U_α *belongs to* $H^s_{\text{loc}}(U_\alpha)$.

That this is a correct definition follows from (25.11) and from Proposition 25.8.

Remark 25.4. We have exploited the compactness of $\mathfrak{M}$ only insofar as we have adopted the notation $H^s(\mathfrak{M})$ instead of $H^s_{\text{loc}}(\mathfrak{M})$: In the case of a non-compact manifold, the latter should stand for the former—everything being otherwise equal. The compactness of $\mathfrak{M}$ is now going to be exploited in a more substantial way: It will enable us to equip $H^s(\mathfrak{M})$ with a Hilbert space structure—which, however, is not canonical (in general).

As $\mathfrak{M}$ is compact, we may select a *finite* open subcovering $U_{\alpha_1}, \ldots, U_{\alpha_r}$ in the covering $\{U_\alpha\}$. Let then $g_1, \ldots, g_r$ be a C^∞ partition of unity subordinated to the subcovering $U_{\alpha_1}, \ldots, U_{\alpha_r}$: For each $j = 1, \ldots, r$, $g_j \in C_c^\infty(U_{\alpha_j})$; $g_1 + \cdots + g_r = 1$ on $\mathfrak{M}$. We may then define a Hilbert space structure on $H^s(\mathfrak{M})$ by setting

$$(u, v)_s = \sum_{j=1}^{r} ((g_j u) \circ \bar{\varphi}_{\alpha j}^1, (g_j v) \circ \bar{\varphi}_{\alpha j}^1)_s$$

where, on the right-hand side, $(\ ,\ )_s$ denotes the Hermitian product in $H^s(\mathbf{R}^n)$.

It follows at once from (25.11) and from Proposition 25.7, that if we change our choice of the finite subcovering $\{U_{\alpha_j}\}$ or that of the partition of unity $\{g_j\}$, we replace the Hilbert space structure of $H^s(\mathfrak{M})$ by an equivalent one (i.e., the underlying locally convex topological vector is unchanged).

Exercises

25.1. We use the concepts and notation of Exercise 17.10, in particular the "semigroup" e^{-tA}, $t \geqq 0$. For any $\theta \in [0, 1]$, let H^θ denote the image of H^1 under the mapping $u \mapsto A^{1-\theta}u$. Prove that H^θ is a Hilbert space for the norm $\|A^\theta u\|_{H^0}$.

Let u_0 be an arbitrary element of $H^{1/2}$ and write $u(t) = e^{-tA}u_0$, $t \geqq 0$. Prove that we have

(25.12)

$$u(t) \in L^2(0, +\infty; H^1) \cap C^0([0, +\infty[;H^0), \qquad u'(t) \in L^2(0, +\infty; H^0).$$

Prove that, conversely, given any function $u(t)$ satisfying (25.12) we have $u(0) \in H^{1/2}$.

Let T be a continuous linear map $H^0 \to H^0$, whose restriction to H^1 defines a continuous linear map $H^1 \to H^1$. By considering $v(t) = Tu(t)$, with $u(t)$ satisfying (25.12), prove that T defines a continuous linear map $H^{1/2} \to H^{1/2}$.

25.2. Let $\mathscr{B}^\infty$ denote the space of C^∞ functions in $\mathbf{R}^n$ whose derivatives of all orders are bounded in the whole of $\mathbf{R}^n$. Prove that, given any function $\phi \in \mathscr{B}^\infty$, the multiplication mapping $u \mapsto \phi u$ is continuous from H^s to H^s for every integer s ($\geqq 0$ or <0). Apply the "interpolation" result in Exercise 25.1 to derive the same conclusion for any real number s.

25.3. Let $f(t)$ be a distribution on the real line, having a compact support contained in the open interval $0 < t < 2\pi$; let $f(\theta)$ denote the same distribution, viewed as a distribution on the unit circumference S^1, and let c_m be its Fourier coefficients,

$$c_m = \int_0^{2\pi} e^{-im\theta} f(\theta)\, d\theta$$

(where the integral is merely a notation for the duality bracket between distributions and C^∞ functions on S^1). Prove that the distribution f belongs to $H^s(\mathbf{R}^1)$ if and only if

$$\sum_{-\infty}^{+\infty} (1 + m^2)^s |c_m|^2 < +\infty \tag{25.13}$$

(s is any real number).

25.4. Using the result in Exercise 25.3, prove that $H^s(\mathbf{T}^n)$ ($\mathbf{T}^n$: the n-dimensional torus) is the space of distributions u on $\mathbf{T}^n$ whose Fourier coefficients c_α ($\alpha \in \mathbf{Z}^n$) satisfy

$$\sum_{\alpha \in \mathbf{Z}^n} (1 + |\alpha|^2)^s |c_\alpha|^2 < +\infty \tag{25.14}$$

(s is any real number).

25.5. Let $\mathfrak{M}$ be a compact C^∞ manifold, and suppose that $H^0(\mathfrak{M})$, $H^1(\mathfrak{M})$ are both equipped with a Hilbert space structure. Show that there is a linear isometry of $H^1(\mathfrak{M})$ onto $H^0(\mathfrak{M})$, A, which is positive, i.e., such that for some $c_0 > 0$,

$$(25.15) \qquad (Au, u)_0 \geqq c_0 \|u\|_0^2, \qquad \forall u \in H^1(\mathfrak{M}).$$

Prove that we have then necessarily, not just (25.15), but in fact

$$(25.16) \qquad (Au, u)_0 \geqq c_1 \|u\|_{1/2}^2, \qquad \forall u \in H^1(\mathfrak{M}),$$

for a suitable $c_1 > 0$. Prove that $A^{-1} : H^0(\mathfrak{M}) \to H^1(\mathfrak{M})$ can be regarded as a *compact* operator (necessarily self-adjoint) of $H^0(\mathfrak{M})$ into itself.

25.6. Use the same concepts and notation as in Exercise 25.5. Let $\chi_1 \geqq \chi_2 \geqq \cdots$ be the sequence of eigenvalues of $A^{-1} : H^0(\mathfrak{M}) \to H^0(\mathfrak{M})$, with repetitions according to multiplicity, and let $\{\phi_j\}$ ($j = 1, 2, \ldots$) be a complete orthonormal system in $H^0(\mathfrak{M})$ such that, for each j, ϕ_j is an eigenfunction of A^{-1} corresponding to the eigenvalue χ_j.

Prove that, whatever the real number s, $H^s(\mathfrak{M})$ is the space of distributions u in $\mathfrak{M}$, which can be written

$$(25.17) \qquad u = \sum_{j=1}^{+\infty} u_j \phi_j,$$

with coefficients u_j satisfying

$$(25.18) \qquad \sum_{j=1}^{+\infty} \chi_j^{-2s} |u_j|^2 < +\infty.$$

25.7. Let $\mathfrak{M}$ be a compact C^∞ manifold, $ds^2 = \sum_{i,j=1}^n g_{ij}\, dx^i\, dx^j$ ($n = \dim \mathfrak{M}$) a Riemannian metric on $\mathfrak{M}$. Show that ds^2 defines a Hilbert space structure on every H^s ($s \in \mathbf{R}$) and a second-order differential operator Λ on $\mathfrak{M}$ which, in turn, defines an isometry of $H^2(\mathfrak{M})$ onto $H^0(\mathfrak{M})$. Prove that Λ must be elliptic (Definition 19.4).

25.8. Let Ω be an open subset of $\mathbf{R}^n$. Prove that, for any $s > 0$, $C^0(\Omega)$ is *not* contained in $H^s_{\mathrm{loc}}(\Omega)$.

25.9. Let S^k denote the k-dimensional unit sphere ($k = 0, 1, \ldots$); regard S^k as the "equator" in S^{k+1}, and thus, for $k \leqq n - 2$, as a subset of S^{n-1} and therefore as one of $\mathbf{R}^n$. Let $d\sigma_k$ be the canonical measure on S^k, regarded as a Radon measure on $\mathbf{R}^n$, precisely the measure

$$C_c^\infty(\mathbf{R}^n) \ni \phi \mapsto \int_{S^k} (\phi|_{S^k})\, d\sigma_k.$$

What is the lowest upper bound of the real numbers s such that $d\sigma_k$ (viewed as a distribution on $\mathbf{R}^n$) belongs to $H^s(\mathbf{R}^n)$?

25.10. Prove that, given any integer $k \geqq 0$ or <0, the mapping "restriction to the open subset Ω of distributions in $\mathbf{R}^n$," r_Ω, is a continuous linear map of $H^k(\mathbf{R}^n)$ into $H^k(\Omega)$. Show that this mapping has a dense image if $k \leqq 0$. Show that there is a natural injection of $H_0^m(\Omega)$ $(m \geqq 0)$ onto a *closed* linear subspace of $H^m(\mathbf{R}^n)$, and that $H^{-m}(\Omega)$ is canonically isomorphic to a closed linear subspace of $H^{-m}(\mathbf{R}^n)$. Describe the latter and relate its orthogonal, in $H^{-m}(\mathbf{R}^n)$, to the restriction map r_Ω above. Prove that, if Ω is bounded, and if $k < m$ ($k \geqq 0$ or <0), the natural injection of $H_0^m(\Omega)$ into $H^k(\Omega)$ is compact. (This is one of the versions of the classical *Rellich's lemma.*)

25.11. Let Ω be an open neighborhood of the origin in $\mathbf{R}^n$, $P(x, D)$ a linear partial differential operator with C^∞ coefficients in Ω, of order m. Assume that the principal symbol (Definition 19.1) $P_m(x, \xi)$ of the operator $P(x, D)$ does not vanish identically, as a polynomial in ξ, for any x in Ω. Prove the following:

LEMMA 25.3. *Let s, s', s'' be three real numbers such that $s' < s$. Let x_0 be an arbitrary point of Ω, ε any number greater than zero. There is an open neighborhood $U \subset \Omega$ of x_0 and a constant $C > 0$, depending on s, s', s'', ε, and x_0, such that*

$$\|u\|_{s'} \leqq C\|P(x, D)u\|_{s''} + \varepsilon\|u\|_s, \qquad \forall u \in C_c^\infty(U). \tag{25.19}$$

[*Hint*: Assume that for fixed s, s', s'', and ε as in Lemma 25.3, no neighborhood U or constant C exists. Show that this implies that there is a sequence of functions $u_\nu \in C_c^\infty(\Omega)$, with diam(supp u_ν) $\to 0$ and whose support converges to $\{x_0\}$, such that

$$\|u_\nu\|_{s'} > 1, \tag{25.20}$$

whereas

$$\|P(x, D)u_\nu\|_{s''} \leqq 1/\nu, \qquad \|u_\nu\|_s \leqq \varepsilon^{-1}. \tag{25.21}$$

By applying Proposition 25.5, derive from this that the u_ν converge, for the norm in $H^{s'}$, to a nonzero distribution u whose support is reduced to the single point $\{x_0\}$. Then apply Lemma 1.1, Exercise 1.12.] Show that the conclusion in Lemma 25.3 is not valid if we remove the hypothesis that the principal symbol of $P(x, D)$ does not vanish identically at any point of Ω.

25.12. Let ρ_ε denote the mollifiers used in Proposition 25.2′ ($\rho_1 = \rho$). Let $L(\xi)$ be an arbitrary linear form on $\mathbf{R}_n$. Prove that there is a constant $C > 0$ such that, for all $\xi, \eta \in \mathbf{R}_n$,

$$|L(\xi)\hat{\rho}(\varepsilon\xi) - L(\eta)\hat{\rho}(\varepsilon\eta)| \leqq C|\xi - \eta|. \tag{25.22}$$

Derive from this that, given any function $c(x) \in C_c^\infty(\mathbf{R}^n)$, the operators defined by

$$(25.23) \qquad T_\varepsilon u = L(D)(\rho_\varepsilon * (cu) - c(\rho_\varepsilon * u))$$

form (as ε varies on]0, 1[) an equicontinuous set of linear mappings $H^s \to H^s$ ($s \in \mathbf{R}$ arbitrary). Derive from this *Friedrich's lemma*:

LEMMA 25.4. *Let $P(x, D)$ denote a linear partial differential operator of order $m \geqq 0$, with C^∞ coefficients in an open subset Ω of $\mathbf{R}^n$. Let s be an arbitrary real number. Given any distribution $u \in H_c^s(\Omega)$, the differences*

$$(25.24) \qquad P(x, D)(\rho_\varepsilon * u) - \rho_\varepsilon * P(x, D)u$$

converge to zero in $H_{\mathrm{loc}}^{s-m+1}(\Omega)$ as $\varepsilon \to +0$.

[*Hint*: In the proof of (25.22) it is enough to consider the case $n = 1$ and $L(\xi) = \xi$. Observe then that

$$\xi\hat{\rho}(\varepsilon\xi) - \eta\hat{\rho}(\varepsilon\eta) = \int_{-\infty}^{+\infty} \left\{ \int_\eta^\xi \frac{\partial}{\partial t}(te^{-i\varepsilon t x})\, dt \right\} \rho(x)\, dx.$$

The proof of the equicontinuity of the operators T_ε is analogous to that of Proposition 25.1.]

26

Traces in $H^m(\Omega)$

Throughout this section, Ω will be a *bounded* open subset of $\mathbf{R}^n$ whose boundary $\partial\Omega = \Gamma$ is a C^∞ hypersurface. For simplicity, we may assume that Ω is connected and lies entirely "on one side" of Γ (i.e., Ω is one of the bounded components of $\mathbf{R}^n \backslash \Gamma$). But Ω could as well be a *finite* union of such connected bounded open sets. The definition of the spaces $H^s(\mathfrak{M})$ given at the end of the preceding section applies to the case where $\mathfrak{M} = \Gamma$.

We shall prove the following results:

THEOREM 26.1. *Let m be any nonnegative integer. The restrictions to Ω of the functions belonging to $C^\infty(\mathbf{R}^n)$ form a dense subspace of $H^m(\Omega)$.*†

As can be seen, this dense subspace is nothing else but $C^\infty(\bar{\Omega})$, the space of C^∞ functions in Ω whose derivatives of all order extend to the closure of Ω as continuous functions. For such functions we may clearly define their values at the boundary Γ, what is called their *trace* on Γ, and which is simply their restriction to Γ.

THEOREM 26.2. *Let m be any integer strictly greater than zero. The trace on Γ, defined at first on $C^\infty(\bar{\Omega})$, can be extended* (*in a unique manner*) *as a continuous linear map of $H^m(\Omega)$ onto $H^{m-1/2}(\Gamma)$.*

In the proofs of both these statements (as in those of many similar ones), the first step consists of a *localization* and a *flattening of the boundary*. This is done as follows.

We choose a finite covering $\{V_1, \ldots, V_r\}$ of Γ by open subsets of $\mathbf{R}^n$ such that, for each $j = 1, \ldots, r$, there is a C^∞ diffeomorphism φ_j of V_j onto the open unit ball of $\mathbf{R}^n$, $B = \{x \in \mathbf{R}^n;\ |x| < 1\}$, such that

(26.1) *the image of $V_j \cap \Omega$ under φ_j is the open upper half-ball*

$$B^+ = \{x \in \mathbf{R}^n;\ |x| < 1,\ x^n > 0\}.$$

† Theorem 26.1 is well known when $m = 0$. In the remainder of this section, we shall assume that $m \geqq 1$.

This is possible, as one sees by using local coordinates in sufficiently small open subsets of Γ (forming a covering of Γ)—and extending these coordinates to open subsets of $\mathbf{R}^n$ by using the *normal* direction to Γ. We leave the details to the student.

Let V_0 be a relatively compact open subset of Ω, containing

$$K_0 = \Omega \backslash \bigcup_{j=1}^{r} (V_j \cap \Omega),$$

and let $\zeta_0, \zeta_1, \ldots, \zeta_r$ denote $r+1$ elements of $C_c^\infty(\mathbf{R}^n)$ such that $\zeta_0 + \zeta_1 + \cdots + \zeta_r = 1$ in a neighborhood of $\overline{\Omega}$, and that, for each $j = 0, 1, \ldots, r$, $\operatorname{supp} \zeta_j \subset V_j$.

Let then u be an arbitrary element of $H^m(\Omega)$. Suppose that, for each j, $0 \leqq j \leqq r$, we find a sequence of functions $u_{j,\nu} \in C_c^\infty(V_j)$ which converge to $\zeta_j u$ in $H^m(V_j \cap \Omega)$. We then claim that the $u_{j,\nu}$ converge to $\zeta_j u$ in $H^m(\Omega)$: This is simply due to the fact that $\zeta_j u$ and the $u_{j,\nu}$ vanish identically outside of a compact subset of V_j, K_j, hence in a neighborhood of

$$\partial(V_j \cap \Omega) \backslash K_j \cap \Gamma.$$

From this it would follow at once that the restrictions to Ω of the functions $u_\nu = \sum_{j=1}^{r} u_{j,\nu}$ converge to u in $H^m(\Omega)$ and would prove Theorem 26.1.

Suppose now that u is the restriction to Ω of a C^∞ function in $\mathbf{R}^n$ and let us denote by γu its trace on Γ, that is, its restriction to Γ. In order to prove Theorem 26.2, we wish to prove that, for some $C > 0$ independent of u,

$$\|\gamma u\|_{H^{m-1/2}(\Gamma)} \leqq C \|u\|_{H^m(\Omega)}. \tag{26.2}$$

As seen at once, this is equivalent to proving that, for each $j = 1, \ldots, r$, we have, for a suitable constant $C_j > 0$ independent of u,

$$\|\gamma\{\zeta_j u\}\|_{H^{m-1/2}(\Gamma)} \leqq C_j \|\zeta_j u\|_{H^m(\Omega)}. \tag{26.3}$$

Of course, we are using on $H^{m-1/2}(\Gamma)$ one of the Hilbert structures defined at the end of Sect. 25, for instance the one defined by means of the local charts $(V_j \cap \Gamma, \varphi_j|_{V_j \cap \Gamma})$.

If we combine (26.2) with Theorem 26.1, we see that γ can be extended as a continuous linear map $H^m(\Omega) \to H^{m-1/2}(\Gamma)$. It remains to show that this mapping is *onto*.

Let v be an arbitrary element of $H^{m-1/2}(\Gamma)$. It is clear what we mean by $\zeta_j v$ ($1 \leqq j \leqq r$): It is the product of v with the *restriction to* Γ *of* ζ_j. Suppose then that we have proved the following: There is $u_j \in H^m(V_j \cap \Omega)$, whose support is contained in a compact subset of V_j and which, consequently, defines an element (also denoted by u_j) of $H^m(\Omega)$, such that the trace on Γ of this element of $H^m(\Omega)$ is equal to $\zeta_j v$. Then, of course, the trace on Γ of $u_1 + \cdots + u_r$ will be equal to v.

This concludes the argument and shows that, in order to prove Theorems 26.1 and 26.2, we may reason in the sets V_j, $V_j \cap \Omega$, $V_j \cap \Gamma$ (with all the precautions suggested above). This is what we have called the *localization*. The next step is the *flattening of the boundary*. It is done by transferring the whole situation to the unit ball B by means of the C^∞ diffeomorphisms φ_j $(1 \leqq j \leqq r)$. We know (Proposition 24.6) that $u \mapsto u \circ \overset{-1}{\varphi}_j$ defines an isomorphism of $H^m(V^\alpha \cap \Omega)$ onto $H^m(B^+)$ and, by Proposition 25.8, that $v \mapsto v \circ \overset{-1}{\psi}_j$ (where ψ_j is the restriction of φ_j to $V_j \cap \Gamma$) defines an isomorphism of $H^{m-1/2}_{\mathrm{loc}}(V_j \cap \Gamma)$ onto $H^{m-1/2}_{\mathrm{loc}}(\Sigma)$, with

$$\Sigma = \{x \in \mathbf{R}^n;\ |x| < 1,\ x^n = 0\}.$$

At this point we should remember that all the elements of $H^m(V_j \cap \Omega)$ and $H^{m-1/2}_{\mathrm{loc}}(V_j \cap \Gamma)$ used in the earlier localization argument had their supports contained in compact subsets of V_j. The analog will be true for their images under the isomorphisms above: These images will have supports contained in compact subsets of the open unit ball B. In view of this, it will suffice for us to prove the following assertions:

(26.4) *Let $u \in H^m(B^+)$ have its support contained in a compact subset of B. Then u is the limit, in $H^m(B^+)$, of a sequence of functions that are the restrictions to B^+ of elements of $C^\infty_c(B)$.*

(26.5) *Let $u \in C^\infty_c(B)$, γu its trace on Σ. There is a constant $C > 0$, independent of u, such that*

$$\|\gamma u\|_{H^{m-1/2}(\mathbf{R}^{n-1})} \leqq C\|u\|_{H^m(B^+)}.$$

We have identified $\mathbf{R}^{n-1}$ with the hyperplane $x^n = 0$ in $\mathbf{R}^n$.

Assertions (26.4) and (26.5) enable us to extend the trace γ on Σ to all elements $u \in H^m(B^+)$ whose support is contained in a compact subset of B. The third assertion will then be

(26.6) *To every $v \in H^{m-1/2}(\mathbf{R}^{n-1})$ with compact support contained in Σ there is $u \in H^m(B^+)$ with support contained in a compact subset of B such that $\gamma u = v$.*

As a matter of fact, because of the condition that the supports of u and v be contained in compact subsets of B and Σ, respectively, B becomes somewhat irrelevant and can be replaced everywhere by $\mathbf{R}^n$ itself (this is equivalent to extending u and v by zero outside their supports). This leads us to replace Σ by $\mathbf{R}^{n-1}$ and B^+ by $\mathbf{R}^n_+ = \{x \in \mathbf{R}^n;\ x^n > 0\}$. We shall then prove the following results:

THEOREM 26.3. *The restrictions to $\mathbf{R}^n_+$ of the functions belonging to $C^\infty_c(\mathbf{R}^n)$ are dense in $H^m(\mathbf{R}^n_+)$.*

In the statement below, we identify $\mathbf{R}^{n-1}$ with the hyperplane $x^n = 0$. We may define the *trace map* γ on the restrictions to $\mathbf{R}^n_+$ of the functions $f \in C^\infty_c(\mathbf{R}^n)$: $\gamma(f)$ is the restriction of f to that hyperplane.

THEOREM 26.4. *The trace* γ *can be extended as a continuous linear map of* $H^m(\mathbf{R}^n_+)$ *onto* $H^{m-1/2}(\mathbf{R}^{n-1})$.

These two statements easily imply our three assertions above, (26.4), (26.5), and (26.6). It suffices to use cutoff functions $\zeta \in C^\infty_c(B)$ equal to one in neighborhoods (in B) of the supports of u and v. Observe that we have the multiplicative property:

$$\gamma(\varphi u) = \gamma(\varphi)\gamma(u), \qquad \varphi \in C^\infty_c(\mathbf{R}^n), \quad u \in H^m(\mathbf{R}^n_+). \tag{26.7}$$

Indeed, (26.7) is true when u itself is the restriction of an element of $C^\infty_c(\mathbf{R}^n)$, hence for all $u \in H^m(\mathbf{R}^n_+)$ by virtue of Theorems 26.3 and 26.4.

Theorems 26.3 and 26.4 represent the last stage in the transformation of our "trace" problem. They are not difficult to prove if we are willing to make the situation a little bit more abstract. The main point will be to interpret $H^m(\mathbf{R}^n)$ [resp. $H^m(\mathbf{R}^n_+)$] as a space of functions on the real line $\mathbf{R}^1$ (resp., on the positive half-line $\mathbf{R}^1_+$), where the variable is x^n, functions which take their values in suitable spaces of functions of the first $n-1$ variables, $x' = (x^1, \ldots, x^{n-1})$. More precisely, we shall regard $u \in H^m(\mathbf{R}^n)$ as an L^2 function on $\mathbf{R}^1$ with values in $H^m(\mathbf{R}^{n-1})$, with the following additional properties:

$$\text{for each} \quad j = 0, 1, \ldots, m, \qquad D^j_n u \in L^2(\mathbf{R}^1; H^{m-j}(\mathbf{R}^{n-1})). \tag{26.8}$$

We have set, as usual, $D^j_n = -\sqrt{-1}\, \partial/\partial x^n$. We use the notation $L^2(X; E)$ to mean the space of L^2 functions in the measure space (X, dx) with values in the Hilbert space E. That (26.8) constitutes an equivalent definition of $H^m(\mathbf{R}^n)$ is obvious. Similarly, $u \in H^m(\mathbf{R}^n_+)$ can be regarded as an L^2 function on $\mathbf{R}^1_+$ such that

$$\text{for each} \quad j = 0, 1, \ldots, m, \qquad D^j_n u \in L^2(\mathbf{R}^1_+; H^{m-j}(\mathbf{R}^{n-1})). \tag{26.9}$$

Also the norm on $H^m(\mathbf{R}^n)$—any one of the norms on this space we have considered so far—is equivalent to the following one:

$$\left\{\sum_{j=0}^{m} \|D^j_n u\|^2_{L^2(\mathbf{R}^1; H^{m-j}(\mathbf{R}^{n-1}))}\right\}^{1/2}.$$

We have an analogous statement with $\mathbf{R}^n_+$ substituted for $\mathbf{R}^n$ and $\mathbf{R}^1_+$ for $\mathbf{R}^1$.

Proof of Theorem 26.3

Let $u \in H^m(\mathbf{R}^n_+)$. We shall make use of the characterization (26.8). By cutting and regularizing u with respect to the "transversal variables" $x' = (x^1, \ldots, x^{n-1})$ we see that u is the limit in $H^m(\mathbf{R}^n_+)$ of a sequence of functions v having the following property:

(26.10) *There is a compact subset K' of $\mathbf{R}^{n-1}$* (depending on v) *such that, for every s real and every $j = 0, 1, \ldots, m$,*

$$D_n^j v \in L^2(\mathbf{R}^1_+ ; H^s(K')).$$

We have stated (26.10) in order to avoid using the spaces $L^2(\mathbf{R}^1_+ ; C_c^\infty(K'))$ which might have put off the student; (26.10) is interesting, from our viewpoint, when applied for s large, $s \sim +\infty$. It has the advantage of involving only one Hilbert space of values, $E = H^s(K)$ (at least, for any given choice of s). Let us denote by t the variable in $\mathbf{R}^1$ or in $\mathbf{R}^1_+$. We know that $v^{(j)} \in L^2(\mathbf{R}^1_+ ; E)$ for every $j = 0, 1, \ldots, m$. Then $v^{(j)}$ must be absolutely continuous in $\mathbf{R}^1_+$ provided that $j < m$; in fact, for $j < m$, $v^{(j)}$ can be extended as a continuous function to the closure of $\mathbf{R}^1_+$ in $\mathbf{R}^1$, and we may talk about the elements $v^{(j)}(+0)$ of E ($j = 0, \ldots, m-1$). Let $\zeta(t) \in C_c^\infty(\mathbf{R}^1)$ (say with real values), equal to one in a neighborhood of the origin, and let us define a new function $V(t)$ on the whole real line $\mathbf{R}^1$, as follows:

$$V(t) = v(t) \quad \text{if} \quad t > 0; \qquad V(t) = \zeta(t) \sum_{j=0}^{m-1} v^{(j)}(+0) \frac{t^j}{j!} \quad \text{if} \quad t < 0.$$

The (distribution) derivatives of $V(t)$ of order $<m$ are continuous functions; $V^{(m)}$ belongs to $L^2(\mathbf{R}^1; E)$ and is equal to $v^{(m)}$ for $t > 0$. Let then $\rho \in C_c^\infty(\mathbf{R}^1)$, $\int \rho \, dt = +1$, and set $\rho_\varepsilon(t) = \varepsilon^{-1}\rho(t/\varepsilon)$, $\varepsilon > 0$. When $\varepsilon \to +0$, whatever $j = 0, \ldots, m$,

$$\left(\frac{d}{dt}\right)^j (\rho_\varepsilon * V) \quad \text{tends to} \quad V^{(j)} \quad \text{in } L^2(\mathbf{R}^1; E);$$

consequently,

$$\sum_{j=1}^{m} \int_{t>0} \|(\rho_\varepsilon * V)^{(j)} - v^{(j)}\|_E^2 \, dt \to 0.$$

We know that the $\rho_\varepsilon * V$ are C^∞ functions valued in $E = H^s(K)$. But s is arbitrarily large and our definition of V is independent of s, hence the $\rho_\varepsilon * V$ are C^∞ functions of t valued in $C_c^\infty(K)$. By multiplying them with cutoff functions of the variable $t = x^n$ (C^∞ functions with compact support in $\mathbf{R}^1$, equal to one on ever larger intervals), we obtain a sequence of functions belonging to $C_c^\infty(\mathbf{R}^n)$ whose restrictions to $\mathbf{R}_+^n$ converge to v in $H^m(\mathbf{R}_+^n)$.

Q.E.D.

Proof of Theorem 26.4

We shall begin by proving the following:

(26.11) *There is a constant $C > 0$ such that, if u is the restriction to $\mathbf{R}_+^n$ of any function belonging to $C_c^\infty(\mathbf{R}^n)$,*

$$\|\gamma u\|_{H^{m-1/2}(\mathbf{R}^{n-1})} \leqq C\|u\|_{H^m(\mathbf{R}_+{}^n)}. \tag{26.12}$$

Let $\hat{u}(\xi', x^n)$ denote the Fourier transform of $u(x)$ with respect to $x' = (x^1, \ldots, x^{n-1})$. We have

$$|\hat{u}(\xi', 0)|^2 = -2 \operatorname{Re} \int_0^{+\infty} \frac{\partial \hat{u}}{\partial x^n}(\xi', t)\overline{\hat{u}(\xi', t)}\, dt. \tag{26.13}$$

We multiply both sides of (26.13) by $(1 + |\xi'|^2)^{m-1/2}$ and integrate with respect to ξ' over $\mathbf{R}_{n-1}$. By the Cauchy–Schwarz inequality we have

$$\left| \int_0^{+\infty} (1 + |\xi'|^2)^{m-1/2} \frac{\partial \hat{u}}{\partial x^n}(\xi', t)\overline{\hat{u}(\xi', t)}\, dt\, d\xi' \right| \tag{26.14}$$

$$\leqq \left\{ \int_0^{+\infty} (1 + |\zeta'|^2)^{m-1} \left| \frac{\partial \hat{u}}{\partial x^n}(\xi', x^n) \right|^2 d\xi'\, dx^n \right\}^{1/2}$$

$$\times \left\{ \int_0^{+\infty} (1 + |\xi'|^2)^m |\hat{u}(\xi', x^n)|^2\, d\xi'\, dx^n \right\}^{1/2}.$$

If we use the remark at the bottom of page 240 on the norms on $H^m(\mathbf{R}_+^n)$, we see at once that the combination of (26.13) and (26.14) implies (26.12).

We are now going to prove that γ maps $H^m(\mathbf{R}_+^n)$ *onto* $H^{m-1/2}(\mathbf{R}^{n-1})$ by constructing a *right inverse* K to γ, in fact a continuous linear map K of $H^{m-1/2}(\mathbf{R}^{n-1})$ into $H^m(\mathbf{R}_+^n)$ such that $\gamma(Kv) = v$ for all $v \in H^{m-1/2}$. It suffices to construct Kv for $v \in \mathscr{S}(\mathbf{R}^{n-1})$. Let $\hat{v}(\xi')$ denote the Fourier transform of v. Then $Kv(x)$ will be the inverse Fourier transform with respect to ξ' of the function

$$\hat{v}(\xi') \exp(-(1 + |\xi'|^2)^{1/2} x^n). \tag{26.15}$$

In other words,

$$Kv(x) = (2\pi)^{-n+1} \int \exp[i\langle x', \xi'\rangle - (1 + |\xi'|^2)^{1/2} x^n] \hat{v}(\xi')\, d\xi'. \tag{26.16}$$

It is clear that, for $x^n \geqq 0$, the function (26.15) belongs to $\mathscr{S}(\mathbf{R}_{n-1})$. Furthermore, if $j < m$,

$$\int_0^{+\infty} \|D_n^j Kv(x)\|^2_{H^{m-j}(\mathbf{R}^{n-1})}\, dx^n$$

$$= (2\pi)^{-n+1} \iint_0^{+\infty} (1 + |\xi'|^2)^{m-j/2} \exp[-2(1 + |\xi'|^2)^{1/2} x^n] |\hat{v}(\xi')|^2\, d\xi'\, dx^n$$

$$= \tfrac{1}{2}\|v\|^2_{H^{m-(j+1)/2}(\mathbf{R}^{n-1})} \leqq \tfrac{1}{2}\|v\|^2_{H^{m-1/2}(\mathbf{R}^{n-1})},$$

which shows that $K : H^{m-1/2}(\mathbf{R}^{n-1}) \to H^m(\mathbf{R}^n_+)$ is continuous. That $\gamma(Kv) = v$ is obvious in view of (26.16). Q.E.D.

Remark 26.1. From (26.16) we derive that

$$(\partial/\partial x^n)^2 Kv = \left(1 - \sum_{j=1}^{n-1} (\partial/\partial x^j)^2\right) v,$$

hence Kv is a solution of the following Dirichlet problem in $\mathbf{R}^n_+$:

(26.17) $$(1 - \Delta)Kv = 0,$$

(26.18) $$\gamma(Kv) = v.$$

We may now define on $H^m(\mathbf{R}^n_+)$ all the *traces of order* $j < m$. Indeed, $(\partial/\partial x^n)^j$ maps $H^m(\mathbf{R}^n_+)$ into $H^{m-j}(\mathbf{R}^n_+)$ and the trace γ maps the latter onto $H^{m-j-1/2}(\mathbf{R}^{n-1})$. We set

(26.19) $$\gamma^j(u) = \gamma((\partial/\partial x^n)^j u).$$

Thus γ^j is a continuous linear map $H^m(\mathbf{R}^n_+) \to H^{m-j-1/2}(\mathbf{R}^{n-1})$.

Let us now go back to our bounded open set Ω with smooth boundary Γ. If we keep in mind what the localization and boundary flattening meant, we see that the partial differentiation in the direction normal to Γ, $\partial/\partial \nu$, must now replace $\partial/\partial x^n$. Of course, the vector field $\partial/\partial \nu$ may not be well defined far away from the boundary, but when we consider the traces on Γ, what happens far away from the boundary is irrelevant. It therefore makes sense to set

(26.20) $$\gamma^j(u) = \gamma((\partial/\partial \nu)^j u), \qquad u \in H^m(\Omega), \quad 0 \leqq j < m.$$

THEOREM 26.5. *Let Ω be a bounded open subset of $\mathbf{R}^n$ with a smooth boundary Γ, lying on one side of Γ. If $0 \leqq j < m$, the jth trace γ^j is a continuous linear map of $H^m(\Omega)$ onto $H^{m-j-1/2}(\Gamma)$.*

It is, of course, the "onto" part of this statement that we must prove. By localization and flattening of the boundary, we are easily reduced to prove the analogous statement in $\mathbf{R}^n_+$:

THEOREM 26.6. *If $0 \leqq j < m$, γ^j maps $H^m(\mathbf{R}^n_+)$ onto $H^{m-j-1/2}(\mathbf{R}^{n-1})$.*

We have once more identified $\mathbf{R}^{n-1}$ with the hyperplane $x^n = 0$.

Proof. Let $v \in H^{m-j-1/2}(\mathbf{R}^{n-1})$. Set

(26.21)

$$K_j u(x) = \frac{(-1)^j}{(2\pi)^{n-1}} \int \exp[i\langle x', \xi'\rangle - (1 + |\xi'|^2)^{1/2} x^n](1 + |\xi'|^2)^{-j/2} \hat{v}(\xi')\, d\xi'.$$

Clearly, $(\partial/\partial x^n)^j K_j u = v$ when $x^n = 0$. That $K_j u \in H^m(\mathbf{R}^n_+)$ can be seen by the same method used in the case $j = 0$ on page 243. Q.E.D.

We conclude this section by proving a few illuminating properties of $H^m(\Omega)$ and $H^m_0(\Omega)$. Without further mention, we assume that Ω *is bounded, has a smooth boundary, and lies on one side of it.*

THEOREM 26.7. *The restriction to Ω maps $H^m(\mathbf{R}^n)$ onto $H^m(\Omega)$.*

Theorem 26.7 is of course true also when $m = 0$, although in the proof we shall assume $m > 0$. It states that every $f \in H^m(\Omega)$ can be extended to $\mathbf{R}^n$ as an element of $H^m(\mathbf{R}^n)$. By localization and flattening of the boundary, the proof reduces to that of the analogous statement in $\mathbf{R}^n_+$:

THEOREM 26.8. *The restriction to $\mathbf{R}^n_+$ maps $H^m(\mathbf{R}^n)$ onto $H^m(\mathbf{R}^n_+)$.*

Proof. We shall content ourselves with sketching it. Let $u \in H^m(\mathbf{R}^n_+)$, and $\gamma^j(u)$ be its traces of order $j = 0, \ldots, m-1$. Let v be any element of $H^m(\mathbf{R}^n_-)$ such that $\gamma^j(u) = \gamma^j(v)$ for every $j = 0, \ldots, m-1$ (of course, $\mathbf{R}^n_- = \{x \in \mathbf{R}^n;\ x^n < 0\}$; here trace means trace on the hyperplane $x^n = 0$). Let $\tilde{u}$ denote the function equal to u for $x^n > 0$ and to v for $x^n < 0$. If we view $\tilde{u}$ as a function of $x^n \in \mathbf{R}^1$ valued in the space of distributions with respect to $x' \in \mathbf{R}^{n-1}$, we can easily check that it has $(m-1)$ continuous derivatives and that, for any k, $0 \leqq k \leqq m$, $(\partial/\partial x^n)^k \tilde{u}$ is the function equal to $(\partial/\partial x^n)^k u$ for $x^n > 0$ and to $(\partial/\partial x^n)^k v$ for $x^n < 0$. We conclude from this that $(\partial/\partial x^n)^k \tilde{u}$ is an L^2 function of x^n with values in $H^{m-k}(\mathbf{R}^{n-1}_{x'})$. Q.E.D.

Remark 26.2. If we know that $\gamma^j(u) = 0$ for every $j < m$, in the preceding proof, we may choose $v \equiv 0$. Then $\tilde{u}$ is the extension of u by zero in $\mathbf{R}^n_-$.

Let $\tilde{\gamma}$ be the mapping $u \mapsto (\gamma^0(u), \ldots, \gamma^{m-1}(u))$ of $H^m(\Omega)$ into

$$\prod_{j=0}^{m-1} H^{m-j-1/2}(\Gamma).$$

THEOREM 26.9. *The kernel of $\tilde{\gamma}$ in $H^m(\Omega)$ is exactly equal to $H^m_0(\Omega)$.*

Since $C_c^\infty(\Omega) \subset \operatorname{Ker} \tilde{\gamma}$, we certainly have $H_0^m(\Omega) \subset \operatorname{Ker} \tilde{\gamma}$. We must prove the converse inclusion. By localization and flattening of the boundary, we are reduced to proving the analogous statement for $\mathbf{R}_+^n$:

THEOREM 26.10. *The kernel of* $\tilde{\gamma}$ *in* $H^m(\mathbf{R}_+^n)$ *is exactly* $H_0^m(\mathbf{R}_+^n)$.

Proof. Let $u \in \operatorname{Ker} \tilde{\gamma}$. By virtue of Remark 26.2, the extension $\tilde{u}$ of u to $\mathbf{R}^n$ by setting $\tilde{u} = 0$ in the complement of $\mathbf{R}_+^n$ belongs to $H^m(\mathbf{R}^n)$. On the other hand, in $H^m(\mathbf{R}_+^n)$, u is the limit, as $\varepsilon \to +0$, of the functions u_ε defined by $u_\varepsilon(x) = \tilde{u}(x', x^n - \varepsilon)$. Since the supports of the latter stay at a distance $\geqq \varepsilon$ from the boundary of $\mathbf{R}_+^n$, they belong to $H_0^m(\mathbf{R}_+^n)$ and therefore so does u.

Q.E.D.

Remark 26.3. Let $u \in H^m(\Omega)$ and let $\tilde{u}$ denote the extension of u to $\mathbf{R}^n$ by zero in the complement of Ω. Naturally $\tilde{u} \in L^2(\mathbf{R}^n)$. But, in general, $\tilde{u}$ does not belong to $H^m(\mathbf{R}^n)$. We have seen that it does when $u \in H_0^m(\Omega)$. Actually, this is the only case where it does (Exercise 26.1).

Appendix Extension to $\mathbf{R}^n$ of Elements of $H^{m,p}(\Omega)$

Although the trace mapping theorems of this section (Theorems 26.2, 26.4 to 26.6, 26.9, and 26.10) have no simple analogs for spaces $H^{m,p}$ with $p \neq 2$, some of the other results, such as Theorem 26.1 for instance, remain valid for $p \neq 2$. We shall briefly indicate how to prove them. Throughout this Appendix, p will be any positive number, $1 \leqq p < +\infty$, m any integer, $m \geqq 0$. It is convenient to state the results in a slightly more general form than those proved in this section. More precisely, we shall not assume that $\Gamma = \partial\Omega$ is a C^∞ hypersurface, but only that it is C^m. This could also have been done in this section, as inspection of the proofs will easily convince the reader (we have not done it, for the sake of simplicity and also in order to have statements that would be valid for any m).

THEOREM 26.A.1. *Let m be a nonnegative integer, p a number such that $1 \leqq p < +\infty$, Ω a bounded open subset of $\mathbf{R}^n$ whose boundary $\partial\Omega$ is a C^m hypersurface. Then the restrictions to Ω of the functions belonging to $C_c^m(\mathbf{R}^n)$ are dense in $H^{m,p}(\Omega)$.*

Proof. By localization and flattening of the boundary, we are reduced to proving a statement of the kind (26.4). It must be kept in mind that, in order to flatten the boundary, we can use diffeomorphisms which are merely C^m and not, in general, C^∞. The analog of (26.4) reads:

(26.A.1) *Let $u \in H^{m,p}(B^+)$ have its support contained in a compact subset of B. Then u is the limit, in $H^{m,p}(B^+)$, of a sequence of functions which are the restrictions to B^+ of elements of $C_c^\infty(B)$.*

Of course, u may be viewed as an element of $H^{m,p}(\mathbf{R}^n_+)$ whose support is contained in a compact subset of B. By duplicating the proof of Theorem 26.3 we can prove the following statement, which obviously implies (26.A.1):

THEOREM 26.A.2. *Let m be any nonnegative integer, p any number such that $1 \leqq p < +\infty$. The restrictions to $\mathbf{R}^n_+$ of functions belonging to $C^\infty_c(\mathbf{R}^n)$ are dense in $H^{m,p}(\mathbf{R}^n_+)$.*

The proof of Theorem 26.A.2 is identical to that of Theorem 26.3, except that we must replace L^2 everywhere by L^p. In particular, we must replace the space $H^s(K')$ by $H^{s,p}(K') = \{w \in H^{s,p}(\mathbf{R}^{n-1});\ \text{supp}\ w \subset K'\}$. Now s is an arbitrarily large integer; $H^{s,p}(K')$ is a separable Banach space. Otherwise nothing is to be changed in the argument.

The restrictions to Ω of functions belonging to $C^m_c(\mathbf{R}^n)$ clearly belong to $C^m(\overline{\Omega})$, the space of C^m functions in Ω whose derivatives of order $\leqq m$ can be extended as continuous functions to the closure $\overline{\Omega}$ of Ω. As a matter of fact, the restriction mapping to Ω, r_Ω, maps $C^m_c(\mathbf{R}^n)$ onto $C^m(\overline{\Omega})$—provided that the hypotheses of Theorem 26.A.1 are satisfied. Furthermore, it is not too difficult, then, to construct a right inverse $\varepsilon_\Omega : C^m(\overline{\Omega}) \to C^m_c(\mathbf{R}^n)$ of r_Ω, which is continuous when the first space is equipped with the norm of $H^{m,p}(\Omega)$ and the second one, with that of $H^{m,p}(\mathbf{R}^n)$. Let φ be an arbitrary element of $C^m(\overline{\Omega})$. After localizing and flattening the boundary, we may assume that $\Omega = B^+$ and that the support of φ is contained in a compact subset of B. We set, for $x^n < 0$,

$$\tilde{\varphi}(x) = \sum_{j=0}^{m} \lambda_j \varphi(x', -\mu_j x^n), \qquad \mu_j > 0, \quad j = 0, \ldots, m,$$

and choose the real numbers λ_j, μ_j in such a way that each trace of order k, $0 \leqq k \leqq m$, of $\tilde{\varphi}$ on the hyperplane $x^n = 0$ is equal to the trace of the same order of φ. This will be true if we require

$$\sum_{k=0}^{m} \lambda_k \mu_k^j = (-1)^j, \qquad j = 0, \ldots, m. \tag{26.A.2}$$

The advantage over the extension by means of the Taylor expansion of φ is that we may now estimate the norm of $\tilde{\varphi}$ in $H^{m,p}(\mathbf{R}^n_-)$ in terms of the norm of φ in $H^{m,p}(\mathbf{R}^n_+)$. Indeed, for some constant C depending only on n, p, and on the choice of the λ_j and μ_j, we have

$$\|\tilde{\varphi}\|_{H^{m,p}(\mathbf{R}^n)} \leqq C\|\varphi\|_{H^{m,p}(\mathbf{R}_+{}^n)}.$$

We shall then set $\Phi = \varphi$ for $x^n > 0$, $\Phi = \tilde{\varphi}$ for $x^n < 0$. We know then that the support of Φ is a compact subset of B, that $\Phi \in C^m_c(B)$, and that

$$\|\Phi\|_{H^{m,p}(\mathbf{R}^n)} \leqq C\|\varphi\|_{H^{m,p}(\mathbf{R}_+{}^n)}.$$

We may now revert to the initially given open set Ω. After some patching up, we obtain the following:

THEOREM 26.A.3. *Let m, p, Ω be as in Theorem 26.A.1. Let Ω_0 be any open subset of $\mathbf{R}^n$ containing the closure $\overline{\Omega}$ of Ω. There is a linear map*

$$\varepsilon_\Omega : C^m(\overline{\Omega}) \to C_c^m(\Omega_0)$$

such that $r_\Omega \varepsilon_\Omega = I$, the identity mapping of $C^m(\overline{\Omega})$, which is continuous when $C^m(\overline{\Omega})$ and $C_c^m(\Omega_0)$ carry the norms of $H^{m,p}(\Omega)$ and $H^{m,p}(\Omega_0)$, respectively.

From Theorem 26.A.1 it follows that $C^m(\overline{\Omega})$ is dense in $H^{m,p}(\Omega)$. Consequently,

COROLLARY 26.A.1. *The mapping ε_Ω in Theorem 26.A.3 can be extended as a continuous linear map $H^{m,p}(\Omega) \to H_0^{m,p}(\Omega_0)$. The map r_Ω, the restriction to Ω, maps $H_0^{m,p}(\Omega_0)$ onto $H^{m,p}(\Omega)$.*

Remark 26.A.1. Inspection of the manner in which the extension mapping ε_Ω is constructed shows that it may be chosen independently of p, $1 \leqq p < +\infty$, and so as to be continuous from $H^{k,p}(\Omega)$ into $H_0^{k,p}(\Omega_0)$ for every $k = 0, 1, \ldots, m$.

Exercises

26.1. Let Ω be a bounded open subset of $\mathbf{R}^n$ with C^∞ boundary Γ; we assume that Ω lies on one side of Γ. Let m be an arbitrary integer at least zero. Prove that $H_0^m(\Omega)$ is the image, under the restriction mapping r_Ω (restriction to Ω of distributions in $\mathbf{R}^n$), of the subspace of $H^m(\mathbf{R}^n)$ consisting of the functions in this space whose support is contained in $\overline{\Omega}$.

26.2. Let Ω be a bounded subset of $\mathbf{R}^n$, open and *star-shaped* about the origin (this means that if x_0 is any point in Ω, the closed straight line segment joining 0 to x_0 is contained in Ω). Prove that the restrictions to Ω of C^∞ functions in $\mathbf{R}^n$ are dense in $H^m(\Omega)$.

26.3. Let $\Omega = \{(x, y) \in \mathbf{R}^2;\ x^2 + y^2 < 1\}$, $\Omega_0 = \Omega \backslash \{0\}$. Prove that the mapping restriction to Ω_0 maps $H^m(\Omega)$ onto a closed linear subspace of $H^m(\Omega_0)$.

26.4. Use the same notation as in Exercise 26.3. Equip $H^1(\Omega_0)$ with the inner product $D(u, v) = \int_\Omega (\operatorname{grad} u) \cdot (\operatorname{grad} \bar{v})\, dx$ (this defines a Hilbert space structure equivalent to the standard one). Show that if $u \in H^1(\Omega_0)$ is orthogonal [for $D(u, v)$] to all the restrictions of elements of $H^1(\Omega)$, we must have

$$u = Q(D)(\log r) + h,$$

where h is harmonic in the unit disk Ω, and $Q(D)$ is a differential operator with constant coefficients in $\mathbf{R}^2$. Conclude from this that u must be identically zero, and that the restriction to Ω_0 maps $H^1(\Omega)$, and therefore also $H^1(\mathbf{R}^n)$, onto $H^1(\Omega_0)$.

26.5. Use the same notation as in Exercise 26.4. Prove that the natural injection of $C_c^\infty(\Omega_0)$ into $C_c^\infty(\Omega)$ can be extended as an isometry of $H_0^1(\Omega_0)$ *onto* $H_0^1(\Omega)$. Derive from this that there are C^∞ functions, belonging to $H_0^1(\Omega_0)$, which are identically one in a neighborhood of the origin.

26.6. Let $\mathbf{E}$ be a complex Hilbert space. Show that there is a bounded linear operator $\varepsilon : H^1(\mathbf{R}_+^1; \mathbf{E}) \to H^1(\mathbf{R}^1; \mathbf{E})$ such that the restriction to $\mathbf{R}_+^1$ of any εu, $u \in H^1(\mathbf{R}_+^1; \mathbf{E})$, is equal to u (apply the method in the proof of Theorem 26.A.3). Let then Ω' be an open subset of $\mathbf{R}^{n-1}$. Apply the preceding result to obtain that there is a continuous linear extension mapping of $H^1(\Omega' \times \mathbf{R}_+^1)$ into $H^1(\Omega' \times \mathbf{R}^1)$ (i.e., a continuous right inverse of the restriction mapping). Let $(\mathbf{R}_+)^n$ denote the product of n copies of the open positive half-line. Prove that there is a continuous extension mapping of $H^1((\mathbf{R}_+)^n)$ into $H^1(\mathbf{R}^n)$. Let Ω be a *convex polygon*, i.e., the interior of the convex hull of a finite set of points. Prove that there is a continuous extension mapping $H^1(\Omega) \to H^1(\mathbf{R}^n)$.

26.7. Let Ω be a bounded open subset of $\mathbf{R}^n$ such that the mapping "restriction to Ω of functions in $\mathbf{R}^n$," r_Ω, regarded as a continuous linear map $H^m(\mathbf{R}^n) \to H^m(\Omega)$, has a continuous right inverse ε_Ω (see Theorem 26.A.3). Prove then that, *whatever the integer* $k < m$ (k could be negative), *the natural injection of* $H^m(\Omega)$ *into* $H^k(\Omega)$ *is compact* (this is one of the standard versions of the classical *Rellich's lemma*; another version is Proposition 25.5).

27

Back to the Dirichlet Problem. Regularity up to the Boundary

Let Ω be a bounded open subset of $\mathbf{R}^n$ whose boundary is a C^∞ hypersurface Γ, and which lies on one side of Γ. With the insight gained in Sect. 26 we may now give a neater formulation of the weak form of the (inhomogeneous) Dirichlet problem:

(27.1) *Given any* $f \in H^{-1}(\Omega)$ *and any* $g \in H^{1/2}(\Gamma)$ *find* $u \in H^1(\Omega)$ *such that:*

(27.2)
$$(\lambda - \Delta)u = f \quad \textit{in } \Omega,$$

(27.3)
$$\gamma(u) = g.$$

We recall that γ is the *trace* of u on Γ. By virtue of Theorem 26.2 we know that there is $v \in H^1(\Omega)$ such that $\gamma(v) = u$. Hence, setting $w = u - v$, we transform our problem into that of solving [in $H^1(\Omega)$]

(27.4)
$$(\lambda - \Delta)w = F \quad \textit{in } \Omega, \qquad \gamma(w) = 0,$$

where $F = f - \lambda v + \Delta v$. By Theorem 26.9 we know that $\gamma(w) = 0$ is equivalent to $w \in H_0^1(\Omega)$. If we apply Theorem 23.1, we see that, for real nonnegative λ, (27.4) has a unique solution, as does problem (27.2)–(27.3). We may state

THEOREM 27.1. *Let* Ω *be a bounded open subset of* $\mathbf{R}^n$, *with boundary a* C^∞ *hypersurface* Γ, *lying on one side of* Γ. *Let* λ *be any nonnegative number. The mapping*

(27.5)
$$u \mapsto (\lambda u - \Delta u, \gamma(u))$$

is an isomorphism of $H^1(\Omega)$ *onto* $H^{-1}(\Omega) \times H^{1/2}(\Omega)$.

We are now going to study the following question. Suppose that we strengthen the regularity requirements on the data f, g in problem (27.1); what can we say about the regularity of the solution u? As a partial answer to this question, we shall prove the following result:

THEOREM 27.2. *Under the same hypotheses as in Theorem 27.1, let m be any integer strictly greater than zero. The mapping (27.5) is an isomorphism of* $H^m(\Omega)$ *onto* $H^{m-2}(\Omega) \times H^{m-1/2}(\Gamma)$.

Proof. All we have to do is to prove that if $f \in H^{m-2}(\Omega)$, $g \in H^{m-1/2}(\Gamma)$, then the solution u of (27.2)–(27.3), which we know to belong to $H^1(\Omega)$, does in fact belong to $H^m(\Omega)$. It suffices to prove this when $g = 0$ as we can always replace u by $u - v$ with $v \in H^m(\Omega)$ such that $\gamma(v) = g$ (by Theorem 26.2).

In order to prove our contention we shall use localization and flattening of the boundary. In fact, we shall use the same open covering of $\overline{\Omega}$, $V_0, V_1, \ldots, V_r$, and the same partition of unity, $\zeta_0, \zeta_1, \ldots, \zeta_r$, introduced at the beginning of Sect. 26. Note that, for each $j = 0, 1, \ldots, r$,

$$(\lambda - \Delta)(\zeta_j u) = f_j \quad \text{in } \Omega, \tag{27.6$_j$}$$

where

$$f_j = \zeta_j f + 2\langle \text{grad } \zeta_j, \text{grad } u\rangle + (\Delta\zeta_j)u. \tag{27.7}$$

Suppose that we have proved that $u \in H^k(\Omega)$, $1 \leqq k < m$. Then f_j will belong to $H^{k-1}(\Omega)$. We are going to show that this implies $u \in H^{k+1}(\Omega)$. Since our hypothesis that $u \in H^k(\Omega)$ is verified when $k = 1$, our assertion will follow by induction on k.

The case $j = 0$ is easily settled by the following lemma:

LEMMA 27.1. *Let* $v \in H^k_c(\mathbf{R}^n)(k \geqq 0)$ *such that* $(\lambda - \Delta)v \in H^{k-1}(\mathbf{R}^n)$. *Then* $v \in H^{k+1}(\mathbf{R}^n)$.

Proof. We have $(1 - \Delta)v = (\lambda - \Delta)v + (1 - \lambda)v \in H^{k-1}(\mathbf{R}^n)$ and $1 - \Delta$ induces an isometry of $H^{s+1}(\mathbf{R}^n)$ onto $H^{s-1}(\mathbf{R}^n)$ (whatever the real number s).

From now on we focus our attention on the cases $j > 0$. This means that V_j intersects the boundary hypersurface Γ. Note that we can choose each V_j as small as we wish (this might of course force us to increase the number r of elements in our covering $\{V_j\}$ and also to enlarge the "central" element V_0—but these facts will not affect the argument below).

Thus we shall study the equation (27.6)$_j$ for $j > 0$. As a matter of fact, we shall drop any subscript j. We shall write u instead of $\zeta_j u$, also V instead of V_j, f instead of f_j, and so on. We have then

$$(\lambda - \Delta)u = f \quad \text{in } V \cap \Omega, \tag{27.8}$$

with $u \in H^1_0(V \cap \Omega)$. We must keep in mind, moreover, that now

(27.9) *the closures of* supp u, supp f *in* V *are compact subsets of* V.

The statement which we must prove, and which will imply what we want, is the following one:

(27.10) *if* $1 \leqq k < m$ *and if* $u \in H^k(V \cap \Omega)$, $f \in H^{k-1}(\Omega \cap V)$, *then* $u \in H^{k+1}(V \cap \Omega)$.

We note, once more, that $(1 - \Delta)u = f + (1 - \lambda)u \in H^{k-1}(V \cap \Omega)$ if the hypotheses in (27.10) are satisfied. This means that we may assume λ to be equal to *one*.

We shall begin by flattening the boundary. Such an operation can be performed in any way one wishes, but we shall perform it by taking advantage of the particular properties of the Laplace operator (in particular, of its rotation invariance). This will simplify our exposition, but is by no means essential. The same reasoning applies to all second-order elliptic operators with C^∞ coefficients, whose principal part is a Hermitian quadratic form.

First of all, we may assume that the point x_0 is the origin in $\mathbf{R}^n$. Second, possibly after performing a rotation in $\mathbf{R}^n$, we may assume that the normal to Γ at x_0 is the x^n coordinate axis. In fact, we may then assume that $V \cap \Gamma$ is defined by an equation $x^n = \varphi(x')$ where, as usual, $x' = (x^1, \ldots, x^{n-1})$. We then set $y' = x'$, $y^n = x^n - \varphi(x')$. As we may eventually want to shrink V (at most, finitely many times!), we may point out that φ is defined and C^∞ in a fixed open neighborhood W' of the x' projection of $\overline{V}$. The change of variables above transforms our operator $1 - \Delta_x$ into

$$Q = 1 - \sum_{j=1}^{n-1} \left(\frac{\partial}{\partial y^j} - \frac{\partial \varphi}{\partial x^j}(y')\frac{\partial}{\partial y^n}\right)^2 - \left(\frac{\partial}{\partial y^n}\right)^2.$$

It should be noted that

$$\text{grad}\, \varphi(0) = 0. \tag{27.11}$$

Let then U, F denote the expressions of u, f in the new coordinates y. We have

$$QU = F \quad \text{in } V \cap \Omega, \tag{27.12}$$

and clearly $U \in H_0^1(V \cap \Omega)$. As a matter of fact, both supp U and supp F are contained in compact subsets of V, hence U can be regarded as an element of $H_0^1(\mathbf{R}_+^n)$ while F can be regarded as an element of $H^{k-1}(\mathbf{R}_+^n)$. We must show that if $U \in H^k(\mathbf{R}_+^n)$, then, in fact, $U \in H^{k+1}(\mathbf{R}_+^n)$.

We shall write

$$Q = (1 - \Delta) + R,$$

with

$$R = 2\sum_{j=1}^{n-1} \frac{\partial \varphi}{\partial y^j} \frac{\partial^2}{\partial y^j\, \partial y^n} - |\text{grad}\, \varphi|^2 \left(\frac{\partial}{\partial y^n}\right)^2 + (\Delta\varphi)\frac{\partial}{\partial y^n}.$$

We shall make Q and R act on functions with support contained in compact subsets of V. Thus the fact that R is not defined outside of V will not matter.

Nevertheless, in order to avoid undue complications in the exposition, we shall extend φ as a C_c^∞ function in the whole space $\mathbf{R}^n$ (in any fashion we wish) and thus R will be defined everywhere.

Let us indicate how we intend to prove our assertion (27.10). We know (Theorem 23.1) that $1 - \Delta$ is an isomorphism of $H_0^1(\mathbf{R}_+^n)$ onto $H^{-1}(\mathbf{R}_+^n)$. We shall give an explicit description of its inverse, which we call G_0. This expression will enable us to prove the following statements:

(27.13) *given any* $k = 0, 1, \ldots,$ G_0 *induces a bounded linear operator* $H^{k-1}(\mathbf{R}_+^n) \to H^{k+1}(\mathbf{R}_+^n)$;

(27.14) *given any* $k = 1, 2, \ldots,$ *and any* $\varepsilon > 0$, *there is* $\delta > 0$ *such that, if the diameter of* V *is less than* δ, *then, for every* $w \in H^{k-1}(\mathbf{R}_+^n)$ *whose support is contained in* V,

$$\|RG_0 w\|_{H^{k-1}(\mathbf{R}_+^n)} \leqq \varepsilon \|w\|_{H^{k-1}(\mathbf{R}_+^n)}.$$

These properties of G_0 will imply what we want. Indeed, set

$$U_1 = G_0(I + RG_0)^{-1}F.$$

We know that $F \in H^{k-1}(\mathbf{R}_+^n)$ and that supp $F \subset V$. In (27.14) we choose $\varepsilon < 1$ and accordingly take the diameter of V small ($<\delta$). Then $(I + RG_0)^{-1}F$ makes sense and defines an element of $H^{k-1}(\mathbf{R}_+^n)$. Consequently, by (27.13), U_1 belongs to $H^{k+1}(\mathbf{R}_+^n)$. It satisfies

$$QU_1 = (1 - \Delta)G_0(I + RG_0)^{-1}F + RG_0(I + RG_0)^{-1}F = F.$$

But the uniqueness of the solution to (27.12) [which follows from the uniqueness of the solution to (27.2)–(27.3)] demands that $U = U_1$.

Construction of the Operator G_0

Given any $v \in H^{-1}(\mathbf{R}_+^n)$ we must have $G_0 v \in H_0^1(\mathbf{R}_+^n)$ and

$$(1 - \Delta)G_0 v = v. \tag{27.15}$$

It is natural to perform a Fourier transformation with respect to $y' = (y^1, \ldots, y^{n-1})$. Let $\hat{v}$ denote the corresponding Fourier transform of v. For simplicity let us write

$$\rho = (1 + |\eta'|^2)^{1/2}, \qquad \eta' \in \mathbf{R}_{n-1}.$$

Equation (27.15) can be rewritten as

$$\left(\frac{\partial}{\partial y^n} - \rho\right)\left(\frac{\partial}{\partial y^n} + \rho\right)\widehat{G_0 v} = -\hat{v}. \tag{27.16}$$

We are reduced to solving an ordinary differential equation for $y^n > 0$, depending on the parameter η'. We have two basic requirements on the solution: First it must vanish when $y^n = 0$; this corresponds to the fact that $G_0 v$ must belong to $H_0^1(\mathbf{R}^n)$. Second, it must be tempered with respect to η', which is necessary if $\widehat{G_0 v}$ is to be the Fourier transform with respect to y' of a distribution in $\mathbf{R}^n_+$. As we shall see, it is possible to comply with both demands.

It is convenient to invert separately the two factors

$$L_+ = \frac{\partial}{\partial y^n} + \rho, \qquad L_- = \frac{\partial}{\partial y^n} - \rho,$$

keeping in mind what we have just said. Note that it suffices to define G_0 on a dense subset of $H^{-1}(\mathbf{R}^n_+)$. Thus we may assume that v is C^∞ and vanishes for y^n large. In this case we may set

$$E_- \hat{v}(\eta', y^n) = -\int_{y^n}^{+\infty} \exp[\rho(y^n - t)]\hat{v}(\eta', t)\, dt.$$

It is clear that $L_- E_- v = v$ and that E_- is tempered with respect to η'. We may also set

$$E_+ \hat{v}(\eta', y^n) = \int_0^{y^n} \exp[-\rho(y^n - t)]\hat{v}(\eta', t)\, dt,$$

and then $L_+ E_+ \hat{v} = \hat{v}$; $E_+ \hat{v}$ is tempered with respect to η'. Finally we set

$$\widehat{G_0 v} = -E_+ E_- \hat{v} \tag{27.17}$$

$$= \int_0^{y^n} \int_t^{+\infty} \exp[-\rho(y^n - t) + \rho(t - s)]\hat{v}(\eta', s)\, ds\, dt.$$

Of course,

$$G_0 v(y) = (2\pi)^{1-n} \int e^{i\langle y', \eta'\rangle} \widehat{G_0 v}(\eta', y^n)\, d\eta'. \tag{27.18}$$

Proof that G_0 Maps $H^{k-1}(\mathbf{R}^n_+)$ into $H^{k+1}(\mathbf{R}^n_+)$

Let us call $\mathscr{E}_+ v$ the inverse Fourier transform with respect to η' of $E_+ \hat{v}$; similarly, call $\mathscr{E}_- v$ that of $E_- \hat{v}$. It suffices to prove that $\mathscr{E}_+$ and $\mathscr{E}_-$ map $H^{k-1}(\mathbf{R}^n_+)$ into $H^k(\mathbf{R}^n_+)$. We shall give the proof for $\mathscr{E}_+$ only; the treatment of $\mathscr{E}_-$ is analogous.

Note that

$$E_+ \hat{v}(\eta', y^n) = \int_0^{+\infty} \exp[-\rho(y^n - t)]H(y^n - t)\hat{v}(\eta', t)\, dt$$

where $H(s)$ stands for the Heaviside function. From the standard Hölder's inequalities for convolution we obtain

$$\int_0^{+\infty} |E_+ \hat{v}(\eta', y^n)|^2 \, dy^n \leqq \left(\int_0^{+\infty} e^{-\rho t} \, dt\right)^2 \int_0^{+\infty} |\hat{v}(\eta', \hat{y}^n)|^2 \, dy^n$$

$$= \int_0^{+\infty} \rho^{-2} |\hat{v}(\eta', y^n)|^2 \, dy^n.$$

If we multiply both extreme sides by ρ^{2k} and integrate them with respect to η' over $\mathbf{R}_{n-1}$, we obtain

(27.19) $$\int_0^{+\infty} \|\mathscr{E}_+ v(\cdot, y^n)\|^2_{H^k(\mathbf{R}^{n-1})} \, dy^n \leqq \int_0^{+\infty} \|v(\cdot, y^n)\|^2_{H^{k-1}(\mathbf{R}^{n-1})} \, dy^n.$$

Note, incidentally, that k need not be an integer—it could be any real number.

Next we exploit the fact that

$$(\partial/\partial y^n) E_+ \hat{v} = \rho E_+ \hat{v} + \hat{v},$$

or, equivalently, with obvious notation,

(27.20) $$(\partial/\partial y^n)\mathscr{E}_+ v = (1 - \Delta')^{1/2} \mathscr{E}_+ v + v.$$

Suppose that we know that $v \in H^{k-1}(\mathbf{R}^n_+)$. This means that given any $j = 0, \ldots, k-1$ [we may assume $k \geqq 1$ here, as we know already that G_0 maps $H^{-1}(\mathbf{R}^n_+)$ onto $H^1_0(\mathbf{R}^n_+)$], $(\partial/\partial y^n)^j v$ is an L^2 function of $y^n > 0$ valued in $H^{k-1-j}(\mathbf{R}^{n-1})$. Suppose then that we have already proved that

(27.21) $$(\partial/\partial y^n)^j (\mathscr{E}_+ v) \in L^2(\mathbf{R}^n_+ ; H^{k-j}(\mathbf{R}^{n-1})).$$

This implies

$$(1 - \Delta')^{1/2} (\partial/\partial y^n)^j (\mathscr{E}_+ v) \in L^2(\mathbf{R}^n_+ ; H^{k-j-1}(\mathbf{R}^{n-1})).$$

If we apply $(\partial/\partial y^n)^j$ to both members in (27.20), we obtain that

$$(\partial/\partial y^n)^{j+1} (\mathscr{E}_+ v) \in L^2(\mathbf{R}^n_+ ; H^{k-(j+1)}(\mathbf{R}^{n-1})).$$

Since (27.21) has been proved for $j = 0$, our induction argument works and proves it for all $j = 0, 1, \ldots, k$. But then it means that $\mathscr{E}_+ v$ belongs to $H^k(\mathbf{R}^n_+)$ (cf. p. 240).

Proof of (27.14)

In what follows w will always denote an arbitrary element of $H^{k-1}(\mathbf{R}^n_+)$ with support in V. Let V_1 be an open neighborhood of $\bar{V}$ and $\zeta \in C^\infty_c(V_1)$ be equal to one in a neighborhood of $\bar{V}$. We have

$$(1 - \Delta)\{(1 - \zeta) G_0 w\} = 2\langle \operatorname{grad} \zeta, \operatorname{grad}(G_0 w)\rangle + (\Delta\zeta) G_0 w.$$

The right-hand side, which we denote by h, belongs to $H^k(\mathbf{R}^n_+)$. On the other hand, we know that $G_0 w$ belongs to $H^1_0(\mathbf{R}^n_+)$ and therefore the same is true of $(1 - \zeta)G_0 w$. By the uniqueness part of Theorem 23.1 we conclude that $(1 - \zeta)G_0 w = G_0 h \in H^{k+2}(\mathbf{R}^n_+)$. Moreover there is a constant $C > 0$, depending on ζ but not on w, such that

$$\|(1 - \zeta)G_0 w\|_{H^{k+2}(\mathbf{R}_+{}^n)} \leqq C\|w\|_{H^{k-1}(\mathbf{R}_+{}^n)}. \tag{27.22}$$

Note that

$$RG_0 w = R[\zeta G_0 w] + R[(1 - \zeta)G_0 w]$$

and that $R : H^{k+1}(\mathbf{R}^n_+) \to H^{k-1}(\mathbf{R}^n_+)$. Suppose that we have proved the following:

(27.23) *to every* $\varepsilon_1 > 0$ *there is* $\eta > 0$ *such that if* diam $V_1 < \eta$, *then, for every* $v \in H^{n+1}(\mathbf{R}^n_+)$ *with support in* V_1,

$$\|Rv\|_{H^{k-1}(\mathbf{R}_+{}^n)} \leqq \varepsilon_1\|v\|_{H^{k+1}(\mathbf{R}_+{}^n)}.$$

If we apply this with $v = \zeta G_0 v$, we obtain

$$\begin{aligned}\|RG_0 w\|_{H^{k-1}(\mathbf{R}_+{}^n)} &\leqq \varepsilon_1\|\zeta G_0 w\|_{H^{k+1}(\mathbf{R}_+{}^n)} + C_1\|(1 - \zeta)G_0 w\|_{H^{k+1}(\mathbf{R}_+{}^n)}\\ &\leqq \varepsilon_1\|G_0 w\|_{H^{k+1}(\mathbf{R}_+{}^n)} + (C_1 + \varepsilon_1)\|(1 - \zeta)G_0 w\|_{H^{k+1}(\mathbf{R}_+{}^n)}\\ &\leqq \varepsilon_1 C_0\|w\|_{H^{k-1}(\mathbf{R}_+{}^n)} + C_2\|w\|_{H^{k-2}(\mathbf{R}_+{}^n)}\end{aligned}$$

by virtue of (27.22) applied with $k - 1$ substituted for k†. It is important to note that whereas C_2 depends on ζ and hence on η, C_0 is simply the norm of the bounded linear operator $G_0 : H^{k-1}(\mathbf{R}^n_+) \to H^{k+1}(\mathbf{R}^n_+)$. We shall eventually prove the following statement:

(27.24) *to every* $\varepsilon_2 > 0$ *there is* $\delta > 0$ *such that, if* diam $V < \delta$, *then, for every* $w \in H^{k-1}(\mathbf{R}^n_+)$ *with support in* V,

$$\|w\|_{H^{k-2}(\mathbf{R}_+{}^n)} \leqq \varepsilon_2\|w\|_{H^{k-1}(\mathbf{R}_+{}^n)}.$$

Let now $\varepsilon > 0$ be arbitrary. Choose η in (27.23) corresponding to $\varepsilon_1 = \varepsilon/2C_0$. From now on we keep ζ unchanged. This enables us to find the constant C_2 (which depends on η). Select then δ in (27.24) corresponding to $\varepsilon_2 = \varepsilon/2C_2$. We see that we have thus proved (27.14).

All that is left for us to do is to prove the two statements (27.23) and (27.24). This is done by standard Sobolev space techniques. We recall that

† We recall that, in the present argument, k is an integer strictly greater than zero.

$k \geqq 1$: This is going to be important. We begin by proving (27.23). For this we go back to the expression of R on page 251. We derive from it:

$$\|Rv\|_{H^{k-1}(\mathbf{R}_+{}^n)} \leqq 2 \sum_{j=1}^{n-1} \left\| \frac{\partial \varphi}{\partial y^j} v \right\|_{H^{k+1}(\mathbf{R}_+{}^n)} + \| |\operatorname{grad} \varphi|^2 v\|_{H^{k+1}(\mathbf{R}_+{}^n)} + C\|v\|_{H^k(\mathbf{R}_+{}^n)} .$$

We are going to avail ourselves of (27.11). We shall prove the following:

LEMMA 27.2. *Let l be any nonnegative integer. Let $\chi \in C_c^\infty(\mathbf{R}^n)$ such that $\chi(0) = 0$. Given any $\varepsilon > 0$ there is $\delta > 0$ such that, if $v \in H^l(\mathbf{R}_+^n)$ has its support contained in the open ball $\{x \in \mathbf{R}^n;\ |x| < \delta\}$,*

$$\|\chi v\|_{H^l(\mathbf{R}_+{}^n)} \leqq \varepsilon \|v\|_{H^l(\mathbf{R}_+{}^n)} . \tag{27.25}$$

Proof. There is a constant $C' > 0$ such that, for all $v \in H^l(\mathbf{R}_+^n)$,

$$\|\chi v\|_{H^l(\mathbf{R}_+{}^n)} \leqq C' \left\{ \sum_{|\alpha| \leqq l} \|\chi D^\alpha v\|_{L^2(\mathbf{R}_+{}^n)} + \|v\|_{H^{l-1}(\mathbf{R}_+{}^n)} \right\} .$$

We select $\delta > 0$ such that the maximum of $|\chi(x)|$ for $|x| \leqq \delta$ does not exceed $\varepsilon/2C'$. Then Lemma 27.2 will follow from

LEMMA 27.3. *Given any $\varepsilon > 0$ there is $\delta > 0$ such that, if $v \in H^l(\mathbf{R}_+^n)$ has its support in $\{x \in \mathbf{R}^n;\ |x| < \delta\}$,*

$$\|v\|_{H^{l-1}(\mathbf{R}_+{}^n)} \leqq \varepsilon \|v\|_{H^l(\mathbf{R}_+{}^n)}. \tag{27.26}$$

Proof. Suppose first that $l = 0$. Then $C_c^\infty(\mathbf{R}_+^n)$ is dense in both $H^l(\mathbf{R}_+^n) = L^2(\mathbf{R}_+^n)$ and $H^{-1}(\mathbf{R}_+^n)$ (the latter by Corollary 23.1). We may therefore take $v \in C_c^\infty(\mathbf{R}_+^n)$. But then the assertion is a particular case of Proposition 25.6.

Suppose now that l is greater than zero. We write, for $l' = l$ or $l' = l - 1$,

$$\|v\|^2_{H^{l'}(\mathbf{R}_+{}^n)} \sim \sum_{j=1}^{l'} \int_0^\delta \left\| \left(\frac{\partial}{\partial x^n} \right)^j v \right\|^2_{H^{l'-j}(\mathbf{R}^{n-1})} dx^n.$$

But if we denote by E any one of the spaces $H^s(\mathbf{R}^{n-1})$,

$$\|v(\cdot, x^n)\|_E^2 \leqq \left(\int_{x_n}^\delta \left\| \frac{\partial v}{\partial x^n}(\cdot, t) \right\|_E dt \right)^2 \leqq |\delta - x^n| \int_0^\delta \left\| \frac{\partial v}{\partial x^n}(.\,, t) \right\|_E^2 dt$$

by the Cauchy–Schwarz inequality. If we integrate both extreme sides with respect to x^n from 0 to δ, we obtain

$$\int_0^\delta \|v(\cdot, x^n)\|_E^2\, dx^n \leqq \frac{\delta^2}{2} \int_0^\delta \left\| \frac{\partial v}{\partial x^n}(\cdot, x^n) \right\|_E^2 dx^n. \tag{27.27}$$

Substituting $(\partial/\partial x^n)^j v$ for v in (27.27) and making the appropriate choice for E yields at once what we wanted.

End of the Proof of Theorem 27.2. Lemma 27.3 not only enables us to complete the proof of Lemma 27.2 but also implies (27.24). Furthermore, if we return to the estimate of $\| Rv \|_{H^{k-1}(\mathbf{R}_+^n)}$ which precedes the statement of Lemma 27.2, we see, by applying Lemma 27.2 to the first two terms on the right-hand side and Lemma 27.3 to the third one, that we have proved the assertion (27.23). Thus the proof of Theorem 27.2 is complete.

COROLLARY 27.1. *We make the same hypotheses as in Theorems 27.1 and 27.2. If* $f \in C^\infty(\bar{\Omega})$ *and* $g \in C^\infty(\Gamma)$, *the solution* u *to (27.2)–(27.3) belongs to* $C^\infty(\bar{\Omega})$.

This simply follows from the fact that

(27.28)
$$C^\infty(\bar{\Omega}) = \bigcap_{m=0}^{+\infty} H^m(\Omega).$$

Exercises

27.1. Let $\Omega = \{(x, y) \in \mathbf{R}^2; x^2 + y^2 < 1\}$. Use Fourier series expansion for functions defined (almost everywhere) in Ω,

(27.29)
$$u(r, \theta) = \sum_{m=-\infty}^{+\infty} c_m(r) e^{im\theta}.$$

State and prove necessary and sufficient conditions on the Fourier coefficients $c_m(r)$ in order that $u(r, \theta) \in H^k(\Omega)$ (k a positive integer). Using the characterization of $H^s(\mathbf{T}^1)(s \in \mathbf{R})$ in Exercise 25.4, and the found necessary and sufficient conditions, prove the following assertions:

(1) given any $k \geqq 1$ and any $u \in H^k(\Omega)$, if $0 \leqq j < k$, the trace of $(\partial/\partial r)^j u$ on $\partial\Omega$ belongs to $H^{k-j-1/2}(\partial\Omega)$;
(2) given any $k \geqq 1$ and any element g of $H^{k-1/2}(\Omega)$, there is a unique harmonic function in Ω, belonging to $H^k(\Omega)$, of which g is the trace on $\partial\Omega$.

27.2. Let Ω be a bounded open subset of $\mathbf{R}^n$, having a C^∞ boundary $\partial\Omega$ and lying on one side of it. Let m be an integer, $m \geqq 1$, and denote by $\mathcal{N}^m(\Omega)$ the space of harmonic functions in Ω which belong to $H^m(\Omega)$. Prove that $H^m(\Omega)$ is the topological direct sum of $\mathcal{N}^m(\Omega)$ and $H^m(\Omega) \cap H_0^1(\Omega)$.

27.3. Let Ω be the unit disk $x^2 + y^2 < 1$ in $\mathbf{R}^2$. Give an example of a function $u \in H_0^1(\Omega) \cap H^2(\Omega)$ whose extension $\tilde{u}$ to $\mathbf{R}^2$, obtained by setting $\tilde{u} = 0$ in $\mathbf{R}^2 \backslash \Omega$, is such that $\Delta\tilde{u} \notin L^2(\mathbf{R}^2)$.

27.4. Let $\Omega \subset \mathbf{R}^n$ be open, bounded, have a smooth boundary, and lie on one side of it. Show that given any set of functions

$$f \in H^{-1}(\Omega), \qquad g_j \in H^{2(k-j)-3/2}(\Omega) \qquad (j = 0, \ldots, k-1),$$

there is a unique function $u \in H^{2k-1}(\Omega)$ such that

$$\Delta^k u = f \quad \text{in } \Omega, \tag{27.30}$$

$$\gamma(\Delta^j u) = g_j, \qquad j = 0, \ldots, k-1, \tag{27.31}$$

where γ is the trace on $\partial\Omega$.

Give a reasonable condition on the polynomial $P(X)$ in one variable X, with real coefficients, in order that the same conclusion be valid after replacing Eq. (27.30) by

$$P(\Delta)u = f \quad \text{in } \Omega. \tag{27.32}$$

28

A Weak Maximum Principle

In this section we take the first step in the transition from the L^2 theory of the weak, or generalized, Dirichlet problem, to the socalled classical theory. In the latter, the regularity assumptions on the data, the open set Ω, its boundary $\Gamma = \partial\Omega$, the right-hand side f, and the boundary value g are relaxed or, more accurately, are different from what they were in the L^2 theory. For instance, we must be able to handle the case where the open set Ω is a square in the plane, or a cube in $\mathbf{R}^3$. Another feature will be that the regularity of the boundary value g will not be describable by means of the space $H^{1/2}(\Gamma)$. For instance, g could merely be a continuous function. Similarly, the right-hand side f will not necessarily belong to $H^{-1}(\Omega)$, and so on.

For the time being, we drop any smoothness requirement on $\partial\Omega$. We shall only assume that Ω is a *bounded* open subset of $\mathbf{R}^n$. As a matter of fact, in many statements, even the boundedness assumption can be dropped.

In this section, *all* functions and distributions will be *real*-valued. This is a technical difference with little bearing on the conclusions: The operator under study, $\lambda - \Delta$, $\lambda \geqq 0$, transforms real functions into real functions, and the complex function $(\lambda - \Delta)u = f$ can be decomposed into two real equations of the same kind. The spaces $H^m(\Omega)$, $C^\infty(\Omega)$, etc., will all consist of real functions and distributions; $H^m(\Omega)$ will be regarded as a real Hilbert space.

We begin by introducing a new notion of (partial) ordering on elements of $H^1(\Omega)$. We recall that $C^\infty(\Omega) \cap H^1(\Omega)$ is dense in $H^1(\Omega)$ (Proposition 24.1).

Definition 28.1. Let E be any subset of $\overline{\Omega}$, u any element of $H^1(\Omega)$. We say that u is nonnegative on E in the sense of $H^1(\Omega)$ if there is a sequence $\{u_\nu\}$ in $C^\infty(\Omega) \cap H^1(\Omega)$ which converges to u in $H^1(\Omega)$ and such that the following is true, for each ν,

(28.1) *there is an open neighborhood U_ν of E in $\mathbf{R}^n$ such that $u_\nu > 0$ in $U_\nu \cap \Omega$.*

We shall then write $u \succcurlyeq 0$ on E.

If $-u \succcurlyeq 0$ on E, we write $u \preccurlyeq 0$ on E; we say then that u is *nonpositive on E in the sense of* $H^1(\Omega)$. If we have both $u \succcurlyeq 0$ and $u \preccurlyeq 0$ on E, we write $u \approx 0$ on E. We may then compare any two elements u, v of $H^1(\Omega)$ on E by checking whether we have $u - v \succcurlyeq 0$ or $u - v \preccurlyeq 0$ on E. Then, of course, we write $u \succcurlyeq v$ or $u \preccurlyeq v$ on E, respectively, and so forth. We say that u is *bounded on E in the sense of* $H^1(\Omega)$ if $u \preccurlyeq M$ on E for some real number M. The infimum of those numbers M will be called the *maximum* of u on E and denoted by $\max_E u$. We can also define the *minimum* of u on E; we shall denote it by $\min_E u$. Let us also observe that the set of functions $u \in H^1(\Omega)$ such that $u \succcurlyeq 0$ on E is a *closed convex cone* in $H^1(\Omega)$.

One may wish to compare the notion "$u \succcurlyeq 0$ on E" with the more customary one, "$u \geqq 0$ almost everywhere (a.e.) in E." The following statement is easy to prove and will be used in the sequel:

PROPOSITION 28.1. *If $u \geqq 0$ a.e. in Ω, then $u \succcurlyeq 0$ on Ω.*
If E is a subset of Ω and if $u \succcurlyeq 0$ on E, then $u \geqq 0$ a.e. in E.

Proof. The proof of the first assertion is practically evident if we go back to the proof of Proposition 24.1: The functions v_ν and v constructed in that proof are $\geqq 0$ in Ω, and $v + \varepsilon$, which is >0 in Ω, converges to u in $H^1(\Omega)$ as ε goes to $+0$.

Next we prove the second assertion. Let $u_\nu \in C^\infty(\Omega) \cap H^1(\Omega)$ converge to u in $H^1(\Omega)$ and for each ν, let U_ν be an open neighborhood of E in Ω in which u_ν is >0. Let F denote the intersection of all the U_ν's; $E \subset F$ and F is measurable. Since the u_ν converge to u in $L^2(\Omega)$ there is a subsequence of the u_ν's which converges to u a.e. in Ω. As the elements of this subsequence are all >0 on F, we reach the conclusion that u itself must be $\geqq 0$ a.e. in F. Q.E.D.

Let us now consider a (real-valued) function $G(t)$ on the real line, which is *uniformly Lipschitz continuous*, i.e., such that, for some constant $K > 0$,

$$|G(t) - G(t')| \leqq K|t - t'| \qquad \text{for all} \quad t, t' \in \mathbf{R}^1. \tag{28.2}$$

It is clear that the distribution derivative G' of G belongs to $L^\infty(\mathbf{R}^1)$ and that its L^∞ norm is bounded by K. Indeed, given any $\varphi \in C_c^\infty(\mathbf{R}^1)$,

$$\langle G', \varphi \rangle = -\int G(t)\varphi'(t)\,dt = \lim_{\substack{h\to 0\\ h\neq 0}} \int G(t)\frac{1}{h}[\varphi(t) - \varphi(t+h)]\,dt$$

$$= \lim_h \int \frac{1}{h}[G(t) - G(t-h)]\varphi(t)\,dt,$$

hence, by (28.2), $|\langle G', \varphi \rangle| \leqq K\|\varphi\|_{L^1}$, which proves our assertion [note that the converse is also true: if $G' \in L^\infty(\mathbf{R}^1)$, then (28.2) holds with $K = \|G'\|_{L^\infty}$]. We have

LEMMA 28.1. *Let G be uniformly Lipschitz in* $\mathbf{R}^1$ *and let* G' *denote a bounded representative of its distribution derivative. Then* $u \mapsto G(u)$ *maps* $H^1(\Omega)$ *into itself, and*

$$(28.3) \qquad (\partial/\partial x^j)G(u) = G'(u)(\partial/\partial x^j)u, \qquad j = 1, \ldots, n,$$

with the agreement that the right-hand side vanishes wherever any one of its factors does. If moreover $G(0) = 0$, *there* $u \mapsto G(u)$ *maps* $H_0^1(\Omega)$ *into itself.*

Proof. Let $\{u_\nu\}$ be a sequence in $C^\infty(\Omega) \cap H^1(\Omega)$ converging to u in $H^1(\Omega)$. We shall apply (28.2) repeatedly: First of all, for each ν, $|G(u_\nu(x))| \leqq |G(u_\nu(x_0))| + K|u_\nu(x_0)| + K|u_\nu(x)|$, where x is a variable point in Ω whereas x_0 is a fixed, arbitrarily chosen point of Ω. This shows that each $G(u_\nu)$ belongs to $L^2(\Omega)$ (observe that all these functions are continuous in Ω). Next we have $\|G(u_\nu) - G(u_\mu)\|_{L^2(\Omega)} \leqq K\|u_\nu - u_\mu\|_{L^2(\Omega)}$, which shows that the $G(u_\nu)$ form a Cauchy sequence in $L^2(\Omega)$. Next, for every $x \in \Omega$, $|G(u(x)) - G(u_\nu(x))| \leqq K|u_\nu(x) - u(x)|$, which shows that $G(u)$ belongs to $L^2(\Omega)$ and that

$$(28.4) \qquad \|G(u_\nu) - G(u)\|_{L^2(\Omega)} \leqq K\|u_\nu - u\|_{L^2(\Omega)};$$

thus the $G(u_\nu)$ converge to $G(u)$ in $L^2(\Omega)$. Formula (28.3) is true, as seen by direct computation [of the distribution derivatives of $G(u)$]. In particular, it implies, for every ν,

$$(28.5) \qquad \|(\partial/\partial x^j)G(u_\nu)\|_{L^2(\Omega)} \leqq K\|u_\nu\|_1.$$

It follows from (28.4) and (28.5) that the $G(u_\nu)$ form a bounded sequence in $H^1(\Omega)$; hence there is a subsequence of it which converges weakly in $H^1(\Omega)$, necessarily to $G(u)$, which thus belongs to $H^1(\Omega)$.

Assume now that $G(0) = 0$ and that $u \in H_0^1(\Omega)$. We may take the sequence $\{u_\nu\}$ given above in $C_c^\infty(\Omega)$. Then the $G(u_\nu)$ have compact support contained in Ω, and therefore belong to $H_0^1(\Omega)$ (Proposition 24.3). On the other hand, we have seen that $G(u)$ belongs to the weakly closed convex hull of the $G(u_\nu)$. By a standard consequence of the Hahn–Banach theorem, this is the same as their strongly closed convex hull, which is contained in the closed linear subspace $H_0^1(\Omega)$. Q.E.D.

PROPOSITION 28.2. *Let u belong to* $H^1(\Omega)$ [*resp., to* $H_0^1(\Omega)$]. *Then* $|u|$, $u^+ = \sup(u, 0)$, $u^- = \inf(u, 0)$ *also do. Let v be another element of* $H^1(\Omega)$ [*resp., of* $H_0^1(\Omega)$]. *Then* $\sup(u, v)$ *and* $\inf(u, v)$ *also belong to* $H^1(\Omega)$ [*resp., to* $H_0^1(\Omega)$].

Proof. For the statement about $|u|$ it suffices to apply Lemma 28.1 with $G(t) = |t|$. Then $u^+ = \frac{1}{2}(u + |u|)$, $u^- = u - u^+$, $\sup(u, v) = u + (v - u)^+$, $\inf(u, v) = u + (v - u)^-$. Q.E.D.

We are going to need an extension of the Lax–Milgram theorem (Lemma 23.1). Let H be a real Hilbert space [for us, H will be $H_0^1(\Omega)$ or, in Corollary 28.1, $H = H^1(\Omega)$]; let K be a *closed convex* subset of H. We denote by H' the dual of H, and by $\langle\ ,\ \rangle$ the duality bracket between H and H'. We set, for any $u \in K$,

$$K_u = \{v \in H;\ \exists \rho > 0 \text{ such that } u + \rho v \in K\}. \tag{28.6}$$

When K is an *affine* subvariety (i.e., the parallel of a linear subspace), we have $K_u = K - u$, which is a linear subspace. In general, K_u is a convex cone; we have $K_u = H$ if and only if u belongs to the *interior* of K.

LEMMA 28.2. *Let $a(u, v)$ be a continuous bilinear form on $H \times H$. Suppose that a is coercive on H* (Definition 23.1), *symmetric, and that the associated quadratic form $a(v, v)$ is nonnegative.*

Given any $f \in H'$ there is a unique element u of K such that

$$a(u, v) \geqq \langle f, v\rangle \qquad \textit{for every} \quad v \in K_u. \tag{28.7}$$

We obtain the Lax–Milgram theorem (Lemma 23.1), in the case of real Hilbert spaces, by taking $K = H$ in Lemma 28.2. In this case, $K_u = H$ and $v \in K_u$ if and only if $-v \in K_u$, hence we must have simultaneously $a(u, v) \geqq$ and $\leqq \langle f, v\rangle$. This argument remains valid whenever K_u is a union of whole straight lines (and not merely of half-lines), e.g., when K_u is a linear subspace (i.e., when K is an affine subvariety).

Proof of Lemma 28.2. Set

$$I(u) = a(u, u) - 2\langle f, u\rangle.$$

We are going to show that there is a number $d > -\infty$ such that $I(u) \geqq d$ for every $u \in K$ and, for a *unique* $u \in K$, $I(u) = d$. The argument is the same as that proving the existence of an orthogonal projection into a closed convex set in a Hilbert space of which, in fact, Lemma 28.2 is a restatement.

First of all, by the coerciveness of $a(u, v)$,

$$I(u) \geqq c^2\|u\|_H^2 - 2\|f\|_{H'}\|u\|_H = \left(c\|u\|_H - \frac{1}{c}\|f\|_{H'}\right)^2 - c^{-2}\|f\|_{H'},$$

hence $d = \inf_{u \in K} I(u) > -\infty$.

For each $j = 1, 2, \ldots$, let $K_j = \{w \in K;\ I(w) \leqq d + 1/j\}$. If w_1, w_2 belong to K_j, we have $\frac{1}{2}(w_1 + w_2) \in K$ since K is convex. Moreover,

$$\begin{aligned}\tfrac{1}{2}a(w_1 - w_2, w_1 - w_2) &= a(w_1, w_1) + a(w_2, w_2) - 2a\left(\frac{w_1 + w_2}{2}, \frac{w_1 + w_2}{2}\right)\\ &= I(w_1) + I(w_2) - 2I\left(\frac{w_1 + w_2}{2}\right) \leqq 2\left(d + \frac{1}{j}\right) - 2d = \frac{2}{j}.\end{aligned}$$

If we use the coerciveness of $a(u, v)$ once more, we see that the diameter of K_j is $\leqq 2/(c\sqrt{j})$. Since K is closed, the intersection of the K_j consists of a single point, $u \in K$. Of course, $I(u) = d$.

Now let $v \in H$ be such that $u + \rho v \in K$ for some $\rho > 0$. Since K is convex, we have $u + tv \in K$ for all t, $0 \leqq t \leqq \rho$, and we know that the function $I(u + tv) - I(u) = t^2 a(v, v) + t\{a(u, v) - \langle f, v\rangle\} \geqq 0$ on $(0, \rho)$. This is possible if and only if (28.7) holds.

Suppose that (28.7) were also true with $u_1 \in K$ in place of u. Observe that we have $v = u_1 - u \in K_u$ whereas $-v \in K_{u_1}$. We derive from (28.7):

$$a(u, v) \geqq \langle f, v\rangle, \qquad -a(u_1, v) \geqq -\langle f, v\rangle,$$

from which, by adding, $a(v, v) \leqq 0$. Since $a(v, v) \geqq c^2\|v\|_H^2$, we must have $u = u_1$. Q.E.D.

COROLLARY 28.1. *Let $a(u, v)$ be a continuous symmetric bilinear functional on $H \times H$, coercive on a closed linear subspace H_0 of H.*

Let K be a closed convex subset of H such that, for some $h^0 \in H$, $K \subset H_0 + h^0$. Then there exists a unique element u of K such that

$$a(u, v) \geqq 0 \qquad \textit{for every} \quad v \in K_u. \tag{28.8}$$

Proof. Set $K^0 = K - h^0$. Clearly K^0 is a closed convex subset of H_0. Furthermore, $u \in K \Leftrightarrow u - h^0 \in K^0$ and

$$u - h^0 + \rho v \in K^0 \quad \Leftrightarrow \quad u + \rho v \in K,$$

hence $K_u = \{v \in H_0;\ \exists \rho > 0$ such that $u - h^0 + \rho v \in K^0\}$. We may apply Lemma 28.2 with H_0 substituted for H, K^0 for K, and with $f : h \mapsto -a(h^0, h)$. We conclude that there is a unique $u - h^0 \in K^0$ such that

$$a(u - h^0, v) \geqq -a(h^0, v) \qquad \text{for every} \quad v \in K_u.$$

Definition 28.2. We say that a distribution in Ω, u, is a subsolution of $\lambda - \Delta$ if $(\Delta - \lambda)u$ is a positive Radon measure in Ω. We say that u is a supersolution if $-u$ is a subsolution.

When $\lambda = 0$, i.e., when the operator under study is minus the Laplace operator, subsolutions are called *subharmonic* distributions (or functions), supersolutions are called *superharmonic* distributions (thus u is subharmonic if Δu is a *positive* Radon measure).

It is a classical theorem of L. Schwartz that any positive distribution is a positive Radon measure; therefore the claim that u is a subsolution of $\lambda - \Delta$ means that

$$\langle(\lambda - \Delta)u, \varphi\rangle = \langle u, (\lambda - \Delta)\varphi\rangle \leqq 0 \tag{28.9}$$
$$\text{for every} \quad \varphi \in C_c^\infty(\Omega) \quad \text{such that} \quad \varphi \geqq 0 \quad \text{in } \Omega.$$

If we know furthermore that $u \in H^1(\Omega)$, we may restate (28.9) in the following manner:

$$a_\lambda(u, \varphi) = \lambda \int_\Omega u\varphi \, dx + \sum_{j=1}^{n} \int_\Omega \frac{\partial u}{\partial x^j} \frac{\partial \varphi}{\partial x^j} \, dx \leqq 0 \tag{28.10}$$

$$\text{for all} \quad \varphi \in C_c^\infty(\Omega), \qquad \varphi \geqq 0.$$

THEOREM 28.1. *Let u, $v \in H^1(\Omega)$ be two subsolutions of $-\Delta + \lambda$. Then* $\sup(u, v)$ *is also a subsolution of* $-\Delta + \lambda$.

Proof. Set $w = \sup(u, v)$ and let K denote the set of $f \in H^1(\Omega)$ such that

$$f \preccurlyeq w \quad \text{on } \Omega, \qquad f - w \in H_0^1(\Omega). \tag{28.11}$$

As the intersection of a closed convex cone and of a closed affine subvariety of $H^1(\Omega)$, K is a closed convex subset of $H^1(\Omega)$. By virtue of Corollary 28.1, applied to $a_\lambda(u, v)$, we obtain that there is a unique $\eta \in K$ such that

$$a_\lambda(\eta, g) \geqq 0 \qquad \text{for all} \quad g \in K_\eta. \tag{28.12}$$

At this point we observe that K_η contains all functions $\varphi \in C_c^\infty(\Omega)$ such that $\varphi \leqq 0$ in Ω. Indeed, if $\eta \preccurlyeq w$, we also have $\eta + \rho\varphi \preccurlyeq w$ whatever $\rho > 0$, and if $\eta - w$ belongs to $H_0^1(\Omega)$, so does $\eta + \rho\varphi - w$. If we replace g by such a function φ in (28.12), we obtain that η is a subsolution of $-\Delta + \lambda$.

Let us now set $\zeta = \sup(u, \eta) = u + (\eta - u)^+ = \eta - (\eta - u)^-$. Clearly (cf. Proposition 28.2), $\zeta \preccurlyeq w$ on Ω. We know that $\eta - w$ is the limit in $H^1(\Omega)$ of a sequence $\{\varphi_\nu\}$ of elements of C_c^∞. If one goes back to the definition of the supremum of two elements of $H^1(\Omega)$ (cf. Proposition 28.1), one derives that $\zeta - w$ belongs to the closed convex hull of the set of functions $\sup(u - w, \varphi_\nu)$, and the latter vanishes outside of the support of φ_ν (since $u \leqq w$ everywhere), hence has a compact support and therefore belongs to $H_0^1(\Omega)$. As a consequence, we see that $\zeta - w \in H_0^1(\Omega)$. In summary, $\zeta \in K$ and, of course, $\zeta - \eta \in K_\eta$. From (28.12) we derive

$$a_\lambda(\eta, \zeta - \eta) \geqq 0. \tag{28.13}$$

From our definition of ζ it follows that $(\zeta - u)(\zeta - \eta) = -(\eta - u)^+(\eta - u)^-$ vanishes identically in Ω. On the other hand, the first-order distribution derivatives of $(\zeta - u)$ and of $(\zeta - \eta)$ are locally integrable functions. It follows from this that

$$\left(\frac{\partial}{\partial x^j}\right)(\zeta - u) \cdot \left(\frac{\partial}{\partial x^k}\right)(\zeta - \eta) \equiv 0 \quad \text{in } \Omega \tag{28.14}$$

(we recall that, in our terminology, a function is always viewed as a distribution and we say that it vanishes "identically" when any one—and every one—

of its representatives vanishes almost everywhere). From (28.14) it follows at once that

$$a_\lambda(\zeta - u, \zeta - \eta) = 0,$$

i.e.,

$$a_\lambda(\zeta, \zeta - \eta) = a_\lambda(u, \zeta - \eta).$$

We shall now use the fact that $\zeta - \eta \in H_0^1(\Omega)$ and that $\zeta \geqq \eta$ in Ω. By cutting and regularizing we see that $\zeta - \eta$ is the limit, in $H^1(\Omega)$, of a sequence of non-negative elements of $C_c^\infty(\Omega)$, hence, by (28.10), we have $a_\lambda(u, \zeta - \eta) \leqq 0$. Consequently,

$$a_\lambda(\zeta, \zeta - \eta) \leqq 0. \tag{28.15}$$

By subtracting (28.13) from (28.15) we see that $a_\lambda(\zeta - \eta, \zeta - \eta) \leqq 0$ hence, since a_λ is coercive, $\zeta = \eta$.

Thus we have $u \leqq \eta \leqq w$ almost everywhere. Similarly, we have $v \leqq \eta \leqq w$ almost everywhere. By modifying, or rather choosing appropriately the representatives of u, v, and η, we may assume that those inequalities are valid everywhere. Then, for any given $x \in \Omega$, if $w(x) = u(x)$, we have $\eta(x) = w(x)$; and the same conclusion is valid if $w(x) = v(x)$. We reach the conclusion that $w = \eta$. As we have already shown that η is a subsolution, the proof of Theorem 28.1 is complete.

COROLLARY 28.2. *If u, $v \in H^1(\Omega)$ are subharmonic, the same is true of* $\sup(u, v)$.

COROLLARY 28.3. *If $u \in H^1(\Omega)$ is subharmonic, the same is true of*

$$\{u\}_\theta = \sup(u, \theta)$$

whatever the real number θ. In particular, $u^+ = \sup(u, 0)$ is subharmonic.

COROLLARY 28.4. *Let λ be $\geqq 0$, $u \in H^1(\Omega)$ be a subsolution of $-\Delta + \lambda$. Then, given any number $\theta \leqq 0$, $\{u\}_\theta$ is a subsolution of $-\Delta + \lambda$.*

Indeed, $(\Delta - \lambda)\theta = -\lambda\theta \geqq 0$.

We come now to the main result of this section:

THEOREM 28.2. *Let $u \in H^1(\Omega)$ be a subsolution of $-\Delta + \lambda (\lambda \geqq 0)$. If $u \geqq 0$ almost everywhere, we have*

$$\max_\Omega u \leqq \max_{\partial\Omega} u. \tag{28.16}$$

Remark 28.1. By virtue of Proposition 28.2, the inequality (28.16) means that $u \leqq \max_{\partial\Omega} u$ almost everywhere in Ω.

Proof of Theorem 28.2. We assume $M = \max_{\partial\Omega} u < +\infty$. We need the following lemma.

LEMMA 28.3. *Let $u \in H^1(\Omega)$ be bounded on $\partial\Omega$ in the sense of $H^1(\Omega)$. If θ is any real number strictly greater than $\max_{\partial\Omega} u$, $(u-\theta)^+ \in H_0^1(\Omega)$.*

Proof. By hypothesis we may find a sequence $\{\varphi_\nu\}$ in $C^\infty(\Omega) \cap H^1(\Omega)$ converging to u in $H^1(\Omega)$ and such that, for each ν, $\varphi_\nu < \theta$ on a set $U_\nu \cap \Omega$, where U_ν is an open neighborhood of $\partial\Omega$ in $\mathbf{R}^n$. But (cf. proof of Lemma 28.1) $(u-\theta)^+$ belongs to the closed convex hull in $H^1(\Omega)$ of the set of functions $(\varphi_\nu - \theta)^+$; for each ν, $(\varphi_\nu - \theta)^+$ vanishes in $U_\nu \cap \Omega$, and therefore belongs to $H_0^1(\Omega)$ (Proposition 24.3) and so does $(u-\theta)^+$. Q.E.D.

Suppose now that $u \geqq 0$ a.e. in Ω. If we had $-\infty \leqq M < 0$, we would derive from Lemma 28.3 that $(u-\theta)^+ = u + |\theta| \in H_0^1(\Omega)$ for any θ, $M < \theta < 0$. But this is incompatible, as the student may easily ascertain, with the fact that $u + |\theta| \geqq |\theta| > 0$ a.e. in Ω. Thus we must have $M \geqq 0$.

Suppose now, in addition to $u \geqq 0$ a.e. in Ω, that u is a subsolution of $-\Delta + \lambda$. Let then θ be any number greater than M. Clearly, $u - \theta$ is a subsolution of $-\Delta + \lambda$, and so is $(u-\theta)^+$ by Corollary 28.4. Consequently, for all $\varphi \in C_c^\infty(\Omega)$, $\varphi \geqq 0$ in Ω,

$$a_\lambda((u-\theta)^+, \varphi) \leqq 0. \tag{28.17}$$

But (28.17) extends by continuity to all $\varphi \in H_0^1(\Omega)$ such that $\varphi \geqq 0$ a.e. in Ω. By Lemma 28.3 we may therefore take $\varphi = (u-\theta)^+$, and by the coercivity of a_λ we conclude that $(u-\theta)^+ = 0$, i.e., $u \leqq \theta$ a.e. This, of course, implies $u \leqq M$ a.e. Q.E.D.

COROLLARY 28.5. *Let $u \in H^1(\Omega)$ be a subsolution of $-\Delta + \lambda (\lambda \geqq 0)$. Then*

$$\max_{\Omega} u \leqq \max_{\partial\Omega} u^+. \tag{28.18}$$

Indeed, $u \leqq u^+$ in Ω, and we may apply (28.16) to u^+ substituted for u (by virtue of Corollary 28.4).

COROLLARY 28.6. *Let $u \in H^1(\Omega)$ be subharmonic. Then*

$$\max_{\Omega} u \leqq \max_{\partial\Omega} u. \tag{28.19}$$

Proof. Call M the right-hand side of (28.19), assumed to be $< +\infty$, and let θ be a real number strictly greater than M. Then $u - \theta$ is obviously subharmonic (here $\lambda = 0$), hence by (28.18) we know that $u - \theta \leqq \max_{\partial\Omega}(u-\theta)^+$ a.e. in Ω. But (Lemma 28.3) $(u-\theta)^+ \in H_0^1(\Omega)$ hence $\max_{\partial\Omega}(u-\theta)^+ = 0$. Thus $u \leqq \theta$ a.e. in Ω and therefore $u \leqq M$ a.e. in Ω. Q.E.D.

COROLLARY 28.7. *Let $u \in H^1(\Omega)$ be a supersolution of $-\Delta + \lambda(\lambda \geqq 0)$, $u \leqq 0$ almost everywhere in Ω. Then*

(28.20)
$$\min_{\partial\Omega} u \leqq \min_{\Omega} u.$$

Proof. u is a subsolution of $-\Delta + \lambda$, a.e. nonnegative, therefore (28.16) is valid for $-u$ replacing u.

Q.E.D.

COROLLARY 28.8. *Let $u \in H^1(\Omega)$ be such that $(-\Delta + \lambda)u = 0$ in Ω. Then*

(28.21)
$$\max_{\Omega} |u| \leqq \max_{\partial\Omega} |u|.$$

Proof. Note that u and $-u$ are subsolutions of $-\Delta + \lambda$, therefore this is also true of u^+ and of $-u^- = \sup(-u, 0)$. We may apply (28.16) to u^+ and to $-u^-$. We immediately obtain (28.21). Q.E.D.

COROLLARY 28.9. *Let $u \in H^1(\Omega)$ be a subsolution of $-\Delta + \lambda$, $h \in H^1(\Omega)$ such that $(-\Delta + \lambda)h = 0$ in Ω. If $u - h \in H_0^1(\Omega)$, $u \leqq h$ a.e. in Ω.*

Proof. Clearly $u - h$ is a subsolution of $(-\Delta + \lambda)$, and so is $(u - h)^+$. The latter belongs to $H_0^1(\Omega)$ by Proposition 28.1, hence, by virtue of Theorem 28.2, $(u - h)^+ \leqq 0$ a.e. in Ω. Q.E.D.

COROLLARY 28.10. *Let $u \in H_0^1(\Omega)$ be a subsolution of $-\Delta + \lambda$. Then $u \leqq 0$ a.e. in Ω.*

Remark 28.2. The entire argument in this section would have worked equally well, had we been dealing with more general second-order elliptic operators than $-\Delta + \lambda$. We could have applied it to any operator L of the form (23.6) whose associated bilinear form $a(u, v)$ [see (23.7)] is *symmetric*, *continuous* on $H^1(\Omega) \times H^1(\Omega)$ and *coercive* on $H_0^1(\Omega)$. The symmetry requirement could even be dropped—but then we need a further generalization of the Lax–Milgram theorem: In Lemma 28.2, the condition that $a(u, v)$ be symmetric should be shown to be superfluous. This can be done, but the corresponding proof is considerably more difficult than that of Lemma 28.2.

When dealing with general operators L of the form (23.6) one might ask whether there is an analog to the results which concern the Laplacian Δ, such as Corollaries 28.2, 28.3, and 28.6. Inspection of their proofs shows that the only fact which was used was that the constant functions where subsolutions (of $-\Delta$). Therefore, analogous statements will be valid whenever the zero-order term $c(x)$ in (23.6) is identically zero. In general, in the proofs of the statements about subsolutions when $c(x)$ is not identically zero, it is necessary that the nonpositive constant functions be subsolutions: This demands that $c(x)$ be $\geqq 0$ a.e. in Ω.

29

Application: Solution of the Classical Dirichlet Problem

We use the same notation as in Sect. 28: Ω is a *bounded* open subset of $\mathbf{R}^n$, where we study $\lambda - \Delta$, with $\lambda \geqq 0$. Unless otherwise specified, all functions and distributions are supposed to be *real*-valued.

Let g be any continuous function on the boundary $\partial\Omega$. We can extend g as a continuous function $\tilde{g}$ with compact support in $\mathbf{R}^n$, and construct a sequence of C^∞ functions, say polynomials, in $\mathbf{R}^n$, which converge to g uniformly on its support. Let then $\{g_j\}(j = 0, 1, \ldots)$ be their restrictions to $\bar{\Omega}$. For each j, there is a unique function $u_j \in H^1(\Omega)$ such that

$$(29.1) \qquad (\lambda - \Delta)u_j = 0 \quad \text{in } \Omega,$$

$$(29.2) \qquad u_j - g_j \in H_0^1(\Omega).$$

By the weak maximum principle (Corollary 28.8) we have

$$(29.3) \qquad \max_{\Omega} |u_j - u_{j'}| \leqq \max_{\partial\Omega} |u_j - u_{j'}| = \max_{\partial\Omega} |g_j - g_{j'}|.$$

This implies at once that the u_j converge uniformly in Ω; let $u \in C^0(\Omega)$ be their limit. Of course, we have $(\lambda - \Delta)u = 0$ in Ω. It is also clear that u is independent of the choice of the sequence $\{g_j\}$ (and therefore of the sequence u_j), as one checks by comparing u to the limit of a different such sequence.

Definition 29.1. The function u just given is called the generalized solution of the classical Dirichlet problem,

$$(29.4) \qquad (\lambda - \Delta)u = 0 \quad \textit{in } \Omega,$$

$$(29.5) \qquad u = g \quad \textit{on } \partial\Omega,$$

and will be denoted by $H_\lambda(g)$.

Suppose that there is $u \in C^0(\bar{\Omega})$ satisfying (29.4)–(29.5) or else that g can be extended to Ω in such a way as to belong to $C^0(\bar{\Omega}) \cap H^1(\Omega)$, and that there is

$u \in H^1(\Omega)$ satisfying (29.4) and such that $u - g \in H_0^1(\Omega)$. In both those cases, $u = H_\lambda(g)$ (Exercise 29.1).

Let us denote by $\mathscr{B}^0(\Omega)$ the space of bounded continuous functions in Ω, equipped with the norm $\sup_\Omega |f|$. From (29.3) and the fact that $\max_\Omega |u_j| \leqq \max_{\partial\Omega} |g_j|$, we derive

PROPOSITION 29.1. *H_λ defines a continuous linear map of $C^0(\partial\Omega)$ into $\mathscr{B}^0(\Omega)$.*

Let then x be a fixed, but arbitrary, point of Ω. The linear map

$$g \mapsto H_\lambda(g)(x) \tag{29.6}$$

is continuous on $C^0(\partial\Omega)$ and therefore defines a Radon measure m_x on $\partial\Omega$:

$$H_\lambda(g)(x) = \int_{\partial\Omega} g(y)\, dm_x(y), \qquad x \in \Omega. \tag{29.7}$$

The case $\lambda = 0$ deserves special mention. In this case, the measure m_x is called the *harmonic measure* on $\partial\Omega$. Because the unique harmonic function in Ω, identically equal to one on $\partial\Omega$, is the function identically equal to one in Ω, we see that the *total mass* of m_x is *one*. Because of the "minimum principle," we have

$$\int_{\partial\Omega} g(y)\, dm_x(y) \geqq 0, \qquad \forall\, x \in \Omega,$$

whatever the *nonnegative* continuous function g on $\partial\Omega$, i.e., m_x is a positive Radon measure on $\partial\Omega$. In summary, we see that m_x is a *probability* on $\partial\Omega$. This fact leads to a new interpretation of the Dirichlet problem (for the Laplace equation), in the context of probability theory. A brief description of this approach (or, rather, of its first steps) can be found in Sect. 31.

It is also clear that the integration theory for Radon measures enables us now to extend the domain of definition of the functional H_λ:

Definition 29.2. A real-valued function g in $\partial\Omega$ (not necessarily continuous) is said to be resolutive if it is integrable with respect to m_x for every x in Ω. Formula (29.7) defines also in this case the generalized solution of the Dirichlet problem (29.4)–(29.5).

We may also define the generalized solution of the inhomogeneous Dirichlet problem:

$$(\lambda - \Delta)u = f \quad \text{in } \Omega, \tag{29.8}$$

$$u = g \quad \text{on } \partial\Omega, \tag{29.9}$$

for $f \in C^0(\overline{\Omega})$, $g \in C^0(\partial\Omega)$.

Let $\tilde{f}$ be equal to f in $\bar{\Omega}$ and to zero in $\mathbf{R}^n \backslash \bar{\Omega}$. Let E be any fundamental solution of $\lambda - \Delta$. We shall apply the following:

LEMMA 29.1. *The convolution $E * \tilde{f}$ is Lipschitz continuous in $\mathbf{R}^n$ and to every compact subset K of $\mathbf{R}^n$ there is a constant $C > 0$ such that*

$$\sup_K |(E * \tilde{f})(x)| \leqq C\|f\|_{L^\infty}, \tag{29.10}$$

$$\sup_{x, x' \in K} |(E * \tilde{f})(x) - (E * \tilde{f})(x')| \leqq C|x - x'| \, \|f\|_{L^\infty}. \tag{29.11}$$

Proof. Since it does not matter what fundamental solution of $\lambda - \Delta$ we deal with, we may take $E = G_\lambda$, the distribution considered in Exercise 9.7. It follows at once from (9.27) and (9.28) that E and grad E are locally integrable in $\mathbf{R}^n$ and that, consequently, $f \mapsto E * \tilde{f}$, $f \mapsto \text{grad } E * \tilde{f}$ are continuous linear mappings of $L^\infty(\Omega)$ into $L^\infty_{\text{loc}}(\mathbf{R}^n; \mathbf{R})$ and into $L^\infty_{\text{loc}}(\mathbf{R}^n; \mathbf{R}^n)$, respectively.

Q.E.D.

Let v then denote the restriction of $E * \tilde{f}$ to Ω, $\dot{v}$ its restriction to $\partial\Omega$. Both restrictions are continuous. We may then take, in (29.8)–(29.9),

$$u = v + H_\lambda(g - \dot{v}). \tag{29.12}$$

It is interesting to consider the case where $g = 0$. Then (29.12) can be written more explicitly, taking (29.7) into account:

$$u(x) = \int_\Omega E(x - x')f(x')\,dx' - \int_{\partial\Omega} \left(\int_\Omega E(y - x')f(x')\,dx' \right) dm_x(y), \tag{29.13}$$

that is,

$$u(x) = \int_\Omega G(x, x')f(x')\,dx', \tag{29.14}$$

where

$$G(x, x') = E(x - x') - \int_{\partial\Omega} E(y - x')\,dm_x(y). \tag{29.15}$$

It is evident that $G(x, x')$ is a distribution in $\Omega \times \Omega$; it is a C^∞ function in the complement of the diagonal. In fact, $G(x, x') - E(x - x')$ is a C^∞ function in $\Omega \times \Omega$. It is the generalized solution of (29.4)–(29.5) when $g(x) = E(x - x')$.

The function $G(x, x')$ does not depend on the choice of the fundamental solution E. Indeed, any other fundamental solution differs from E by a function which is a solution of $(\lambda - \Delta)h = 0$ in the whole space. But then

$$h(x - x') = \int_{\partial\Omega} h(y - x')\,dm_x(y), \qquad x, x' \in \bar{\Omega}.$$

Definition 29.3. The distribution $G(x, x')$ in $\Omega \times \Omega$ is called the Green function of the Dirichlet problem for $\lambda - \Delta$ in Ω.

We have

$$(29.16) \quad (\lambda - \Delta_x)G(x, x') = (\lambda - \Delta_{x'})G(x, x') = \delta(x - x') \quad \text{in } \Omega \times \Omega,$$

$$(29.17) \quad G(x, x') = 0 \quad \text{for} \quad x \in \partial\Omega \quad (x' \in \Omega),$$

in a generalized sense akin to that of Definition 29.1.

For "general" f and g, the generalized solution of (29.8)–(29.9), given in (29.12), can be written

$$(29.18) \quad u(x) = \int_\Omega G(x, x')f(x')\, dx' + \int_{\partial\Omega} g(y)\, dm_x(y), \qquad x \in \Omega.$$

It is important to underline the fact that, even when g is a continuous function in $\partial\Omega$, and x_0 an arbitrary point of $\partial\Omega$, $u(x)$ does not necessarily converge to $g(x_0)$ as $x \in \Omega$ tends to x_0. We wish now to take a look at the points where this occurs [that $u(x) \to g(x_0)$ as $x \to x_0$].

Definition 29.4. A point x_0 of the boundary $\partial\Omega$ is called regular if, whatever the continuous function g on $\partial\Omega$, $H_\lambda(g)(x)$ converges to $g(x_0)$ as $x \in \Omega$ converges to x_0.

It is not difficult to obtain a *necessary* condition for a point to be regular. First of all,

PROPOSITION 29.2. *Let $x_0 \in \partial\Omega$ be regular for $\lambda - \Delta$. Then, whatever the point x' in Ω, $G(x, x')$ tends to zero as $x \in \Omega$ tends to x_0, and whatever the continuous function f in $\overline{\Omega}$, $\int_\Omega G(x, x')f(x')\, dx' \to 0$.*

This is evident by (29.13) through (29.15).

PROPOSITION 29.3. *Let $x_0 \in \partial\Omega$ be regular for $\lambda - \Delta$. There is a function $\beta \in H^1(\Omega)$ having the following properties:*

$$(29.19) \quad (\lambda - \Delta)\beta = -1 \quad \textit{in } \Omega;$$

$$(29.20) \quad \beta < 0 \textit{ in } \Omega \textit{ and, whatever } y \in \partial\Omega,\ y \neq x_0,\ \overline{\lim_{\Omega \ni x \to y}}\, \beta(x) < 0;$$

$$(29.21) \quad \beta(x) \to 0 \textit{ as } x \in \Omega \textit{ tends to } x_0.$$

Proof. Consider $w(x) = 1 - \exp(M|x - x_0|^2)$. It is easily checked that $(\lambda - \Delta)w > \lambda \geqq 0$ as soon as $2nM > \lambda$. Let β be the unique solution of (29.19) in $H^1(\Omega)$ such that $\beta - w \in H_0^1(\Omega)$. Since β is a subsolution and w a supersolution, we derive from Corollary 28.10 that $\beta < w$ in Ω, whence (29.20). By

Definition 29.1, we have $\beta(x) = -\int_\Omega G(x, x')\, dx' + \int_{\partial\Omega} w(y)\, dm_x(y)$, and by the hypothesis that x_0 is regular, taking into account Proposition 29.2, we get that $\beta(x) \to w(x_0) = 0$ as $x \in \Omega$ tends to x_0. Q.E.D.

We shall prove a kind of converse to Proposition 29.3. First we introduce the classical definition:

Definition 29.5. Let x_0 be a point in the boundary $\partial\Omega$. A continuous function β in Ω will be called a barrier at x_0 (for $\lambda - \Delta$) if β is a subsolution of $\lambda - \Delta$ and satisfies (29.20) and (29.21).

Before stating and proving the main result of this section (Theorem 29.1) let us observe that the following function in $\partial\Omega$,

$$\beta^*(y) = \overline{\lim_{\Omega \ni x \to y}}\ \beta(x), \qquad y \in \partial\Omega, \tag{29.22}$$

is *upper semicontinuous*. In particular, it reaches its *maximum* on any compact subset of $\partial\Omega$.

THEOREM 29.1. *For a point $x_0 \in \partial\Omega$ to be regular, it is necessary and sufficient that there be a barrier β at x_0 which belongs to $H^1(\Omega)$.*

Proof. The necessity follows from Proposition 29.3 and we shall now prove the sufficiency. Let g be an arbitrary continuous function on $\partial\Omega$ and set $u = H_\lambda(g)$. We are going to show that given any $\varepsilon > 0$, there is an open neighborhood $U(x_0)$ of x_0 in $\mathbf{R}^n$ such that

$$|u(x) - g(x_0)| < \varepsilon \qquad \text{for every} \quad x \in U(x_0) \cap \Omega. \tag{29.23}$$

If we go back to the definition of $H_\lambda(g)$ at the beginning of this section (by means of the sequences g_j and u_j), we see that it suffices to prove our assertion when g can be extended as a C^∞ function in $\mathbf{R}^n$; we continue to denote the extension by g.

Let η, τ be two numbers greater than zero (η will be small, τ large). Let h_+(resp. h_-) be a function in $\mathbf{R}^n$ satisfying $(\lambda - \Delta)h_\pm = 0$ and such that

$$h_\pm(x_0) = g(x_0) \pm \eta.$$

We set

$$v = h_- + \tau\beta, \qquad w = h_+ - \tau\beta;$$

v is a subsolution, w a supersolution of $\lambda - \Delta$. We are going to show that $v < g < w$ in Ω, provided that τ is large enough. Indeed, in a sufficiently small neighborhood $N(x_0)$ of x_0 in $\mathbf{R}^n$, we have

$$h_- < g < h_+,$$

hence, since $\beta < 0$ in Ω,

(29.24) $$v(x) < g(x) < w(x), \qquad \forall x \in N(x_0) \cap \Omega.$$

Let us assume that $N(x_0)$ is open and let Γ_0 be the complement of $N(x_0) \cap \partial\Omega$ with respect to $\partial\Omega$. According to our preliminary remark, if β^* is the function on $\partial\Omega$ defined in (29.22), there is a number $c > 0$ such that $\beta^*(y) < -c$ for all $y \in \Gamma_0$. But this implies at once that, if we decrease c a little bit, we will have

$$\beta < -c \quad \text{in } \Omega \backslash N(x_0).$$

Of course this means that we may choose τ so large as to have

(29.25) $$v < \inf g - \eta, \qquad \sup g + \eta < w \quad \text{in } \Omega\backslash N(x_0).$$

The conjunction of (29.24) and (29.25) implies $v < g < w$ in Ω.

Set $V = v - u$; V obviously belongs to $H^1(\Omega)$ and is a subsolution of $\lambda - \Delta$ in Ω. Similarly, $W = w - u \in H^1(\Omega)$ and it is a supersolution in Ω. We have, by Corollary 28.5,

(29.26) $$\max_{\Omega} V \leqq \max_{\partial\Omega} V^+.$$

Observe that

$$V^+ = (v - g - (u - g))^+ = \sup(v - g, u - g) - (u - g).$$

We know $u - g \in H_0^1(\Omega)$. Let $\{\phi_j\}$, $j = 1, 2, \ldots$, be a sequence of functions in $C_c^\infty(\Omega)$ which converge to $u - g$ in $H^1(\Omega)$. We know that $\sup(v - g, u - g)$ belongs to the closed convex hull spanned by the functions $\sup(v - g, \phi_j)$ (cf. the proof of Lemma 28.1) and since $v < g$ in Ω, we have $\sup(v - g, \phi_j) = 0$ where $\phi_j = 0$. We conclude that every function $\sup(v - g, \phi_j)$ belongs to $H_0^1(\Omega)$ and that this is also true of $\sup(v - g, u - g)$ and therefore of V^+. But then $\max_{\partial\Omega} V^+ = 0$ and, by virtue of (29.26), $V \leqq 0$ a.e. in Ω. But v is continuous and u analytic in Ω, hence $v \leqq u$ everywhere in Ω. Similarly, $u \leqq w$ in Ω. By using the definition of v, w and the continuity of $h_\pm$ together with the property (29.21), we see that we may choose η and $U(x_0)$ sufficiently small, so as to satisfy (29.23). Q.E.D.

We shall conclude this section by giving a sufficient condition in order that there be a barrier at a given point x_0 of the boundary $\partial\Omega$. This criterion is classical; it should be noted that it is independent of $\lambda \geqq 0$.

PROPOSITION 29.4. *If there is a closed ball $\bar{B}$ in $\mathbf{R}^n$ such that*

(29.27) $$\{x_0\} = \bar{\Omega} \cap \bar{B},$$

then there exists a distribution β in $\mathbf{R}^n$ such that

(29.28) $$\Delta\beta = 1 \quad \textit{in the complement of the center of } B;$$

(29.29) $$\beta(x_0) = 0, \ \beta(x) < 0 \textit{ for all } x \in \bar{\Omega}, \ x \neq x_0.$$

The condition (29.27) is often expressed by saying that there is an *osculating* sphere to $\bar{\Omega}$ at x_0.

Proof. We may assume that the center of the ball B is the origin; let R be its radius. We take β to be a function of $s = |x|^2$ alone; then

$$\Delta\beta = 4s(\partial/\partial s)^2\beta + 2n(\partial/\partial s)\beta.$$

$$\beta(x) = \frac{1}{2n}(|x|^2 - R^2) + C(|x|^{2-n} - R^{2-n}) \qquad \text{if} \quad n > 2,$$

$$\beta(x) = \tfrac{1}{4}(|x|^2 - R^2) - C\log(|x|/R) \qquad \text{if} \quad n = 2.$$

Whatever the constant $C > 0$, $\Delta\beta = 1$ in $\mathbf{R}^n\backslash\{0\}$ (in fact, $\Delta\beta = 1 + C'\delta$ in $\mathbf{R}^n$). Furthermore, $\beta(x) = 0$ for $|x| = R$, in particular $\beta(x_0) = 0$. It remains to choose C large enough so as to have, for every $x \in \bar{\Omega}$, $x \neq x_0$,

$$|x|^2 - R^2 < 2nC(R^{2-n} - |x|^{2-n}) \qquad \text{if} \quad n > 2,$$

$$|x|^2 - R^2 < 2C\log(|x|^2/R^2) \qquad \text{if} \quad n = 2.$$

Since Ω is bounded it suffices to verify this when $|x| = R + \varepsilon$, with $\varepsilon > 0$ small, in which case it is practically evident. Q.E.D.

We may say that *the classical Dirichlet problem for $\lambda - \Delta$ is well posed in Ω if to every $f \in C^0(\bar{\Omega})$ and to every $g \in C^0(\partial\Omega)$ there is $u \in C^0(\bar{\Omega})$ satisfying $(\lambda - \Delta)u = f$ in Ω, $u = g$ in $\partial\Omega$.* By Proposition 29.3 we see that this is the case if and only if there is a barrier belonging to $H^1(\Omega)$ at every point of $\partial\Omega$. By Proposition 29.4 we see that large classes of open sets have this property. For instance: (i) all convex sets; (ii) all sets Ω whose boundary is a C^1 hypersurface (and which lie on one side of it). It is easy to make up more examples.

Exercises

29.1. Prove the assertions immediately following Definition 29.1.

29.2. Give an example of a bounded open subset Ω of the plane $\mathbf{R}^2$ which is not convex, whose boundary is not a C^1 curve, and which nevertheless has the property that, given any point x_0 in its boundary, there is a closed disk in $\mathbf{R}^2$ intersecting $\bar{\Omega}$ only at x_0 (cf. Proposition 29.4).

29.3. Prove that the set of points x_0 in the boundary $\partial\Omega$ of a bounded open subset Ω of $\mathbf{R}^n$, such that there is an open ball B in $\mathbf{R}^n\backslash\Omega$ having the property (29.27), is dense in $\partial\Omega$.

29.4. Let Ω be any bounded open subset of $\mathbf{R}^n$, $G(x, x')$ the Green function of the operator $\lambda - \Delta$ in Ω ($\lambda \geqq 0$). Prove that G is a *symmetric* distribution in $\Omega \times \Omega$ [this means that

$$(29.30)\qquad \int_\Omega G(x, x')\phi(x')\,dx' = \int_\Omega G(x', x)\phi(x')\,dx', \qquad \forall\, \phi \in C_c^\infty(\Omega)].$$

29.5. Let Ω and $G(x, x')$ be as in Exercise 29.4. Prove that, in $\Omega \times \Omega$, when $n \geqq 3$,

$$(29.31)\qquad G(x, x') \sim \frac{(n-2)}{|S^{n-1}|}\,|x - x'|^{-n+2} \qquad \text{as} \quad |x - x'| \sim 0,$$

whereas, when $n = 2$,

$$(29.32)\qquad G(x, x') \sim \frac{1}{2\pi}\log|x - x'| \qquad \text{as} \quad |x - x'| \sim 0.$$

Give a precise meaning to these equivalences.

29.6. Prove that the harmonic measure, relative to the open ball $|x| < R$ in $\mathbf{R}^n$, is equal to

$$(29.33)\qquad dm_x(y) = \frac{1}{|S^{n-1}|R}\,\frac{R^2 - |x|^2}{|x - y|^n}\,d\sigma_y,$$

where $d\sigma_y$ is the canonical area measure on the sphere $|y| = R$. Let us now denote by x^* the inverse of x with respect to the sphere $|y| = R$:

$$x^* = \frac{R^2}{|x|^2}\,x.$$

Prove that the Green function for the Laplace operator, relative to the open ball $|x| < R$, is equal, when $n \geqq 3$, to

$$(29.34)\qquad G(x, x') = C_n\left\{\frac{1}{|x - x'|^{n-2}} - \left(\frac{R}{|x|}\right)^{n-2}\frac{1}{|x^* - x'|^{n-2}}\right\},$$

where $C_n = (n-2)/|S^{n-1}|$, whereas, when $n = 2$, it is equal to

$$(29.35)\qquad G(x, x') = \frac{1}{2\pi}\log\left(\frac{|x|}{R}\,\frac{|x^* - x'|}{|x - x'|}\right).$$

29.7. Let Ω be a bounded open subset of $\mathbf{R}^n$ whose boundary is C^1, and which lies on one side of it. Let $G(x, x')$ be the Green function of $\lambda - \Delta$ ($\lambda > 0$) relative to Ω, and $dm_x(y)$ the "metaharmonic" measure, defined in (29.7). Derive from Green's formula (10.8) that

$$(29.36)\qquad dm_x(y) = \frac{dG}{\partial\nu}(x, y), \qquad x \in \Omega, \quad y \in \partial\Omega,$$

where $\partial/\partial\nu$ is the normal derivative, with respect to the y variable, in the *interior* normal direction to $\partial\Omega$.

29.8. Check formula (29.36) when $\lambda = 0$ and Ω is the open ball of radius $R > 0$, centered at the origin.

29.9. Let Ω be a bounded open subset of $\mathbf{R}^n$, $G(x, x')$ the Green function of $\lambda - \Delta$ in Ω. Prove that the mapping

$$C_c^\infty(\Omega) \ni \phi \mapsto \left(x \mapsto \int_\Omega G(x, x')\phi(x')\, dx'\right)$$

can be extended as an isomorphism of $H^{-1}(\Omega)$ onto $H_0^1(\Omega)$.

29.10. Let us denote by x, y, z the coordinates in $\mathbf{R}^3$, and set $r^2 = x^2 + y^2$. Denote by Ω the open set in $\mathbf{R}^3$ defined by

$$(29.37) \qquad r^2 + z^2 < 1, \qquad r > \exp(-1/2z) \qquad \text{if} \quad z > 0.$$

Describe Ω and show that every point of its boundary different from zero is regular (Definition 29.4).

Let μ be the positive Radon measure $\phi \mapsto \int_0^1 \phi(0, 0, z) z\, dz$ and call u the restriction to Ω of the convolution $E * \mu$, where $E = (4\pi)^{-1}(x^2 + y^2 + z^2)^{-1/2}$. Show that, up to a constant factor,

$$(29.38) \quad \begin{aligned} u = {} & (r^2 + (z-1)^2)^{1/2} - (r^2 + z^2)^{1/2} \\ & + z \log|(r^2 + z^2)^{1/2} + z|\ |(r^2 + (z-1)^2)^{1/2} + 1 - z| \\ & - 2z \log r. \end{aligned}$$

Prove that u is harmonic in Ω and that the restriction of u to the boundary $\partial\Omega$ is continuous on $\partial\Omega$.

Compute the limit of u as $(x, y, z) \to 0$ along the surfaces

$$r = \exp(-\alpha/2z), \qquad z > 0,$$

for various values of α, $0 < \alpha < 1$. Conclude from this that the origin is *not* regular (relatively to the Laplace equation and to the open set Ω; a point of $\partial\Omega$ like the origin in this exercise is called a *Lebesgue spine*).

29.11. Let Ω be a bounded open subset in the plane $\mathbf{R}^2$, z_0 a point in its boundary such that there is a straight line segment $[z_0, z_1]$ entirely contained in $\mathbf{R}^2 \backslash \Omega$ ($z_0 \neq z_1$). Prove that z_0 is then regular for the Laplace equation in Ω. [*Hint*: Consider a branch of $\log(z - z_0)$ adapted to the situation and study the function $-\mathrm{Re}\{1/\log(z - z_0)\}$.]

29.12. Let $a < b$ be two (finite) real numbers; set $L = b - a$. Show that, given any complex number λ which is not *one* of the numbers $-k^2\pi^2/L^2$, k an integer ≥ 1, there is a Green function $G_\lambda(x, x')$ for the differential operator $\lambda - d^2/dx^2$ in the open interval $]a, b[$. Setting

$$E_\lambda(x) = \frac{1}{2\sqrt{\lambda}}\{H(x)e^{-x\sqrt{\lambda}} + H(-x)e^{x\sqrt{\lambda}}\},$$

where $\sqrt{z}$ is the branch of the square root which is positive for z real positive, give the explicit expression of

$$G_\lambda(x, x') - E_\lambda(x - x').$$

Describe what happens when a tends to $-\infty$ or when b tends to $+\infty$, and also what happens when λ tends to one of the critical values $-k^2\pi^2/L^2$, $k = 1, 2, \ldots$ (while a and b remain finite).

30

Theory of the Laplace Equation: Superharmonic Functions and Potentials

In the preceding two sections, we "solved" the Dirichlet problem when the data were merely continuous by adapting the classical *Perron's method* and combining it with the use of the weak maximum (see Sect. 28). The advantages of this approach are twofold: first of all, it is a natural extension of the variational method, which has led to the weak solution; second, and more important, although we have applied it only to the typical operator $\lambda - \Delta$, it extends to a very wide class of second-order elliptic equations, equations whose coefficients are not even required to be continuous (see [2]). It should be noted, however, that in the study of the Laplace operator proper, $-\Delta$, we dispose of more refined information than that available in the more general situation. Specifically, we have used subsolutions (and supersolutions) which obey the maximum (or the minimum) principle. In the case of the Laplace equation, these are subharmonic (or superharmonic) functions, which satisfy a more precise inequality, namely that the value of the function at the center of a sphere is at most equal to its average on the sphere itself (at least equal, if we deal with superharmonic functions). This fact has deep implications on their regularity, as we shall see in this section.

The importance of superharmonic functions is that they include the so-called potentials (as we are going to see, roughly speaking every superharmonic function can be represented as the sum of a potential and of a harmonic function). Potentials are of course very important in the applications to gravitation theory, electrostatics and magnetism, heat transfer, and so on. Historically (in the eighteenth and nineteenth centuries), the Laplace equation came to play a fundamental role in the process by which these theories were established on a firm mathematical basis. The type of phenomena it has been used to describe conforms to a well-defined and fairly simple pattern: They relate to the "states" of a homogeneous medium (which fills a certain region $\mathscr{R}$ in space), and they are stationary, i.e., they do not vary with time. There is a *field* in the medium, which for us means that we are given a vector field X in $\mathscr{R}$. The field has two basic properties: There is no circulation, and the flux is conserved. In common language, the first one means that there are no eddies. Mathematically speaking, this means that curl $X = 0$, which also means (at least locally) that X is the gradient of a scalar function U, the potential. The flux of X across a piece of surface σ is mathematically defined as the integral over σ of the normal component of X. Its conservation means that the incoming flux, at one cross-sectional end of a tube of force, equals the outgoing one, at the other end. Again, in mathematical terms, this means that the field X is divergence free: div $X = 0$. If we combine this with the fact that $X = \text{grad } U$, we obtain precisely that $\Delta U = 0$.

However, the field must be created by something: the somethings are the "charges" or the "masses" (depending on which theory we are considering). Their presence in a (compact) region of space, $\mathfrak{K}$, is detected by the fact that the flux of the field X across any closed surface Σ enclosing $\mathfrak{K}$ is not zero; this flux does not depend on the surface, since div $X = 0$ in the complement of $\mathfrak{K}$ (provided of course that Σ is not too far away from $\mathfrak{K}$). Its value can be taken as a measure of the total charge, or mass, carried by $\mathfrak{K}$. By virtue of the conventions agreed upon by physicists, in what concerns gravitation and electricity, it is the custom to equate this total charge to the *incoming* flux across Σ, that is, to

$$-\int_{\Sigma} (X \cdot \nu)\, d\sigma,$$

where ν is the *outer* normal to Σ. By formula (10.6), this is equal to the integral of $-\Delta U$ over the interior of Σ, that is, over $\mathfrak{K}$. If we assume that there is "creation" inside Σ (as opposed to absorption), we see that the total mass of $-\Delta U$ is at least zero. If we push this argument to the infinitesimal level, we see that $-\Delta U$ must be a positive (Radon) measure; i.e., U must be a superharmonic distribution. Therein lies much of the motivation for our taking a closer look at this kind of distribution.

All functions and distributions dealt with in this section are *real-valued*, unless otherwise specified. Given a function f defined on the closed ball $|x - x_0| \leq r$ (and sufficiently regular there), we introduce its average over the sphere $|x - x_0| = r$,

$$f^{\#}(x_0; r) = |S^{n-1}|^{-1} \int_{S^{n-1}} f(x_0 + r\dot{x})\, d\dot{x}, \tag{30.1}$$

and also its average over the ball,

(30.2) $$f^{@}(x_0; r) = |B^n|^{-1} \int_{B^n} f(x_0 + ry)\, dy,$$

where B^n is the unit ball in $\mathbf{R}^n$ and $|B^n|$ is its "volume."

Let Ω be an open subset of $\mathbf{R}^n$. In what follows we deal with functions, defined in Ω, whose values are either (finite) real numbers or else $\pm\infty$. Such a function is said to be *lower semicontinuous* at a point x_0 of Ω if to every (finite) real number $\alpha < f(x_0)$ there is a neighborhood U_α of x_0 in Ω such that $f(x) > \alpha$ for all x in U_α; f is lower semicontinuous in a subset A of Ω if f is lower semicontinuous at every point of A. It is easily seen that, for a function f to be lower semicontinuous in Ω, it is necessary and sufficient that f be the upper envelope of a family of continuous functions in Ω.

Definition 30.1. *A function f in Ω is called hyperharmonic if it is lower semicontinuous, if $f(x) > -\infty$ for every $x \in \Omega$, and if*

(30.3) $$f^{\#}(x; r) \leqq f(x), \qquad \forall x \in \Omega, \quad \forall r < d(x, \complement\Omega).$$

By integrating both sides of (30.3) with respect to r we immediately obtain

(30.3′) $$f^{@}(x; r) \leqq f(x), \qquad \forall x \in \Omega, \quad \forall r < d(x, \complement\Omega).$$

The terminology "hyperharmonic" is justified by the following fact, whose proof we leave as an exercise to the student:

(30.4) *If f is hyperharmonic and u is harmonic in Ω, and if, whatever the point y of the boundary of Ω (or at infinity, if Ω is unbounded),*

$$\varliminf_{x \to y} (f(x) - u(x)) \geqq 0,$$

then $f \geqq u$ in Ω.

We have denoted by $\varliminf$ the lower limit. In particular, if B is any open ball whose closure is contained in Ω, and if I_f^B denotes the *Poisson integral* of f with respect to B [defined in formula (10.36): the integration over ∂B can obviously be extended to semicontinuous functions], then $f \geqq I_f^B$ everywhere in B.

The hyperharmonic functions in Ω form a *convex cone*: Any finite linear combination of such functions with positive coefficients is again such a function. Two other important, but almost evident, properties of hyperharmonic functions are the following ones:

(30.5) *The upper envelope of any increasing oriented set of hyperharmonic functions in Ω is hyperharmonic in Ω*

["increasing oriented" means that, given any two elements f, g in the set, there is a third one, h, such that $h \geqq \sup(f, g)$].

(30.6) *If f, g are hyperharmonic in Ω, so is* $\inf(f, g)$.

The next property is slightly less obvious:

PROPOSITION 30.1. *Let f be a hyperharmonic function in Ω. In each connected component of Ω, either f is identically equal to $+\infty$ or else it is locally Lebesgue-integrable.*

Proof. The subset A of Ω consisting of points in a neighborhood of which f is integrable is obviously open. Let $x_0 \in \Omega \backslash A$. By lower semicontinuity, there is a number, r, $0 < r < d(x_0, \complement\Omega)$, and a finite constant M such that $f(x) + M \geqq 0$ for every $x \in B_r(x_0)$, i.e., for every x such that $|x - x_0| < r$ (we are using here the property that f is nowhere equal to $-\infty$). If $|x_1 - x_0| < r/2$, then $x_0 \in B_{r/2}(x_1) \subset \Omega$, and $f(x) + M$ cannot be integrable over $B_{r/2}(x_1)$, otherwise f itself would be integrable over a neighborhood of x_0. By (30.3′) we have $(f + M)(x_1) \geqq (f + M)^{@}(x_1, r/2) = +\infty$. This shows that $\Omega \backslash A$ is also open and that $f \equiv +\infty$ in this set. Q.E.D.

A function g in Ω is called *hypoharmonic* if $-g$ is hyperharmonic. By the mean value theorem (Theorem 10.1) we see that harmonic functions are both hyperharmonic and hypoharmonic; as a matter of fact, they are the only such functions (Corollary 30.1 below).

Let us introduce now the classical terminology:

Definition 30.2. Any hyperharmonic function which is locally integrable in Ω is called superharmonic.

A function f is called *subharmonic* if $-f$ is superharmonic.

If f is harmonic in Ω, $-|f|$ is superharmonic. An important class of superharmonic functions is the following:

Example 30.1 Newton's Potential

This is the function

(30.7) $$G(x) = \frac{1}{(n-2)|S^{n-1}|} \frac{1}{r^{n-2}},$$

where n is assumed to be $\geqq 3$ and $r = |x|$. If we go back to the definition of the fundamental solution E of the Laplace equation in Sect. 9, we see that

(30.8) $$-\Delta G = \delta, \quad \text{the Dirac measure,} \quad \text{in } \mathbf{R}^n.$$

When $n = 2$, we take G to be the *logarithmic potential*:

$$G(x) = \frac{1}{2\pi} \log \frac{1}{r}. \tag{30.9}$$

It is clear that G is locally integrable in $\mathbf{R}^n$; we shall regard it as a "true" function, not as an equivalence class of functions a.e. equal; we shall agree that $G(0) = +\infty$.

Let us then introduce the standard mollifiers $\rho_\varepsilon(\varepsilon > 0) : \rho_\varepsilon(x) = \varepsilon^{-n}\rho(x/\varepsilon)$; $\rho \in C_c^\infty(\mathbf{R}^n)$, $\rho \geqq 0$ everywhere, $\int \rho\, dx = +1$. Then the regularizations $G * \rho_\varepsilon$ are solutions of $-\Delta T = \rho_\varepsilon$, hence are C^∞ functions in $\mathbf{R}^n$, which satisfy (Exercise 10.1) $(G * \rho_\varepsilon)^\#(x_0; r) \leqq (G * \rho_\varepsilon)(x_0)$ for all $x_0 \in \mathbf{R}^n$, $r > 0$. By going to the limit as $\varepsilon \to +0$, we obtain

$$G^\#(x_0; r) \leqq G(x_0), \qquad \forall x_0 \in \mathbf{R}^n, \quad \forall r > 0 \tag{30.10}$$

(noting that the restriction of G to any sphere is integrable over that sphere). Of course, we then also have

$$G^@(x_0; r) \leqq G(x_0), \qquad \forall x_0 \in \mathbf{R}^n, \quad r > 0. \tag{30.11}$$

It is clear that $G(x)$ is lower semicontinuous.

Example 30.2 Potentials in $\mathbf{R}^n$

Definition 30.3. Let μ be a distribution with compact support in $\mathbf{R}^n$, G the Newton potential (30.7) if $n \geqq 3$, the logarithmic potential (30.9) if $n = 2$.

The convolution $G * \mu$ is called the potential of μ and denoted by U^μ.

Potentials of the kind defined above are "global," that is, they are defined in the whole Euclidean space. Later on we shall introduce the Green potentials, which are defined in a given open subset Ω of $\mathbf{R}^n$.

The potential U^μ of the distribution $\mu \in \mathscr{E}'$ satisfies the inhomogeneous Laplace equation

$$-\Delta U^\mu = \mu \quad \text{in } \mathbf{R}^n. \tag{30.12}$$

We recall (Weyl's lemma, Theorem 9.1) that U^μ is an analytic function in the complement of supp μ.

PROPOSITION 30.2. *If $\mu \in \mathscr{E}'$, its potential U^μ is the unique solution of (30.12) which tends to zero at infinity.*

Proof. The uniqueness is obvious, since if there were another solution of (30.12) decaying at infinity, V^μ, the difference $U^\mu - V^\mu$ would be a harmonic function in the whole space, decaying at infinity, hence identically zero by the maximum principle. We must prove that U^μ itself decays at infinity. It suffices to note that, given some bounded open subset Ω of $\mathbf{R}^n$ containing the support of μ, as $d(x, \Omega) \to +\infty$, $y \mapsto G(x-y)$ converges to zero in $C^\infty(\Omega)$ and therefore $U^\mu(x) = \langle \mu_y, G(x-y) \rangle \to 0$. Q.E.D.

In the sequel we shall limit our attention to the case where μ is a *positive* Radon measure, at first with compact support. The convolution of such a measure with a lower semicontinuous function such as G is lower semicontinuous. Moreover,

PROPOSITION 30.3. *If μ is a positive Radon measure with compact support in $\mathbf{R}^n$, its potential U^μ is locally Lebesgue-integrable in $\mathbf{R}^n$.*

Proof. Let us first consider the case $n \geqq 3$. Let f be the characteristic function of an arbitrary compact subset of $\mathbf{R}^n$. By Fubini's theorem for Radon measures and by a straightforward application of Lemma 29.1, we obtain

$$\int U^\mu f \, dx = \int (G * f) \, d\mu \leqq \text{const} \sup_{x \in \operatorname{supp} \mu} |(G * f)(x)| < +\infty.$$

In more than two dimensions, $G \geqq 0$ everywhere, and therefore $U^\mu \geqq 0$ everywhere. This implies at once what we wanted.

Suppose now $n = 2$. Since the support of μ is compact, given any compact subset K of the plane, there is a number $R > 0$ such that

$$U^\mu = (\chi_R G) * \mu = (\chi_R G + \tau \mathbf{1}) * \mu - \tau \int d\mu \quad \text{in } K,$$

where χ_R stands for the characteristic function of the disk $|x| < R$, $\mathbf{1}$ for the constant function equal to one, τ for a large positive number. It suffices to show that $(\chi_R G + \tau \mathbf{1}) * \mu$ is locally L^1. For τ large enough, $\chi_R G + \tau \mathbf{1} \geqq 0$ everywhere, and the same argument as in the case $n \geqq 3$ applies. Q.E.D.

Since $y \mapsto G(x-y)$ is lower semicontinuous for every x in $\mathbf{R}^n$, we may form the *potential of an arbitrary positive Radon measure* μ (not necessarily with compact support),

$$U^\mu(x) = \int G(x-y) \, d\mu(y). \tag{30.13}$$

Of course $U^\mu(x)$ can be infinite. Noting that μ is the (weak) limit of an increasing sequence of positive Radon measures μ_α with compact support [for instance, the measures $\chi_R(x)\mu$ as $R \to +\infty$], we see that U^μ is equal to the upper envelope of an increasing sequence of lower semicontinuous functions (at least when $n \geqq 3$; when $n = 2$ one must add suitable constant functions, as in the proof of Proposition 30.3), hence it is lower semicontinuous.

We may compute the average of both members of (30.13) (on the right-hand side, under the integral sign) over any sphere $|x - x_0| = r$. The inequality (30.10) at once implies

(30.14) $$(U^\mu)^\#(x_0; r) \leqq U^\mu(x_0), \qquad \forall x_0 \in \mathbf{R}^n, \quad \forall r > 0.$$

We may summarize our findings as follows:

PROPOSITION 30.4. *Let μ be a positive Radon measure in $\mathbf{R}^n$. Its potential U^μ is a hyperharmonic function in $\mathbf{R}^n$. If the support of μ is compact, U^μ is a superharmonic function in $\mathbf{R}^n$.*

It is obvious that U^μ might be superharmonic, that is, locally Lebesgue-integrable, even in cases where the support of μ is not compact.

We recall that a *superharmonic distribution* in Ω is a distribution T such that $-\Delta T$ is a positive Radon measure in Ω.

PROPOSITION 30.5. *Let T be a superharmonic distribution in Ω. Given any relatively compact open subset Ω' of Ω, there is a positive Radon measure μ with compact support in $\mathbf{R}^n$ such that $T - U^\mu$ is a harmonic function in Ω'.*

Proof. Any positive Radon measure μ with compact support equal to $-\Delta T$ in Ω' satisfies the requirements of the lemma.

The next statement justifies the introduction of Definitions 30.1 and 30.2:

PROPOSITION 30.6. *For any distribution T in Ω the following properties are equivalent*:

(30.15) *T is a superharmonic distribution in Ω;*

(30.16) *T is a function (in the sense of distribution theory) and one, and only one, of its representatives is a superharmonic function in Ω.*

Proof.

I. (30.15) implies (30.16). This follows at once from Proposition 30.5 [the uniqueness of the representative in (30.16) is a consequence of the fact that two superharmonic functions which are a.e. equal are everywhere equal].

II. (30.16) implies (30.15). For $d > 0$, let Ω_d denote the subset of Ω consisting of the points x such that $d(x, \complement\Omega) > d$. It is clear that, if $f \in L^1_{\text{loc}}(\Omega)$,

$$(\rho_\varepsilon * f)(x) = \int \rho_\varepsilon(x - y) f(y)\, dy$$

is well defined, and C^∞, in Ω_ε. Let us take for f the representative of T which is a superharmonic function in Ω. For $d > \varepsilon$, $x_0 \in \Omega_d$, and $r < d - \varepsilon$, we derive from (30.3),

$$(\rho_\varepsilon * f)^\#(x_0; r) \leqq (\rho_\varepsilon * f)(x_0).$$

We apply formula (10.32) to $u = \rho_\varepsilon * f$ and conclude that $-\Delta u = -\rho_\varepsilon * \Delta f \geqq 0$ in Ω_d. It suffices then to go to the limit, first as $\varepsilon \to +0$, then as $d \to +0$. Q.E.D.

COROLLARY 30.1. *Any function in Ω which is both hyperharmonic and hypoharmonic is harmonic.*

Indeed, f is continuous and by Proposition 30.6, $\Delta f = 0$ in Ω.

Example 30.3 Green's Potentials

As usual, let Ω denote a bounded open subset of $\mathbf{R}^n$, and let $G(x, x')$ denote the Green function of $-\Delta$ in Ω. We have

$$G(x, x') = G(x - x') - \int_{\partial\Omega} G(y - x')\, dm_x(y), \tag{30.17}$$

where dm_x is the harmonic measure on $\partial\Omega$ and we may take $G(x)$ to be either Newton's potential if $n \geqq 3$ (30.7), or else the logarithmic potential (30.9) if $n = 2$ [see (29.15)].

The first term on the right-hand side of (30.17) is a superharmonic function in $\mathbf{R}^n$; the second term is a harmonic function in Ω. Therefore, if μ is a positive Radon measure in Ω, we may form

$$U^\mu(x) = \int_\Omega G(x, x')\, d\mu(x'), \tag{30.18}$$

which is called the *Green potential* of μ. Observe that the potential of μ in $\mathbf{R}^n$,

$$V^\mu(x) = \int_\Omega G(x - x')\, d\mu(x'), \tag{30.19}$$

possibly infinite (notice that μ does not necessarily extend as a measure on $\mathbf{R}^n$), is certainly hyperharmonic. We note that

$$\int_{\partial\Omega} V^\mu(y)\, dm_x(y)$$

is harmonic, though possibly constant and equal to $+\infty$ in some connected component of Ω. In a generalized sense, it takes the value V^μ at the boundary.

From the fact that the total mass of the harmonic measure m_x is one, we derive easily that $\int_{\partial\Omega} V^\mu(y)\, dm_x(y) \leqq V^\mu(x)$ for every $x \in \Omega$. We have

$$U^\mu(x) = V^\mu(x) - \int_{\partial\Omega} V^\mu(y)\, dm_x(y), \qquad x \in \Omega. \tag{30.20}$$

If then ν is another positive Radon measure in Ω, by Fubini's theorem for Radon measures we obtain the *reciprocity formula*

$$\int_\Omega U^\mu\, d\nu = \int_\Omega U^\nu\, d\mu. \tag{30.21}$$

We know (by Proposition 30.1) that, in each connected component of Ω, the Green potential U^μ is either identically equal to $+\infty$ or else it is a superharmonic function. A sufficient condition for the latter to happen is easy to obtain:

PROPOSITION 30.7. *If the total mass of the positive Radon measure μ in Ω is finite, its Green potential U^μ is a superharmonic function in Ω.*

Proof. To say that the total mass of μ is finite is equivalent to saying that μ extends as a continuous linear functional on $\mathscr{B}^0(\Omega)$, the space of bounded continuous functions in Ω. Whatever $x \in \Omega$.

$$h(x, x') = \int_{\partial\Omega} G(y - x')\, dm_x(y)$$

is a bounded function of x' in Ω, according to Proposition 29.1, and therefore $\int_\Omega h(x, x')\, d\mu(x') < +\infty$. On the other hand, μ can be extended as a positive Radon measure $\tilde{\mu}$ on $\mathbf{R}^n$, setting $\tilde{\mu} = 0$ in $\mathbf{R}^n \backslash \bar{\Omega}$; and then we know that $G * \tilde{\mu}$, which is the potential of $\tilde{\mu}$ in $\mathbf{R}^n$, is equal to $\int_\Omega G(x - x')\, d\mu(x')$; it is a superharmonic function. Q.E.D.

Remark 30.1. To say that the total mass of the positive Radon measure μ is finite is equivalent to saying that μ can be extended as a Radon measure on $\mathbf{R}^n$, supported by $\bar{\Omega}$ (or "carried" by Ω, since the boundary of Ω is necessarily of measure zero, in the sense of the extended measure). In these circumstances the notation U^μ is ambiguous, for we must distinguish between the Green potential of μ (in Ω) and the potential of the extended measure in $\mathbf{R}^n$. For our limited intent here, this ambiguity will not be harmful.

Of course, one can easily define the Green potential of an arbitrary Radon measure μ in Ω. By linearity, we write $\mu = \mu^+ - \mu^-$, where μ^+ and μ^- are positive measures, and set

$$U^\mu = U^{\mu^+} - U^{\mu^-}.$$

Potentials with Finite Energy

Definition 30.4. Let μ be a Radon measure in the bounded open set Ω, U^μ its Green potential. We say that μ and U^μ have finite energy if U^μ is locally integrable in Ω and if its Dirichlet integral,

$$\int_\Omega |\text{grad } U^\mu(x)|^2 \, dx \tag{30.22}$$

is finite. It is then called the energy of U^μ (or of μ).

The energy of U^μ, (30.22), will be denoted by $|U^\mu|_e^2$, also by $|\mu|_e^2$. If ν is another Radon measure in Ω, we set

$$(\mu, \nu)_e = (U^\mu, U^\nu)_e = \int_\Omega (\text{grad } U^\mu) \cdot (\text{grad } U^\nu) \, dx. \tag{30.23}$$

Similar definitions can be introduced when $\Omega = \mathbf{R}^n$.

We restrict ourselves, in the sequel, to the case where Ω is bounded. We know then that $-\Delta$ defines an isomorphism of $H_0^1(\Omega)$ onto $H^{-1}(\Omega)$. Its inverse, when restricted to $C_c^\infty(\Omega)$, or even to $C_c^0(\Omega) \subset H^{-1}(\Omega)$, is identical with the Green kernel mapping:

$$\phi \mapsto \int_\Omega G(\cdot, x')\phi(x') \, dx' \tag{30.24}$$

(Exercise 29.9). Let μ be a Radon measure belonging to $H^{-1}(\Omega)$. By cutting and mollifying we may find a sequence of functions $\phi_j \in C_c^\infty(\Omega)$ which converges to μ in $H^{-1}(\Omega)$, such that

$$\int_\Omega G(x, x')\phi_j(x') \, dx' \to U^\mu(x) \qquad \text{for each} \quad x \in \Omega.$$

This implies that U^μ is the pointwise limit of $(-\Delta)^{-1}\phi_j$; the latter converge to $(-\Delta)^{-1}\mu$ in $H_0^1(\Omega)$. This means that the equivalence class of U^μ (in the sense of Lebesgue integration theory) belongs to $H_0^1(\Omega)$. In the language of distribution theory we may state

PROPOSITION 30.8. *The potentials with finite energy in Ω belong to $H_0^1(\Omega)$.*

The potentials with finite energy form a dense, but *not* closed, subset of $H_0^1(\Omega)$: Indeed, on one hand their set contains the preimage under $(-\Delta)$ of the dense subspace $C_c^0(\Omega)$ of $H^{-1}(\Omega)$; on the other hand, recalling that $n > 1$, there are elements in $H^{-1}(\Omega)$ which are not Radon measures. Noting, however, that any positive distribution is a positive Radon measure and that the limit of a sequence of such distributions is also a positive Radon measure, we derive

PROPOSITION 30.9. *The cone of the Green potentials of positive Radon measures in Ω, with finite energy, is closed in $H_0^1(\Omega)$. It is identical to the set of all nonnegative superharmonic functions belonging to $H_0^1(\Omega)$.*

Now taking Proposition 30.8 into account and the duality between $H_0^1(\Omega)$ and $H^{-1}(\Omega)$, we see that, if μ and ν are two Radon measures in Ω, belonging to $H^{-1}(\Omega)$,

$$(\mu, \nu)_e = (U^\mu, U^\nu)_e = \int_\Omega U^\mu \, d\nu. \tag{30.25}$$

Of course, $\|\mu\|_e^2 = \int_\Omega U^\mu \, d\mu$. The quantity (30.25) is called the *mutual energy* of the measures (or of the "charges," or of the "masses") μ, ν. Thus, under the present hypotheses (of finiteness of the energy), the reciprocity formula (30.21) merely expresses the symmetry of the inner product $(\mu, \nu)_e$. Note that, since Ω is bounded, the energy norm $\|U^\mu\|_e$ is equivalent to the norm $\|U^\mu\|_1$ on $H_0^1(\Omega)$, and $\|\mu\|_e$ is equivalent to the norm $\|\mu\|_{-1}$ on $H^{-1}(\Omega)$. Thus the inner product $(\ , \ )_e$ defines, on the set of Radon measures belonging to $H^{-1}(\Omega)$, a structure of pre-Hilbert space (Hausdorff but not complete). Note also that if μ and ν are positive Radon measures, their mutual energy is $\geqq 0$. From this it follows that, if $\mu \geqq \nu$, then $\|\mu\|_e^2 \geqq \|\nu\|_e^2$.

The preceding argument does not extend to $\mathbf{R}^n$, as $-\Delta$ is not an isomorphism of $H_0^1 = H^1$ onto H^{-1}.

The Riesz Representation of Superharmonic Functions

Let f be a superharmonic function in an open subset Ω of $\mathbf{R}^n$ and suppose that there is at least one subharmonic function v in Ω, $v(x) > -\infty$ for every $x \in \Omega$, such that $v \leqq f$ throughout Ω. Let f^* denote the upper envelope of all such subharmonic functions v. We contend that f^* is *harmonic* in Ω. Indeed, if $v \leqq f$ is subharmonic and if B is any open ball with closure contained in Ω, the function $\tilde{v}$ equal to v in $\Omega \backslash B$ and to the Poisson integral I_v^B in B is also subharmonic [as one sees by checking that $\tilde{v}(x) \leqq \tilde{v}^\#(x; r)$ for every $x \in \Omega$ and every r sufficiently small] and smaller than v, hence, than f, and therefore f^* is equal to the upper envelope of these functions $\tilde{v}$, which is harmonic in B. The function f^* is called the *greatest harmonic minorant* of f.

Suppose now that Ω is bounded and $f \geqq 0$. Given any $\varepsilon > 0$ we denote by Ω_ε the set of $x \in \Omega$ such that $d(x, \complement\Omega) > \varepsilon$. The function $x \mapsto f^@(x; \varepsilon)$ is well defined, and obviously superharmonic in Ω_ε. Moreover, $f^@(x; \varepsilon) \leqq f(x)$ there and $f^@(x; \varepsilon)$ is a *continuous* function of $x \in \Omega_\varepsilon$. When $\varepsilon \to +0, f^@(x; \varepsilon) \to f(x)$. For every ε, let Ω'_ε be a relatively compact open subset of Ω_ε, containing

$\Omega_{2\varepsilon}$, whose boundary is a C^∞ hypersurface and such that Ω'_ε lies on one side of it (locally). Let h_ε denote the generalized solution of the homogeneous Dirichlet problem for $-\Delta$ in the open set $\Omega\backslash\Omega'_\varepsilon$, with boundary data $f^{@}(\cdot;\varepsilon)$ on $\partial\Omega'_\varepsilon$ and zero on $\partial\Omega$. Since all the points of $\partial\Omega'_\varepsilon$ are regular, the function equal to $f^{@}(\cdot;\varepsilon)$ in Ω'_ε and to h_ε in $\Omega\backslash\Omega'_\varepsilon$ is continuous in the whole of Ω. We shall denote it by f_ε; it is superharmonic, and $\mu_\varepsilon = -\Delta f_\varepsilon$ has compact support (contained in $\overline{\Omega}'_\varepsilon$). Moreover $f_\varepsilon \leqq f$ and f_ε converges pointwise to f (in fact, if $2\varepsilon < \varepsilon'$, $f_\varepsilon \geqq f_{\varepsilon'}$).

If U^{μ_ε} denotes the Green potential of μ_ε (in Ω), we note that $f_\varepsilon - U^{\mu_\varepsilon}$ is harmonic in Ω and tends to zero at $\partial\Omega$, in a generalized sense: i.e., it is the uniform limit (over Ω) of a sequence of functions $u_j \in H^1(\Omega)$, harmonic in Ω, such that $u_j - g_j \in H_0^1(\Omega)$, with $g_j \in C^0(\overline{\Omega}) \cap H^1(\Omega)$ tending to zero uniformly on $\overline{\Omega}$. We conclude easily from this that it vanishes identically. In other words,

$$f^\varepsilon = U^{\mu_\varepsilon} = \int_\Omega G(\cdot, x')\, d\mu_\varepsilon(x'). \tag{30.26}$$

But the measure μ_ε is equal to $\mu = -\Delta f$ in Ω'_ε and therefore the right-hand side in (30.26) is not less than

$$\int_{\Omega'_\varepsilon} G(\cdot, x')\, d\mu(x'),$$

which converges to U^μ as $\varepsilon \to +0$ [indeed, $G(x, x')$ is integrable with respect to $d\mu(x')$ and nonnegative]. By going to the limit as $\varepsilon \to +0$, we conclude that $U^\mu \leqq f$. Of course we know that $f - U^\mu$ is harmonic in Ω (note that it is $\geqq 0$ and $\leqq f$, therefore locally integrable).

Since $f \geqq 0$ it has a greatest harmonic minorant, f^*, which is also $\geqq 0$. We may apply the preceding argument to $f - f^*$ instead of f; U^μ remains the same, since μ does. We conclude that $f - U^\mu - f^* = f_1^*$ is harmonic and $\geqq 0$ in Ω. But then $f^* + f_1^*$ is harmonic, and $\leqq f$ in Ω. In view of the maximal character of f^*, we conclude that $f_1^* \equiv 0$, and we obtain the *representation formula* for superharmonic functions, due to Riesz:

$$f(x) = \int_\Omega G(x, x')\, d\mu(x') + f^*(x), \tag{30.27}$$

where $\mu = -\Delta f$ and f^* is the greatest harmonic minorant of f (which, we recall, is nonnegative).

We derive from (30.27):

PROPOSITION 30.10. *Let Ω be a bounded open subset of $\mathbf{R}^n$. In order that a nonnegative superharmonic function in Ω be the Green potential of a positive Radon measure in Ω, it is necessary and sufficient that its greatest harmonic minorant be identically zero.*

Let $\mathscr{F}$ denote the space of locally integrable functions f in Ω whose Dirichlet integral is finite (i.e., such that grad f is square-integrable). To say that $\|f\|_e = 0$ is equivalent to saying that f is locally constant in Ω. Let $\dot{\mathscr{F}}$ denote the quotient of $\mathscr{F}$ modulo the locally constant functions in Ω, and $\|\dot{f}\|_e$ the norm on $\dot{\mathscr{F}}$ associated with the energy norm. It is a Hilbert norm, but of course $\dot{\mathscr{F}}$ is not complete. However, $H_0^1(\Omega)$ can be regarded as a closed subspace of $\dot{\mathscr{F}}$ (using again the fact that Ω is bounded) and the cone of Green potentials of positive Radon measures in Ω with finite energy can be regarded as a (closed convex) cone in $\dot{\mathscr{F}}$, which we denote by Γ. Although $\dot{\mathscr{F}}$ is not complete, in view of the fact that Γ is, we may introduce the *orthogonal projection* [for the inner product $(\ ,\ )_e$] of $\dot{\mathscr{F}}$ onto Γ. Let π denote this projection. Let f be any nonnegative superharmonic function in Ω, $\dot{f}$ its class modulo locally constant functions. We know that $\dot{f} - \pi\dot{f}$ is orthogonal, for $(\ ,\ )_e$, to the potentials $U^\nu \in H_0^1(\Omega)$ where ν is a positive Radon measure; but then this is also true when ν is not necessarily positive. We know that the U^ν form a dense linear subspace of $H_0^1(\Omega)$, and therefore $\dot{f}^* = \dot{f} - \pi\dot{f}$ is orthogonal to the latter subspace. This implies that it is harmonic in Ω. But of course,

$$-\Delta(\pi\dot{f}) = -\Delta f = \mu, \qquad \text{i.e.,} \quad \pi\dot{f} = U^\mu,$$

and there is a representative f^* of $\dot{f}^*$ in $\mathscr{F}$ such that

$$f = U^\mu + f^*, \tag{30.28}$$

which is the representation formula (30.27) under the present particular circumstances. Thus (30.27) can be viewed as a generalization of an *orthogonal* decomposition formula, associated with the energy norm.

Capacity Potential and Capacity

Let Ω be a bounded open subset of $\mathbf{R}^n$, K a compact subset of Ω. We shall denote by W_K the lower envelope (i.e., the infimum) of all superharmonic functions v in Ω, such that $v \geqq 0$ in Ω and $v \leqq 1$ on K. It is seen at once that $W_K(x) \geqq W_K^\#(x;\ r)$, $0 < r < d(x, \mathbf{R}^n\backslash\Omega)$. However, W_K might fail to be a superharmonic function, because it might fail to be lower semicontinuous. We denote then by V_K the upper envelope (i.e., the supremum) of all lower semicontinuous functions w in Ω such that $w \leqq W_K$. The following facts are easy to ascertain:

$$0 \leqq V_K \leqq W_K \leqq 1 \quad \text{in } \Omega; \tag{30.29}$$

$$V_K = W_K = 1 \quad \text{in the interior of } K; \tag{30.30}$$

$$V_K = W_K \quad \text{in } \Omega\backslash K \quad \text{and both are harmonic in } \Omega\backslash K. \tag{30.31}$$

PROPOSITION 30.11. *The function V_K is the Green potential (in Ω) of a positive measure μ_K carried by K.*

Proof. It is easy to derive from its definition that V_K is superharmonic in Ω (it is of course lower semicontinuous and all one has to check is that it majorizes its average on spheres, which is easy). Suppose, for the sake of simplicity, that Ω is connected (otherwise reason in each connected component separately). Fix $x_0 \in \Omega$ arbitrarily; the Green function $G(x, x_0)$ is a potential [that of the Dirac measure $\delta(x - x_0)$] and is greater than zero everywhere in Ω, therefore it has a minimum $c > 0$ on K, hence $0 \leqq V_K \leqq c^{-1}G(\cdot, x_0)$. The greatest harmonic minorant of cV_K is a nonnegative harmonic minorant of $G(\cdot, x_0)$. We apply Proposition 30.10 twice: once to conclude that, since $G(\cdot, x_0)$ is a potential, the greatest harmonic minorant of V_K is zero, the second time to draw from this that V_K itself is a potential. Since V_K is harmonic in $\Omega \backslash K$, $\mu_K = -\Delta V_K$ is carried by K. Q.E.D.

Definition 30.5. The potential V_K is called the capacity potential of K, the measure μ_K is called the capacity distribution of K, and the total mass of μ_K, which is $\mu_K(K)$, is the capacity of K (relative to Ω). The capacity of K is denoted by $\mathscr{C}(K)$.

We leave the proof of the next two assertions as an exercise to the student:

(30.32) *The capacity potential V_K is the greatest Green potential U^μ of positive Radon measures μ, supported by K, such that $U^\mu \leqq 1$ in Ω.*

(30.33) *The capacity $\mathscr{C}(K)$ is the supremum of the total masses $\mu(K)$, as μ ranges over the set of all positive Radon measures supported by K whose Green potential does not exceed 1 in Ω.*

The notion of capacity plays a truly fundamental role in the modern potential theory and that is why we have decided to include its definition and basic properties here (see below), even though we do not give the proofs. As can be guessed, it originated with electrostatics: It is a measure of the maximum charge that can be loaded, at equilibrium, on a conductor (occupying the compact set K) surrounded by a medium void of charges (here, $\Omega \backslash K$) and itself bounded by a grounded surface (here $\partial\Omega$: "grounded" means that the potential V_K on it is maintained at zero). The requirement that V_K not exceed one in Ω is a "normalizing" condition, enabling us to compare capacities.

PROPOSITION 30.12. *Let Ω be a bounded open subset of $\mathbf{R}^n$, $\mathscr{C}(K)$ the capacity with respect to Ω of an arbitrary compact subset of Ω. The following is true:*

(30.34) $\mathscr{C}(\varnothing) = 0$ *and* $K_1 \subset K_2 \Rightarrow \mathscr{C}(K_1) \leqq \mathscr{C}(K_2)$;

(30.35) $\mathscr{C}(K_1 \cup K_2) + \mathscr{C}(K_1 \cap K_2) \leqq \mathscr{C}(K_1) + \mathscr{C}(K_2)$;

(30.36) *if $\{K_j\}$ $(j = 1, 2, \dots)$ is a decreasing sequence of compact subsets of Ω, with intersection K, $\mathscr{C}(K) = \lim_{j \to +\infty} \mathscr{C}(K_j)$.*

For a proof, see [*H*, Theorem 7.20].

Based on the properties (30.34) to (30.36) one can extend the capacity (regarded as a function of sets), somewhat in the manner one extends a Radon measure. One can define the *inner capacity* of an arbitrary subset A of Ω as the supremum of $\mathscr{C}(K)$, when K ranges over the collection of compact sets contained in A, and the *outer capacity* as the infimum of the inner capacities of all open sets containing A. A set is then said to be *capacitable* if its inner and outer capacities are equal (they are then called its capacity) and so on. An interesting result, due to H. Cartan, states that there is identity between polar sets (see Exercise 30.6, Definition 30.6) and sets with zero capacity (the surrounding Ω is here irrelevant).

Exercises

30.1. Let f be a superharmonic function $> -\infty$ in an open neighborhood of a closed ball $|x - x_0| \leqq R$ ($x_0 \in \mathbf{R}^n$, $R > 0$). Derive from the remark which follows (30.4) and from the Poisson integral formula (10.36) that the restriction of f to the sphere $|x - x_0| = R$ is integrable (with respect to the area measure).

30.2. Let ϕ be a *convex* mapping of an interval $]a, b[$ $(-\infty \leqq a < b \leqq +\infty)$ of the real line, into $\mathbf{R}$ [convex means that $\phi(\alpha t_1 + \beta t_2) \leqq \alpha\phi(t_1) + \beta\phi(t_2)$ if $\alpha + \beta = 1$, $\alpha, \beta \geqq 0$].

Let Ω be an open subset of $\mathbf{R}^n$, h a harmonic function in Ω such that $h(x) \in\ ]a, b[$ for all $x \in \Omega$. Prove that $\phi \circ h$ is subharmonic in Ω.

Show, moreover, that if ϕ is monotone increasing, then $\phi \circ f$ is subharmonic in Ω whatever the subharmonic function f in Ω whose image lies in the interval $]a, b[$.

30.3. Let f be a superharmonic function in $\mathbf{R}^n$, $f^{@}(\cdot\,; r)$ its average on the ball of radius $r > 0$ (centered at the point under consideration). Show that $f^{@}(\cdot\,; r)$ is a *continuous* superharmonic function in $\mathbf{R}^n$. What is $f^{@}$ when f is the fundamental solution of $-\Delta$, given in (30.7) or (30.9)? For this choice of f, compute the spherical average $f^{\#}$ [see (30.1) and (30.2)].

Derive from the property of $f^{@}$ above that, given any integer $m \geqq 0$, f is the pointwise limit of an *increasing* sequence of C^m superharmonic functions in $\mathbf{R}^n$.

30.4. Let f be a superharmonic function in an open subset Ω of $\mathbf{R}^n$. By using the ball averages $f^{@}(\cdot\,; r)$ defined in (30.2), show that f is the pointwise limit of an increasing sequence of C^2 superharmonic functions in Ω.

30.5. Let Ω be a bounded open subset of $\mathbf{R}^n$, $G(x, x')$ the Green function of $-\Delta$ in Ω. Let μ_f be the measure $f(x)\,dx$ in Ω, where $f \in L^1(\Omega)$. Prove that the Green potential of μ_f [see (30.20)] is a *continuous* function in Ω. [*Hint*:

Reduce the proof to the case where the support of f is contained in a closed ball $\bar{B} \subset \Omega$ and then exploit the properties of the Green function for the ball, (29.34) or (29.35).]

30.6. This exercise concerns the following important notion:

Definition 30.6. A subset S of $\mathbf{R}^n$ is called polar if there is an open neighborhood U of S and a superharmonic function f in U which is equal to $+\infty$ on S.

Prove that a polar set has (Lebesgue) measure zero and that its intersection with an arbitrary sphere $|x - x_0| = R$ has (area) measure zero.

30.7. Let (x, y, z) denote the variable point in $\mathbf{R}^3$ and J, the segment $x = y = 0, 0 \leqq z \leqq 1$. Consider the measure $\mu : \phi \mapsto \int_0^1 \phi(0, 0, z)\, dz$ (carried by J), and its potential in $\mathbf{R}^3$, U^μ (Definition 30.3). Show that $U^\mu\,(0, 0, z) = +\infty$ if $0 \leqq z \leqq 1$ and conclude from this that every straight line segment in $\mathbf{R}^n$, $n \geqq 3$, is polar (Definition 30.6).

30.8. Let r be a positive number, and denote by Π_r the piece of hyperplane

$$|x^j| < r \quad (1 \leqq j \leqq n-1), \qquad x^n = 0, \tag{30.37}$$

in $\mathbf{R}^n$. Suppose that there were an open neighborhood of Π_r, U, and a superharmonic function f in U, equal to $+\infty$ on Π_r. By rotation and translations (which leave superharmonicity invariant), show that there would be a number r', $0 < r' < r$, a function F, defined and superharmonic in an open neighborhood of the hypercube $\{x \in \mathbf{R}^n;\ |x^j| < r', j = 1, \ldots, n\}$, equal to $+\infty$ on the boundary of this hypercube. Derive from the minimum principle for superharmonic functions that this is not possible and conclude that Π_r cannot be polar (Definition 30.6).

30.9. Let Ω be the open ball of radius $R > 0$ in $\mathbf{R}^n$ ($n \geqq 2$), centered at the origin, K the closed ball of radius R', $0 < R' < R$, also centered at the origin. Compute the capacity (Definition 30.5) of K with respect to Ω. What are the capacity potential and distribution of K relative to Ω? What are the answers to the same questions if we now take K to be the sphere $|x| = R'$?

30.10. Prove the following result, due to Hartog: Let $\{v_k\}$ ($k = 1, 2, \ldots$) be a sequence of superharmonic functions in an open set $\Omega \subset \mathbf{R}^n$, such that:

(i) given any compact subset K of Ω, there is a constant $M_K > 0$ such that $v_k(x) \geqq -M_K$ for every index k and every $x \in K$;

(ii) there exists a constant $C > 0$ such that, whatever $x \in \Omega$, the *lower limit* of the numbers $v_k(x)$, as $k \to +\infty$, is $\geqq -C$.

Then, given any compact subset K of Ω and any number $\varepsilon \geqq 0$, there is an index k_0 such that

$$v_k(x) \geqq -C - \varepsilon, \qquad \forall k \geqq k_0, \quad \forall x \in K. \tag{30.38}$$

31

Laplace Equation and the Brownian Motion

31.1 The Discrete Case

Let $\mathbf{Z}^N$ denote the lattice of points $x = (x^1, \ldots, x^N)$ in $\mathbf{R}^N$ whose coordinates x^j are integers (≥ 0 or < 0). We say that two points x and y in $\mathbf{Z}^N$ are *neighbors* if their Euclidean distance is exactly equal to one. Every point in $\mathbf{Z}^N$ has exactly $2N$ neighbors. A *path* in $\mathbf{Z}^N$ is a mapping of an interval in $\mathbf{Z}$ into $\mathbf{Z}^N$ such that the image of two neighboring integers is a pair of neighboring points. If the interval in which this map is defined has a finite lower limit, the path has a *starting point*; if the interval has a finite upper limit, it has an *arriving point*.

We shall now look at a "particle" moving over the lattice $\mathbf{Z}^N$ (the reader may prefer to think of a "piece" moving on an infinite N-dimensional checker-board). The motion of the particle is effected as follows: Before each move we throw a dice having $2N$ faces, on each one of which is labeled an N-tuple, precisely the set of coordinates of one of the neighbors of the origin. By translating the result from the origin to the point where the particle happens to be at present, we discover at what neighboring point it must be next. Starting with a position x, successive dice throwing (or *trials*) yield the path followed by the particle. Its motion, under the rules above, is called *random walk*. An important feature of random walk is that each trial is independent of the preceding ones. Let us denote by $x(0) = x, x(1), \ldots, x(n)$, the successive positions of the particle. We may introduce the *transition probabilities* $p(n, x, B)$: B is a subset of $\mathbf{Z}^N$ and $p(n, x, B)$ is the probability that the particle starting at $x(0) = x$ finds itself in the set B at the nth step, i.e. that $x(n) \in B$. If B consists of a single point y, we write $p(n, x, y)$. It is natural to define $p(0, x, y)$ as one if $x = y$ and as zero otherwise. Note that

(31.1) $p(1, x, y) = 1/2N$ if x and y are neighbors, $\quad 0$ otherwise.

The function $B \mapsto p(n, x, B)$ defines a positive measure, with total mass one, on $\mathbf{Z}^N$: It is indeed a probability. If the distance between x and B exceeds n, we have $p(n, x, B) = 0$; in other words, $y \mapsto p(n, x, y)$ has compact (finite) support (easy to describe).

Let Ω be a subset of $\mathbf{Z}^N$. We shall assume Ω to be *connected*, which means that any two points in Ω can be joined by a path entirely contained in Ω. A point of Ω will be called an *interior point* if all its neighbors also belong to Ω, a *boundary point* if this is not so.

A (say, vector-valued) function f in Ω is said to be *harmonic* if its value at any interior point $x \in \Omega$ is equal to its average over the set of neighbors of x:

$$f(x) = \frac{1}{2N} \sum_{|x-y|=1} f(y). \tag{31.2}$$

Note that the right-hand side of (31.2) is nothing else but

$$\mathrm{P}f(x) = \sum_{y \in \mathbf{Z}^N} p(1, x, y) f(y). \tag{31.3}$$

The operator $f \mapsto \mathrm{P}f$ is often called the (one-step) *shift* or *averaging* operator. A real-valued function f is called *superharmonic* if $f(x) \geqq \mathrm{P}f(x)$ at every interior point x; nonnegative superharmonic functions are called *excessive functions* in the theory of Markoff processes.

It is obvious that the *maximum principle* holds for complex-valued harmonic functions in Ω. Also, any harmonic function in the whole lattice $\mathbf{Z}^N$ which is bounded must be a constant.

We may set

$$\Delta = -2(I - P), \tag{31.4}$$

where P is the shift operator (31.3) (and I the identity); Δ is the *discrete Laplacian*. By definition, a function f is harmonic if $\Delta f = 0$.

Suppose now that the set Ω is *bounded*, and let $\partial\Omega$ denote its boundary. Let g be a *real-valued* function defined in $\partial\Omega$. The *Dirichlet problem* in the discrete case is the problem of solving

$$\Delta u = 0 \quad \text{in the interior points of } \Omega, \tag{31.5}$$

$$u = g \quad \text{on } \partial\Omega. \tag{31.6}$$

This problem has a remarkable probabilistic interpretation, and so does its unique solution! Indeed, we may interpret g as a *payoff*: If our particle, starting from a point x in the interior, hits the boundary for the first time at the point y, we have winnings or losses equal to $g(y)$ [whether it is a gain or a loss depends on the sign of $g(y)$]. The value of the solution u at x will then be the value of the *expected winnings* (or *losses*) at x, as we are now going to show.

Let B be an arbitrary subset of the boundary $\partial\Omega$. If x is an interior point, we define $m_x(B)$ as the probability that the random walk of the particle, starting at x, hits the boundary for the first time at a point y belonging to the set B. It is immediately seen that m_x defines a positive measure on $\partial\Omega$, with total mass $\leqq 1$. Let us fix B and consider the nonnegative function $x \mapsto m_x(B)$; it is defined at the interior points of Ω but we may extend its definition to boundary points x by setting it equal there to one if $x \in B$, to zero otherwise. It is clear that we have, if x is an interior point,

$$m_x(B) = \sum_{|x-y|=1} m_y(B)p(1, x, y) = \frac{1}{2N} \sum_{|x-y|=1} m_y(B), \tag{31.7}$$

which shows that $m_x(B)$ is harmonic. Then take $B = \partial\Omega$: When x is a boundary point, we have $m_x(\partial\Omega) = 1$, and therefore $m_x(\partial\Omega)$ is a harmonic function in Ω, identically equal to one on the boundary. By the maximum principle, it must be identically equal to one throughout Ω. This implies that the total mass of the measure m_x on $\partial\Omega$ is one, for every $x \in \Omega$.

Now, in order to obtain the solution u of (31.5)–(31.6) for arbitrary g, it suffices to take

$$u(x) = \sum_{y \in \partial\Omega} g(y) m_x(\{y\}). \tag{31.8}$$

Indeed, u is harmonic as a finite sum of harmonic functions, and it takes the value $g(y)$ at a boundary point y, because of the properties of the harmonic measure m_x [cf. (29.7)]. Because of our definition of $m_x(\{y\})$ (the probability that the particle first hits the boundary at the point y), (31.8) defines $u(x)$ exactly as the *mathematical expectation* at the point x of the payoff g.

We may interpret the values of any function f on $\mathbf{Z}^N$ as coefficients of a Fourier series. Let us set, indeed,

$$\tilde{f}(\theta) = \sum_{y \in \mathbf{Z}^N} f(y) e^{i\theta \cdot y}, \tag{31.9}$$

where θ is the variable in the N-dimensional torus $\mathbf{T}^N : \theta = (\theta_1, \ldots, \theta_N)$, $0 \leqq \theta_j \leqq 2\pi$ for every j, and $\theta \cdot y = \theta_1 y^1 + \cdots + \theta_N y^N$. The series (31.9) certainly converges if f has finite support (and also, of course, under broader conditions). We have then

$$f(y) = (2\pi)^{-N} \int_{\mathbf{T}^N} e^{-i\theta \cdot y} \tilde{f}(\theta)\, d\theta. \tag{31.10}$$

We may apply this to the function $p(n, x, y)$ for fixed x and n. We note that

$$\tilde{p}(0, x, \theta) = e^{i\theta \cdot x}, \tag{31.11}$$

and on the other hand, by the total probability formula, for $n \geqq 1$,

(31.12)

$$p(n, x, y) = \sum_{|x-x'|=1} p(1, x, x')p(n-1, x', y) = \frac{1}{2N} \sum_{|x-x'|=1} p(n-1, x', y),$$

whence

(31.13) $$\tilde{p}(n, x, \theta) = \frac{1}{2N} \sum_{|x-x'|=1} \tilde{p}(n-1, x', \theta).$$

Taking $n = 1$ and applying (31.11) yields, since $|x - x'| = 1$ implies $x' \cdot \theta = x \cdot \theta \pm \theta_j$ for some j, $1 \leqq j \leqq N$,

(31.14) $$\tilde{p}(1, x, \theta) = e^{i\theta \cdot x}\Phi(\theta),$$

where we have set

(31.15) $$\Phi(\theta) = \frac{1}{N} \sum_{j=1}^{N} \cos \theta_j .$$

If we return to (31.13) and apply induction on $n = 1, 2, \ldots$, we immediately obtain

(31.16) $$\tilde{p}(n, x, \theta) = e^{i\theta \cdot x}\Phi^n(\theta),$$

and (31.11) shows that this is also true for $n = 0$. Applying now the Fourier inversion formula (31.10) yields

(31.17) $$p(n, x, y) = (2\pi)^{-N} \int_{\mathbf{T}^N} e^{i\theta \cdot (x-y)}\Phi^n(\theta)\, d\theta$$

Let us then consider the sum of all the powers Φ^n, $n = 0, 1, \ldots, (1 - \Phi)^{-1}$; it is clear that it has a pole at $\theta = 0$; the order of $(1 - \Phi(\theta))^{-1}$ at that pole is 2, in the sense that

(31.18)

$$\text{for} \quad \theta \sim 0, \qquad (1 - \Phi(\theta))^{-1} \sim 2N|\theta|^2 \qquad [|\theta| = (\theta_1^2 + \cdots + \theta_N^2)^{1/2}]$$

We derive from this that, *provided* $N \geqq 3$, *the function* $(1 - \Phi)^{-1}$ *on the torus* $\boldsymbol{T}^N$ *is integrable*, and therefore the function

(31.19) $$G(z) = \sum_{n=0}^{+\infty} p(n, z, 0) = (2\pi)^{-N} \int_{\mathbf{T}^N} e^{i\theta \cdot z}(1 - \Phi(\theta))^{-1}\, d\theta$$

is bounded. If we apply the total probability formula (31.12), we immediately obtain

(31.20) $$-\tfrac{1}{2}\Delta G(x) = \delta(x),$$

where δ is the analog of the Dirac measure: It is equal to one if $x = 0$ and to zero otherwise. Thus G is harmonic in the complement of the origin. Let us also denote by G the convolution operator

$$G : \mu \mapsto G * \mu(x) = \sum_y G(x - y)\mu(y) \tag{31.21}$$

(defined, say, on functions—which are the same thing as measures, on $\mathbf{Z}^N$—with compact support). Then (31.20) can be rewritten in the form

$$(I - P)G = I, \qquad \text{the identity operator,} \tag{31.22}$$

i.e., $G = (I - P)^{-1}$. If we look back at (31.15) and (31.19), we see that $\Phi(\theta)$ is the Fourier transform of the shift operator P, and $(1 - \Phi(\theta))^{-1}$ is the one of G. The usefulness of (31.19) is that it shows readily that $G(z)$ is a bounded function when $N \geqq 3$. One can also easily derive from it that

$$\text{when} \quad |x| \sim +\infty, \qquad G(x) \sim c_N |x|^{2-N} \tag{31.23}$$

[we leave the derivation as an exercise for the student: The basic idea, of course, is to exploit (31.18) and to make a change of variable $\theta = |z|^{-1}\theta'$ in the integral in (31.19)].

The properties just given show that G is the analog of the Newton potential (see Example 30.1). The presence of the factor $\frac{1}{2}$ in (31.20) is merely due to our presently adopting a different convention (which, actually, is the one commonly adopted in potential and probability theories). We may of course define the potential in $\mathbf{Z}^N$ of a measure (i.e., function) with compact support, μ, by the formula

$$U^\mu = G * \mu \qquad \text{(cf. Definition 30.3).} \tag{31.24}$$

It is seen at once (cf. Proposition 30.2) that U^μ is the unique solution of the inhomogeneous Laplace equation

$$-\tfrac{1}{2}\Delta U^\mu = \mu \qquad (\text{in } \mathbf{Z}^N) \tag{31.25}$$

which tends to zero at infinity.

If we rewrite (31.24) in the following fashion:

$$U^\mu(x) = \sum_{n=0}^{\infty} \left(\sum_{y \in \mathbf{Z}^N} p(n, x, y)\mu(y) \right), \tag{31.26}$$

we obtain the probabilistic interpretation of the potential. Again, we must regard the given function $\mu(y)$ as a payoff. The nth term,

$$\sum_y p(n, x, y)\mu(y), \tag{31.27}$$

is the mathematical expectation of winnings μ at the nth trial. Equation (31.26) says therefore that $U^\mu(x)$ is the total mathematical expectation (or, equivalently, the mean value) of the payoff during a random walk of the particle starting at the point x.

We conclude the first half of this section by computing the quantity (31.27) when $\mu(y) = |y|^2$, the square of the Euclidean norm of $y \in \mathbf{Z}^N$. The result will give us the mean value of the square of the distance (from the origin) reached by a particle in a random walk started at x. Let us denote it by $d^2(n, x)$. For $n \geqq 1$, we have

$$d^2(n, x) = \frac{1}{2N} \sum_{|x-x'|=1} d^2(n-1, x'). \tag{31.28}$$

Of course, we have $d^2(0, x) = |x|^2$. By induction on n, we derive from (31.28) that

$$d^2(n, x) = |x|^2 + n. \tag{31.29}$$

Indeed, (31.28) implies

$$d^2(n, x) = n - 1 + (2N)^{-1} \sum_{|x-x'|=1} |x'|^2.$$

But because of the obvious symmetry,

$$\sum_{|x-x'|=1} |x'|^2 = N(|x + \mathbf{e}_1|^2 + |x - \mathbf{e}_1|^2) = 2N(|x|^2 + 1),$$

where $\mathbf{e}_1$ is the unit vector along the x^1-axis.

In §31.2, we shall make use of (31.29) or rather of its consequence (which is also a direct consequence of the *central limit theorem*) that the mean value of the distance reached at the nth step by a particle in a random walk is of the order of $\sqrt{n}$ (for large n).

31.2 The Continuous Case

Instead of dealing with the lattice $\mathbf{Z}^N$ in §3.1, we could have dealt with any one of its homothetical images $h\mathbf{Z}^N$ (the points of $h\mathbf{Z}^N$ are obtained by multiplying those of $\mathbf{Z}^N$ by h). Then, by going to the limit as $h \to +0$, we should arrive at the concept of a *continuous random walk*. However in the process of going to the limit, we should not lose sight of the fact that the notion of distance covered during the random walk must not be allowed to dissolve! Specifically, the distance covered during a random walk in $\mathbf{Z}^N$ must be essentially the same as the one covered in a walk "of same duration" in $h\mathbf{Z}^N$. But we cannot, really, talk of the distance covered during a random walk, we can only talk of its mean value which, by the remark concluding §31.1, we know to be of the order of $\sqrt{n}$. Thus, if we wish to divide the length of the steps by h^{-1}, we must divide the "duration" of each step by h^{-2}; i.e., we must multiply the number of steps effected in a given time by h^{-2}. The switch to a "temporal" terminology is not accidental: the variable n, the number of steps, in §31.1, is now going to be replaced by the time variable t, i.e., the duration of the walk. For reasons of convenience in the notation, we shall divide the duration of each step, not by h^{-2}, but by Nh^{-2}.

Thus we return to the total probability formula (31.12), but now in the lattice $h\mathbf{Z}^N$ and according to the preceding remarks, we rewrite it as follows:

$$p(t + h^2/N, x, y) = (2N)^{-1} \sum_{|x-x'|=h} p(t, x', y) \tag{31.30}$$

[thus we have written t instead of $(n-1)h^2/N$]. Let us denote by $\mathbf{e}_j$ the unit vector along the x^j-axis ($j = 1, \ldots, N$). We subtract $p(t, x, y)$ from both members in (31.30) and divide the result by h^2/N. We obtain

$$\begin{aligned} &\frac{N}{h^2}\{p(t + h^2/N, x, y) - p(t, x, y)\} \\ &\qquad = \frac{1}{2}\sum_{j=1}^{N} \frac{1}{h^2}\{p(t, x + h\mathbf{e}_j, y) + p(t, x - h\mathbf{e}_j, y) - p(t, x, y)\}. \end{aligned} \tag{31.31}$$

Now we go to the limit as $h \to +0$. We easily obtain

$$\left(\frac{\partial}{\partial t} - \tfrac{1}{2}\Delta\right) p(t, x, y) = 0 \qquad (\text{for } \ t > 0). \tag{31.32}$$

We see that the transition probability $p(t, x, y)$ is a solution of the heat equation, in the half-space $t > 0$ (with $\frac{1}{2}\Delta$ substituted for Δ). For fixed $t > 0$, we must have $p(t, x, y) \to 0$ as $|x - y| \to +\infty$, since this is true in the discrete case. Furthermore,

$$p(0, x, y) = \delta(x - y), \qquad \text{the Dirac measure (on the diagonal).} \tag{31.33}$$

These conditions determine $p(t, x, y)$ uniquely (see §6.1, also Example 44.1). We have $p(t, x, y) = p(t, x - y)$, where

$$p(t, z) = (2\pi t)^{-N/2} \exp(-|z|^2/2t). \tag{31.34}$$

Here $t > 0$; when $t \to +0$, $p(t, z)$ converges (in the distributions, or in the measures sense) to the Dirac measure at the origin δ. If B is a Lebesgue measurable subset of $\mathbf{R}^N$, the probability that the particle, presently (i.e., at some given time which we take to be the origin) at the position x, will be, at time $t > 0$, at some point belonging to B is equal to

$$P_x[x(t) \in B] = \int_B p(t, x - y)\, dy. \tag{31.35}$$

A probability distribution of the kind (31.34) is called *normal* or *Gaussian*. It is spherically symmetric, indicating the perfect isotropy of the continuous random walk (which is called *Brownian motion*, in honor of the botanist Robert Brown who observed the "rapid oscillatory motion" of pollen grains suspended in water, in 1827). The graph of $p(t, z)$ (for fixed $t > 0$) is the usual bell-shaped surface, with maximum at $z = 0$. The fact that, for arbitrary x, the probability distribution is equal to $p(t, x - y)$ emphasizes the fact that the motion is *translation-invariant*. Note also that, according to (31.35), past events have no influence on the future motion. One notable difference with the discrete case is that the probability that the particle finds itself at any given individual point, at some time $t > 0$, is always zero. The particle, presently at the point x, will at random pick up a path—according to the law (31.35). Paths are simply continuous mappings of $\overline{\mathbf{R}}_+ = \{t \in \mathbf{R};\, t \geqq 0\}$ into $\mathbf{R}^N$; with every "starting point" x we may associate a probability distribution, i.e., a positive measure with total mass one, on the space of all paths originating at x, but we do not go into this here.

The probabilistic approach to the Dirichlet problem can now easily be guessed. Let Ω be a bounded open set in $\mathbf{R}^N$, $\partial\Omega$ its boundary. We introduce the *time of first exit* τ of a particle starting at $x \in \Omega$: τ is the minimum of all numbers t such that $x(t) \in \mathbf{R}^N \backslash \Omega$. Next we define the *escape probability* m_x: It is a measure on the boundary $\partial\Omega$. To every (say, Borel) subset A of $\partial\Omega$ it assigns the number which is the probability that $x(\tau) \in A$. If now g is an arbitrary continuous function on $\partial\Omega$, and if we set

$$u(x) = \int_{\partial\Omega} g(y)\, dm_x(y), \tag{31.36}$$

we shall have

$$\Delta u = 0 \quad \text{in } \Omega, \tag{31.37}$$

$$u = g \quad \text{on } \partial\Omega. \tag{31.38}$$

A few facts have to be established and the meaning of (31.38) must be made clear. First of all, we must show that the time of escape is *finite*, or rather (more in keeping with the standpoint adopted here) that the particle escapes from the set Ω with probability 1. Actually, it suffices to prove that the particle escapes from an arbitrary ball, $B_R(x)$, centered at x, with probability 1. Let us show that the probability that the particle remains inside the ball forever is zero. Suppose it remains inside $B_R(x)$ until time $n \in \mathbf{Z}_+$; the increments $x(j) - x(j-1)$, $j = 1, \ldots, n$, must all have norm $\leqq 2R$. Because of the probability law presiding over the motion, we have

$$P[|x(j) - x(j-1)| \leqq 2R] = (2\pi)^{-N/2} \int_{|z|<2R} \exp(-|z|^2/2)\, dz = \alpha_R < 1,$$

and by the formula of total probability,

$$P_x[\tau > n] \leqq \prod_{j=1}^{n} P[|x(j) - x(j-1)| \leqq 2R] = \alpha_R^n$$

which converges to zero as $n \to +\infty$. (We have denoted by $P[e]$ the probability of the event e.)

A very simple probabilistic argument shows that the function (31.36) is harmonic. Let $B_r(x)$ be any open ball, centered at x, whose closure is contained in Ω. Before hitting the boundary the particle starting at x must hit the boundary $S_r(x)$ of the ball. But then, if we want to compute the probability of hitting (at the time of first exit from Ω, τ) some subset of $\partial\Omega$, we may as well forget what happened before the time of first exit from the ball $B_r(x)$, τ'. In other words, we may assume that the particle starts at time τ' from a point of $S_r(x)$. But we must specify the probability that the particle finds itself (at the time τ') in a given subset of $S_r(x)$, A'. This must be necessarily equal to the probability that the particle hit A' at τ' when coming from x. The latter cannot be anything but uniform on $S_r(x)$, by virtue of the symmetry of the Gaussian law. In mathematical terms, the analog of the measure m_x for $B_r(x)$ is the area measure on the sphere $S_r(x)$, normalized, i.e., divided by the total area of $S_r(x)$. What we have just said means exactly that

$$\text{(31.39)} \qquad m_x(A) = |S^{n-1}|^{-1} \int_{S^{n-1}} m_{x+r\dot{x}}(A)\, d\dot{x} \qquad \text{[cf. (30.1)]},$$

where A is a Borel subset of $\partial\Omega$. In other words, the measure m_x is harmonic (its measurability with respect to x is easy to verify) and this implies (31.37).

Let us now look at the boundary condition (31.38). We know, by the considerations of Sect. 29, which led us to the notion of regular point (Definition 29.4), that things are not simple. Suppose, for instance, that Ω is the complement of the origin in the unit disk in the plane $\mathbf{R}^2$. Then it can be shown, whatever the point x in Ω, that is, however near x is to 0, that the probability that the particle, starting from x, will exit for the first time from Ω at the origin is equal to zero. The notion of regularity of a boundary point has the following probabilistic interpretation: *If we denote by τ the time of first exit from Ω of a particle starting at x, a point x_0 of $\partial\Omega$ is regular if and only if*

$$\text{(31.40)} \qquad \lim_{x \to x_0} P_x[\tau > h] = 0, \qquad \textit{for every} \quad h > 0.$$

(in standard terminology, τ converges to zero *in probability* as $x \to x_0$). We shall not prove this assertion (the proof is not very difficult and the student might try to find it). It is a consequence of the following assertion: *x_0 is regular if and only if*

$$\text{(31.41)} \qquad \forall r > 0, \qquad \lim_{x \to x_0} P_x[x(\tau) \in \partial\Omega \cap B_r(x_0)] = 1.$$

Exercises

31.1. Let $G(z)$ be the potential in $\mathbf{Z}^N$ given by (31.19). Show that, when $N = 2$,

$$G(0) = \pi^{-2} \int_{-\pi}^{\pi} \int_{-\pi}^{\pi} \frac{d\theta_1\, d\theta_2}{4 - (\cos\theta_1 + \cos\theta_2)^2} = +\infty$$

but that $G(0) < +\infty$ if $N \geqq 3$. [*Hint*: In the case $N = 2$, observe that $p(2k + 1, 0, 0) = 0$ for all $k = 0, 1, \ldots$.]

31.2. This exercise deals with the important concept introduced by

Definition 31.1. A subset B of $\mathbf{Z}^N$ *is called recurrent if, given any point x of* $\mathbf{Z}^N$, *the probability that a particle, starting at x, eventually hits B is equal to 1.*

If $x \in \mathbf{Z}^N$, denote by $\pi_B(x)$ the probability that a particle, starting at x, hits B at least once during its random walk, by $\pi_B^*(x)$ the probability that it hits B infinitely often. Prove the following assertion:

(31.42) *if B is recurrent*, $\pi_B^*(x) = 1$; *if B is nonrecurrent*, $\pi_B^*(x) = 0$.

[*Hint*: Express $\pi_B(x)$ as the sum $\sum_{n=0}^{+\infty} q_B(n, x, y)$, where $q_B(n, x, y)$ is the probability that the particle, starting at x, hits the set B for the first time on the nth step, at the point $y \in B$, and show that $\pi_B^*(x) = \pi_B(x)\pi_B^*(x)$.]

31.3. Prove that if $N \leqq 2$, every subset of $\mathbf{Z}^N$ is recurrent (Definition 31.1), whereas, if $N \geqq 3$, every bounded subset of $\mathbf{Z}^N$ is nonrecurrent. Prove that the "plane" $x^3 = 0$ in $\mathbf{Z}^3$ is recurrent.

31.4. Prove the *Riesz decomposition* formula [see (30.27)] in the discrete case. That is, prove that given any *excessive* function f in $\mathbf{Z}^N$ (i.e., any nonnegative superharmonic function), we have

$$f = U^\mu + f^*, \tag{31.43}$$

where $U^\mu = G * \mu$ is the potential of $\mu = -\frac{1}{2}\Delta f$, and f^* is a harmonic function in $\mathbf{Z}^N$. Show that f^* is the greatest harmonic minorant of f.

31.5. Let $B \in \mathbf{Z}^N$, and π_B, π_B^* be the functions so denoted in Exercise 31.2. Prove that π_B is excessive and that π_B^* is the greatest harmonic minorant of π_B. Prove that $\mu_B = -\frac{1}{2}\Delta\pi_B$ is supported by B and that B is nonrecurrent if and only if π_B is a potential.

31.6. Suppose that B is *nonrecurrent* (see Definition 31.1). Prove that $\mu_B = -\frac{1}{2}\Delta\pi_B$ is the greatest charge carried by B whose potential is $\leqq 1$. [Then the total mass $\mathscr{C}(B) = \sum_{y \in B} \mu_B(y)$ is called the (discrete) *capacity* of B.]

31.7. Prove that, if $N \geqq 3$, the capacity of a set B consisting of k points ($k < +\infty$) is equal to $k/G(0)$, $G(z)$ being the potential (31.19), and that the capacity of any infinite nonrecurrent set is infinite.

31.8. Let Ω be a bounded open subset of $\mathbf{R}^N$. We shall denote by σ the first *strictly positive* time of exit from Ω of a particle starting at x. (Note that if $x \in \complement\Omega$, the first time of exit τ is obviously zero, but if x belongs to the boundary of Ω, the particle might reenter Ω at once and then reexit from Ω at a later time.) Let h be an arbitrary number greater than zero. Show that, given any point x,

$$P_x[\tau > h] \leqq P_x[\sigma > h]. \tag{31.44}$$

Denote then by $A(t_0)$ the event that, in the time interval $t_0 \leqq t \leqq h\,(t_0 > 0)$, the trajectory of the particle lies entirely in Ω, and denote by $P_x[A(t_0)]$ the probability of $A(t_0)$ for a particle starting at x. Prove that $P_x[A(t_0)]$, for fixed t_0, $0 < t_0 < h$, is a continuous function of x in $\mathbf{R}^N$. [*Hint*: Prove that

$$P_x[A(t_0)] = \int_{\mathbf{R}^N} P_y[\tau > h - t_0]p(t_0, x - y)\,dy.] \tag{31.45}$$

Show that

$$P_x[\sigma > h] = \lim_{t_0 \searrow +0} P_x[A(t_0)], \tag{31.46}$$

and conclude from this that $P_x[\sigma > h]$ is upper semicontinuous with respect to x, and that, by using the necessary and sufficient condition (31.40) for the regularity of $x_0 \in \partial\Omega$, together with (31.44), x_0 is regular if

$$P_{x_0}[\sigma > 0] = 0. \tag{31.47}$$

31.9. Use the same notation as in Exercise 31.8. Let h, t_1 be two numbers such that $0 < h < t_1$ and let now $A(t_1)$ denote the event that, in the time interval $h \leqq t \leqq t_1$, the trajectory of the particle lies entirely inside Ω. Prove that, if x is any point in $\mathbf{R}^n$,

$$P_x[\sigma > 0, A(t_1)] = P_x[\sigma > 0]P_x[A(t_1)], \tag{31.48}$$

i.e., for $h > 0$, the events $\sigma > 0$ and $A(t_1)$ are independent (observe that the occurrence of $\sigma > 0$ is decided in an infinitely small time interval $0 < t < \varepsilon$). Derive from (31.48) that

$$P_x[\sigma > 0] = (P_x[\sigma > 0])^2, \tag{31.49}$$

whence the *zero–one law*: $P_x[\sigma > 0]$ is either equal to zero or to one.

31.10. Let Ω be a bounded open subset of $\mathbf{R}^N$, x_0 a point of its boundary. Suppose that there is a truncated spherical cone Γ, contained in $\mathbf{R}^N \backslash \Omega$, with vertex at x_0. By combining the sufficient condition of regularity (31.47) with the zero–one law (31.49), show that if x_0 were irregular, there would be $t_0 > 0$ such that the probability that the trajectory lie entirely inside Ω during the time interval $0 < t < t_0$ would be 1. Thus for $0 < t < t_0$, the trajectory would *certainly* lie outside the cone Γ. Use then the rotation invariance of the Gaussian law to conclude that, during the open interval $0 < t < t_0$, the trajectory would lie outside an open ball, with strictly positive radius, centered at x_0, and derive from this that x_0 must be regular, contrary to the assumption.

31.11. Derive from the Gaussian law the mean value of the distance covered, during a time interval $0 \leqq t \leqq T$, in a continuous random walk.

32

Dirichlet Problems in the Plane. Conformal Mappings

In this section we are concerned with the relationship between harmonic functions of two variables (x, y) in the plane $\mathbf{R}^2$ (or in open subsets of $\mathbf{R}^2$) and analytic functions of the complex variable $z = x + iy$ $\left(i = \sqrt{-1}\right)$. The relationship is based on the fact that

$$(32.1) \qquad \Delta = \left(\frac{\partial}{\partial x}\right)^2 + \left(\frac{\partial}{\partial y}\right)^2 = 4\,\frac{\partial^2}{\partial z\,\partial \bar{z}},$$

where $\partial/\partial\bar{z} = \frac{1}{2}(\partial/\partial x + i\,\partial/\partial y)$ is the Cauchy–Riemann operator (cf. Sect. 5) and $\partial/\partial z$ its "complex conjugate."

The first consequence of (32.1) is that the Laplace equation is invariant under a large class of transformations of the plane, well beyond the orthogonal group in two variables. For if $z \mapsto Z = Z(z)$ is a *holomorphism* of an open subset Ω of $\mathbf{R}^2$ onto another such set Ω' [$Z(z)$ is holomorphic in Ω, $Z'(z)$ does not vanish at any point of Ω, and the mapping $z \mapsto Z$ is a bijection of Ω onto Ω'], then $h(Z) \mapsto h(Z(z))$ constitutes a bijection (and in fact, an isomorphism for the natural topologies) of the space of harmonic functions in Ω' onto that of harmonic functions in Ω. This is simply because

$$(32.2) \qquad \Delta_{x,y} = |Z'|^2\,\Delta_{X,Y},$$

where $X = \operatorname{Re} Z$, $Y = \operatorname{Im} Z$.

The fact that the Laplace equation in the plane is invariant under holomorphisms, or conformal mappings as they are more often called, enables us to transfer the Dirichlet problem from an open set onto one of its biholomorphic images, preferably a simpler set. Perhaps the simplest of all bounded open subsets of $\mathbf{C}$ is the unit disk $\{z;\ |z| < 1\}$, which we shall denote by $\mathscr{D}$. If Ω is a bounded *domain* in $\mathbf{C}$ (that is, a bounded open *connected* and *simply connected* set), we know, by the Riemann mapping theorem, that there is a holomorphism $z \mapsto w = w(z)$ of Ω onto $\mathscr{D}$. Moreover, given any point z_0 of Ω, we may select the function w so as to have $w(z_0) = 0$. Let us furthermore assume that the boundary $\partial\Omega$ is a C^1 curve and that the mapping $z \mapsto w$ extends as a C^1 homeomorphism (meaning that its derivative exists, is continuous, and nowhere vanishes) of $\partial\Omega$ onto the unit circumference $\partial\mathscr{D}$. It is then easy to obtain an expression of the harmonic measure on $\partial\Omega$. We are here looking at the homogeneous Dirichlet problem:

(32.3) $$\Delta u = 0 \quad \text{in } \Omega,$$

(32.4) $$u = g \quad \text{on } \partial\Omega,$$

under reasonable regularity assumptions on the boundary datum g. Let us set $f(w) = g(z)$ with $w = w(z)$. Thus f is a function of w, $|w| = 1$. Consider then the solution v of the Dirichlet problem

(32.5) $$\Delta v = 0 \quad \text{in } \mathscr{D},$$

(32.6) $$v = f \quad \text{on } \partial\mathscr{D}.$$

The mean value formula (10.17) yields

$$v(0) = \frac{1}{2\pi}\int_0^{2\pi} f(e^{i\theta})\, d\theta = \frac{1}{2\pi i}\oint_{|w|=1} f(w)\frac{dw}{w}.$$

Let us then set

(32.7) $$M(z_0, z) = \frac{w'(z)}{w(z)},$$

recalling that the choice of w depends on that of z_0. We obtain

(32.8) $$u(z_0) = \frac{1}{2\pi i}\oint_{\partial\Omega} g(z) M(z_0, z)\, dz.$$

In particular, we may apply this formula when Ω is the unit disk itself, and $z \mapsto w$ is a holomorphism of $\mathscr{D}$ onto itself, extending into a C^1 homeomorphism of the unit circumference onto itself, and mapping z_0 onto zero. It is standard to take

(32.9) $$w = (z - z_0)/(1 - \bar{z}_0 z),$$

and an easy computation shows that if $|z| = 1$,

(32.10) $$M(z_0, z) = \frac{1 - |z_0|^2}{|1 - \bar{z} z_0|^2}\frac{1}{z}.$$

If we put this into (32.8), we obtain the *Poisson formula* (10.28) in the present particular circumstances:

$$(32.11)\quad u(z_0) = \frac{1}{2\pi}\int_0^{2\pi} g(e^{i\theta})\frac{(1 - r_0^2)\,d\theta}{1 - 2r_0\cos(\theta - \theta_0) + r_0^2},$$

$$z_0 = r_0 \exp(i\theta_0), \quad r_0 < 1.$$

We continue to assume that Ω is a bounded domain. If f is a holomorphic function in Ω, it is harmonic and therefore so are its real and imaginary parts. Conversely, let u be a real-valued harmonic function in Ω. The differential form

$$\omega = -\frac{\partial u}{\partial y}\,dx + \frac{\partial u}{\partial x}\,dy$$

is closed in Ω, and therefore (since Ω is simply connected) it is equal to the differential dv of a function in Ω. We may take

$$v(x, y) = \int_{(x_0,y_0)}^{(x,y)} \omega,$$

where (x_0, y_0) is an arbitrary (but fixed) point of Ω, and the integral is performed over any reasonably smooth (e.g., piecewise linear) path in Ω, joining (x_0, y_0) to (x, y) (since ω is closed, the choice of the path is immaterial). Since $\omega = dv$, we have the Cauchy–Riemann equations:

$$(32.12)\qquad \frac{\partial v}{\partial x} = -\frac{\partial u}{\partial y}, \qquad \frac{\partial v}{\partial y} = \frac{\partial u}{\partial x},$$

which imply that v also is harmonic in Ω. It is customary to say that u and v are *conjugate harmonic functions.* Any other function satisfying the relations (32.12) differs from v by a constant; and this of course corresponds to a change in the choice of the point (x_0, y_0) in the integral defining v.

An interesting remark is that the relations (32.12) imply

$$(32.13)\qquad (\text{grad } u)\cdot(\text{grad } v) = 0.$$

This orthogonality has an important interpretation. Suppose that Ω represents an isotropic (and uniform) medium and that u is the temperature inside that medium; suppose that the values of u are maintained constant (in time) at each point of the boundary of Ω. Then the curves $u = \text{const}$ in Ω are the *isothermic lines.* It should be noted that they are not necessarily true "curves": for instance, if the value of the temperature at the boundary is constant, say equal to one at every point of $\partial\Omega$, then $u \equiv 1$ in Ω, and there is only one isothermic line, identical with the set Ω itself. But let us assume, for the sake of the argument, that the isothermic lines form a bona fide one-parameter family of smooth curves. Then the orthogonal trajectories of these curves are the level curves of v: When u is the temperature, these curves would be the *lines of heat flow*: the vector grad u is tangent to the curves $v = \text{const}$ and grad u is proportional to the flux.

It is also clear that if we perform a conformal mapping of Ω onto another domain in $\mathbf{C}$, the whole configuration will be conformally transformed: The isothermic lines will be mapped into isothermic lines and the lines of heat flow into lines of heat flow, in the new domain.

Let f be a continuous function in the *closed* unit disk $\bar{\mathscr{D}}$, holomorphic in the interior $\mathscr{D}$, and set $f = u + iv$. Noting that, if $|z| = 1$,

$$\frac{1 - |z_0|^2}{|1 - \bar{z}z_0|^2} = \text{Re}\left(\frac{z + z_0}{z - z_0}\right),$$

we have, by (32.8) and (32.10),

$$u(z_0) = \frac{1}{2\pi i}\oint_{|z|=1} u(z)\,\text{Re}\left\{\frac{z + z_0}{z - z_0}\right\}\frac{dz}{z},$$

and therefore

$$v(z_0) = \frac{1}{2\pi i}\oint_{|z|=1} u(z)\,\text{Im}\left\{\frac{z + z_0}{z - z_0}\right\}\frac{dz}{z} + C,$$

since the integral at the right is a harmonic conjugate of $u(z_0)$ (both being regarded as functions of z_0 in Ω). We conclude that

$$f(z_0) = \frac{1}{2\pi i}\oint_{|z|=1} u(z)\left\{\frac{z + z_0}{z - z_0}\right\}\frac{dz}{z} + C.$$

But it is clear that we may apply this formula with $-if$ substituted for f and $v = \text{Re}(-if)$ for u, whence

$$f(z_0) = \frac{1}{2\pi i}\oint_{|z|=1} iv(z)\,\frac{z + z_0}{z - z_0}\,\frac{dz}{z} + C'.$$

By adding these two formulas and modifying the definition of the constant C we obtain

$$2f(z_0) = \frac{1}{2\pi i}\oint_{|z|=1} f(z)\,\frac{z + z_0}{z - z_0}\,\frac{dz}{z} + C.$$

Taking $z_0 = 0$ in the preceding expression and applying the mean value theorem shows that

$$C = f(0) = \frac{1}{2\pi i}\oint_{|z|=1} f(z)\,\frac{dz}{z},$$

hence that

$$f(z_0) = \frac{1}{2\pi i}\oint_{|z|=1} f(z)\frac{1}{2}\left\{\frac{z+z_0}{z-z_0}+1\right\}\frac{dz}{z} = \frac{1}{2\pi i}\oint_{|z|=1} f(z)\frac{dz}{z-z_0},$$

which shows that Cauchy's formula in the unit disk [cf. (5.13)] is a consequence of the Poisson formula (32.11). This is not unexpected and it should be noted that the Poisson formula is a consequence of the mean value theorem, itself an immediate consequence of Cauchy's formula.

We continue to consider the Dirichlet problem (32.3)–(32.4) in the case where $\Omega = \mathscr{D}$, the unit disk. We assume that the boundary values g are sufficiently regular, which could mean here $g \in L^2$, or $g \in L^p(1 \leqq p \leqq +\infty)$. As a matter of fact, most of what we are now going to say remains valid if we only assume that g is a *distribution* on the unit circumference $\partial\mathscr{D}$. The important thing, in what follows, is that g should have a *Fourier expansion*:

$$g(e^{i\theta}) = \sum_{m=-\infty}^{+\infty} g_m e^{im\theta}. \tag{32.14}$$

In the case where all the Fourier coefficients g_m vanish for $m < 0$ the solution of (32.3)–(32.4) (in some suitably generalized sense) is readily obtained: It is a holomorphic function of z, $|z| < 1$; its *Taylor expansion* is simply

$$\sum_{m=0}^{+\infty} g_m z^m. \tag{32.15}$$

If all the coefficients g_m for $m > 0$ were to vanish, the solution would be given by

$$\sum_{m=0}^{-\infty} g_m \bar{z}^{-m}. \tag{32.16}$$

In the general case, therefore, we may take

$$u(x, y) = u^+(z) + u^-(z) - g_0, \tag{32.17}$$

where u^+ and u^- are the functions (32.15) and (32.16), respectively. As one easily checks by means of the Cauchy formula applied to u^+ and of the anti-Cauchy formula applied to u^-, the solution u coincides (under the appropriate assumptions on g) with that given by the Poisson formula.

It should also be noted that the Poisson kernel in (32.11) can be regarded as a convolution kernel on the unit circumference (identified with the group of rotations in the plane). For $r_0 < 1$, it is C^∞ and, as a consequence, we have the right to apply it to any distribution on the unit circumference: The result is then "the solution" of the homogeneous Dirichlet problem in the unit disk, having that distribution as boundary value. One can even show that the solution does indeed "take" the boundary values: Indeed, u is a smooth function of $r_0 > 0$, valued in the space of distributions on the circumference $|z| = r_0$; as $r_0 \nearrow 1$, $u(r_0 e^{i\theta})$, regarded as a distribution in the angular variable θ, converges to $g(e^{i\theta})$. This assertion can be checked at once, for instance on the expression of u given by (32.17) (using Fourier series).

Finally we note that the Fourier series method can be used in any annulus $R_1 < |z| < R_2$ $(0 < R_1 < R_2)$ (see Exercise 32.2). Many "doubly" connected open sets can be mapped onto such an annulus by means of a holomorphism; the solution of the Dirichlet problem in the annulus can then be transferred to such an open set.

One may want to solve the Dirichlet problem in an unbounded open set, for instance in the exterior of the unit circumference. It is then necessary to impose "conditions at infinity," for instance the rate of decay of the solution one seeks, and therein lies a certain amount of ambiguity (cf. Exercises 32.4 to 32.7).

Exercises

32.1. Suppose that the function of θ, $0 \leqq \theta \leqq 2\pi$, $g(e^{i\theta})$ is square-integrable and let u be the function in the open unit disk given by the Poisson formula (32.11). Prove that, as $r_0 > 0$ converges to 1,

$$\int_0^{2\pi} |u(r_0 e^{i\theta}) - g(e^{i\theta})|^2 \, d\theta \to 0$$

[*Hint*: Use the Fourier series expression of u.]

32.2. Let g_j be a distribution on the circumference $|z| = r_j$ $(j = 1, 2;$ $0 < r_1 < r_2)$. By the Fourier series method, show that there is a harmonic function $u(r, \theta)$ in the annulus $r_1 < r < r_2$ having the following property: Given any C^∞ function $\phi(e^{i\theta})$ on the unit circumference,

$$\int_0^{2\pi} u(r, \theta)\phi(e^{i\theta}) \, d\theta \to \int_0^{2\pi} g_j(e^{i\theta})\phi(e^{i\theta}) \, d\theta$$

as $r \to r_j$ $(j = 1, 2)$.

32.3. Prove that the harmonic function u in Exercise 32.2 is unique. [*Hint*: Show that, in the case where the boundary data g_1, g_2 vanish identically, every convolution

$$\int_0^{2\pi} u(r, \theta_0 - \theta)\phi(e^{i\theta}) \, d\theta, \qquad \phi \in C_\theta^\infty([0, 2\pi]),$$

is also a solution, and that it is continuous up to the boundary.]

32.4. Let $g(e^{i\theta})$ be a continuous function on the unit circumference. What is the condition on g in order that there be a holomorphic function $f(z)$ in the region $|z| > 1$, equal to g when $|z| = 1$? Show that, if f exists, it is *not* unique. Show that there is only one such holomorphic function f if we impose the additional condition that it vanish at infinity. Derive from this that, for an arbitrary continuous function $g(e^{i\theta})$, there is a unique harmonic function in the region $|z| > 1$, equal to g when $|z| = 1$, which tends to zero at infinity.

32.5. Let $g(e^{i\theta})$ be an arbitrary continuous function on the unit circumference, $u(x, y)$ the unique harmonic function in the region $x^2 + y^2 > 1$, equal to g when $x^2 + y^2 = 1$, and tending to zero at infinity. Show that u can be expressed as a convolution on the unit circumference, similar to the Poisson formula (32.11) (but now with $r_0 > 1$). Compute the kernel acting on $g(e^{i\theta})$ in this convolution. [*Hint*: Perform an inversion on the variable z_0 in the Poisson formula.]

32.6. Let $g(t)$ denote an arbitrary continuous function, with compact support, in the real line. Let us set

$$g(z) = (2\pi i)^{-1} \int_{-\infty}^{+\infty} g(t)(t - z)^{-1}\, dt,$$

$$u(x, y) = g(z) - g(\bar{z}) \qquad (z = x + iy, \quad i = \sqrt{-1}).$$

Writing

$$u(x, y) = \int_{-\infty}^{+\infty} K(x, y, t) g(t)\, dt,$$

show that the kernel $K(x, y, t)$ (often called the *Poisson kernel* relative to the upper half-plane) can be regarded as a distribution in (x, t) depending smoothly (i.e., in a C^∞ fashion) on the parameter $y > 0$ and satisfying:

$$\lim_{y \to +0} K(x, y, t) = \delta(x - t) \qquad (\delta, \text{ the Dirac measure}). \tag{32.18}$$

Prove that u is the unique harmonic function in the upper half-plane which satisfies both

(32.19) $\lim_{y \to +0} u(x, y) = g(x)$, *uniformly with respect to x in compact subsets of the real line*;

and

$$\lim_{z \to \infty} u(x, y) = 0, \tag{32.20}$$

but that it is *not* the only one which satisfies (32.19) alone.

32.7. Let $g(t)$ be a continuous function on the real line, periodic with period 2π. Show that there is a unique harmonic function u in the open upper half-plane, satisfying

$$\lim_{y \to +0} u(x, y) = g(x), \tag{32.21}$$

$$\lim_{y \to +\infty} u(x, y) = (2\pi)^{-1} \int_0^{2\pi} g(t)\, dt, \tag{32.22}$$

where the limits are uniform with respect to x in $\mathbf{R}^1$. Can u be expressed in terms of g by means of the Poisson kernel for the upper half-plane (see Exercise 32.6)?

32.8. Consider the complex differential equation

$$\frac{dw}{dz} = (1 - z^2)^{-1/2}(1 - k^2 z^2)^{-1/2} \qquad (0 < k < 1), \tag{32.23}$$

where the radicals are assumed to be equal to $+1$ when $z = 0$. Show that there is a unique holomorphic solution $w(z)$ of (32.23) for Re $z > 0$, equal to zero when $z = 0$ (it is clear that w will be holomorphic in a neighborhood of the origin). Show that $z \mapsto w(z)$ is a biholomorphic mapping of the open upper half-plane onto the *rectangle*

$$0 < \operatorname{Re} z < K, \qquad 0 < \operatorname{Im} z < K', \tag{32.24}$$

where K and K' are two real constants, which the student should prove to be equal to $w(1)$ and to $(1/i)w(1/k) - w(1)$.

By using the inverse mapping $w \mapsto z = \operatorname{sn}(w)$ (sn is Jacobi's elliptic function) and the fact that

$$z \mapsto u = e^{i\theta} \frac{z - \alpha}{z - \bar{\alpha}} \qquad (0 \leqq \theta \leqq 2\pi, \quad \alpha \in \mathbf{C}, \quad \operatorname{Im} \alpha > 0) \tag{32.25}$$

is a conformal mapping of the half-plane Im $z > 0$ onto the open unit disk, give the expression of the harmonic measure relative to the rectangle (32.24).

32.9. Solve the weak Dirichlet problem in the annulus $0 < |z| < 1$ by means of Fourier series. Explain what you find and compare with what one seeks in the classical Dirichlet problem.

32.10. Let u and v be conjugate harmonics in the unit disk $|z| < 1$. Writing $z = re^{i\theta}$, consider the Fourier expansions of u and of v with respect to θ and state the relation between the Fourier coefficients $u_m(r)$ and $v_m(r)$ of u and v, respectively ($m \in \mathbf{Z}$).

33

Approximation of Harmonic Functions by Harmonic Polynomials in Three Space. Spherical Harmonics

Let us look provisionally at the Laplace equation in two independent variables, x and y, and try to determine all *polynomials* $P(x, y)$ which are solutions of it. Let us write $P(x, y) = P_m(x, y) + P_{m-1}(x, y) + \cdots$, where $P_d(x, y)$ is homogeneous of degree d. Since the $P_j(x, y)$ are linearly independent, each one of them must be a solution of the Laplace equation and we may therefore assume that $P = P_m$ is homogeneous of degree m. By going to polar coordinates r, θ, we may write

$$(33.1) \qquad P_m(x, y) = r^m \sum_{\nu=-m}^{+m} c_\nu e^{i\nu\theta}.$$

But clearly the exponentials $\exp(i\nu\theta)$ are also linearly independent, and therefore each term $r^m e^{i\nu\theta}$ must be harmonic. By using the expression of the Laplace operator in polar coordinates, we see that we must have

$$(33.2) \qquad \left\{ r\frac{\partial}{\partial r}\left(r\frac{\partial}{\partial r}\right) + \frac{\partial^2}{\partial \theta^2}\right\}(r^m e^{i\nu\theta}) = 0,$$

which implies at once $m^2 = \nu^2$, i.e., $m = \pm\nu$. In other words,

$$(33.3) \qquad P_m(x, y) = Az^m + B\bar{z}^m.$$

Let now $f(x, y)$ be an arbitrary harmonic function in the disk $r < R$, say continuous up to the boundary. By using the Fourier series of its boundary value, we have seen that it can be represented by a Fourier series whose coefficients are multiples of powers of r, specifically:

$$(33.4) \qquad f(x, y) = \sum_{\nu=-\infty}^{+\infty} C_\nu r^\nu e^{i\nu\theta} = C_0 + f^+(z) + f^-(\bar{z}),$$

where f^+ and f^- are both holomorphic in the disk $r < R$ and vanish at the origin. Thus we see that the harmonic polynomials are dense in the space of harmonic functions in the disk (this remains true if the disk is replaced by any *domain* in $\mathbf{R}^2$). This approximation property has many applications. Foremost among them is the possibility of solving the Dirichlet problem in disks and annuli, as shown in Sect. 32, by exploiting the series expansions (33.4). It is therefore natural to ask whether a similar approximation property is valid in higher dimensions and whether it leads to reasonably simple series representations which would enable us to solve "explicitly" the Dirichlet problem in balls (or in "shells"). It should be pointed out, however, that we cannot expect the same rewards in dimension $\geqq 3$ as those reaped in the plane: Precisely, we shall not have at our disposal the analog of the Riemann mapping theorem, which tells us that we may transfer the Dirichlet problem from any bounded domain in $\mathbf{R}^2$ to the unit disk, and therefore we shall not be able to transfer the Dirichlet problem to the ball (from even simple open subsets of $\mathbf{R}^n$, $n \geqq 3$).

If Ω is an open subset of $\mathbf{R}^n$ whose complement has no bounded connected components, it is true that every harmonic function in Ω is the limit (say for the uniform convergence on compact subsets of Ω) of harmonic polynomials. We shall not prove this theorem in the present book. We shall only look at the case of dimension 3 and show that relatively simple expansions, akin to (33.4), indeed exist. Furthermore, they exhibit spherical symmetries which make them very handy if one desires to solve the Dirichlet problem in balls.

As in the case of two variables, we shall look at a homogeneous polynomial of degree m, here $P_m(x, y, z)$, and switch to spherical coordinates r, ϕ, θ (ϕ is the *longitude*, θ the *colatitude*):

$$x = r\cos\phi\sin\theta, \qquad y = r\sin\phi\sin\theta, \qquad z = r\cos\theta. \tag{33.5}$$

An easy computation (cf. Exercise 33.1) shows that, in these coordinates,

$$r^2\Delta = r^2\frac{\partial^2}{\partial r^2} + 2r\frac{\partial}{\partial r} + (\sin\theta)^{-2}\Lambda_{\phi,\theta}, \tag{33.6}$$

where

$$\Lambda_{\phi,\theta} = \frac{\partial^2}{\partial\phi^2} + (\sin\theta)^2\frac{\partial^2}{\partial\theta^2} + (\cos\theta)\sin\theta\frac{\partial}{\partial\theta}. \tag{33.7}$$

Here takes place an automatic separation of variables, since $P_m(x, y, z) = r^m P_m(\cos\phi\sin\theta, \sin\phi\sin\theta, \cos\theta)$; hence we must have

$$\left(r^2\frac{d^2}{dr^2} + 2r\frac{d}{dr}\right)r^m = kr^m, \tag{33.8}$$

$$(\Lambda_{\phi,\theta} + k\sin^2\theta)h(\phi, \theta) = 0, \tag{33.9}$$

representing $P_m(\cos\phi \sin\theta, \sin\phi \sin\theta, \cos\theta)$ as a linear combination of linearly independent functions $h(\phi, \theta)$. The first equation immediately gives

$$k = m(m+1). \tag{33.10}$$

On the other hand, because of the linear independence of the exponentials $\exp[i(\alpha\phi + \beta\theta)]$ we may also use separation of variables in (33.9). This gives, if we write $h(\phi, \theta) = v(\phi)u(\theta)$,

$$v'' + k_1 v = 0, \tag{33.11}$$

$$(\sin\theta)^2 \frac{d^2u}{d\theta^2} + (\cos\theta)(\sin\theta)\frac{du}{d\theta} + k(\sin\theta)^2 u - k_1 u = 0, \tag{33.12}$$

and (33.11) shows at once that $k_1 = n^2$, where n is an integer (positive, negative, or zero). Thus

$$v = Ae^{in\phi} + Be^{-in\phi}. \tag{33.13}$$

We must now concentrate on (33.12). It is convenient to make the change of variables $t = \cos\theta$. It transforms (33.12) into

$$\frac{d}{dt}\left\{(1-t^2)\frac{du}{dt}\right\} + \left(m(m+1) - \frac{n^2}{1-t^2}\right)u = 0. \tag{33.14}$$

When $n = 0$, (33.14) reduces to the *Legendre equation*,

$$\frac{d}{dt}\left\{(1-t^2)\frac{du}{dt}\right\} + m(m+1)u = 0. \tag{33.15}$$

The solutions of (33.15) are called the *Legendre functions.* Here we shall only consider the case of m (and also n) taking nonnegative integral values, but the theory of these equations has been developed for all complex values of m and n. As we shall see later, the interesting solutions of (33.14) can be expressed in terms of the (interesting) solutions of (33.15). We shall therefore look first at the latter. Of course there is no need to restrict the variation of t to the interval $[-1, +1]$ [although we were led to (33.14) by setting $t = \cos\theta$ in (33.12)].

In order to solve (33.15) it is natural to try a power series expansion

$$u(t) = \sum_{j=0}^{+\infty} u_j t^j.$$

This leads to the following recursion formula for the coefficients u_j:

$$(j+2)(j+1)u_{j+2} = -(m-j)(m+j+1)u_j. \tag{33.16}$$

Formula (33.16) shows at once the existence of a polynomial solution to (33.15): If m is even, we take all u_j with j odd equal to zero, and all u_j with j even $> m$ also equal to zero; (33.16) enables us then to determine the remaining

u_j with j even in terms of the first one of them. The same argument holds when m is odd—except that one must exchange even for odd throughout. When $m = 0$, this polynomial solution is a constant. For $m = 1$, it is of the form Ct with C an arbitrary constant. For any $m = 0, 1, \ldots$, the polynomial solutions are all multiples of one of them, i.e., span a one-dimensional linear space. It is not difficult to find a generator of this linear space. Let us set $u = D^m v$ ($D = d/dt$) and integrate (33.15) from 0 to t. We get

$$(33.17) \quad (1 - t^2)D^{m+1}v + m(m+1)D^{m-1}v = D^m[(1 - t^2)Dv + 2mtv] = 0.$$

We obtain at once a solution v of (33.17) by taking $v = (1 - t^2)^m$. Thus the polynomial solutions of (33.15) are all of the form

$$(33.18) \qquad u(t) = C\left(\frac{d}{dt}\right)^m (1 - t^2)^m.$$

Integration by parts shows at once that

$$(33.19) \qquad \int_{-1}^{+1} t^h \left(\frac{d}{dt}\right)^m (1 - t^2)^m \, dt = 0 \qquad \text{if} \quad h < m,$$

which implies that any two functions (33.18), corresponding to two different values of m, are necessarily orthogonal in $L^2(]-1, +1[)$. On the other hand,

$$\begin{aligned}\int_{-1}^{+1} \left|\left(\frac{d}{dt}\right)^m (1 - t^2)^m\right|^2 dt &= \int_{-1}^{+1} (1 - t^2)^m \left(\frac{d}{dt}\right)^{2m} (t^2 - 1)^m \, dt \\ &= (2m)! \int_{-1}^{+1} (1 - t^2)^m \, dt \\ &= m! \int_{-1}^{+1} \left\{\left(\frac{d}{dt}\right)^m (1 + t)^{2m}\right\} (1 - t)^m \, dt \\ &= (m!)^2 \int_{-1}^{+1} (1 + t)^{2m} \, dt = 2^{2m+1}(m!)^2/(2m + 1).\end{aligned}$$

Definition 33.1. The polynomials

$$(33.20) \qquad P_m(t) = \frac{1}{2^m m!}\left(\frac{d}{dt}\right)^m [(t^2 - 1)^m], \qquad m = 0, 1, 2, \ldots,$$

are called Legendre's polynomials.

We have shown that

$$(33.21) \qquad (m + \tfrac{1}{2})^{1/2}(m' + \tfrac{1}{2})^{1/2} \int_{-1}^{1} P_m(t)P_{m'}(t) \, dt = \delta_{m, m'}$$

(Kronecker's index).

THEOREM 33.1. *The sequence of functions*

$$(m + \tfrac{1}{2})^{1/2} P_m(t), \qquad m = 0, 1, \ldots, \tag{33.22}$$

forms a Hilbert basis (i.e., a complete orthonormal system) in $L^2(]-1, +1[)$.

Proof. In view of (33.21) it suffices to prove that the Legendre polynomials span a dense linear subspace of $L^2(]-1, +1[)$. As a matter of fact, $P_0(t), \ldots, P_m(t)$ constitutes a basis of the linear space of polynomials in one variable of degree $\leq m$. The proof of the latter is by induction on m, using the observation that

$$P_m(t) - 2^{-m} \frac{(2m)!}{(m!)^2} t^m \quad \text{is of degree} \quad m - 1. \qquad \text{Q.E.D.}$$

We know that the solutions of (33.15), in the open interval $|t| < 1$, form a two-dimensional linear space. For $m = 0$ we have

$$\frac{du}{dt} = C(1 - t^2)^{-1} = \frac{C}{2}\left\{\frac{1}{1 - t} + \frac{1}{1 + t}\right\}. \tag{33.23}$$

The constant solutions of (33.23) are multiples of $P_0(t) \equiv 1$. The "other" solutions are multiples of

$$Q_0(t) = \frac{1}{2} \log \frac{1 + t}{1 - t}. \tag{33.24}$$

It is easy to show that the general solution of (33.15), for any $m = 0, 1, \ldots$, is

$$A\{\alpha_0 P_0(t) + \cdots + \alpha_{m-1} P_{m-1}(t)\} + AP_m(t) \frac{1}{2} \log \frac{1 + t}{1 - t} + BP_m(t), \tag{33.25}$$

where the α_j's ($j = 0, \ldots, m - 1$) are well-determined numbers and A, B are arbitrary constants. The presence of the logarithmic term excludes that solutions (33.25) with $A \neq 0$ enter in the expression of the *spherical harmonics* (which we are seeking) as we know that the latter are polynomials with respect to $t = \cos\theta$ and $\sin\theta = (1 - t^2)^{1/2}$. We shall therefore restrict ourselves to multiples of the Legendre polynomials. With this in mind we look now at Eq. (33.14) for $n \neq 0$. A change of unknown function

$$u(t) = (1 - t^2)^{n/2} v(t)$$

transforms it into

$$(1 - t^2)v'' - 2(n + 1)tv' + [m(m + 1) - n(n + 1)]v = 0. \tag{33.26}$$

On the other hand, if we differentiate n times the Legendre equation (33.15) for h, we obtain precisely (33.26) for $v = h^{(n)}$. In view of the preceding considerations about the Legendre equation, we see that we obtain a solution of (33.14) for all $n = 0, 1, 2, \ldots$, by taking u equal to

$$(-1)^n(1-t^2)^{n/2}\left(\frac{d}{dt}\right)^n P_m(t), \tag{33.27}$$

or any scalar multiple of this function. The functions (33.27) are often called the *associated Legendre functions* (of the first kind) and denoted by $P_m^n(t)$. When n is odd, the square root in (33.27) must be understood as the *positive square root* (for $|t| < 1$). According to (33.20) we have

$$P_m^n(t) = \frac{(-1)^{m+n}}{2^m m!}(1-t^2)^{n/2}\left(\frac{d}{dt}\right)^{m+n}(1-t^2)^m \qquad (m, n = 0, 1, \ldots). \tag{33.28}$$

Observe that $P_m^n \equiv 0$ if $n > m$. Q.E.D.

The proof of the next statement is very similar to that of Theorem 33.1 and we leave it to the reader:

THEOREM 33.2. *For each fixed* $n = 0, 1, \ldots$, *the sequence of functions*

$$(m+\tfrac{1}{2})^{1/2}\left[\frac{(m-n)!}{(m+n)!}\right]^{1/2} P_m^n(t), \qquad m = n, n+1, \ldots, \tag{33.29}$$

forms a Hilbert basis in $L^2(]-1, +1[)$.

We return now to our original problem, that of determining the harmonic polynomials in three variables. According to (33.28) we have

$$P_m^n(\cos\theta) = \sum_{2k \leq m-n} \alpha_{m,k}^n (\sin\theta)^n (\cos\theta)^{m-n-2k}. \tag{33.30}$$

We derive from this:

$$r^m e^{in\phi} P_m^n(\cos\theta) = \sum_{2k \leq m-n} \alpha_{m,k}^n (x+iy)^n z^{m-n-2k}(x^2+y^2+z^2)^k, \tag{33.31}$$

which means that the left-hand side is a homogeneous polynomial of degree m in the variables x, y, z. We have shown that it is harmonic. Since n can only take the values $0, 1, \ldots, m$, and since we can obviously substitute $-\phi$ for ϕ (which is equivalent to replacing $x + iy$ by $x - iy$) in (33.31), we have thus obtained $2m + 1$ linearly independent harmonic polynomials which are homogeneous of degree m (in x, y, z).

Let $\mathscr{P}^d_{\text{hom}, N}$ denote the space of homogeneous polynomials of degree $d \geq 0$ in N variables. It has dimension $(N + d - 1)!/(N - 1)!\, d!$. We identify it with its own dual by means of the duality bracket

$$\langle f, g\rangle = \sum_{|\alpha|=d} \frac{1}{\alpha!} f^{(\alpha)} g^{(\alpha)}, \tag{33.32}$$

where $\alpha = (\alpha_1, \ldots, \alpha_N)$ is an N-tuple of nonnegative integers. Let us denote by $x^j, j = 1, \ldots, N$, the variables and set $D_j = \partial/\partial x^j$. Then D_j is a linear map of $\mathscr{P}^d_{\text{hom}, N}$ into $\mathscr{P}^{d-1}_{\text{hom}, N}$; its transpose ${}^tD_j$ maps $\mathscr{P}^{d-1}_{\text{hom}, N}$ into $\mathscr{P}^d_{\text{hom}, N}$. We have

$$\langle f, D_j g\rangle = \sum_{|\alpha|=d-1} \frac{1}{\alpha!} f^{(\alpha)}(D_j g)^{(\alpha)} = \sum_{|\beta|=d} \frac{1}{\beta!}(x_j f)^{(\beta)} g^{(\beta)},$$

which shows that the transpose of $\partial/\partial x^j$ is multiplication by x^j. Now let $P(D) = P(D_1, \ldots, D_N)$ be a homogeneous polynomial in the D_j, of degree m. We may state:

LEMMA 33.1. *The transpose of the linear mapping*

$$P(D) : \mathscr{P}^d_{\text{hom}, N} \to \mathscr{P}^{d-m}_{\text{hom}, N} \tag{33.33}$$

is the multiplication of elements of $\mathscr{P}^{d-m}_{\text{hom}, N}$ *by* $P(x)$.

COROLLARY 33.1. *The mapping (33.33) is surjective and the dimension of its kernel is equal to*

$$\frac{(N + d - 1)!}{(N - 1)!\, d!} - \frac{(N + d - m - 1)!}{(N - 1)!\,(d - m)!}. \tag{33.34}$$

If we apply this to the Laplace operator (thus $m = 2$) in $N = 3$ variables, we see that the following is true:

THEOREM 33.3. *The linear space of homogeneous harmonic polynomials of degree* $m \geqq 0$ *in three variables* x, y, z, *has dimension exactly equal to* $2m + 1$. *The polynomials (33.31), together with their complex conjugates, form a linear basis of this space.*

It is also clear how we can exploit Theorem 33.3 and the properties established earlier, for the Legendre functions, in order to solve the Dirichlet problem in the open ball

$$B_R = \{(x, y, z) \in \mathbf{R}^3;\ x^2 + y^2 + z^2 < R^2\} \qquad (R > 0).$$

As a matter of fact, because of radial homogeneity, we may as well reason in the case $R = 1$; B_R is then the unit ball, B^3, and its surface is the unit sphere, S^2. Suppose we are given any square-integrable function (for the Lesbegue measure on the unit sphere), $g(\phi, \theta)$, on S^2. We may of course expand it in a Fourier series with respect to ϕ:

$$g(\phi, \theta) = g_0(\theta) + \sum_{n=1}^{+\infty} \{g_n^+(\theta)e^{in\phi} + g_n^-(\theta)e^{-in\phi}\}. \tag{33.35}$$

By applying now Theorem 33.2 we may expand each one of the coefficients g_0, $g_n^\pm$ in terms of the Legendre functions with respect to $\cos\theta$ [we view the

Fourier coefficients in (33.35) as functions of $\cos\theta$: we recall that θ has to vary only from 0 to π, since ϕ varies from 0 to 2π]. Finally we obtain an expansion in $L^2(S^2)$:

$$g(\phi, \theta) = \sum_{m=0}^{+\infty} g_{m,0} P_m(\cos\theta) + \sum_{n=1}^{+\infty} \sum_{m=n}^{+\infty} \{g_{m,n}^{+} P_m^n(\cos\theta) e^{in\phi} + g_{m,n}^{-} P_m^n(\cos\theta) e^{-in\phi}\}. \tag{33.36}$$

It is now very easy to construct the harmonic function u in the open ball B_1 which takes (in some suitable sense) the boundary value g on S^2. Let us denote by $S_{m,n}(x, y, z)$ the spherical harmonic (33.31). The solution u is then given by

$$u = \sum_{m=0}^{+\infty} g_{m,0} S_{m,0} + \sum_{n=1}^{+\infty} \sum_{m=n}^{+\infty} (g_{m,n}^{+} S_{m,n} + g_{m,n}^{-} \bar{S}_{m,n}), \tag{33.37}$$

where $\bar{S}$ denotes the complex conjugate of S.

Exercises

33.1. Let $r, \theta_1, \ldots, \theta_{n-1}$ denote spherical coordinates in $\mathbf{R}^n$. Write the expression of the Laplace operator Δ in these coordinates.

33.2. Show that, given any integer $n = 0, 1, \ldots$, if a polynomial in one variable x, of degree $\leqq n$, is orthogonal in $L^2(]-1, +1[)$ to every polynomial of degree $\leqq n-1$, it is necessarily a multiple of a Legendre polynomial. (Do not use computation, only linear algebra.)

If $P_n(x)$ denotes the nth Legendre polynomial, show that

$$(1 - 2xt + t^2)^{-1/2} = \sum_{n=0}^{+\infty} P_n(x) t^n. \tag{33.38}$$

Prove the following formulas:

$$2^n P_n(x) = \frac{1}{2\pi i} \oint_\gamma \frac{(w^2 - 1)^n}{(w - x)^{n+1}}\, dw \qquad \text{(Schläfli's formula)}, \tag{33.39}$$

where γ is a circle centered at x;

$$P_n(x) = \frac{1}{\pi} \int_0^\pi \{x + (x^2 - 1)^{1/2} \cos\theta\}^n \, d\theta \qquad \text{(Laplace's formula)}; \tag{33.40}$$

$$(n + 1)P_{n+1} - (2n + 1)xP_n + nP_{n-1} = 0, \tag{33.41}$$

$$nP_n = xP_n' - P_{n-1}', \qquad nP_{n-1} = P_n' - xP_{n-1}'. \tag{33.42}$$

33.3. Let $P_n(x)$ denote the nth Legendre polynomial. Prove that the roots of P_n (which, of course, number n) are all real, pairwise distinct, and that they all belong to the interval $-1 < x < +1$.

34

Spectral Properties and Eigenfunction Expansions

We return now to a second-order differential operator $A = A(x, \partial/\partial x)$ whose coefficients (when it is written in the variational form below) are L^∞ functions in the open set $\Omega \subset \mathbf{R}^n$:

$$A = -\sum_{j,k=1}^{n} \frac{\partial}{\partial x^j} a^{jk}(x) \frac{\partial}{\partial x^k} + \sum_{j=1}^{n} b^j(x) \frac{\partial}{\partial x^j} + c(x). \tag{34.1}$$

As we have said, we assume that a^{jk}, b^j $(1 \leqq j \leqq n)$, c belong to $L^\infty(\Omega)$. In addition to this, we are going to assume that A is *formally self-adjoint*:

$$\langle Au, \bar{v} \rangle = \langle u, \overline{Av} \rangle, \qquad \forall u, v \in H_0^1(\Omega), \tag{34.2}$$

where $\langle\ ,\ \rangle$ is the bracket of the duality between $H_0^1(\Omega)$ and $H^{-1}(\Omega)$. It is not difficult to state necessary and sufficient conditions in order that (34.2) holds. The principal part of A,

$$-\sum_{j,k=1}^{n} \frac{\partial}{\partial x^j} a^{jk}(x) \frac{\partial}{\partial x^k}$$

must be (formally) self-adjoint, and this is so if and only if

$$a^{jk}(x) = \overline{a^{kj}(x)} \qquad \text{for almost every} \quad x \in \Omega, \quad 1 \leqq j, k \leqq n. \tag{34.3}$$

The part of order $\leqq 1$ in A must also be self-adjoint, which implies

$$b^j(x) = -\overline{b^j(x)}, \quad \text{a.e. in } \Omega, \qquad 1 \leqq j \leqq n, \tag{34.4}$$

$$c(x) - \overline{c(x)} = \sum_{j=1}^{n} \frac{\partial b^j}{\partial x^j}(x) \quad \text{a.e. in } \Omega. \tag{34.5}$$

In particular, the divergence of the vector $(b^1, \ldots, b^n)$ must belong to $L^\infty(\Omega)$.

Condition (34.2) is equivalent to the property that the sesquilinear form on $H_0^1(\Omega) \times H_0^1(\Omega)$ associated with A,

$$a(u, v) = \sum_{j,k=1}^{n} \int_\Omega a^{jk}(x) \frac{\partial u}{\partial x^k} \frac{\partial \bar{v}}{\partial x^j} \, dx + \sum_{j=1}^{n} \int_\Omega b^j(x) \frac{\partial u}{\partial x^j} \bar{v} \, dx + \int_\Omega c(x) u \bar{v} \, dx,$$

is *Hermitian*, i.e., $a(v, u) = \overline{a(u, v)}$.

We shall also make the *strong ellipticity* assumption:

(34.6) *for suitable* $c_0 > 0$, *all* $\zeta \in \mathbf{C}^n$, *and almost all* $x \in \Omega$,

$$\sum_{j,k=1}^{n} a^{jk}(x) \zeta_j \bar{\zeta}_k \geqq c_0 |\zeta|^2.$$

As usual, this leads to the coercivity of the form

$$\lambda \int_\Omega |u|^2 \, dx + a(u, u),$$

provided that the real number λ is sufficiently large:

(34.7) *There is* $\lambda_0 \in \mathbf{R}$ *such that, for all* $\lambda > \lambda_0$ *and all* $u \in H_0^1(\Omega)$,

$$\langle (\lambda I + A)u, \bar{u} \rangle \geqq c_1 \|u\|^2_{H^1(\Omega)} . \tag{34.8}$$

The constant c_1 in (34.8) is strictly greater than zero and independent of u (but depends on λ).

The coercivity inequality (34.8) implies that, for $\lambda > \lambda_0$, $\lambda I + A$ defines an isomorphism of $H_0^1(\Omega)$ onto $H^{-1}(\Omega)$. Let us denote by $G(\lambda)$ the inverse isomorphism. Throughout this section we make the hypothesis that

(34.9) Ω *is bounded.*

Then if J denotes the natural injection of $H_0^1(\Omega)$ into $H^{-1}(\Omega)$, we know, for instance by Proposition 25.5 (cf. also Exercise 25.10), that J is *compact*. It follows from this:

PROPOSITION 34.1. *If* $\lambda > \lambda_0$ *and if* Ω *is bounded,* $G(\lambda)J$ *is a compact operator of* $H_0^1(\Omega)$ *into itself.*

Thus, under the hypothesis (34.9), we may apply the classical theorem, due to Riesz, on the spectral decomposition of a *positive compact* operator, in a Hilbert space. Its proof can be found in any text on the subject (see, e.g., [TVS, D&K, Theorem 48.1]). An operator of this kind has a *discrete* spectrum, which means that every point in the spectrum is isolated and is an eigenvalue of the operator. Furthermore, the eigenspace corresponding to any given eigenvalue χ, which we may denote by V_χ, is finite-dimensional. If χ, χ' are two distinct eigenvalues, the corresponding eigenspaces $V_\chi, V_{\chi'}$ are orthogonal. The eigenvalues, which are (of course) strictly positive numbers, form a decreasing sequence converging to zero. The orthogonal sum of the eigenspaces V_χ, as χ ranges over the spectrum, is dense in the orthogonal of the kernel of the operator, hence in the whole Hilbert space if the operator happens to be injective, as is the case for $G(\lambda)J$ when $\lambda > \lambda_0$. Thus, in the injective case, we may find a complete orthonormal system, in our Hilbert space, consisting of eigenfunctions.

Let h be an eigenfunction of $G(\lambda)J$ in $H_0^1(\Omega)$, corresponding to the eigenvalue χ; then χ is strictly greater than zero. We have

$$Ah = (\chi^{-1} - \lambda)h, \tag{34.10}$$

which shows the relation between the spectrum of A and the one of $G(\lambda)$. Since λ is somewhat arbitrary, we shall in the sequel focus on the spectrum of A, $S(A, \Omega)$: It consists of a sequence of real numbers converging to $+\infty$, which we may order as an increasing sequence

$$\lambda_1 \leqq \lambda_2 \leqq \cdots \leqq \lambda_j \leqq \cdots, \tag{34.11}$$

with repetitions according to the multiplicity of each eigenvalue.

We may view the *resolvent* of A,

$$(zI - A)^{-1} = (z - \lambda)^{-1}G(\lambda)(G(\lambda) - (z - \lambda)^{-1}I)^{-1}, \tag{34.12}$$

as a holomorphic function of z in $\mathbf{C}\backslash S(A, \Omega)$ valued in the space of bounded linear operators $H^{-1}(\Omega) \to H_0^1(\Omega)$. Note that it enables us to solve the Dirichlet problem:

$$(A - zI)u = f \quad \text{in } \Omega, \tag{34.13}$$

$$u = g \quad \text{on } \partial\Omega, \tag{34.14}$$

whenever $z \notin S(A, \Omega)$, for sufficiently regular f and g: f must belong to $H^{-1}(\Omega)$ and g must be the trace on $\partial\Omega$ of an element w of $H^1(\Omega)$ (assuming that the boundary $\partial\Omega$ is regular enough, so that the trace of w can be defined). Then the solution u of (34.13)–(34.14) is given by

$$u = w - (zI - A)^{-1}[f - (A - zI)w]. \tag{34.15}$$

Suppose now that z is an eigenvalue of A. The range $\mathscr{R}$ of $A - zI$ is canonically isomorphic to the orthogonal $\mathscr{M}$ of its kernel $\mathscr{K}$ in $H_0^1(\Omega)$, since A is self-adjoint (and therefore z is real). We may regard $A - zI$ as an isomorphism of $\mathscr{M}$ onto $\mathscr{R}$, whose inverse we may still denote by $(A - zI)^{-1}$. With such notation, and provided that $f - (A - zI)w \in \mathscr{R}$, (34.15) still gives a solution of (34.13)–(34.14), but now we may obtain a whole affine space of solutions, whose dimension is $\geqq 1$, by adding to the right-hand side in (34.15) an arbitrary element of $\mathscr{K}$. If one wishes to select in a unique manner one solution among these, one must adjoin a sufficient number of conditions —a familiar situation, which goes with the *Fredholm alternative.*

We are now going to look at *Hilbert space bases,* i.e., complete orthonormal systems, consisting of eigenfunctions of A. It is convenient to redefine the inner product in $H^1(\Omega)$, in a manner better suited to the study of A. We shall set

$$((u, v)) = a(u, v) + \kappa \int_\Omega u\bar{v}\, dx, \tag{34.16}$$

with $\kappa > \lambda_0$ [see (34.7)]. We may then select a sequence of eigenfunctions of A, $\{U_j\}$ $(j = 1, 2, \ldots)$ belonging to $H_0^1(\Omega)$, such that [cf. (34.11)]:

$$AU_j = \lambda_j U_j, \qquad j = 1, 2, \ldots, \tag{34.17}$$

$$((U_i, U_j)) = \delta_{i,j} \qquad \text{(Kronecker's index)}, \qquad i, j = 1, 2, \ldots. \tag{34.18}$$

By the Riesz theorem we know that the U_j form a Hilbert space basis of $H_0^1(\Omega)$. Noting that $\kappa + \lambda_j > 0$, let us now set

$$E_j = (\kappa + \lambda_j)^{1/2} U_j \qquad (j = 1, 2, \ldots). \tag{34.19}$$

PROPOSITION 34.2. *Suppose Ω is bounded. Then the functions E_j [resp. $(\kappa + \lambda_j)^{1/2} E_j$] form a Hilbert space basis in $L^2(\Omega)$ [resp. in $H^{-1}(\Omega)$, equipped with the Hilbert space structure dual of the one defined by $((\ ,\ ))$ on $H_0^1(\Omega)$]. A distribution u in Ω belongs to $H_0^1(\Omega)$ [resp. to $L^2(\Omega)$, resp. to $H^{-1}(\Omega)$] if and only if*

$$u = \sum_{j=1}^{+\infty} u_j E_j, \tag{34.20}$$

where the series converges in $\mathscr{D}'(\Omega)$ and where

$$\sum_{j=1}^{+\infty} (\kappa + \lambda_j) |u_j|^2 < +\infty \tag{34.21}$$

[*resp., where*

$$\sum_{j=1}^{+\infty} |u_j|^2 < +\infty, \tag{34.22}$$

resp., where

$$\sum_{j=1}^{+\infty} (\kappa + \lambda_j)^{-1} |u_j|^2 < +\infty \tag{34.23}$$

].

Proof. We have, by virtue of (34.16), (34.18), and (34.19),

$$\int_\Omega E_i \bar{E}_j \, dx = \langle (A + \kappa)G(\kappa)JE_i, \bar{E}_j \rangle = ((G(\kappa)JE_i, E_j))$$

$$= (\kappa + \lambda_i)^{-1}((E_i, E_j)) = \left(\frac{\kappa + \lambda_j}{\kappa + \lambda_i}\right)^{1/2} ((U_i, U_j)) = \delta_{i,j}.$$

On the other hand,

$$(E_i, E_j)_{H^{-1}(\Omega)} = \langle E_i, \overline{G(\kappa)JE_j} \rangle = (\kappa + \lambda_j)^{-1} \int_\Omega E_i \bar{E}_j \, dx,$$

from which the first part of the statement follows. As for the last part, it follows from

LEMMA 34.1. *Let $\mathscr{H}$ be a Hilbert space of distributions in Ω, $\{\varepsilon_j\}$ $(j = 1, 2, \ldots)$ a Hilbert space basis in $\mathscr{H}$. For a distribution u in Ω to belong to $\mathscr{H}$ it is necessary and sufficient that there be a sequence (c_j) in l^2 such that the series $\sum_j c_j \varepsilon_j$ converges to u in $\mathscr{D}'(\Omega)$. Then the sequence (c_j) is unique.*

We recall that the space l^2 is the Hilbert space of square-summable sequences of complex numbers.

Proof. The mapping $(c_j)_{j=1,2,\ldots} \mapsto \sum_j c_j \varepsilon_j$, where the convergence is to be understood in the sense of $\mathscr{H}$, is an isometry of l^2 onto $\mathscr{H}$. Let $\tilde{\mathscr{H}}$ denote the space of distributions u which can be represented as series $\sum_j c_j \varepsilon_j$ converging in $\mathscr{D}'(\Omega)$ [with $(c_j) \in l^2$]. There is a natural mapping of $\tilde{\mathscr{H}}$ onto $\mathscr{H}$: To each series as above we assign the same series, but now converging in $\mathscr{H}$. Since the topology of $\mathscr{H}$ is finer than the one induced by $\mathscr{D}'(\Omega)$, this mapping must be the identity. Q.E.D.

Remark 34.1. The analogy with the global situation, where $\Omega = \mathbf{R}^n$ and $A = -\Delta$, is worth stressing. Instead of the eigenfunction expansion (34.20) we have, in the global situation, the Fourier inversion formula: $\exp(-ix \cdot \xi)$ is obviously an eigenfunction of $-\Delta$ (corresponding to the eigenvalue $|\xi|^2$ which varies over $\mathbf{R}_+$; the difference, of course, is that the spectrum is not discrete; note also that the eigenfunctions do not belong to H^1). The "weight" $(\kappa + \lambda_j)$ should be replaced by $(\kappa + |\xi|^2)$, where κ is any number greater than zero (usually one takes $\kappa = 1$). The analog of Proposition 34.2 would then state that $u \in H^j(\mathbf{R}^n)$ $(j = 1, 0, -1)$ if and only if

$$|\hat{u}(\xi)|^2(1 + |\xi|^2)^j \in L^1,$$

which is the definition of the Sobolev spaces (via the Plancherel formula).

Remark 34.2. The eigenfunction expansion (34.20) enables us to "insert" $H_0^1(\Omega)$, $H^0(\Omega)$, and $H^{-1}(\Omega)$ into a one-parameter family of distribution spaces, which we could denote by $\tilde{H}_0^s(\Omega; A)$ ($s \in \mathbf{R}$): a distribution u in Ω belongs to the latter space if and only if it has a series expansion (34.20) with coefficients u_j satisfying

$$\sum_{j=1}^{+\infty} (\kappa + \lambda_j)^s |u_j|^2 < +\infty. \tag{34.24}$$

In particular, for $|s| \leqq 1$, we obtain a family of *interpolation spaces* between $H_0^1(\Omega)$ and $H^{-1}(\Omega)$.

Example 34.1

We shall determine the eigenfunctions E_j ($j = 1, 2, \ldots$) in the case of one space variable, denoted by x, when $A = -(d/dx)^2$ and when Ω is a finite interval $a < x < b$. We have

$$-E_j'' = \lambda_j E_j, \tag{34.25}$$

and since E_j must belong to $H_0^1(\Omega)$,

$$E_j(a) = E_j(b) = 0. \tag{34.26}$$

The general solution of (34.25) is given by

$$E_j = u \cos(x\sqrt{\lambda_j}) + v \sin(x\sqrt{\lambda_j}),$$

and the boundary conditions (34.26) require that the vector $(u, v) \in \mathbf{R}^2$ be orthogonal to the two straight lines through the origin whose angles with the u-axis are $a\lambda_j^{1/2}$ and $b\lambda_j^{1/2}$, respectively. Since $u^2 + v^2 \neq 0$, this is only possible if the two lines in question are identical, that is, setting $L = b - a$,

$$L\lambda_j^{1/2} = j\pi, \qquad j = 1, 2, \ldots. \tag{34.27}$$

This condition determines the eigenvalues of $-(d/dx)^2$ in our present setup; observe that these eigenvalues must be strictly positive, by Proposition 23.4. They are the numbers

$$\lambda_j = (j\pi/L)^2, \qquad j = 1, 2, \ldots. \tag{34.28}$$

By taking into account (34.26) and the fact that the E_j must form an orthonormal system in $L^2(\Omega)$, we immediately obtain

$$E_j = \left(\frac{2}{L}\right)^{1/2} \sin\left(j\pi \frac{x-a}{b-a}\right), \qquad j = 1, 2, \ldots. \tag{34.29}$$

Observe that the multiplicity of each eigenvalue is *one*.

Example 34.2

We continue to look at $A = -\Delta$, but this time in two independent variables, denoted by x and y. We take Ω to be the open disk $x^2 + y^2 < R^2$, $R > 0$. It is convenient, therefore, to switch to polar coordinates r, θ and to expand the prospective solution of $(\Delta + \lambda)h = 0$ into a Fourier series (in θ):

$$h(r, \theta) = h_0(r) + \sum_{m=1}^{+\infty} (h_m(r)e^{im\theta} + h_{-m}(r)e^{-im\theta}). \tag{34.30}$$

We use the expression of the Laplacian in polar coordinates:

$$\Delta = \left(\frac{\partial}{\partial r}\right)^2 + \frac{1}{r}\frac{\partial}{\partial r} + \frac{1}{r^2}\left(\frac{\partial}{\partial \theta}\right)^2.$$

Because of the fact that any two functions $v(r)e^{im\theta}$, $w(r)e^{im'\theta}$ with $m \neq m'$, are orthogonal in $L^2(\Omega)$, in order that h, given by (34.30), be a solution of $(\Delta + \lambda)h = 0$, it is necessary and sufficient that each term $h_m(r)e^{im\theta}$, $m \in \mathbf{Z}$, be one. We must therefore study the equation

$$h_m'' + \frac{1}{r}h_m' + \left(\lambda - \frac{m^2}{r^2}\right)h_m = 0, \qquad m = 0, \pm 1, \pm 2, \ldots. \tag{34.31}$$

We require $h \in H_0^1(\Omega)$, hence

$$h_m(R) = 0, \qquad \forall m \in \mathbf{Z}. \tag{34.32}$$

Notice that we know a priori that there will be nontrivial solutions to (34.31)–(34.32) only if λ is an eigenvalue of the problem, and then there must be solutions only for finitely many m [otherwise the space of the functions $h \in H_0^1(\Omega)$ satisfying $(\Delta + \lambda)h = 0$ would be infinite-dimensional].

If we set $s = \sqrt{\lambda}\, r$ in (34.31) and denote now by primes the differentiations with respect to s, that equation is transformed into the mth *Bessel equation*

$$h'' + \frac{1}{s}h' + \left(1 - \frac{m^2}{s^2}\right)h = 0. \tag{34.33}$$

The space of the solutions that are regular at the origin is one-dimensional, it is spanned by the mth *Bessel function* (supposing now that $m \geqq 0$):

$$J_m(s) = \left(\frac{s}{2}\right)^m \sum_{p=0}^{+\infty} \frac{(-1)^p}{\Gamma(p+1)\Gamma(m+p+1)}\left(\frac{s}{2}\right)^{2p}, \tag{34.34}$$

where Γ denotes the Euler gamma function. The way to find the expression (34.34) is straightforward: One writes that the solution h in (34.33) is a power series in the variable s and determines the coefficients of this series by the relations between them derived from (34.33). If m is a negative integer one defines $J_m(s)$ as $(-1)^{-m}J_{-m}(s)$; it is obviously also a solution of (34.33)!

The question, of course, is whether one can also achieve

$$J_m(\sqrt{\lambda}\, R) = 0. \tag{34.35}$$

In answering this question we invoke classical results of the theory of Bessel functions (to be found, e.g., in [W]). We may as well suppose $m \geqq 0$ (most of the statements below are valid even when m is not an integer, but we are solely interested in this case, of course).

(34.36) *For any* $m \geqq 0$, $J_m(s)$ *has only real zeros.*

(34.37) *Let* $z_{m,j}$ $(j = 1, 2, \ldots)$ *denote the positive zeros of* J_m, *arranged in increasing order. Then* (interlacing of the zeros of the Bessel functions)

$$0 < z_{m,1} < z_{m+1,1} < z_{m,2} < z_{m+1,2} < \cdots.$$

Property (34.37) makes it important to know something about the distribution of zeros of J_0:

(34.38) J_0 *has an infinity of positive zeros; they lie in the intervals*

$$[(k + \tfrac{3}{4})\pi, (k + \tfrac{7}{8})\pi], \qquad k = 0, 1, \ldots.$$

Finally, let us emphasize the fact [implicit in (34.37)] that any two functions J_m $(m = 0, 1, \ldots)$ have no common zeros except the origin (when J_0 is not one of them).

If we take all this into account, we conclude that we have a (double) sequence of eigenvalues:

$$\lambda = (z_{m,j}/R)^2, \qquad m = 0, 1, \ldots, \quad j = 1, 2, \ldots, \tag{34.39}$$

and that the eigenspace corresponding to each one of them is two-dimensional —it is spanned by $J_m(z_{m,j}\, r/R)e^{im\theta}$ and by $J_m(z_{m,j}\, r/R)e^{-im\theta}$ [except for $m = 0$, in which case the eigenspace is one-dimensional and spanned by $J_0(z_{0,j}\, r/R)$].

Exercises

34.1. This exercise concerns the meaning of the *smallest eigenvalue*. Let $V \subsetneq H$ be two complex Hilbert spaces and assume that V is dense in H and that its injection into H is compact. By composing the injections of V into H and that of H into the antidual $\overline{V}'$ of V (we are identifying H with its own antidual) we obtain an injection J of V into $\overline{V}'$. Show that this enables us to introduce *positive* continuous linear mappings of V into $\overline{V}'$, and if A is a positive isomorphism of V onto $\overline{V}'$, we may define its square root, $A^{1/2}$. Prove that $A^{1/2}$ is an isomorphism of V onto H. Let $a(u, v) = (A^{1/2}u, A^{1/2}v)_H$. Show that the smallest eigenvalue of A is equal to the largest number $c > 0$ such that

(34.40) $$c\|u\|_H^2 \leqq a(u, u), \qquad \forall u \in V.$$

34.2. Apply the abstract description in Exercise 34.1 to the case where $V = H_0^1(\Omega)$, $H = L^2(\Omega)$ [hence $\overline{V}' = H^{-1}(\Omega)$] when Ω is the open subset of $\mathbf{R}^n$ defined by

(34.41) $$-\infty < a_j < x_j < b_j < +\infty, \qquad j = 1, \ldots, n.$$

Taking A to be minus the Laplace operator, $-\Delta$, compute its smallest eigenvalue.

34.3. Let $J_m(z)$ be the mth Bessel function, defined in (34.34) ($m \in \mathbf{Z}$). Show that

(34.42) $$e^{iz\sin\theta} = \sum_{m=-\infty}^{+\infty} J_m(z)e^{im\theta}.$$

Derive from this that, for all *real* x,

(34.43) $$|J_m(x)| \leqq \frac{1}{\sqrt{2}} \quad \text{if} \quad m \neq 0, \qquad |J_0(x)| \leqq 1;$$

Prove the following formulas:

(34.44) $$J_m(z) = \frac{1}{\pi}\int_0^\pi \cos(m\theta - z\sin\theta)\, d\theta \qquad \text{(Bessel's formula);}$$

(34.45) $$J_m(z) = \Gamma(m + \tfrac{1}{2})^{-1}\Gamma(\tfrac{1}{2})^{-1}\left(\frac{z}{2}\right)^m \int_{-1}^{+1} e^{izt}(1 - t^2)^{m-1/2}\, dt;$$

(34.46) $$J_{m-1} + J_{m+1} = \frac{2m}{z} J_m, \qquad J_{m-1} - J_{m+1} = 2J_m';$$

(34.47) $$J_m(y + z) = \sum_{\nu=-\infty}^{+\infty} J_\nu(y)J_{m-\nu}(z).$$

Explain the relation between (34.47) and the properties of Fourier series related to convolution (in $\mathbf{Z}$ and on the torus $\mathbf{T}$).

34.4. Formula (34.34) is used to define the Bessel function $J_m(s)$ for any complex number m which is not an integer <0, in particular when $m = p + \frac{1}{2}$ with p an integer. Let $u(x) \in L^1(\mathbf{R}^n)$ be rotation-invariant; set $u^\#(r) = u(x)$ if $|x| = r$. Show that the Fourier transform of u is also rotation-invariant and that, if $p = |\xi|$, it is equal to

(34.48) $$\hat{u}(\xi) = \rho^{1-n/2}\int_0^{+\infty} u^\#(r)J_{n/2-1}(r\rho)r^{n/2}\, dr.$$

34.5. Prove that the Laplace operator Δ can be regarded as an unbounded self-adjoint operator (which we shall denote here by $-A$) on $L^2(\mathbf{R}^n)$ with domain $H^2(\mathbf{R}^n)$. Prove that the spectrum of A is exactly equal to $\mathbf{R}_+$ but that there is no eigenfunction of A belonging to $L^2(\mathbf{R}^n)$, hence no eigenvalue of A. Prove that given any number $\lambda \geqq 0$, there is a function $f_\lambda \in L^\infty(\mathbf{R}^n) \cap C^\infty(\mathbf{R}^n)$ such that $-\Delta f_\lambda = \lambda f_\lambda$ (f_λ not identically zero).

34.6. Let λ be an arbitrary number $\geqq 0$. Let Δ be the Laplace operator on $\mathbf{R}^2$. Prove that all the *tempered* distributions u which satisfy the homogeneous equation

$$(\Delta + \lambda)u = 0 \quad \text{in } \mathbf{R}^2, \tag{34.49}$$

are given by a series expansion

$$u(z) = \sum_{m=-\infty}^{+\infty} c_m J_m(\sqrt{\lambda}\, r)e^{im\theta} \qquad (z = re^{i\theta}), \tag{34.50}$$

where $\{c_m\}_{m \in \mathbf{Z}}$ is a sequence of complex numbers growing slowly at infinity [i.e., $\exists C > 0$ such that $|c_m| \leqq C(1 + |m|)^C$, $\forall m \in \mathbf{Z}$]. Prove that the series on the right in (34.50) converges in $C^\infty(\mathbf{R}^n)$ (J_m stands for the mth Bessel function).

34.7. How do you reconcile the fact that, given a bounded open subset of $\mathbf{R}^n$, for suitable values of $\lambda > 0$ the homogeneous equation $(\Delta + \lambda)h = 0$ has nontrivial solutions in $H_0^1(\Omega)$, with the fact that no linear PDE with constant coefficients can have compactly supported distribution solutions different from zero (the latter assertion can be established by means of Fourier transformation, applying the Paley–Wiener theorem)?

35

Approximate Solutions to the Dirichlet Problem. The Finite Difference Method

Conformal mappings and eigenfunction expansions provide us with the means of approximating the solution to the Dirichlet problem

$$(35.1) \qquad (\lambda - \Delta)u = f \qquad \text{(in } \Omega\text{, an open subset of } \mathbf{R}^n\text{)},$$

$$(35.2) \qquad u = g \quad \text{on } \partial\Omega \qquad \text{(the boundary of } \Omega\text{)}.$$

Unfortunately, their use is clearly limited to highly "symmetric" configurations: That of conformal mappings presumes that we are essentially in a two-dimensional situation, and the use of spherical harmonics, of cylindrical, ellipsoidal, etc., harmonics, is possible only when the domain Ω exhibits spherical, cylindrical, ellipsoidal, etc., symmetries, respectively. In practice these are excessive limitations. Thus quite early (at the beginning of this century), physicists and engineers were led to devise a simple and practical method of approximating the solution of (35.1)–(35.2). This method is closely related to the notion of discrete harmonic functions described in Sect. 31. It goes under the name of *finite difference method.* A crucial advantage of this method is that it is particularly well adapted to treatment by computer. Another method, which has been in favor recently, is the *Galerkin method*; it has a certain similarity to eigenfunction expansion, but much greater flexibility. Both the finite difference and the Galerkin method are based on the approximation of Eq. (35.1) by linear equations in *finite*-dimensional spaces. One solves the latter and shows that their solutions converge to that of (35.1) as the finite-dimensional spaces suitably "increase." We give here a very brief description of the two methods, from the standpoint of the *variational* approach, currently the standpoint which is most often adopted.

It pays to look at a setup which is general enough to cover both the finite difference and Galerkin's method. We shall deal with a *separable* Hilbert space V, i.e., with a Hilbert space having a countable dense subset or, if one prefers, a countable orthonormal complete system. In the application given here V will be $H_0^1(\Omega)$, and the separability assumption is certainly satisfied—as it always is, in practice [usually, V is a closed subspace of $H^m(\Omega)$].

We shall now describe an *external approximation* of the Hilbert space V. At first this concept may seem a bit too abstract, but the examples will help, we hope, to understand the reason for the abstraction. We shall deal with spaces and mappings indexed by a parameter h; h will run over a sequence (or sometimes over a net with a countable basis—like an interval in the real line) which we assume to converge; the limits with respect to h will always be meant as h tends to the limit of that sequence. But first of all we introduce a monomorphism of V into another Hilbert space F, J, i.e., a continuous linear injection whose image is closed but not necessarily all of F (by the closed graph theorem, this implies that J is a homeomorphism into). For each h, we introduce a Hilbert space V_h which, in the applications, will be finite-dimensional, and a pair of continuous linear mappings:

(35.3) $$r_h : V \to V_h, \qquad p_h : V_h \to F$$

(the r_h are often called the *restriction* mappings, the p_h the *prolongation* or *extension* mappings).

Definition 35.1. The system (V_h, p_h, r_h, F, J) *is called a convergent external approximation of* V *if both following conditions are satisfied*:

(35.4) $\forall u \in V$, *$p_h r_h u$ converges to Ju in F* (for the norm of F);

(35.5) *suppose that, for each h, we are given an element v_h of V_h such that $p_h v_h$ converges weakly in F to some v; then necessarily $v = Ju$, $u \in V$.*

Furthermore the approximation is said to be stable if the norms of the mappings r_h and p_h are bounded independently of h.

Example 35.1. Let $V = F$ and J be the identity mapping of V. Let h vary over the sequence $1, 1/2, \ldots, 1/k, \ldots$, and let $\{V_h\}$ be an increasing sequence of finite-dimensional subspaces of V whose union is dense in V (recalling that V is separable). For instance, we may be given a complete orthonormal system $\{e_1, \ldots, e_k, \ldots\}$ in V and let V_h be the subspace spanned by $e_1, \ldots, e_k$ where $k = 1/h$. We may then take for p_h the natural injection of V_h into V and for r_h, the orthogonal projection of V onto V_h. The approximation thus obtained is convergent and stable; it is called a *Galerkin approximation* of V. We shall simply denote it by $\{V_h\}$.

Let us momentarily go back to Sects. 22, 23, and 26 and recall that, under suitable assumptions on Ω, on $\partial\Omega$, and on the boundary datum g, we may reduce the problem (35.1)–(35.2) to the same one, but where $g = 0$. We must now assume that $f \in H^{-1}(\Omega)$ and the variational approach consists in seeking u in $H_0^1(\Omega)$ satisfying, for all $v \in H_0^1(\Omega)$, the equation

$$a(u, v) = \langle f, \bar{v} \rangle, \tag{35.6}$$

where $\langle\ ,\ \rangle$ is the bracket of the duality between $H_0^1(\Omega)$ and $H^{-1}(\Omega)$, and

$$a(u, v) = \lambda \int_\Omega u\bar{v}\, dx + \sum_{j=1}^{n} \int_\Omega \frac{\partial u}{\partial x^j} \frac{\partial \bar{v}}{\partial x^j}\, dx. \tag{35.7}$$

(We assume throughout that λ is $\geqq 0$; when Ω is unbounded, we assume $\lambda > 0$.)

In the abstract situation, we assume that we are given a continuous sesquilinear form $a(u, v)$ on $V \times V$ which is *coercive* (Definition 23.1) and a continuous *antilinear* form ϕ on V; we want to approximate the solution u, which belongs to V (Lemma 23.1), of the equation

$$a(u, v) = \phi(v) \qquad \text{for all} \quad v \in V. \tag{35.8}$$

[When dealing with real-valued functions and distributions, hence with real Hilbert spaces V, H, and V', the form a is assumed to be bilinear, and the form ϕ linear. Note that we do not assume the form a to be Hermitian—or symmetric, in the real case, as it is in the particular case (35.7).]

Let us consider an external approximation of V, (V_h, p_h, r_h, F, J), as in Definition 35.1, and assume that we are given, for each h, a continuous sesquilinear form a_h on $V_h \times V_h$ and a continuous antilinear functional ϕ_h on V_h, submitted to the following conditions:

(35.9) *there is a constant* $c_0 > 0$, *independent of* h, *such that*

$$|a_h(u_h, u_h)| \geqq c_0 \|u_h\|_{V_h}^2, \qquad \forall u_h \in V_h;$$

(35.10) *there is a constant* $C > 0$, *independent of* h, *such that*

$$\|\phi_h\|_{V_h'} \leqq C.$$

Of course, we must somehow relate the data a_h, ϕ_h to a and ϕ. This is done by requiring that a_h converge to a and ϕ_h to ϕ in a reasonable sense, which is made precise by the following *consistency conditions*:

(35.11) *for all* v, $w \in V$, *if the sequence* $p_h v_h$, $v_h \in V_h$, *converges weakly to* Jv *in* F, *then*:

$$\lim_h a_h(v_h, r_h w) = a(v, w), \qquad \lim_h a_h(r_h w, v_h) = a(w, v); \tag{35.12}$$

$$\lim_h \phi_h(v_h) = \phi(v). \tag{35.13}$$

Clearly, when a_h is Hermitian for every h, one of the two conditions (35.12) is redundant.

Example 35.2. Let $\{V_h\}$ be a Galerkin approximation of V (Example 35.1). If we take for a_h the restriction of a to $V_h \times V_h$ and for ϕ_h the restriction of ϕ to V_h, we see at once that conditions (35.9) to (35.11) are satisfied.

By virtue of the Lax–Milgram lemma (Lemma 23.1), for every h the equation

$$(35.14)_h \qquad a_h(u_h, v_h) = \phi_h(v_h), \qquad \forall v_h \in V_h,$$

has a unique solution u_h in V_h. We may state and prove:

THEOREM 35.1. *Let (V_h, p_h, r_h, F, J) be a convergent and stable external approximation of the Hilbert space V. Let $a(u, v)$ be a continuous coercive sesquilinear form on $V \times V$, ϕ a continuous antilinear form on V. For every h, let a_h be a continuous sesquilinear coercive form on $V_h \times V_h$ and ϕ_h a continuous antilinear form on V_h, such that (35.9), (35.10), and (35.11) hold.*

If $u \in V$ denotes the unique solution of (35.8) and $u_h \in V_h$, for each h, the one of $(35.14)_h$, $p_h u_h$ converges to Ju in F (for the norm).

Proof. By applying $(35.14)_h$ with $v_h = u_h$ and applying (35.9) and (35.10) we immediately obtain

$$(35.15) \qquad \|u_h\|_{V_h} \leqq C/c_0 .$$

If $\sup_h \|p_h\| = M < +\infty$ (since the approximation is stable, Definition 35.1),

$$(35.16) \qquad \|p_h u_h\|_F \leqq MC/c_0 .$$

We may extract a subsequence, $\{p_{h'} u_{h'}\}$, which converges weakly in F, to some element, of the form Ju_1, $u_1 \in V$, by (35.5).

Let us arbitrarily fix $v \in V$ and apply now $(34.14)_h$ with $v_h = r_h v$. We apply (35.4) and see that the sequence $\{v_h\}$ is of the kind considered in (35.11). We derive from (35.13) that $\phi_h(v_h)$ converges to $\phi(v)$. On the other hand, let $\tilde{u}_h = u_{h'}$ if $h = h'$ belongs to the set of indices of the weakly convergent subsequence $p_{h'} u_{h'}$ above, and $\tilde{u}_h = r_h u_1$ otherwise. We see that $p_h \tilde{u}_h$ converges weakly to Ju_1 in F, hence, by (35.12), $a_h(u_h, v_h)$ converges to $a(u_1, v)$. But $a_{h'}(u_{h'}, v_{h'}) = \phi_{h'}(v_{h'}) \to \phi(v)$, hence u_1 satisfies (35.8), which demands that $u_1 = u$. This shows that

$$(35.17) \qquad \{p_h u_h\} \quad \text{converges weakly to} \quad Ju \quad \text{in } F.$$

Let us then consider

$$\begin{aligned} T &= a_h(u_h - r_h u, u_h - r_h u) \\ &= a_h(u_h, u_h) + a_h(r_h u, r_h u) - a_h(u_h, r_h u) - a_h(r_h u, u_h). \end{aligned}$$

By (35.9) we have

$$\|u_h - r_h u\|_{V_h}^2 \leqq c_0^{-1} T. \tag{35.18}$$

If we apply (35.12) [with $v_h = r_h u$ and recall (35.4)], we see that

$$a_h(r_h u, r_h u), \quad a_h(u_h, r_h u), \quad a_h(r_h u, u_h) \quad \text{converge to} \quad a(u, u).$$

On the other hand, by (35.13),

$$a_h(u_h, u_h) = \phi_h(u_h) \qquad \text{converges to} \quad \phi(u) = a(u, u).$$

Combining these assertions yields that T converges to zero, and therefore

$$\lim_h \|u_h - r_h u\|_{V_h} = 0. \tag{35.19}$$

By using the fact that the norms of the p_h are bounded independently of h, and also (35.4), we conclude that $p_h u_h$ converges to Ju for the norm in F.

Q.E.D.

The property (35.19) is sometimes expressed by saying that the u_h converge *discretely* to u (from which arises the notion of *discrete convergence*).

Inspection of the proof of Theorem 35.1 yields an estimate of the *error*. Actually there are several notions of error:

(1) the *error between u and u_h*, which is the value of $\|Ju - p_h u_h\|_F$ (thus this error depends on the choice of J);
(2) the *discrete error* between u and u_h, which is the value of $\|u_h - r_h u\|_{V_h}$;
(3) the *cutoff error* on u, $\|Ju - p_h r_h u\|_F$.

It is on the size of the latter that we might best be able to have an influence, as it depends solely on the choice of the approximation (V_h, p_h, r_h, F, J) and not on the sesquilinear form $a(u, v)$. We shall determine an upper bound for the error between u and u_h (1), in terms of the cutoff error (3). First we shall determine an upper bound for the discrete error (2), in terms of an "extension" α of a to $F \times F$, that is, of

(35.20) *a continuous sesquilinear form α on $F \times F$ such that $\alpha(Jv, Jw) = a(v, w)$ for all $v, w \in V$.*

Then let u be the solution of (35.8). We *define* the following continuous linear functional on F:

$$\Phi(g) = \alpha(Ju, g), \qquad g \in F. \tag{35.21}$$

It is verified at once that $\Phi(v) = \phi(Jv)$ for all $v \in V$.

Remark 35.1. Observing that $F = (JV) \oplus (JV)^\perp$ [$(JV)^\perp$ is the orthogonal of JV in F], we see that forms such as α always exist: We may take α to be the only extension of a such that $\alpha(f, g) = 0$ for all $f \in F$, $g \in (JV)^{\perp}$. But other extensions might be more suited to the problem under study, as shown in the following:

Example 35.3. Let $V = H^1(\Omega)$ [or $H_0^1(\Omega)$] and take $F = (L^2(\Omega))^{n+1}$,

$$J : v \mapsto \left(v, \frac{\partial v}{\partial x^1}, \ldots, \frac{\partial v}{\partial x^n}\right). \tag{35.22}$$

We know that J is an *isometry* of V into F. The extension of the form (35.7) is evident:

$$\alpha(f, g) = \int_\Omega \left(\lambda f_0 \bar{g}_0 + \sum_{j=1}^n f_j \bar{g}_j\right) dx, \qquad f, g \in (L^2(\Omega))^{n+1}. \tag{35.23}$$

Observe that (35.23) is certainly different from the extension described in Remark 35.1. Indeed, the latter is never coercive when J is not onto, whereas (35.23) is obviously coercive (when $\lambda > 0$) although J is certainly not onto.

We introduce now the following two quantities:

$$\varepsilon_h(\phi) = \| {}^t p_h \Phi - \phi_h \|_{V_h'}, \tag{35.24}$$

$$\eta_h(u) = \sup_{0 \neq v_h \in V_h} |\alpha(Ju, p_h v_h) - a_h(r_h u, v_h)| / \|v_h\|_{V_h}. \tag{35.25}$$

Concerning (35.24), note that p_h is a continuous linear map $V_h \to F$, hence ${}^t p_h$ is a continuous linear map $F' \to V_h'$.

THEOREM 35.2. *Use the same hypotheses and same notation as in Theorem 35.1. We have*

$$\|u_h - r_h u\|_{V_h} \leqq c_0^{-1}[\varepsilon_h(\phi) + \eta_h(u)], \tag{35.26}$$

where c_0 is the positive coercivity constant in (35.9).

Proof. It suffices to reexamine the quantity T introduced at the end of the proof of Theorem 35.1, and rewrite it as follows:

$$\begin{aligned} T &= a_h(u_h, u_h - r_h u) - a_h(r_h u, u_h - r_h u) \\ &= \phi_h(u_h - r_h u) - \alpha(Ju, p_h(u_h - r_h u)) + \alpha(Ju, p_h(u_h - r_h u)) \\ &\quad - a_h(r_h u, u_h - r_h u) \\ &= \phi_h(u_h - r_h u) - \Phi(u_h - r_h u) + \alpha(Ju, p_h(u_h - r_h u)) - a_h(r_h u, u_h - r_h u), \end{aligned}$$

whence

$$|T| \leqq [\varepsilon_h(\phi) + \eta_h(u)] \|u_h - r_h u\|_{V_h}. \tag{35.27}$$

By combining (35.18) and (35.27) we get (35.26). Q.E.D.

COROLLARY 35.1. *We have the same hypotheses as in Theorem 35.2. Let M be an upper bound for the norms of the extension operators p_h. Then*

$$\|Ju - p_h u_h\|_F \leqq \|Ju - p_h r_h u\|_F + \frac{M}{c_0}\,[\varepsilon_h(\phi) + \eta_h(u)]. \tag{35.28}$$

The Finite Difference Method

In this second part of the section we apply the general considerations of the first part and describe the approximation of the solution to the Dirichlet problem one obtains by the finite difference method. We assume throughout that the open set Ω is bounded. We deal with diagonal matrices h whose diagonal entries $h_1, \ldots, h_n$ are strictly greater than zero (and converge to zero). If $x \in \mathbf{R}^n$, hx denotes the vector $(h_1 x^1, \ldots, h_n x^n)$; $h\mathbf{Z}^n$ is the image of $\mathbf{Z}^n$, the lattice in $\mathbf{R}^n$ of points with integral coordinates (>0 or $\leqq 0$), under the mapping $x \mapsto hx$. In what follows a matrix h will be our parameter (so denoted in the first part) and will range over a sequence of (strictly positive) diagonal matrices converging to the zero matrix. The sequence will not be specified and will simply be denoted by $\mathfrak{h}$ (of course, in actual computations one must make a choice of the sequence $\mathfrak{h}$).

Given $x_0 \in \mathbf{R}^n$ and $h \in \mathfrak{h}$, we write

$$\sigma_h(x_0) = \prod_{j=1}^{n} [x_0^j - \tfrac{1}{2}h_j,\, x_0^j + \tfrac{1}{2}h_j[; \tag{35.29}$$

(35.30) *if r is a nonnegative integer,*

$$\sigma_h(x_0, r) = \bigcup_{\substack{\alpha \in \mathbf{Z}^n \\ |\alpha| \leq r}} \sigma_h(x_0 + \tfrac{1}{2}h\alpha).$$

Three examples of sets $\sigma_h(x_0, r)$ in the plane (i.e., when $n = 2$) can be found in Fig. 35.1.

We denote by w_{h, x_0} the characteristic function of the set $\sigma_h(x_0)$ and by $\delta_{h, j}$ ($j = 1, \ldots, n$) the *finite difference operator*

$$\delta_{h, j} F(x) = \frac{1}{h_j}\,[F(x + \tfrac{1}{2}h_j \mathbf{e}_j) - F(x - \tfrac{1}{2}h_j \mathbf{e}_j)], \tag{35.31}$$

where $\mathbf{e}_j$ stands for the unit vector along the x^j-axis.

Returning to the open set Ω, we shall use the notation

$$\mathring{\Omega}_h^r = \{x \in h\mathbf{Z}^n;\ \sigma_h(x, r) \subset \Omega\}. \tag{35.32}$$

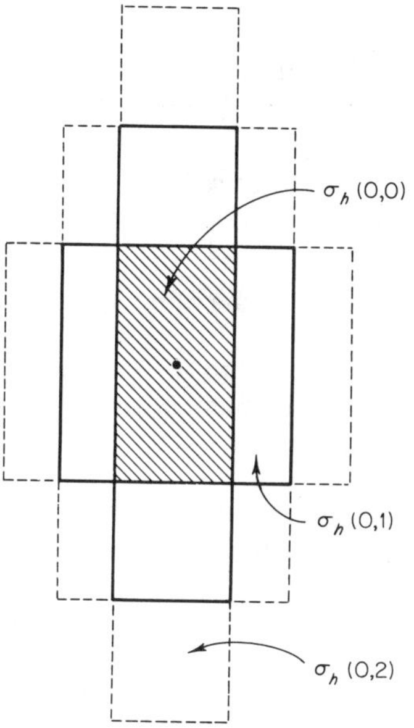

FIG. 35.1

When dealing with second-order elliptic operators, one makes use of these sets $\mathring{\Omega}_h^r$ only for $r = 0, 1$. If the order of the operator under study is $2m$, one will customarily make use of the same sets but for $r = 0, \ldots, m$.

We describe now the external approximation (or *difference scheme*) we are going to use. As we have said, $V = H_0^1(\Omega)$. We take $F = (L^2(\Omega))^{n+1}$ and J to be the isometric injection (35.22). For each $h \in \mathfrak{h}$, the space V_h will be the linear span of the characteristic functions $w_{h,y}$ as y ranges over the set (35.32) where $r = 1$, i.e., V_h will be the space of step functions:

(35.33) $$v_h(x) = \sum_{y \in \mathring{\Omega}_h{}^1} v_h(y) w_{h,y}(x).$$

The notation is consistent since $w_{h,y}\, w_{h,y'} \equiv 0$ whenever y, y' are different elements of $\mathring{\Omega}_h^1$. Observe that the dimension of V_h is exactly equal to the number of such elements.

We equip V_h with the inner product

(35.34) $$(v_h, w_h)_{h,1} = \int \{v_h \bar{w}_h + \sum_{j=1}^{n} (\delta_{h,j} v_h)(\delta_{h,j} \bar{w}_h)\}\, dx, \qquad v_h, w_h \in V_h.$$

The extension operator p_h will be the mapping

(35.35) $$v_h \mapsto (v_h, \delta_{h,1} v_h, \ldots, \delta_{h,n} v_h),$$

which is obviously an isometry of V_h into F.

As for the restriction operator $r_h : V \to V_h$, it assigns to every $v \in H_0^1(\Omega)$ the step function (35.33) defined by

(35.36) $$v_h(y) = |\sigma_h(y)|^{-1} \int_{\sigma_h(y)} v(x)\, dx, \qquad y \in \mathring{\Omega}_h^1,$$

where

(35.37) $$|\sigma_h(y)| = \text{volume of } \sigma_h(y) = h_1 \cdots h_n.$$

We are going to prove that the difference scheme described above is stable and convergent (Definition 35.1), but under a restrictive condition involving the open set Ω and the sequence $\mathfrak{h}$.

Definition 35.2. Let $\mathfrak{h}$ be a sequence of strictly positive diagonal matrices converging to zero. We say that the open set $\Omega \subset \mathbf{R}^n$ is $\mathfrak{h}$-regular if there is a finite collection of vectors in $\mathbf{R}^n$, $\theta_1, \ldots, \theta_p$, and a constant $C > 0$, independent of $j = 1, \ldots, n$, and of $h \in \mathfrak{h}$, such that the following is true:

(35.38) *given any point $y \in \mathring{\Omega}_h^1$ such that either $y + h_j \mathbf{e}_j$ or $y - h_j \mathbf{e}_j$ does not belong to $\mathring{\Omega}_h^1$, to every $z \in \sigma_n(y)$ there is ρ, $0 < \rho < C$, such that, for some q, $1 \leqq q \leqq p$, $z + \rho h_j \theta_q \notin \Omega$.*

THEOREM 35.3. *Let $\mathfrak{h}$ be a sequence of strictly positive diagonal matrices, converging to zero, Ω a bounded open subset of $\mathbf{R}^n$.*

If Ω is $\mathfrak{h}$-regular, the external approximation (V_h, p_h, r_h, F, J) of $V = H_0^1(\Omega)$ described above is stable and convergent.

Proof. Since the mappings p_h are isometries, in order to prove that the approximation is stable, it suffices to prove that

(35.39) $$\|r_h v\|_{V_h} \leqq \text{const } \|v\|_V$$

for some constant independent of h and of $v \in V$. Of course it suffices to prove (35.39) for v in a dense subset of V; we shall take $v \in C_c^\infty(\Omega)$.

First of all, recalling that the sets $\sigma_h(y)$, $y \in h\mathbf{Z}^n$, are pairwise disjoint, we see that

$$\int_\Omega |v_h(x)|^2\, dx = \sum |v_h(y)|^2 |\sigma_h(y)| = (h_1 \cdots h_n)^{-1} \sum \left| \int_{\sigma_h(y)} v(x)\, dx \right|^2$$

$$\leqq \sum \int_{\sigma_h(y)} |v(z)|^2\, dz \qquad \text{(by Cauchy–Schwarz)}$$

$$\leqq \int_\Omega |v(x)|^2\, dx,$$

where all summations are performed with respect to $y \in \mathring{\Omega}_h^1$ [cf. (35.32)]. Thus,

$$\|v_h\|_{L^2(\Omega)} \leqq \|v\|_{L^2(\Omega)}. \tag{35.40}$$

In order to show that the difference scheme (V_h, p_h, r_h, F, J) is *stable*, we must estimate the L^2 norm of $\delta_{h,j} v_h$ in terms of the H^1 norm of v. Let us fix j, $1 \leqq j \leqq n$, and let us denote, for any point x,

$$x^+ = x + \tfrac{1}{2}h_j \mathbf{e}_j, \qquad x^- = x - \tfrac{1}{2}h_j \mathbf{e}_j.$$

We have

$$v_h(x^+) = \sum_y v_h(y) w_{h,y}(x^+) = \sum_{y^-} |\sigma_h(y^-)|^{-1} \int_{\sigma_h(y^-)} v(z^+)\,dz\, w_{h,y^-}(x),$$

that is,

$$v_h(x^+) = (h_1 \cdots h_n)^{-1} \sum_{y^-} w_{h,y^-}(x) \int_{\sigma_h(y^-)} v(z^+)\,dz, \tag{35.41}$$

the summation being performed over the image of $\mathring{\Omega}_h^1$ under the translation $y \mapsto y^-$. Similarly,

$$v_h(x^-) = (h_1 \cdots h_n)^{-1} \sum_{y^+} w_{h,y^+}(x) \int_{\sigma_h(y^+)} v(z^-)\,dz. \tag{35.42}$$

It follows from the definition of $\mathring{\Omega}_h^1$ that y^+ and y^- belong to Ω as long as y remains in $\mathring{\Omega}_h^1$. Note also that, if $y' = y - h_j \mathbf{e}_j$, then $y'^+ = y^-$. If there are two points y, y' in $\mathring{\Omega}_h^1$ such that $y'^+ = y^-$, we shall denote the latter point by y^0. Otherwise we write y^*; thus $y^* = y^+$ (resp. y^-) whenever

(35.43) $y + h_j \mathbf{e}_j$ *(resp.* $y - h_j e_j$*) does not belong to* $\mathring{\Omega}_h^1$.

We derive from (35.41)–(35.42):

$$\begin{aligned}(h_1 \cdots h_n)[v_h(x^+) - v_h(x^-)] &= \sum_{y^\circ} w_{h,y^\circ}(x) \int_{\sigma_h(y^\circ)} [v(z^+) - v(z^-)]\,dz \\ &\quad + \sum_{y^*} \pm w_{h,y^*}(x) \int_{\sigma_h(y^*)} v(z^\pm)\,dz.\end{aligned} \tag{35.44}$$

We have [cf. proof of (35.40)]

$$\begin{aligned}\int_\Omega |v_h(x^+) - v_h(x^-)|^2\,dx &\leqq \sum_{y^\circ} \int_{\sigma_h(y^\circ)} |v(z^+) - v(z^-)|^2\,dz \\ &\quad + \sum_{y^*} \int_{\sigma_h(y^*)} |v(z^\pm)|^2\,dz.\end{aligned} \tag{35.45}$$

In the terms corresponding to points y° we use the fact that

$$(35.46)\qquad |v(z^+) - v(z^-)|^2 \leqq h_j^2 \int_{-1/2}^{+1/2} \left|\frac{\partial v}{\partial x^j}(z + th_j \mathbf{e}_j)\right|^2 dt.$$

As for the terms corresponding to points y^*, we observe that, if $y^* = y^+$ for some $y \in \mathring{\Omega}_h^1$, the corresponding term in (35.45) is

$$(35.47)\qquad \int_{\sigma_h(y^+)} |v(z^-)|^2\, dz = \int_{\sigma_h(y)} |v(z)|^2\, dz.$$

At this point we exploit the $\mathfrak{h}$-regularity of Ω (Definition 35.2) and property (35.43) of y. For each $z \in \sigma_h(y)$ we may select $\rho = \rho(z)$, $0 < \rho < C$, such that $z + \rho h_j \theta_q \notin \Omega$ for some q, $1 \leqq q \leqq p$. Since $v \in C_c^\infty(\Omega)$, we have

$$(35.48)\qquad \begin{aligned} |v(z)|^2 &= |v(z + \rho h_j \theta_q) - v(z)|^2 \\ &\leqq \rho h_j^2 \int_0^\rho |\langle \theta_q, \operatorname{grad} v\rangle(z + th_j\theta_q)|^2\, dt \\ &\leqq CC_1^2 h_j^2 \int_0^C \sum_{q=1}^p |\operatorname{grad} v(z + th_j\theta_q)|^2\, dt, \end{aligned}$$

where $C_1 = \sup_q |\theta_q|$. Since the sets $\sigma_h(y)$ and $\sigma_h(y^\circ)$ are all contained in Ω [by (35.32)] we derive from (35.45), (35.46), and (35.48):

$$h_j^{-2} \int_\Omega |v_h(x^+) - v_h(x^-)|^2\, dx \leqq \int_{-1/2}^{+1/2} \int_\Omega \left|\frac{\partial v}{\partial x^j}(z + th_j\mathbf{e}_j)\right|^2 dz\, dt$$
$$+ CC_1^2 \sum_{q=1}^p \int_0^C \int_\Omega |\operatorname{grad} v(z + th_j\theta_q)|^2\, dz\, dt.$$

After obvious changes of variables in the z integrals we obtain

$$(35.49)\qquad \int |\delta_{h,j} v(x)|^2\, dx \leqq (1 + pC^2C_1^2) \int |\operatorname{grad} v(x)|^2\, dx,$$

which, together with (35.40), implies (35.39), and hence the stability of the scheme we are looking at. In order to complete the proof of Theorem 35.3 we must show that it is convergent.

First we show that, for each $v \in V$, $p_h r_h v$ converges to Jv in F. By the Banach–Steinhaus theorem (i.e., the principle of uniform boundedness), thanks to the fact that the norms of p_h and r_h are bounded independently of h, it suffices to prove the assertion when $v \in C_c^\infty(\Omega)$. Then it suffices to show that $v_h \to v$ and $\delta_{h,j} v \to (\partial/\partial x^j)v$ in $L^2(\Omega)$, which is quite obvious.

Next, for each $h \in \mathfrak{h}$, let v_h be an element of V_h such that the sequence $\{p_h v_h\}$ converges weakly in F. This implies that the v_h and the $\delta_{h,j} v_h$ converge in the distribution sense, and that their limits, respectively v and $(\partial/\partial x^j)v$, belong to $L^2(\Omega)$, i.e., $v \in H^1(\Omega)$. But since the supports of the v_h and of the $\delta_{h,j} v_h$ are compact subsets of Ω, it is not difficult to see that $v \in H_0^1(\Omega)$.

Q.E.D.

From Theorem 35.3 we can derive an approximation of the weak solution u to the Dirichlet problem

$$(\lambda - \Delta)u = f \in H^{-1}(\Omega), \qquad u \in H_0^1(\Omega), \tag{35.50}$$

that is, the solution u of Eq. (35.8) where a is given by (35.7) and

$$\phi(v) = \langle f, \bar{v} \rangle, \qquad v \in V = H_0^1(\Omega). \tag{35.51}$$

We introduce the following approximations of a and ϕ, respectively:

$$a_h(v_h, w_h) = \int_\Omega \left(\lambda v_h \bar{w}_h + \sum_{j=1}^n (\delta_{h,j} v_h)(\delta_{h,j} \bar{w}_h) \right) dx, \tag{35.52}$$

$$\phi_h(v_h) = \int_\Omega \left(f_0 \bar{v}_h - \sum_{j=1}^n f_j \, \delta_{h,j} \bar{v}_h \right) dx, \tag{35.53}$$

assuming that $v_h, w_h \in V_h$ and that we have

$$f_0 + \sum_{j=1}^n \frac{\partial f_j}{\partial x_j} = f. \tag{35.54}$$

[If we have at our disposal additional information about f, for instance that $f \in L^2(\Omega)$, the solution of Eq. (35.54) can become very easy; for instance, $f_0 = f$, $f_j = 0$ for $j > 0$.] Condition (35.9) is satisfied with $c_0 = \inf(1, \lambda)$; if $\lambda = 0$, we may still be able to satisfy (35.9) by modifying the inner product in V_h [usually given by (35.34)]. Condition (35.10) is satisfied by

$$C = \sum_{j=0}^n \int_\Omega |f_j|^2 \, dx.$$

It is seen at once that condition (35.11) is satisfied (taking into account what has been established in the proof of Theorem 35.3).

We consider the approximate problem

$$a_h(u_h, v_h) = \phi_h(v_h), \qquad \forall v_h \in V_h. \tag{35.55}$$

It suffices to verify (35.55) when v_h ranges over a basis of V_h, in particular when $v_h = w_{h,y}$, $y \in \mathring{\Omega}_h^1$. We may set, for any two points y, y' in $\mathring{\Omega}_h^1$,

$$\tilde{a}_h(y, y') = a_h(w_{h,y}, w_{h,y'}), \qquad \tilde{\phi}_h(y) = \phi_h(w_{h,y}). \tag{35.56}$$

In order to find the solution of (35.55),

$$u_h(x) = \sum_{y \in \mathring{\Omega}_h{}^1} u_h(y) w_{h,y}, \tag{35.57}$$

we must solve the system of linear equations

$$(35.58) \qquad \sum_{y \in \mathring{\Omega}_h^1} \tilde{a}_h(y, y')u_h(y) = \tilde{\phi}_h(y'), \qquad y' \in \mathring{\Omega}_h^1.$$

Calling $n(h)$ the number of points in $\mathring{\Omega}_h^1$, i.e., $n(h) = \dim V_h$, we see that (35.58) is a system of $n(h)$ linear equations, in $n(h)$ unknowns, of the complex numbers $u_h(y)$.

By Theorems 35.1 and 35.3, we know that, when h tends to zero along $\mathfrak{h}$, the solution u_h of (35.55), given by (35.57), converges to the solution u of (35.50).

The requirement that Ω be $\mathfrak{h}$-regular depends, of course, on the choice of $\mathfrak{h}$. Let us assume that, when $h \in \mathfrak{h}$ tends to zero, the "components" h_k remain comparable, that is, that there is a constant $\gamma > 0$ such that

$$(35.59) \quad \gamma \leqq h_j/h_k \leqq \gamma^{-1} \qquad \text{for all} \quad h \in \mathfrak{h} \quad \text{and all} \quad j, k = 1, \ldots, n.$$

Then, by choosing enough vectors θ_q (for instance, enough to ensure that every convex cone with solid angle greater than a given value $\omega_0 > 0$ will contain at least one of them), one can easily see that every *convex* bounded open set Ω is $\mathfrak{h}$-regular, or that every bounded open set Ω whose boundary is C^1 (and which lies on one side of $\partial\Omega$) is $\mathfrak{h}$-regular.

Finally, we wish to underline the fact that the application of the finite difference method presented above is probably the most primitive one can give, and that the method is amenable to great elaboration and diversification. It can be applied to a variety of problems besides the Dirichlet problem, for instance to the Neumann problem, the oblique derivative problem, (cf. Sect. 37), and also to higher order equations (cf. Sects. 36 and 38). One may obtain, under suitable circumstances, convergence properties other than the one contemplated in Theorem 35.1, for instance convergence in the L^∞ norm. Also, rather than using simple sets such as the $\sigma_h(y)$ introduced in (35.29) and the associated $\sigma_h(y, r)$, $\mathring{\Omega}_h^1$, it might be convenient to use more complicated, curvilinear sets. Also the method can be extended to elliptic equations with coefficients which are not smooth, and to nonlinear equations. On all of this we refer to the texts on approximation of solutions to elliptic boundary value problems, such as [A] and [FS]. The range of application of the finite difference method is not limited to elliptic boundary value problems. With appropriate modifications it has been successfully applied to a number of hyperbolic and parabolic evolution equations and mixed problems (on these problems, see Chapter IV; a very simple application of the Galerkin method to hyperbolic evolution equations can be found in the proof of Theorem 47.1).

Exercises

35.1. All the functions considered in this exercise will be *real*-valued. Let V_h be the linear space of step functions (35.33), v_h any element of V_h, M an arbitrary number at least zero. Prove that $(v_h - M)^+ = \sup(v_h - M, 0)$ also belongs to V_h.

Let $f \in L^\infty(\Omega)$ and consider the linear functional on V_h [cf. (35.53)] $\int_\Omega f v_h \, dx$. Let λ be strictly greater than zero and choose $M > 0$ such that $\|f\|_{L^\infty(\Omega)} \leqq \lambda M$. Let $a_h(v_h, w_h)$ be the bilinear form (35.52) and u_h the only element of V_h such that

$$a_h(u_h, v_h) = \int_\Omega f v_h \, dx, \qquad \forall v_h \in V_h. \tag{35.60}$$

Derive from (35.60) that if $w_h = u_h - M$,

$$a_h(w_h, w_h^+) = \int_\Omega (f(x) - \lambda M) w_h^+(x) \, dx. \tag{35.61}$$

Prove that

$$(\delta_{h,j} w_h^+(x))^2 \leqq \delta_{h,j} w_h(x) \, \delta_{h,j} w_h^+(x), \qquad \forall x \in \mathbf{R}^n, \tag{35.62}$$

and derive from (35.61) that $\int_\Omega (w_h^+(x))^2 \, dx = 0$, hence that

$$u_h(x) \leqq M, \qquad \forall x \in \Omega. \tag{35.63}$$

[Similarly one would obtain $u_h(x) \geqq -M$, $\forall x \in \Omega$.] Since $L^\infty(\Omega)$ is the dual of $L^1(\Omega)$, conclude that there is a subsequence of the u_h which converges weakly, i.e., for the weak dual topology on $L^\infty(\Omega)$, $\sigma(L^\infty(\Omega), (L^1\Omega))$, to u, and that we have

$$u \in L^\infty(\Omega), \qquad \|u\|_{L^\infty(\Omega)} \leqq \frac{1}{\lambda} \|f\|_{L^\infty(\Omega)}. \tag{35.64}$$

35.2. Let $\Omega_h = \mathring{\Omega}_h^0$ in the notation of (35.32). Call V_h the space of functions

$$v_h(x) = \sum_{y \in \Omega_h} v_h(y) w_{h,y}(x), \tag{35.65}$$

where $w_{h,y}$ is the characteristic function of $\sigma_h(y)$. Let p_h denote the natural injection of V_h into $F = L^2(\Omega)$ and r_h the mapping of $V = L^2(\Omega)$ into V_h which assigns the functions (35.65) to $v \in V$ such that

$$v_h(y) = (h_1 \cdots h_n)^{-1} \int_{\sigma_h(y)} v(x) \, dx \qquad (y \in \Omega_h). \tag{35.66}$$

Prove that (V_h, p_h, r_h, F) is a stable and convergent external approximation of $L^2(\Omega)$.

35.3. Let Ω_h^1 be the set of points $y \in h\mathbf{Z}$ such that $\sigma_h(y)$ intersects Ω; define V_h as the space of functions

$$v_h(x) = \sum_{y \in \Omega_h{}^1} v_h(y) w_{h,y}(x) \qquad \text{[cf. (35.33)]} \tag{35.67}$$

equipped with the inner product (35.34). Take $F = (L^2(\Omega))^{n+1}$ and the operator p_h

$$v_h \mapsto (v_h, \delta_{h,1} v_h, \ldots, \delta_{h,n} v_h)|_\Omega \tag{35.68}$$

where $|_\Omega$ denotes, as usual, the restriction (of each component) to Ω. Suppose that Ω is bounded and regular enough so that there is a continuous extension mapping ε_Ω: $H^1(\Omega) \to H^1(\mathbf{R}^n)$ (cf. Appendix to Sect. 26). Take $r_h v$, $v \in H^1(\Omega)$, to be the step function (35.67) such that

$$v_h(y) = (h_1 \cdots h_n)^{-1} \int_{\sigma_h(y)} \varepsilon_\Omega v(x)\, dx \qquad (y \in \Omega_h^1). \tag{35.69}$$

Prove that (V_h, p_h, r_h, F) is a stable and convergent external approximation of $V = H^1(\Omega)$.

35.4. Let Ω be a bounded open subset of $\mathbf{R}^n$, lying on one side of its boundary Γ; the latter is assumed to be a C^1 hypersurface. Exploit the approximation of $H^1(\Omega)$ described in Exercise 35.4 to obtain an approximation of the solution of the *Neumann problem* in Ω, for $-\Delta + \lambda$ [$\lambda > 0$; see (37.1)–(37.2) and §37.1].

36

Gårding's Inequality. Dirichlet Problem for Higher Order Elliptic Equations

In this section we show how the variational method can be used to obtain weak solutions to the Dirichlet problem for higher order strongly elliptic equations. Basic ingredients will be an open subset Ω of the Euclidean space $\mathbf{R}^n$, which need not be bounded but which we shall suppose so, in order to streamline the statements, and an elliptic differential operator in Ω, of *even* order $2m > 0$,

(36.1) $$P(x, D) = \sum_{|\alpha| \leq 2m} a_\alpha(x) D^\alpha.$$

Here $\alpha = (\alpha_1, \ldots, \alpha_n)$ is an n-tuple of integers $\geqq 0$; $|\alpha| = \alpha_1 + \cdots + \alpha_n$, and $D^\alpha = D_1^{\alpha_1} \cdots D_n^{\alpha_n}$, $D_j = -\sqrt{-1}\, \partial/\partial x^j$. We assume that the coefficients a_α are C^∞ functions in Ω, and that the operator $P(x, D)$ is *uniformly strongly elliptic* in Ω. This is a condition bearing on its *principal symbol* (Definition 19.1),

(36.2) $$P_{2m}(x, \xi) = \sum_{|\alpha| = 2m} a_\alpha(x) \xi^\alpha.$$

The condition is that, for a suitable constant $c_0 > 0$,

(36.3) $$\operatorname{Re} P_{2m}(x, \xi) \geqq c_0 |\xi|^{2m}, \qquad \forall x \in \Omega, \quad \forall \xi \in \mathbf{R}_n$$

[such a positivity condition requires that the order of $P(x, D)$ be even]. We shall also make the following hypothesis:

(36.4) *The coefficients a_α and all their derivatives of order $\leqq m$ are bounded functions in the whole of Ω.*

We shall now state and prove the main result of this section, the celebrated *Gårding's inequality*:

THEOREM 36.1. *Under the preceding hypotheses, there are two positive constants* C_o, C_1 *such that*

$$\|u\|_m^2 \leqq C_o \operatorname{Re}(P(x, D)u, u)_0 + C_1\|u\|_{m-1}^2, \qquad \forall u \in C_c^\infty(\Omega). \tag{36.5}$$

We have denoted by $(\, , \,)_k$ and $\| \; \|_k$ the inner product and the norm in the Sobolev space $H^k(\mathbf{R}^n)$ [or in the Sobolev space $H^k(\Omega)$].

Proof. Let x_o be an arbitrary point of Ω. By virtue of (36.3) we have

$$c_o|\xi|^{2m}|\hat{u}(\xi)|^2 \leqq \operatorname{Re}\{P_{2m}(x_o, \xi)|\hat{u}(\xi)|^2\}, \tag{36.6}$$

where $\hat{u}$ stands for the Fourier transform of $u \in C_c^\infty(\Omega)$. We derive at once from (36.6) and from Plancherel's formula, that

$$c_o\|u\|_m^2 \leqq \operatorname{Re}(P_{2m}(x_o, D)u, u)_0 + c_o\|u\|_{m-1}^2. \tag{36.7}$$

Suppose now that the support of u lies in the open ball

$$B_r(x_0) = \{x \in \mathbf{R}^n; \; |x - x_0| < r\}.$$

There is a constant $B_0 > 0$, depending only on the mth first partial derivatives of the coefficients a_α, $|\alpha| = 2m$, such that

$$|(P_{2m}(x, D)u, u)_0 - (P_{2m}(x_0, D)u, u)_0| \leqq B_0 r\|u\|_m^2 + B_0\|u\|_m \|u\|_{m-1}. \tag{36.8}$$

In order to prove (36.8) it suffices to consider the case where $P_{2m}(x, \xi) = a(x)\xi^\alpha$, $|\alpha| = 2m$. It is clear that

$$a(x)D^\alpha = D^{\alpha'}a(x)D^{\alpha''} + \sum_{\beta', \beta''} D^{\beta'}a_{\beta', \beta''}(x)D^{\beta''}, \tag{36.9}$$

where $|\alpha'| = |\alpha''| = m$, and where the summation ranges over a set of pairs of n-tuples β', β'' of length at most m and $m - 1$, respectively. Consequently,

(36.10)

$$|(\{a(x) - a(x_o)\}D^\alpha u, u) \pm (\{a(x) - a(x_o)\}D^{\alpha''}u, D^{\alpha'}u)| \leqq B_1\|u\|_m\|u\|_{m-1},$$

where the constant B_1 depends solely on the $a_{\beta', \beta''}$, which are suitable derivatives of $a(x)$. But if the gradient of $a(x)$ is bounded in Ω, we see that, for a suitable constant B_2 depending only on the bound of grad a in Ω, we have

$$|(\{a(x) - a(x_o)\}D^\alpha u, u)| \leqq B_2 r\|u\|_m^2 + B_1\|u\|_m\|u\|_{m-1}, \tag{36.11}$$

which implies what we want, that is, (36.8).

If we combine (36.7) and (36.8), and choose $r \leqq c_o/2B_o$, we obtain

$$\tfrac{1}{2}c_o \|u\|_m^2 \leqq \mathrm{Re}(P_{2m}(x, D)u, u)_0 + B_3 \|u\|_m \|u\|_{m-1}, \quad \forall u \in C_c^\infty(B_r(x_o)). \tag{36.12}$$

In view of the hypothesis (36.4) and by an argument similar to that which led to (36.10), we see that there is a constant $B_4 > 0$ such that

$$|(\{P(x, D) - P_{2m}(x, D)\}u, u)_0| \leqq B_4 \|u\|_m \|u\|_{m-1}, \quad \forall u \in C_c^\infty(\Omega).$$

If we combine this with (36.12) and choose the constant B_5 appropriately, we obtain

$$c_0 \|u\|_m^2 \leqq 2\,\mathrm{Re}(P(x, D)u, u)_0 + B_5 \|u\|_m \|u\|_{m-1}, \quad \forall u \in C_c^\infty(B_r(x_0)). \tag{36.13}$$

It should be emphasized that the constants c_0 and B_5 are independent of x_0.

The last step consists of "patching together" the local estimates (36.13). In order to do this, we cover the closure of Ω, $\bar{\Omega}$, with a finite number of open balls $B_r(x_i)$, $i = 1, \ldots, J$, and introduce an equal number of nonnegative test functions ϕ_i $(i = 1, \ldots, J)$ with the following properties: For each i, supp $\phi_i \subset B_r(x_i)$; $\phi_1^2 + \cdots + \phi_J^2 = 1$ in Ω (thus $0 \leqq \phi_i \leqq 1$ for every i). We consider then an arbitrary test function u in Ω and apply (36.13) to each $\phi_i u$ (in place of u). We add the corresponding inequalities and observe that

$$|(P(x, D)(\phi_i u) - \phi_i P(x, D)u, \phi_i u)_0| \leqq B_6 \|u\|_{m-1} \|u\|_m.$$

Finally, by using the fact that, given any $\eta > 0$,

$$\|u\|_m \|u\|_{m-1} \leqq \tfrac{1}{2}\left(\eta \|u\|_m^2 + \frac{1}{\eta} \|u\|_{m-1}^2\right),$$

and by choosing η small enough, one easily obtains (36.5). Q.E.D.

Remark 36.1. The hypotheses in Theorem 34.1 could have been relaxed in several ways. For instance, the result remains valid even when the open set Ω is *unbounded.* In this case we must exploit the uniformity in the hypotheses (36.3)–(36.4) with greater care: We must construct a partition of unity, $\{\phi_i^2\}$, as in the end of the proof, but now *infinite*, though having *uniform* properties (for instance, consisting of the translates at the points of a suitable lattice of a fixed test function). Also the regularity requirements on the coefficients of $P(x, D)$ could have been made less stringent: By inspection of the proof one sees easily that all one needs is that the coefficient of D^α in $P(x, D)(|\alpha| \leqq 2m)$ have $\sup(|\alpha| - m, 0)$ derivatives in $L^\infty(\Omega)$.

Remark 36.2. We have stated and proved a global version of Gårding's inequalities which is well suited for the study of boundary value problems. But it is clear that we could have as well stated and proved a local version: If we drop the uniformity requirements in (36.3)–(36.4), we could have proved (36.5) for all u with support in any given relatively compact open subset of Ω, Ω', and with constants C_o, C_1 depending on Ω'.

Remark 36.3. We could have replaced the norm $\| \; \|_{m-1}$ on the right-hand side of (36.5) by any norm $\| \; \|_j$, $j < m$, in particular by $\| \; \|_0$. It suffices to observe that, given any $\eta > 0$, there is a constant $C(\eta) > 0$ such that

$$(1 + |\xi|^2)^{m-1} \leqq \eta(1 + |\xi|^2)^m + C(\eta), \qquad \forall \xi \in \mathbf{R}_n,$$

hence

$$\|u\|_{m-1}^2 \leqq \eta \|u\|_m^2 + C(\eta)\|u\|_0^2, \qquad \forall u \in C_c^\infty. \tag{36.14}$$

By choosing $\eta \leqq 1/2C_1$ and increasing C_o we derive from (36.5) and (36.14):

$$\|u\|_m^2 \leqq C_o \operatorname{Re}(P(x, D)u, u)_0 + C_1'\|u\|_0^2, \qquad \forall u \in C_c^\infty(\Omega). \tag{36.15}$$

We shall now make the additional hypothesis that (36.15) is valid with $C_1' = 0$:

$$\|u\|_m^2 \leqq C_o \operatorname{Re}(P(x, D)u, u)_0, \qquad \forall u \in C_c^\infty(\Omega). \tag{36.16}$$

Remark 36.4. If $P(x, D)$ satisfies (36.15), $P(x, D) + \lambda$ satisfies (36.16) as soon as $\operatorname{Re} \lambda$ is sufficiently large.

We observe then that

$$(P(x, D)u, v)_0 \tag{36.17}$$

can be extended as a sesquilinear form on $H_0^m(\Omega) \times H_0^m(\Omega)$ and that (36.16) implies that this form is *coercive* (Definition 23.1). We may apply the Lax–Milgram lemma (Lemma 23.1) and derive (cf. Theorem 23.2)

THEOREM 36.2. *If (36.16) holds, the operator $P(x, D)$ defines an isomorphism of $H_0^m(\Omega)$ onto $H^{-m}(\Omega)$.*

Remark 36.5. It is clear that we could have assumed the coercivity of (36.17) instead of formulating the hypothesis (36.16). But "in practice" it is easier to check whether (36.16) holds, rather than whether the form under study is coercive.

Theorem 36.2 enables us to show that the (inhomogeneous) Dirichlet problem, relative to Ω and $P(x, D)$, has a unique solution. But let us first state this problem.

As usual we shall make rather stringent regularity assumptions on Ω: We shall assume that its boundary Γ is a smooth (that is, C^∞) hypersurface and that Ω lies on one side of it. We refer now to the theory of traces developed at the end of Sect. 26. We introduce the traces of order $j = 0, 1, \ldots, m-1$, on Γ, defined in (26.20), and the "multiple" trace $\tilde{\gamma}$ considered in Theorem 26.9:

$$\tilde{\gamma} : u \mapsto (\gamma^0(u), \ldots, \gamma^{m-1}(u)), \qquad \gamma^j(u) = \frac{\partial^j u}{\partial \nu^j}\Big|_\Gamma .$$

The inhomogeneous Dirichlet problem we consider here is that of finding u in $H^m(\Omega)$ satisfying

(36.18) $\quad P(x, D)u = f \in H^{-m}(\Omega) \qquad (\text{in } \Omega),$

(36.19) $\quad \gamma^j(u) = g^j \in H^{m-j-1/2}(\Gamma) \qquad (\text{on } \Gamma), \qquad j = 0, \ldots, m-1.$

It suffices to combine Theorem 36.2 with Theorems 26.5 and 26.9 to obtain

THEOREM 36.3. *If (36.16) holds,*

(36.20) $$u \mapsto (P(x, D)u, \tilde{\gamma}(u))$$

is an isomorphism of $H^m(\Omega)$ *onto* $H^{-m}(\Omega) \times \prod_{j=0}^{m-1} H^{m-j-1/2}(\Gamma)$.

Exercises

36.1. Let $P_{2m}(x, \xi)$ be a homogeneous polynomial of degree $2m$ in the variables $\xi_1, \ldots, \xi_n$, with coefficients in $C^\infty(\bar{\Omega})$ [cf. (36.2)]. Let N be an arbitrary nonzero vector in $\mathbf{R}_n$ and let Π_N be a hyperplane of $\mathbf{R}_n$ supplementary to N. Give a necessary and sufficient condition on the roots of the polynomial in the variable z, $P_{2m}(x, \theta + zN)$, as x ranges over $\bar{\Omega}$ and θ ranges over Π_N, in order that (36.3) be verified.

36.2. Let Ω be a bounded open subset of $\mathbf{R}^n$, $P(x, D)$ a uniformly strongly elliptic differential operator of order $2m$ in Ω [thus (36.3) holds]. Assume, furthermore, that all the coefficients of $P(x, D)$ belong to $\mathscr{B}^\infty(\Omega)$ (i.e., they are complex C^∞ functions in Ω, and all their derivatives, of any order, are bounded in Ω). Prove that, given any nonnegative integer s, there are two positive constants $C_0(s)$, $C_1(s)$ such that

(36.21) $\quad \|u\|_{m+s}^2 \leqq C_0(s) \operatorname{Re}(P(x, D)u, u)_s + C_1(s)\|u\|_s^2, \qquad \forall u \in C_c^\infty(\Omega).$

36.3. Let Ω and $P(x, D)$ be as in Exercise 36.2, but suppose now that all the coefficients of $P(x, D)$ can be extended as C^∞ functions, say with compact support, in $\mathbf{R}^n$. Prove that, for a suitable choice of $C_0(s)$, $C_1(s)$, the estimate (36.21) can be extended to all real values of s.

36.4. Let Ω be an arbitrary bounded open subset of $\mathbf{R}^n$, $P(x, D)$ an elliptic operator of order m (Definition 19.4) with C^∞ coefficients in Ω. Prove that every point $x_0 \in \Omega$ has an open neighborhood $U \subset \Omega$ such that, for some constant $C > 0$,

$$\|u\|_m \leqq C\|P(x, D)u\|_0, \qquad \forall u \in C_c^\infty(U). \tag{36.22}$$

Derive from the analogous estimate for the transpose of $P(x, D)$ that the equation

$$P(x, D)u = f \in H^{-m}(U) \qquad (\text{in } U) \tag{36.23}$$

has a solution $u \in L^2(U)$ and, in fact, that u can be taken to depend linearly and continuously on f.

36.5. Let Ω and $P(x, D)$ be as in Exercise 36.4. Prove that, for all positive real numbers s, there is an open neighborhood $U_s \subset \Omega$ of x_0 and a constant $C_s > 0$ such that

$$\|u\|_{m+s} \leqq C_s\|P(x, D)u\|_s, \qquad \forall u \in C_c^\infty(U_s). \tag{36.24}$$

From the estimate analogous to (36.24) for the transpose of $P(x, D)$ derive that, given any point x_0 of Ω and any real number s, there is an open neighborhood $V \subset \Omega$ of x_0 such that, for every $f \in H^s(V)$, there is a $u \in H^{s+m}(V)$ satisfying $P(x, D)u = f$ in V.

36.6. Let Ω and $P(x, D)$ be as in Exercise 36.4. Prove that, given any real number s and any compact subset K of Ω, there is a constant $C > 0$ such that

$$\|u\|_{m+s} \leqq C(\|P(x, D)u\|_s + \|u\|_{m-1+s}), \qquad \forall u \in C_c^\infty(K). \tag{36.25}$$

By combining this with Lemma 25.3 (Exercise 25.11) prove that, for an appropriate choice of U_s and C_s, the inequality (36.24) is also valid when s is a negative real number.

36.7. Let Ω and $P(x, D)$ be as in Exercise 36.4. Let u be a distribution with compact support in Ω, such that $f = P(x, D)u \in H^{s_0}$ ($s_0 \in \mathbf{R}$). Let s be a real number such that $u \in H^s$ and that $s - m \leqq s_0$. Apply the Friedrichs lemma (Lemma 25.4, Exercise 25.12) to derive from the inequality (36.25) (where $s + 1$ is substituted for $s + m$) that the functions $\rho_\varepsilon * u$ converge weakly in H^{s+1}, and therefore that $u \in H^{s_0+m}$.

By using cutoff functions derive from this the following result:

THEOREM 36.4. *Let $P(x, D)$ be an elliptic operator of order m, with C^∞ coefficients in the open subset Ω of $\mathbf{R}^n$.*

Every distribution u in Ω such that $P(x, D)u \in H_{\text{loc}}^{s_0}(\Omega)$ belongs to $H_{\text{loc}}^{s_0+m}(\Omega)$.

COROLLARY 36.1. *Every elliptic linear partial differential operator with C^∞ coefficients in Ω is hypoelliptic in Ω* (Definition 2.1).

36.8. Let $P(x, D)$ be an elliptic operator of order m, with C^∞ coefficients in the open set $\Omega \subset \mathbf{R}^n$. Let K be any compact subset of Ω. Prove, by exploiting the estimate (36.25), that the set of C^∞ functions h with support contained in K, satisfying the homogeneous equation $P(x, D)h = 0$, forms a finite-dimensional linear space.

36.9. Let $P(x, D)$ be a uniformly strongly elliptic differential operator of order $2m$, with C^∞ coefficients, in a bounded open subset Ω of $\mathbf{R}^n$. Assume that the coefficients of $P(x, D)$ and all their derivatives are bounded in Ω. Prove that, given any complex number ζ, the following is true:

(36.26) *the kernel of* $P(x, D) + \zeta$ *in* $H_0^m(\Omega)$ *is a finite* (perhaps, zero) *dimensional linear subspace* V_0, *and* $P(x, D) + \zeta$ *induces an isomorphism of the quotient space* $H_0^m(\Omega)/V_0$ *onto a subspace* V' *of* $H^{-m}(\Omega)$ *whose codimension is finite* (and possibly zero); V' *is the orthogonal* (for the duality) *of the kernel of the adjoint operator* $P(x, D)^* + \bar{\zeta}$.

Prove furthermore that the values of ζ for which dim V_0 or codim V' are at least one form a sequence converging to infinity in the complex plane, and all lying in a half-plane Re $\zeta \leqq C < +\infty$. [*Hint*: Use Rellich's lemma (see Exercise 25.10) and the basic properties of compact operators (see, e.g., [Y, Chap. X, §5, Theorem 2]).]

37

Neumann Problem and Other Boundary Value Problems (Variational Form)

Throughout this section Ω is a bounded open subset of $\mathbf{R}^n$, with a smooth boundary and lying on one side of it; the boundary is denoted by Γ.

Boundary value problems, other than the Dirichlet one, occur frequently in applications. For instance, one may want to study the Neumann problem, say for the equation

$$(\lambda - \Delta)u = f \qquad (\text{in } \Omega), \tag{37.1}$$

which is to say, the problem where the following boundary condition is imposed on the solution:

$$\frac{\partial u}{\partial \nu} = g \quad \text{on } \Gamma. \tag{37.2}$$

As before $\partial/\partial\nu$ stands for the partial differentiation along the normal to Ω, outwardly oriented. The reasons for studying the Neumann problem are obvious: For instance, one might want to study the distribution of temperature inside the body Ω when one only knows the heat flow through the boundary surface Γ, rather than the temperatures on Γ, as in the Dirichlet problem. One might also want to study "mixtures" of the Dirichlet and Neumann problems, such as the following one. Let Γ be represented as the union of two disjoint subsets Γ_0, Γ_1 (which we assume to be fairly regular). We may seek a solution of the "hybrid" problem:

$$(\lambda - \Delta)u = f \quad \text{in } \Omega, \tag{37.3}$$

$$u = g_0 \quad \text{on } \Gamma_0, \qquad \frac{\partial u}{\partial \nu} = g_1 \quad \text{on } \Gamma_1. \tag{37.4}$$

In this section we are going to show that, as long as we are satisfied with weak solutions and select our problems from a relatively restricted list, a unified treatment is available to us, generalizing the one described in Sect. 23 and based on application of the Lax–Milgram lemma (Lemma 23.1). Predictably, it requires that the problem be reformulated in terms of a suitable coercive sesquilinear form on the "right" Hilbert space, which in general, but not always (see 37.1b), is a closed linear subspace of $H^1(\Omega)$, intermediate between the latter and $H_0^1(\Omega)$—or between $H^m(\Omega)$ and $H_0^m(\Omega)$ if we deal with a differential operator of order $2m$.

Before embarking on the discussion of these variational boundary value problems, we must emphasize the fact that many kinds of boundary value problems of great importance are out of reach of the method we use here: nonvariational boundary value problems, which enter in the Lopatinski class of which a brief description is given in Sect. 38; problems like the *oblique derivative* one, to be found in §37.5, but where the oblique derivative is allowed to become tangent to the boundary at some point; or problems for certain operators, or systems of operators, which are *not* strongly elliptic, and are not stable by small perturbation of the same order. The foremost example of the latter is the so-called $\bar{\partial}$-Neumann problem, which is a boundary value problem for the overdetermined system $\partial u/\partial \bar{z}_j = f_j$ ($j = 1, \ldots, n$), of fundamental importance in the theory of holomorphic functions, of which nothing more will be said in this book.

37.1a The (Weak) Neumann Problem for $-\Delta + \lambda$, $\lambda > 0$

The sesquilinear form we are going to use is the same as in the Dirichlet problem:

$$a_\lambda(u, v) = \int_\Omega \{\lambda(u\bar{v}) + (\text{grad } u) \cdot (\text{grad } \bar{v})\}\, dx.$$

Since we assume $\lambda > 0$, $a_\lambda(u, v)$ is coercive (Definition 23.1) on $H^1(\Omega)$ [and therefore also on any closed linear subspace of $H^1(\Omega)$]. Let then $f \in L^2(\Omega)$, $g \in L^2(\Omega)$. We note that

$$v \mapsto \int_\Omega f\bar{v}\, dx + \int_\Gamma g\bar{v}\, d\sigma \tag{37.5}$$

defines a continuous antilinear functional on $H^1(\Omega)$ [in the integral over the boundary Γ, v stands for $\gamma(v)$, the trace of v on Γ; see Sect. 26]. The Lax–Milgram lemma implies that there is a unique element u of $H^1(\Omega)$ such that

$$a_\lambda(u, v) = \int_\Omega f\bar{v}\, dx + \int_\Gamma g\bar{v}\, d\sigma, \qquad \forall v \in H^1(\Omega). \tag{37.6}$$

If we take $v \in C^\infty(\overline{\Omega})$, we may integrate by parts in the left-hand side of (37.6) and rewrite it as

$$\int_\Omega u(\lambda - \Delta)\bar{v}\, dx + \int_\Gamma u \frac{\partial \bar{v}}{\partial \nu}\, d\sigma \tag{37.7}$$

[the definition of the surface measure $d\sigma$ on Γ is precisely such that (37.7) is the result of the integration by parts of $a_\lambda(u, v)$]. By equating (37.7) to the right-hand side of (37.6) and choosing v in $C_c^\infty(\Omega)$, since the surface integral in (37.7) is then zero, we obtain (37.1). We return to arbitrary v in $C^\infty(\overline{\Omega})$, now taking (37.1) into account. Actually, we may proceed as follows.

Let $\{\overline{\Omega}_\varepsilon\}$ $(0 < \varepsilon < 1)$ be a one-parameter family of relatively compact open subsets of Ω, with smooth boundaries Γ_ε, say monotone increasing and exhausting Ω. We have

$$a_\lambda(u, v) = \lim_{\varepsilon \to +0} \int_{\Omega_\varepsilon} \{\lambda u\bar{v} + (\operatorname{grad} u) \cdot (\operatorname{grad} \bar{v})\}\, dx. \tag{37.8}$$

Since $f \in L^2(\Omega)$ we know (and it is not difficult to show) that u belongs to $H^2_{\text{loc}}(\Omega)$. We may therefore integrate by parts in the integrals over Ω_ε [in (37.8)] and thus obtain

$$a_\lambda(u, v) = \lim_{\varepsilon \to +0} \left(\int_{\Omega_\varepsilon} f\bar{v}\, dx + \int_{\Gamma_\varepsilon} \frac{\partial u}{\partial \nu}\, \bar{v}\, d\sigma \right). \tag{37.9}$$

In view of (37.1) this means that

$$\lim_{\varepsilon \to +0} \int_{\Gamma_\varepsilon} \frac{\partial u}{\partial \nu}\, \bar{v}\, d\sigma = \int_\Gamma g\bar{v}\, d\sigma. \tag{37.10}$$

This can be regarded as the meaning of (37.2).

In fact, if we choose the open sets Ω_ε carefully, we can find a diffeomorphism ϕ_ε of Γ_ε onto Γ and transfer the normal derivatives of u on Γ_ε into normal derivatives, on Γ, of a one-parameter family of functions $u_\varepsilon \in H^2(\Omega)$ which converge to u, say in $H^1(\Omega)$, and whose normal derivatives on Γ converge to g. At any rate, we may consider that we have obtained a weak solution of the Neumann problem (37.1)–(37.2) for $\lambda > 0$. If we strengthened the regularity requirements on the data f and g, we would be able to establish better regularity properties for the solution u and, if these were good enough, conclude that the boundary condition (37.2) is satisfied in a "classical" sense.

37.1b The Weak Neumann Problem for the Laplace Equation

It is clear that, when $\lambda = 0$, we cannot expect that the solution u of (37.1)–(37.2), if it exists, will be unique: The constant functions are evidently solutions of the homogeneous problem. The Dirichlet sesquilinear form

$$a(u, v) = \int_\Omega (\text{grad } u) \cdot (\text{grad } \bar{v})\, dx$$

is *not* coercive on $H^1(\Omega)$. The *associated seminorm* is the *Dirichlet integral* seminorm:

$$D(u) = \left(\int_\Omega |\text{grad } u|^2\, dx\right)^{1/2}, \tag{37.11}$$

whose null space (or kernel) is the linear subspace of $H^1(\Omega)$ consisting of the *locally constant* functions in Ω. Let us assume that Ω is connected, for the sake of simplicity. Then $D(u) = 0 \Rightarrow u = \text{const}$. Let us go to the quotient $\dot{H}^1(\Omega)$ of $H^1(\Omega)$ modulo the constant functions (which, of course, form a one-dimensional linear subspace, identifiable with $\mathbf{C}$); we shall denote by $\dot{u}$, $\dot{v}$, etc., the elements of this quotient space. By definition, so to speak, $\dot{u} \mapsto D(\dot{u})$ is a norm of $\dot{H}^1(\Omega)$ and turns it into a Hilbert space; we denote by $a(\dot{u}, \dot{v})$ the associated sesquilinear form, which is of course the inner product on $\dot{H}^1(\Omega)$, and also the transfer of $a(u, v)$ to the quotient. Needless to say, $a(\dot{u}, \dot{v})$ is coercive, and is therefore amenable to the Lax–Milgram treatment.

Now let f be an arbitrary function belonging to $L^2(\Omega)$, g one belonging to $L^2(\Omega)$. We may ask under which conditions it is true that the continuous antilinear functional (37.5) defines a continuous linear functional on the quotient space $\dot{H}^1(\Omega)$. The answer is obvious: It must be orthogonal to the constant functions; that is, it must satisfy

$$\int_\Omega f\, dx + \int_\Gamma g\, d\sigma = 0. \tag{37.12}$$

If this is satisfied, we may proceed as before (in §37.1a) and conclude that there exists a function $u \in H^1(\Omega)$, unique up to constants (if Ω is connected), satisfying (37.1)–(37.2) when $\lambda = 0$, in the weak sense described in §37.1a.

37.2 The "Hybrid" Problem for $-\Delta + \lambda (\lambda > 0)$

We refer here to the problem (37.3)–(37.4). We continue to deal with the sesquilinear form $a_\lambda(u, v)$. The important step is the choice of the subspace V

of $H^1(\Omega)$ on which we consider it. It is defined as follows: It is the closure in $H^1(\Omega)$ of the space of functions $\phi \in C^\infty(\bar{\Omega})$ which vanish in a neighborhood of Γ_0 (in $\bar{\Omega}$). In this case V is the space of functions belonging to $H^1(\Omega)$ which "vanish on Γ_0" [when $\Gamma_0 = \Gamma$, $V = H_0^1(\Omega)$]. Let $f \in L^2(\Omega)$, $g_1 \in L^2(\Gamma_1)$. By applying once again the Lax–Milgram lemma we see that there is a unique element u of V which satisfies

(37.13) $$a_\lambda(u, v) = \int_\Omega f\bar{v}\, dx + \int_{\Gamma_1} g_1 \bar{v}\, d\sigma, \qquad \forall v \in V.$$

By a reasoning quite analogous to the one used for the Neumann problem we may interpret (37.13) as meaning that

(37.14) $$(\lambda - \Delta)u = f \quad \text{in } \Omega,$$

(37.15) $$\frac{\partial u}{\partial \nu} = g_1 \quad \text{on } \Gamma_1.$$

The fact that u belongs to V means that $u = 0$ on Γ_0—in the usual generalized sense. If we wish to satisfy an inhomogeneous condition

(37.16) $$u = g_0 \quad \text{on } \Gamma_0,$$

it is convenient to make some rather strong regularity assumptions about g_0 : For instance, we may assume that g_0 is the restriction to Γ_0 of a function on Γ which belongs to $H^{3/2}(\Gamma)$. For then we may find $h \in H^2(\Omega)$ which is equal to g_0 on Γ_0 (precisely, whose trace on Γ is equal to g_0 on Γ_0). Let then w be the unique solution in V of the problem:

(37.17) $$(\lambda - \Delta)w = f - (\lambda - \Delta)h \quad \text{in } \Omega,$$

(37.18) $$\frac{\partial w}{\partial \nu} = g_1 - \frac{\partial h}{\partial \nu} \quad \text{on } \Gamma_1.$$

It is clear that $u = w + h$ is a solution of (37.14)–(37.15)–(37.16). The question of its uniqueness is more delicate, since we must state precisely within what framework we ask this question. For instance, if g_1 is sufficiently regular, say g_1 is the restriction to Γ_1 of an element of $H^{1/2}(\Gamma)$, we may hope to prove that the solution u belongs to $H^2(\Omega)$ (at least in some cases), and we might ask whether such a solution is then unique—in $H^2(\Omega)$. It will of course suffice to prove that any element h of $H^2(\Omega)$ which satisfies $h = 0$ on Γ_0 actually belongs to V, i.e., can be approximated in $H^1(\Omega)$ by functions which are C^∞ in the closure of Ω and vanish identically in a neighborhood of Γ_0 (in $\bar{\Omega}$). This will certainly be possible in a number of cases (recognizable by sufficiently strong regularity assumptions on Γ_0 and Γ_1).†

† On the subject of the problem (37.3)–(37.4), it is suggested that the reader look at Exercises 37.2 to 37.4 (where the case $\lambda = 0$ is also discussed).

37.3 Weak Neumann Problems for More General Second-Order Elliptic Equations

We might want to know how the Neumann problem should be posed for elliptic equations more general than $\lambda - \Delta$, and first of all for a differential operator of the kind (23.6), that is,

$$L = -\sum_{j,k=1}^{n} \frac{\partial}{\partial x^k} a^{jk}(x) \frac{\partial}{\partial x^j} + \sum_{j=1}^{n} b^j(x) \frac{\partial}{\partial x^j} + c(x), \tag{37.19}$$

assuming that the associated sesquilinear form,

$$a(u, v) = \sum_{j,k=1}^{n} \int_\Omega a^{jk} \frac{\partial u}{\partial x^j} \frac{\partial \bar{v}}{\partial x^k} \, dx + \sum_{j=1}^{n} \int_\Omega b^j \frac{\partial u}{\partial x^j} \bar{v} \, dx + \int_\Omega c u \bar{v} \, dx, \tag{37.20}$$

is coercive on $H^1(\Omega)$. If we proceed exactly as in the case of the operator $\lambda - \Delta$, we reach the conclusion that, given any $f \in L^2(\Omega)$ and any $g \in L^2(\Omega)$, there is a unique function u in $H^1(\Omega)$ satisfying

$$a(u, v) = \int_\Omega f\bar{v} \, dx + \int_\Gamma g\bar{v} \, d\sigma, \qquad \forall v \in H^1(\Omega). \tag{37.21}$$

It remains to interpret this relation. The implication concerning the interior is clear [and is derived by choosing $v \in C_c^\infty(\Omega)$]:

$$Lu = f \quad \text{in } \Omega. \tag{37.22}$$

The implication concerning the boundary is better understood if we assume that the coefficients a^{jk} are real-valued and the matrix $A = (a^{jk})$ is symmetric. It is then positive definite because of the hypothesis that $a(u, v)$ is coercive. By taking v in $C^\infty(\bar{\Omega})$ and integrating by parts in the "leading part" $\tilde{a}(u, v)$ of the form $a(u, v)$, we obtain

$$\begin{aligned} \tilde{a}(u, v) &= \int_\Omega (\text{grad } u) \cdot (A \text{ grad } \bar{v}) \, dx \\ &= \int_\Omega \text{div}\{u(A \text{ grad } \bar{v})\} \, dx - \int_\Omega u \, \text{div} \, (A \text{ grad } \bar{v}) \, dx, \end{aligned}$$

hence, by the standard Stokes-like formula (Lemma 10.1),

$$a(u, v) = \int_\Omega u\overline{L^* v} \, dx + \int_\Gamma u\{\nu \cdot (A \text{ grad } \bar{v})\} \, d\sigma, \tag{37.23}$$

where L^* denotes the *formal adjoint* of L, and ν the unit vector along the outer normal to Γ. We have

$$\nu \cdot A \text{ grad } v = ({}^t A \nu) \cdot \text{grad } v = \frac{\partial v}{\partial \nu_a'},$$

where we use the notation

$$\frac{\partial}{\partial \nu_a'} = \sum_{k=1}^{n} \left(\sum_{j=1}^{n} a^{jk}(x) \nu^j \right) \frac{\partial}{\partial x^k}. \tag{37.24}$$

Let us remark that if u were sufficiently differentiable, up to the boundary, we could have integrated by parts "the other way," and obtained

$$a(u, v) = \int_{\Omega} (Lu)\bar{v}\, dx + \int_{\Gamma} \frac{\partial u}{\partial \nu_a} \bar{v}\, d\sigma, \tag{37.25}$$

where now we use the notation

$$\frac{\partial}{\partial \nu_a} = \sum_{j=1}^{n} \left(\sum_{k=1}^{n} a^{jk}(x) \nu^k \right) \frac{\partial}{\partial x^j}. \tag{37.26}$$

In (37.24) and in (37.26) it is understood that x lies on the boundary Γ. Then $\partial/\partial \nu_a$ is the normal differentiation on Γ, defined by the metric

$$\sum_{j,k} a^{jk}(x)\, dx^j\, dx^k$$

(this metric is defined on $\bar{\Omega}$). In the case where the operator (37.19) is equal to $-\Delta$, with Δ the Laplace operator in n variables, the matrix $A = (a^{jk})$ is equal to I, the $n \times n$ identity matrix. In the general case, we recall that, by the coercivity hypothesis on the form $a(u, v)$, the matrix A is *positive definite*.

By reasoning as in the case of the Neumann problem, we derive from (37.22) and (37.23) that

$$\frac{\partial u}{\partial \nu_a} = g \quad \text{on } \Gamma, \tag{37.27}$$

in a weak sense (and in a classical sense, if sufficiently strong regularity hypotheses are made). The conjunction of (37.22) and (37.27) is the Neumann problem naturally associated with the differential operator L.

37.4 Weak Neumann Problems for Higher Order Elliptic Equations

We shall now describe briefly how the Neumann problem is posed for elliptic equations of order $2m$, with m arbitrary. We shall assume that the differential operator we propose to study is given in the variational form, that is, in the form

$$P(x, D) = \sum_{|p|, |q| \leq m} (-1)^{|q|}\, \partial^q a_{p,q}(x)\, \partial^p, \tag{37.28}$$

where $\partial^p = (\partial/\partial x^1)^{p_1} \cdots (\partial/\partial x^n)^{p_n}$, $|p| = p_1 + \cdots + p_n$, and where the coefficients $a^{p,q}(x)$ belong to $L^\infty(\Omega)$ (and might be required to have stronger regularity properties). In the form (37.28) the operator $P(x, D)$ is naturally associated with a continuous sesquilinear functional on $H^m(\Omega) \times H^m(\Omega)$, namely

$$a(u, v) = \sum_{|p|, |q| \leqq m} \int_\Omega a_{p,q}(x)\, \partial^p u\, \overline{\partial^q v}\, dx, \tag{37.29}$$

and we formulate the hypothesis that $a(u, v)$ is *coercive* [on $H^m(\Omega)$].

We consider then the following continuous linear functional on $H^m(\Omega)$:

$$F(v) = \int_\Omega fv\, dx + \sum_{j=0}^{m-1} \int_\Gamma g_j \left(\frac{\partial}{\partial v}\right)^j v\, d\sigma, \tag{37.30}$$

where, say, $f \in L^2(\Omega)$, $g_j \in L^2(\Gamma)(0 \leqq j \leqq m-1)$. By virtue of the Lax–Milgram lemma we obtain that there is a unique u in $H^m(\Omega)$ satisfying

$$a(u, v) = F(\bar{v}) \qquad \text{for every} \quad v \in H^m(\Omega). \tag{37.31}$$

The problem for us now is to interpret Eq. (37.31). In the interior we have, obviously,

$$P(x, D)u = f \qquad (\text{in } \Omega). \tag{37.32}$$

As for the boundary conditions it is convenient to assume as much regularity on the solution u as we need: We may take u in $H^{2m}(\Omega)$ [and v, say in $C^\infty(\overline{\Omega})$]. Of course there remains the question of finding out when the assumption is satisfied; we do not discuss it here, however.

In order to write down explicitly the boundary conditions implied by (37.31) we shall, as before, integrate by parts—which amounts to using a *Green formula* for the differential operator $P(x, D)$ [given in the variational form (37.28)] and the $m - 1$ first normal derivatives on v.

Let χ_Ω denote the characteristic function of the set Ω (which, we recall, is bounded). Let us extend v (in any admissible manner) as a C^∞ function in $\mathbf{R}^n$. Then

$$\begin{aligned} a(u, v) &= \sum_{|p|, |q| \leqq m} \langle \chi_\Omega a_{p,q}\, \partial^p u, \partial^q \bar{v} \rangle \\ &= \sum_{p,q} (-1)^{|q|} \sum_{r \leqq q} \binom{q}{r} \langle (\partial^r \chi_\Omega)\, \partial^{q-r}\{a_{p,q}\, \partial^p u\}, \bar{v} \rangle \\ &= \langle \chi_\Omega P(x, D)u, \bar{v} \rangle \\ &\quad + \sum_{0 < |r| \leqq m} \sum_{p} \sum_{\substack{q \geqq r \\ |p|, |q| \leqq m}} (-1)^{|q|} \binom{q}{r} \langle \partial^r \chi_\Omega\, \partial^{q-r}\{a_{p,q}\, \partial^p u\}, \bar{v} \rangle, \end{aligned}$$

where we have used the notation $r \leqq q$ to mean $r_j \leqq q_j$ for every $j = 1, \ldots, n$. Now,

$$\text{(37.33)} \qquad \text{if} \quad |r| > 0, \qquad \partial^r \chi_\Omega = c_r(x) \left(\frac{\partial}{\partial \nu}\right)^{|r|-1} \delta_\Gamma,$$

where $c_r(x)$ is a C^∞ function on Γ (cf. Exercise 37.6) and δ_Γ stands for the "surface Dirac measure" associated with Γ,

$$\text{(37.34)} \qquad \langle \delta_\Gamma, \phi \rangle = \int_\Gamma \phi \, d\sigma, \qquad \phi \in C_c^\infty(\mathbf{R}^n).$$

Applying once more Leibniz's formula, we obtain

$$\text{(37.35)} \quad a(u, v) = \int_\Omega \bar{v} P(x, D)u \, dx + \sum_{j=0}^{m-1} \int_\Gamma \left\{ \left(\frac{\partial}{\partial \nu}\right)^j \bar{v} \right\} \mathcal{N}_j(x, D)u \, d\sigma,$$

where the $\mathcal{N}_j$ are differential operators, defined in the neighborhood of Γ, of orders $\leqq 2m - 1 - j$, respectively, and which can be easily calculated (cf. Exercise 37.6).

If we return then to the expression (37.30) of $F(v)$, we see that the boundary conditions implied by the equation (37.31) are

$$\text{(37.36)} \qquad \mathcal{N}_j(x, D)u = g_j \quad \text{on } \Gamma, \qquad j = 0, \ldots, m - 1.$$

In the absence of sufficient regularity (up to the boundary) on the part of the solution u, we can only say that the boundary conditions (37.12) are satisfied in a weak sense.

Remark 37.1. It is very important to observe that the boundary operators $\mathcal{N}_j$ in (37.35) and (37.36) do not only depend on Γ and on $P(x, D)$ but also on the variational form of $P(x, D)$ used, in our case (37.28). Several choices of the variational form are possible for a given differential operator $P(x, D)$, and they may lead to different boundary conditions. This remark applies also to the second-order Neumann problems described in §37.3, and even to simple operators such as $\lambda - \Delta$ (cf. Exercise 37.5).

Remark 37.2. Underlying the treatment of the Neumann problem in the present section is *the generalized Green formula*,

$$\text{(37.37)} \quad \int_\Omega u \overline{P(x, D)^* v} \, dx = \int_\Omega \{P(x, D)u\} \bar{v} \, dx + \sum_{j=0}^{2m-1} \int_\Gamma \{\mathcal{N}_j(x, D)u\} \left(\frac{\partial}{\partial \nu}\right)^j \bar{v} \, d\sigma,$$

valid for u and v, say, in $H^{2m}(\Omega)$. In (37.37), $P(x, D)^*$ stands for the *formal adjoint* of $P(x, D)$ [of which that equation is, in fact, a definition, if we take u and v in $C_c^\infty(\Omega)$], while the $\mathcal{N}_j (j = 0, \ldots, 2m - 1)$ are $2m$ operators defined in a neighborhood of Γ (and which can be computed by integration by parts).

37.5 Radiation Problems and Oblique-Derivative Problems for $-\Delta + \lambda$

Another interesting problem which is amenable to the variational method is that of solving the equation

(37.38) $$(\lambda - \Delta)u = f \quad \text{in } \Omega,$$

under a "radiation" boundary condition:

(37.39) $$\frac{\partial u}{\partial \nu} + \alpha u = g \quad \text{on } \Gamma,$$

where α is a *nonnegative* function belonging to $L^\infty(\Gamma)$ (f and g can be taken as in the Neumann problem). In this case we consider the sesquilinear form

(37.40) $$a(u, v) = \int_\Omega \{\lambda u\bar{v} + (\operatorname{grad} u) \cdot (\operatorname{grad} \bar{v})\}\, dx + \int_\Gamma \alpha\, u\bar{v}\, d\sigma,$$

on $H^1(\Omega)$. Since the trace of an element of $H^1(\Omega)$ belongs to $H^{1/2}(\Gamma)$ (Theorem 26.2), $a(u, v)$ is clearly continuous on $H^1(\Omega) \times H^1(\Omega)$. Since $\alpha \geqq 0$ it is coercive, from which arise the existence and uniqueness of a solution to the variational problem

(37.41) $$a(u, v) = \int_\Omega f\bar{v}\, dx + \int_\Gamma g\bar{v}\, d\sigma.$$

Assuming sufficient regularity on u (or otherwise interpreting it in a weak sense), we obtain, by Green formula,

(37.42) $$\int_\Omega (\lambda - \Delta)u\bar{v}\, dx + \int_\Gamma \left(\frac{\partial u}{\partial \nu} + \alpha u\right)\bar{v}\, d\sigma = \int_\Omega f\bar{v}\, dx + \int_\Gamma g\bar{v}\, d\sigma,$$

which shows that we have solved (37.38)–(37.39)—in a weak sense.

Next we consider a certain type of oblique-derivative problems. We consider a smooth, real vector field T on the boundary Γ, also a smooth, real function α on Γ, *which never vanishes in* Γ. We shall make the following assumption:

(37.43) $\alpha^{-1}T + T^*\alpha^{-1}$ *is a smooth function, everywhere nonnegative on* Γ.

Here T^* means the adjoint of T for the natural Hilbert space structure on $L^2(\Gamma)$ (that defined by means of the measure $d\sigma$ or, if one prefers, by the metric induced by the Euclidean metric on $\mathbf{R}^n$). Under this hypothesis the variational method applies to the *oblique-derivative* problem

(37.44) $$(\lambda - \Delta)u = f \quad \text{in } \Omega,$$

(37.45) $$\left(\alpha \frac{\partial}{\partial \nu} + T\right)u = g \quad \text{on } \Gamma.$$

First of all, after division of Eq. (37.45) by α we may assume that $\alpha \equiv 1$. We consider then the following sesquilinear form.

$$a(u, v) = \int_\Omega \{u\bar{v} + (\text{grad } u) \cdot (\text{grad } \bar{v})\}\, dx + \langle Tu, \bar{v}\rangle_\Gamma, \tag{37.46}$$

where $\langle\ ,\ \rangle_\Gamma$ stands for the duality bracket between $H^{1/2}(\Gamma)$ (to which the trace of v belongs) and $H^{-1/2}(\Gamma)$ (to which Tu belongs). It is seen at once (by virtue of Theorem 26.2) that (37.46) is continuous on $H^1(\Omega) \times H^1(\Omega)$, and it follows from (37.43) (where now $\alpha = 1$) that it is coercive on $H^1(\Omega)$. The Lax–Milgram lemma implies that the problem (37.44)–(37.45) has one, and only one, (weak) solution.

By taking T to be not a vector field but a first-order differential operator on Γ, submitted to (37.43), condition (37.45) would generalize the radiation condition (37.39), as well as the preceding oblique-derivative boundary condition.

Exercises

37.1. Let Ω be a bounded domain (i.e., connected and simply connected open set) in the plane $\mathbf{R}^2$ whose boundary is a C^∞ curve Γ. Let us denote by $\partial/\partial t$ the unit vector field tangent to Γ oriented *clockwise*. Let u be a C^∞ function in $\bar{\Omega}$, harmonic in Ω. Show that if v is another C^∞ function in $\bar{\Omega}$, such that u and v are conjugate harmonics [see (32.12)] in Ω, then v is necessarily a solution of the Neumann problem

$$\Delta v = 0 \quad \text{in } \Omega, \qquad \frac{\partial v}{\partial \nu} = \frac{\partial u}{\partial t} \quad \text{on } \Gamma \tag{37.47}$$

(the normal differentiation $\partial/\partial\nu$ is oriented *outward*).

37.2. Let Ω be a bounded open set in $\mathbf{R}^n$, $n \geqq 2$, with C^∞ boundary Γ, lying on one side of Γ. Let x_0 be an arbitrary point of $\bar{\Omega}$. Prove that the set of C^∞ functions in $\bar{\Omega}$ which vanish in some neighborhood of x_0 is dense in $H^1(\Omega)$. [*Hint*: Let $B_R(x_0)$ be an open ball centered at x_0 with very large radius R; show, by a duality argument, that $H_0^1(B_R(x_0)\backslash\{x_0\}) = H_0^1(B_R(x_0))$ and derive from this the sought result.] Deduce from the preceding result that, when Γ_0 reduces to a single point, the hybrid problem (37.3)–(37.4) reduces to the Neumann problem.

37.3. Let B_R denote the open ball centered at the origin, with radius $R > 0$, in the space $\mathbf{R}^n (n \geqq 2)$. Write $x' = (x^1, \ldots, x^{n-1})$ and let $g(x')$ be an L^∞ function in $\mathbf{R}^{n-1}$. Prove that

$$C_c^\infty(B_R) \ni \phi \mapsto \int_{\mathbf{R}^n} g(x')\phi(x', 0)\, dx'$$

extends as a continuous linear functional on $H_0^1(B_R)$.

Next, consider an open set Ω as in Exercise 37.2 and let Γ_0 be an *open* nonempty subset of its boundary Γ. Prove that there is a continuous linear functional on $H^1(\Omega)$ which does not vanish on the subspace of the constant functions, but which does vanish on any C^∞ functions in $\bar{\Omega}$ which is identically zero in some neighborhood, in $\bar{\Omega}$, of Γ_0. [*Hint*: Use a continuous extension mapping (cf. Appendix to Sect. 26) of $H^1(\Omega)$ into $H_0^1(B_R + x_0)$, where $x_0 \in \Gamma_0$ and R is sufficiently large, and exploit the result proved in the first part of the present exercise.]

37.4. Let Ω and Γ_0 be as in Exercise 37.3. Call V the closure in $H^1(\Omega)$ of the space of C^∞ functions in $\bar{\Omega}$ which vanish in a neighborhood of Γ_0. Let $\Gamma_1 = \Gamma \backslash \Gamma_0$.

Derive from the conclusion of Exercise 37.3 that the Dirichlet sesquilinear functional, $\int_\Omega (\text{grad } u) \cdot (\text{grad } \bar{v})\, dx$, is coercive on V (cf. §37.2b), and that, as a consequence, the problem (37.3)–(37.4) when $\lambda = 0$, has a unique solution u in V for every choice of data, $f \in L^2(\Omega)$, $g_1 \in L^2(\Gamma)$, with both g_0 and g_1 identically zero in Γ_0. Explain how to obtain a solution of the same problem, but this time with g_0 not identically zero in Γ_0.

37.5. Let Ω be a bounded open subset of the plane $\mathbf{R}^2$, lying on one side of its boundary Γ, which is a smooth curve. Let $c(x)$ be a real-valued C^1 function in $\bar{\Omega}$. Describe the boundary value problem associated with the sesquilinear form

(37.48)

$$\int_\Omega \left(\frac{\partial u}{\partial x^1} \frac{\partial \bar{v}}{\partial x^1} + \frac{\partial u}{\partial x^2} \frac{\partial \bar{v}}{\partial x^2} + c \frac{\partial u}{\partial x^1} \frac{\partial \bar{v}}{\partial x^2} - c \frac{\partial u}{\partial x^2} \frac{\partial \bar{v}}{\partial x^1} + \frac{\partial c}{\partial x^2} \frac{\partial u}{\partial x^1} \bar{v} - \frac{\partial c}{\partial x^1} \frac{\partial u}{\partial x^2} \bar{v} \right) dx.$$

What is the restriction of (37.48) to $u, v \in H_0^1(\Omega)$?

37.6. Let Ω be the following elliptic region in the plane,

$$\left\{ (x, y) \in \mathbf{R}^2 ; \frac{x^2}{a^2} + \frac{y^2}{b^2} < 1 \right\},$$

and let χ_Ω denote its characteristic function. Given any two integers $p, q \geqq 0$, compute $(\partial/\partial x)^q (\partial/\partial y)^p \chi_\Omega$[cf. (37.33)]. Give the expressions of the Neumann boundary operators $\mathcal{N}_j$ [defined in (37.35)] relative to Ω and to the sesquilinear form

$$\sum_{p,\, q \leq m} \int_\Omega \left(\frac{\partial^{p+q} u}{\partial x^p \, \partial y^q} \right) \left(\frac{\partial^{p+q} \bar{v}}{\partial x^p \, \partial y^q} \right) dx \qquad (m \in \mathbf{Z}_+).$$

37.7. Let r, θ be the polar coordinates in $\mathbf{R}^2$, $g(e^{i\theta})$ a sufficiently regular function on the unit circumference. Give a complete description, by means of

Fourier series, of the solutions (when they exist) of the boundary value problem

$$\text{(37.49)}\quad \Delta u = 0 \quad \text{if} \quad r < 1, \qquad \alpha \frac{\partial u}{\partial r} + \beta \frac{\partial u}{\partial \theta} + \gamma u = g \qquad \text{when} \quad r = 1,$$

where α, β, γ are complex numbers.

In the remaining exercises, Ω denotes a bounded open subset of $\mathbf{R}^n$, with smooth boundary Γ, lying on one side of Γ.

37.8. Show that, whatever the complex number λ, the sesquilinear form

$$\alpha(u, v) = \int_\Omega \{\lambda u\bar{v} + (\Delta u)(\Delta \bar{v})\}\, dx$$

is *not* coercive on $H^2(\Omega)$(Δ is the Laplace operator on $\mathbf{R}^n$, $n \geqq 2$). Describe the boundary value problem *formally* associated with the equation

$$\alpha(u, v) = \int_\Omega f\bar{v}\, dx, \qquad \forall v \in H^2(\Omega),$$

where $f \in L^2(\Omega)$ [the student may assume that $u \in H^4(\Omega)$].

37.9. Let V denote the subspace of $H_0^1(\Omega)$ consisting of the functions v such that grad $(\Delta v) \in L^2(\Omega; \mathbf{C}^n)$, equipped with the Hilbert norm

$$(\|v\|^2_{H^1(\Omega)} + \|\text{grad}(\Delta v)\|^2_{L^2})^{1/2}.$$

Show that the sesquilinear form

$$a(u, v) = \int_\Omega \{(\text{grad } u) \cdot (\text{grad } \bar{v}) + (\text{grad } \Delta u) \cdot (\text{grad } \Delta \bar{v})\}\, dx$$

is continuous and coercive on V and that, as a consequence, given any $f \in H_0^1(\Omega)$, there is a unique element u of V such that

$$\text{(37.50)}\qquad a(u, v) = \int_\Omega (\text{grad } f) \cdot (\text{grad } \bar{v})\, dx, \qquad \forall \bar{v} \in V.$$

From the fact (Theorem 23.1) that Δ defines an isomorphism of $H_0^1(\Omega)$ onto $H^{-1}(\Omega)$, derive that we have

$$\text{(37.51)}\qquad \Delta(\Delta u) + u = f \quad \text{in } \Omega,$$

$$\text{(37.52)}\qquad \frac{\partial}{\partial \nu}(\Delta u) = 0 \quad \text{on } \Gamma.$$

Show that, in addition to (37.52), u satisfies another, nontrivial boundary condition.

38

Indications on the General Lopatinski Conditions

Throughout this section, Ω is a bounded open subset of $\mathbf{R}^n$ whose boundary is a C^∞ hypersurface; we shall assume that Ω lies on one side of $\partial\Omega$. We shall consider an *elliptic* operator of even order $m = 2m_0$, $P(x, D)$, whose coefficients are complex functions, defined and C^∞ in an open neighborhood of $\overline{\Omega}$. We assume that we are given J boundary differential operators, $B_j(x, D)$ ($j = 1, \ldots, J$): These are differential operators with coefficients defined and C^∞ in an open neighborhood of the boundary $\partial\Omega$, and they determine the boundary conditions in the problem we propose to study, namely that of "finding" and studying the solution of the problem:

$$P(x, D)u = f \quad \text{in } \Omega, \tag{38.1}$$

$$B_j(x, D)u = g_j \quad \text{on } \partial\Omega \ (j = 1, \ldots, J), \tag{38.2}$$

where f and the g_j are, say, smooth functions in $\overline{\Omega}$ and $\partial\Omega$, respectively.

After having encountered a wealth of particular cases, as we have in the earlier sections, it might seem reasonable to ask the following question: Given the operator $P(x, D)$, which are the sets of boundary operators $\{B_j(x, D)\}$ ($j = 1, \ldots, J$) such that the problem (38.1)–(38.2) is well posed? Of course we must state in precise terms what we mean by "well posed." Our choice of the meaning is dictated by the experience acquired on the subject. Without going into the details let us say that the "natural" meaning will not do: The natural meaning is that the problem is well posed if $(f, g_1, \ldots, g_J) \mapsto u$ is a bijection of $C^\infty(\overline{\Omega}) \times \{C^\infty(\partial\Omega)\}^J$ onto $C^\infty(\overline{\Omega})$. We need much stronger properties than this.

Indeed, we have two basic requirements: Our definition should be invariant under diffeomorphism (this is true of the preceding one); it should be stable under "small" perturbations of the operators $P(x, D)$, $B_j(x, D)$ ($1 \leqq j \leqq J$). By small we mean that we may modify each one of these operators by adding to it an operator of the same order, provided that the absolute values of the coefficients, in the latter, and of their derivatives (up to a certain order) are sufficiently small. These two requirements are natural, but it should be said that the reason for setting them is that, without them, the analysis could not be carried through.

Now we observe that, if we deal with isomorphisms of a Fréchet space into another one, we cannot be sure of their stability under small perturbations (in a reasonable sense of the phrase). But this is true if we deal with isomorphisms between Banach spaces. In PDE theory, whenever faced with similar circumstances, one has the recourse of representing the space C^∞ as the union of a family of Banach or of Hilbert spaces, ordered into a scale; the Sobolev spaces H^s are usually the best suited for such an approach. We shall do it again here. We observe that if m_j is the order of $B_j(x, D)$, $u \mapsto B_j(x, D)u|_{\partial\Omega}$ is a continuous linear map of $H^s(\Omega)$ into $H^{s-m_j-1/2}(\partial\Omega)$ (here s is an integer $\geqq m_j + 1$).

Definition 38.1. We say that the boundary problem (38.1)–(38.2) is well posed if, given any $s \geqq \sup(m, m_1 + 1, \ldots, m_J + 1)$,

$$u \mapsto \left(P(x, D)u, B_1(x, D)u\Big|_{\partial\Omega}, \ldots, B_J(x, D)u\Big|_{\partial\Omega}\right) \tag{38.3}$$

is an isomorphism of $H^s(\Omega)$ *onto*

$$H^{s-m}(\Omega) \times \prod_{j=1}^{J} H^{s-m_j-1/2}(\partial\Omega). \tag{38.4}$$

"Isomorphism," in the preceding statement, refers to the locally convex space (not the Hilbert space) structures.

The fact that (38.3) is an isomorphism of $H^s(\Omega)$ onto (38.4) is equivalent to a pair of inequalities, the first of which reads as follows:

$$\|u\|_{H^s(\Omega)} \leqq C\Big\{\|P(x, D)u\|_{H^{s-m}(\Omega)} + \sum_{j=1}^{J} \|\gamma[B_j(x, D)u]\|_{H^{s-m_j-1/2}(\partial\Omega)}\Big\}, \qquad u \in H^s(\Omega). \tag{38.5}$$

We have denoted by γ the trace on $\partial\Omega$ (Sect. 26). The second estimate applies to what is called the *adjoint* of the problem (38.1)–(38.2). Our purpose here is merely to provide an idea of how the analysis goes, and of the results to which it leads, without getting too involved with technicalities, and certainly without giving the proofs. We shall therefore not describe the adjoint problem (cf. Exercise 38.9), nor state what the adjoint estimate is. Observe that the estimate (38.5) leads to the uniqueness and the regularity of the solution, whereas the adjoint inequality leads to its existence.

In principle, we would like to find necessary and sufficient conditions, bearing on the coefficients of all the intervening operators, $P(x, D)$ and the $B_j(x, D)$, and on the open set Ω, in order that the problem (38.1)–(38.2) be well posed. But a moment of reflection shows that this is somewhat hopeless. Think of the extremely simple case where $P(x, D) = \lambda - \Delta$, $J = 1$, $B_1(x, D) = I$, the identity. What are the values of the (complex) number λ for which the corresponding problem (38.1)–(38.2), that is, the Dirichlet problem in Ω for $\lambda - \Delta$, is well posed? A precise answer to such a question is beyond our reach, except in very primitive cases, since it would require the precise knowledge of the eigenvalues of the problem.

Thus we shall settle for somewhat less. Note that the dual of the space (38.4) can be identified with

$$(H^{s-m}(\Omega))^* \times \prod_{j=1}^{J} H^{-s+m_j+1/2}(\partial\Omega). \tag{38.6}$$

On the other hand, since $C^\infty(\bar{\Omega})$ is dense in $H^{s-m}(\Omega)$, the elements of the dual of the latter space can be identified with distributions in $\bar{\Omega}$ (regarded as a manifold with boundary), and of course it makes sense to speak of the elements of (38.6) which are C^∞ functions: It means that they are $(J + 1)$-tuples $(f, g_1, \ldots, g_J)$ where $f \in C^\infty(\bar{\Omega})$ and $g_j \in C^\infty(\partial\Omega)$ for every j.

Definition 38.2. We say that the problem (38.1)–(38.2) is elliptic if the kernel of the mapping (38.3) and its cokernel, i.e., the kernel of its transpose consist solely of C^∞ functions.

Since the kernel of (38.3) is closed, and contained in $C^\infty(\bar{\Omega})$, the topology induced on it by $C^\infty(\bar{\Omega})$ is identical with that induced by $H^s(\Omega)$ [or with that induced by $H^{s+k}(\Omega)$, whatever the integer $k \geqq 0$]. From this and from Rellich's lemma (Theorem 26.11) it follows that this kernel is a locally compact, hence finite-dimensional linear space. A similar argument applies to the transpose of the mapping (38.3):

PROPOSITION 38.1. *If the problem (38.1)–(38.2) is elliptic, the mapping (38.3) has a finite-dimensional kernel and cokernel. In particular, it is a Fredholm operator* (i.e., if N is its kernel and M its cokernel, its *index*, $\dim N - \dim M$, is finite; the ellipticity implies that M and N do not depend on s).

The ellipticity of the problem (38.1)–(38.2) is also equivalent to a pair of inequalities, one bearing on the mapping (38.3) and the other on its adjoint. The estimate relative to (38.3) is obtained by a slight modification of (38.5):

$$\|u\|_{H^s(\Omega)} \leqq C\Big\{\|P(x, D)u\|_{H^{s-m}(\Omega)} + \sum_{j=1}^{J} \|\gamma(B_j(x, D)u)\|_{H^{s-m_j-1/2}(\partial\Omega)} + \|u\|_{H^{s'}(\Omega)}\Big\}, \quad \forall u \in H^s(\Omega), \tag{38.7}$$

where we may take for s' any nonnegative integer $< s$, for instance $s' = s - 1$ (but of course the constant C depends on the choice of s'). We note that the estimate (38.7) is invariant under diffeomorphism, and under small perturbations. The latter now has the following meaning: We may add to $P(x, D)$ or to each one of the $B_j(x, D)$ a differential operator of the same order (m and m_j, respectively) whose principal part has sufficiently small coefficients; the coefficients of the lower order terms are only required to belong to $C^\infty(\overline{\Omega})$.

The most important new feature of (38.7), absent in (38.5), is that such an estimate is *localizable*: Suppose that every point $x_0 \in \Omega$ has an open neighborhood U in $\mathbf{R}^n$ such that the inequality in (38.7) is valid for every $u \in H^s(\Omega)$ whose support (as a distribution in $\mathbf{R}^n$) is contained in U; then it is valid whatever $u \in H^s(\Omega)$. For by using a C^∞ partition of unity in a neighborhood of $\overline{\Omega}$, whose elements have sufficiently small supports, one can piece the local estimates together (the "error terms" resulting from the partition of unity can be absorbed into the term const $\|u\|_{H^{s-1}(\Omega)}$ by increasing the constant).

The preceding remarks enable us to considerably reduce the difficulties of the situation. First of all, we may localize it completely in an open neighborhood of a point x_0 of the boundary. Indeed, in the neighborhood of any point in the interior the estimate (38.7) (or rather, its localization) does not involve the boundary operators $B_j(x, D)$ at all, and it is then a trivial consequence of Gårding's inequality (34.5), applied to

$$P(x, D)^*(1 - \Delta_x)^{s-m}P(x, D)$$

instead of $P(x, D)$ [$P(x, D)^*$ is the formal adjoint of $P(x, D)$; see p. xiv].

Second, we may assume that the neighborhood U of $x_0 \in \partial\Omega$ is an open ball centered at x_0, and that $U \cap \partial\Omega$ lies in the hyperplane $x^n = x_0^n$, as we have done systematically in Sect. 27.

Third, by choosing U sufficiently small and by perturbing $P(x, D)$ and the $B_j(x, D)$, we may assume that, in U, $P(x, D) = p(D)$, $B_j(x, D) = b_j(D)$, where $p(\xi)$ and the $b_j(\xi)$ are homogeneous polynomials in n variables, of degrees m and m_j, respectively.

For the sake of simplicity let us assume that $x^0 = 0$ and that Ω is the upper half-ball $\{x \in U;\ x^n > 0\}$. The argument is of course closely related to, and in fact an elaboration of, the one in Sect. 27. Here we are concerned with the validity of the estimate

$$(38.8) \qquad \|u\|_{H^s(\Omega)} \leqq C\Big\{\|p(D)u\|_{H^{s-m}(\Omega)} + \sum_{j=1}^{J} \|\gamma(b_j(D)u)\|_{H^{s-m_j-1/2}(\partial\Omega)} + \|u\|_{H^{s-1}(\Omega)}\Big\},$$

$$\forall u \in H^s(\Omega), \quad \operatorname{supp} u \Subset U.$$

Just as (38.7) corresponds to the problem (38.1)–(38.2), the estimate (38.8) is associated with the following boundary value problem:

(38.9) $$p(D)u = f \quad \text{in } \Omega,$$

(38.10) $$b_j(D)u = g_j \quad \text{in } U \cap \partial\Omega, \qquad j = 1, \ldots, J.$$

Here $f \in H^{s-m}(\Omega)$, supp $f \Subset U$, and $g_j \in H^{s-m_j-1/2}(\partial\Omega)$, supp $g_j \Subset U \cap \partial\Omega$ (in other words, f and the g_j vanish identically in a neighborhood of the sphere which is the boundary of U). What we want is that the kernel and cokernel of the mapping

(38.11) $$u \mapsto (p(D)u, \gamma(b_1(D)u), \ldots, \gamma(b_J(D)u))$$

consist exclusively of C^∞ functions in $\bar{\Omega}$ [recalling that $u \in H^s(\Omega)$]. Here γ stands for the trace on the $(n-1)$-dimensional open ball $U \cap \partial\Omega$.

As we have done in Sect. 27, we shall avail ourselves of a Fourier transformation with respect to the tangential variables. The problem (38.9)–(38.10) is thus transformed into an "initial value" problem for an ordinary differential equation (in the variable $t = x^n$ normal to the flat part of the boundary of Ω), depending on the parameters $\xi' = (\xi_1, \ldots, \xi_{n-1})$:

(38.12) $$p(\xi', D_t)\hat{u} = \hat{f} \qquad \text{when} \quad t \geqq 0,$$

(38.13) $$b_j(\xi', D_t)\hat{u} = \hat{g}_j \quad (j = 1, \ldots, J) \qquad \text{when} \quad t = 0.$$

We are going to look at the solutions of (38.12) when $f = g_1 = \cdots = g_J = 0$ identically. Since all solutions of a linear ordinary differential equation with constant coefficients (here, depending on the parameters ξ') are easy to construct by means of the "characteristic roots," let us introduce the latter: We shall subdivide them into two sets, the first, $\lambda_1, \ldots, \lambda_p$, made up of those whose imaginary part is greater than zero, the second, $\lambda_{p+1}, \ldots, \lambda_m$, made up of those having a strictly negative imaginary part. Since the operator $P(x,D)$ is elliptic and $p(D) = P_m(x_0, D)$ is its principal part, frozen at a point x_0 of the boundary, no characteristic root is ever real provided that ξ' remains different from zero, which we are assuming throughout the present argument (when $\xi' = 0$ all characteristic roots "collapse" to the value zero).

PROPOSITION 38.2. *If the number n of independent variables is $\geqq 3$, the number of characteristic roots lying in the upper half-plane is exactly equal to* $m_0 = m/2$ (and therefore to the number of roots lying in the lower half-plane).

Proof. If $n > 2$, the complement of the origin in $\mathbf{R}^{n-1}$ is connected and we can find a smooth simple curve γ joining $\xi' \neq 0$ to $-\xi'$. Along γ the roots of $p(\xi', \tau)$, as a polynomial in τ, can be represented as continuous functions of ξ'. Because of the homogeneity of p, it is clear that the number μ of roots in the upper half-plane, at the point ξ', must be equal to the number ν of roots in the lower half-plane, at the point $-\xi'$. But μ must be constant over γ, otherwise some root would have to be real at some point of γ. Thus $\mu = \nu$ and $\mu + \nu = m$. Q.E.D.

From now on we assume that

(38.14) *The number of roots of $p(\xi', \cdot)$ whose imaginary part is greater than zero, for every $\xi' \in \mathbf{R}^{n-1}\backslash\{0\}$, is exactly equal to $m_0 = m/2$.*

Why we require this to be true also when $n = 2$ (or $n = 1$) will be clear in a moment. Let us mention that the operators $P(x, D)$ which satisfy this condition (38.14) at every point of $\bar{\Omega}$ are often called *strongly elliptic*. The Cauchy–Riemann operator in the plane is not strongly elliptic.

The importance of the characteristic roots $\lambda_1, \ldots, \lambda_{m_0}$ belonging to the upper half-plane lies in the fact that, though every one of the functions $\exp(i\lambda_j(\xi')t)$ is a solution of the equation

$$p(\xi', D_t)h = 0, \qquad t > 0, \tag{38.15}$$

only those such that $\operatorname{Im} \lambda_j(\xi') > 0$ can be regarded as Fourier transforms, in the variables x', of smooth functions (or even, for that matter, of distributions). Indeed, the other ones grow much too fast as $|\xi'| \to +\infty$, as soon as $t > 0$.

Let us assume for the moment that the $\lambda_j(\xi')$ $(j = 1, \ldots, m_0)$ are pairwise distinct, for every $\xi' \neq 0$. This restriction will be removed later. In this case all solutions of (38.15) are of the form

$$h(\xi', t) = \sum_{j=1}^{m_0} C_j(\xi') \exp(i\lambda_j(\xi')t). \tag{38.16}$$

If we want this function to satisfy the homogeneous boundary conditions

$$b_j(\xi', D_t)h = 0 \quad \text{when} \quad t = 0, \qquad j = 1, \ldots, J, \tag{38.17}$$

we must have

$$\sum_{k=1}^{m_0} b_j(\xi', \lambda_k(\xi'))C_k(\xi') = 0, \qquad j = 1, \ldots, J. \tag{38.18}$$

Suppose that (38.18) had nontrivial solutions $C(\xi') = (C_1(\xi'), \ldots, C_{m_0}(\xi'))$. Then we would be able to choose the latter to be any function of ξ' we wished, provided that it be valued in the null space of the matrix

$$(b_j(\xi', \lambda_k(\xi'))_{j=1,\ldots,J,\, k=1,\ldots,m_0}. \tag{38.19}$$

For instance, by multiplying it by a suitable scalar $\alpha(\xi')$, we could achieve easily that $h(\xi', 0)$ not be of rapid decay as $|\xi'| \to +\infty$, and thus not be the

Fourier transform of a smooth function of x', contrary to our requirement. We must therefore require that *the matrix (38.19) define an injective mapping of* $\mathbf{C}^{m_0}$ *into* $\mathbf{C}^J$. Similar considerations concerning the *cokernel* of (38.12)–(38.13), i.e., concerning the kernel of the transpose problem, lead us to the requirement that also *the transpose of (38.19) must be injective.* Both conditions can be combined and rephrased as follows:

(38.20) *$J = m_0$ and the determinant of (38.19) is different from zero for every $\xi' \in \mathbf{R}^{n-1}\backslash\{0\}$.*

This formulation has the defect of assuming that the $\lambda_j (1 \leqq j \leqq m_0)$ are distinct. Rather than removing this restriction at the present stage, we shall avail ourselves of it in the construction of the solution of (38.15), that is, of (38.12) when the right-hand side f is identically zero, satisfying the boundary conditions (38.13). In fact, let us denote by $\hat{v}_j(\xi', t)$ the solution of (38.12)–(38.13) when $\hat{f}$ and all $\hat{g}_k$ vanish identically, except $\hat{g}_j$. We may write

(38.21) $$\hat{v}_j(\xi', t) = \hat{g}_j(\xi')\hat{F}_j(\xi', t),$$

where $\hat{F}_j$ is the same as $\hat{v}_j$ except that $\hat{g}_j = 1$ in its definition. The function $\hat{F}_j$ must have the form (38.16), with constants C_j satisfying the system of linear equations

$$\sum_{k=1}^{m_0} b_{j'}(\xi', \lambda_k(\xi'))C_k(\xi') = \delta_{jj'} \qquad (j' = 1, \dots, m_0),$$

where $\delta_{jj'}$ is Kronecker's index. Cramer's rule yields at once

(38.22) $$\hat{F}_j(\xi', t) = B_j(\xi', t)/B(\xi'),$$

where

(38.23) $$B(\xi') = \det[b_j(\xi', \lambda_k(\xi'))],$$

i.e. $B(\xi')$ is the determinant of (38.19), while $B_j(\xi', t)$ is the same determinant, except that the row $\{b_j(\xi', \lambda_1(\xi')), \dots, b_j(\xi', \lambda_{m_0}(\xi'))\}$ has been replaced by $\{\exp(i\lambda_1(\xi')t), \dots, \exp(i\lambda_{m_0}(\xi')t)\}$.

The noteworthy feature of formula (38.22) is that it remains valid even when the characteristic roots are not distinct, as the reader may easily ascertain (cf. Exercise 38.4). Observe that the functions $\hat{F}_j$ are well defined (as functions of ξ'): Indeed, they are symmetric functions of the characteristic roots lying in the upper-plane, and therefore they are functions of the coefficients of the polynomial in the variable τ,

(38.24) $$M^+(\xi', \tau) = \prod_{j=1}^{m_0} (\tau - \lambda_j(\xi')).$$

Since no characteristic root ever crosses the real axis (as long as $\xi' \neq 0$) the coefficients of M^+ are continuous functions of ξ'. In particular we see which hypothesis should replace (38.20) in the general case:

(38.25) *$J = m_0$ and the functions $\hat{F}_j(\xi', t)$ are finite for all $\xi' \in \mathbf{R}^{n-1}$.*

Of course, this is a condition bearing on the ratios B_j/B (and not only on the denominator B: the latter may have zeros, but the factors which vanish must be canceled by factors in the numerator). There is an elegant restatement of (38.25), whose proof we leave to the student (Exercise 38.6):

(38.26) *the polynomials in τ, $b_j(\xi', \tau)$ $(j = 1, \ldots, J = m_0)$, are linearly independent modulo $M^+(\xi', \tau)$.*

It is obvious that there is an "invariant" formulation of the condition (38.31), involving the normal (or, rather, the conormal) to the boundary $\partial\Omega$ of Ω (without assuming that it has been flattened) and the boundary operators $B_j(x, D_x)$ (precisely, their principal parts), as well as the operator $M^+(x, D_x)$, extracted from the principal part $P_m(x, D_x)$ of $P(x, D_x)$ in the manner $M^+(D_{x'}, D_t)$ was extracted from $p(D_{x'}, D_t)$—with the normal direction to the boundary playing the role of the t-axis. We leave this invariant formulation to the reader. The boundary operators which satisfy it are often said to *cover* $P(x, D)$ or to associate with it the *Lopatinski* (or *Lopatinski–Shapiro*) *boundary conditions.*

In order to complete the construction of the solution of (38.12)–(38.13) we have yet to construct the solution, denoted by $\hat{v}$, corresponding to the case of homogeneous boundary conditions, when $\hat{g}_j = 0$ for all $j = 1, \ldots, m_0$. We proceed in a manner analogous to that used to construct $\widehat{G_0 v}$ in Sect. 27. We introduce the polynomial

$$M^-(\xi', \tau) = \prod_{j=m_0+1}^{2m_0} (\tau - \lambda_j(\xi')). \tag{38.27}$$

We introduce the solution $\hat{E}_+(\xi', t)$ of the homogeneous equation

$$M^+(\xi', D_t)\hat{E}_+ = 0, \tag{38.28}$$

satisfying the initial conditions

$$\partial_t^k \hat{E}_+(\xi', 0) = \begin{cases} 0 & \text{if } k < m_0 - 1, \\ 1 & \text{if } k = m_0 - 1. \end{cases} \tag{38.29}$$

Similarly, we denote by $\hat{E}_-$ the analogous solution of $M^-(\xi', D_t)\hat{E}_- = 0$. Of course, when the characteristic roots are distinct, $\hat{E}_+$ is of the form (38.16) with a suitable choice of the constants $C_j(\xi')$. In fact,

$$\hat{E}_+(\xi', t) = \tilde{V}(\lambda_1, \ldots, \lambda_{m_0}; t)/V(\lambda_1, \ldots, \lambda_{m_0}), \tag{38.30}$$

where $V(\lambda_1, \ldots, \lambda_{m_0})$ is the *Vandermonde determinant* $\det\{(\lambda_j^{k-1})_{1 \leq j,k \leq m_0}\}$ whereas the numerator $\tilde{V}(\lambda_1, \ldots, \lambda_{m_0}; t)$ is the same as the Vandermonde determinant, except that the last row, $\{\lambda_1^{m_0-1}, \ldots, \lambda_{m_0}^{m_0-1}\}$, has been replaced by

$$\{\exp(i\lambda_1 t), \ldots, \exp(i\lambda_{m_0} t)\}.$$

The expression (38.30) remains valid even when the roots λ_j "coalesce." We have an analogous expression for E_-, using the λ_j for $j > m_0$.

By the general argument for ODE described in Sect. 4, we see that

$$\hat{f}_1(\xi', t) = -\int_t^{+\infty} \hat{E}_-(\xi', t - t')\hat{f}(\xi', t')\, dt'$$

satisfies the equation $M^-(\xi', D_t)\hat{f}_1 = \hat{f}$ for $t > 0$. Because of the signs of the imaginary parts of the λ_j when $j > m_0$ and because of the integration from t to $+\infty$, $\hat{f}_1$ is tempered with respect to ξ'. Let us then set

$$\hat{w}(\xi', t) = -\int_0^t \int_{t'}^{+\infty} \hat{E}_+(\xi', t - t')\hat{E}_-(\xi', t' - t'')\hat{f}(\xi', t'')\, dt'\, dt'', \tag{38.31}$$

and the same argument, applied now to $\hat{E}_+$, shows that $M^+(\xi', D_t)\hat{w} = \hat{f}_1$ for $t > 0$ and that $\hat{w}$ is tempered with respect to ξ'.

Does $\hat{w}$ satisfy the homogeneous boundary conditions $b_j(\xi', D_t)\hat{w} = 0$ when $t = 0$ for all $j = 1, \ldots, m_0$? If it did, we would take $\hat{w}$ as the solution of (38.12)–(38.13) when the $\hat{g}_j$ all vanish. It is clear that this is so when the orders of the $b_j(\xi', D_t)$ do not exceed $m_0 - 1$ (as in the Dirichlet problem); then the definition of $\hat{E}_+$ implies that all the derivatives of $\hat{w}$ of order $< m_0$ vanish when $t = 0$. But if the orders of the b_j exceed $m_0 - 1$ for some j, this need not be the case. On the other hand, we need not worry: from the argument which precedes, we know how to construct the solution h of (38.15), satisfying $b_j(\xi', D_t)\hat{h} = b_j(\xi', D_t)\hat{w}$ when $t = 0$ $(j = 1, \ldots, m_0)$, and it will suffice to take $\hat{v} = \hat{w} - \hat{h}$. An explicit formula for $\hat{v}$ is easy to obtain, and so is one for the solution $\hat{u} = \hat{v} + \hat{v}_1 + \cdots + \hat{v}_{m_0}$ of the general problem (38.12)–(38.13).

Let us retrace our steps. First of all, after an inverse Fourier transformation with respect to ξ', we must go from the expression of u in terms of $f, g_1, \ldots, g_{m_0}$, to the estimate (38.8) (and to the analogous adjoint estimate). Then we must derive (38.7) from (38.8). We must also devise a concrete realization of the adjoint problem and submit it to a similar treatment. Finally we must show that these estimates, (38.7) and the analog for the adjoint problem imply what we want: existence and uniqueness modulo finite-dimensional kernel and cokernel, and regularity of the solutions up to the boundary. Of course, the latter part is quite technical and is not described here.

Exercises

38.1. Verify that the Dirichlet boundary conditions

$$\left.\frac{\partial^j u}{\partial \nu^j}\right|_{\partial\Omega} = g^j \in H^{m_0-j-1/2}\,(\partial\Omega), \qquad j = 0, \ldots, m_0 - 1, \tag{38.32}$$

"cover" *every* strongly elliptic operator $P(x, D)$ of order $2m_0$ (cf. Theorem 36.3).

38.2. Is it true that the radiation problem and the oblique-derivative problem considered in §37.5 are of Lopatinski's type?

38.3. Consider the oblique-derivative problem (37.44)–(37.45). Prove that if the function $\alpha \in C^\infty(\Gamma)$ vanishes at some point of Γ, that problem is *not* of Lopatinski's type.

38.4. Consider the boundary value problem (38.9)–(38.10), assuming that $m_0 = J = 2$, and that

$$p(\xi', \tau) = (\tau - \lambda(\xi'))^2(\tau - \mu(\xi'))^2 \tag{38.33}$$

with $\operatorname{Im}\lambda > 0$, $\operatorname{Im}\mu < 0$ for any $\xi' \neq 0$ ($\xi' \in \mathbf{R}^{n-1}$). Prove that, under these hypotheses, the function defined in (38.30) is

$$\hat{E}_+(\xi', t) = te^{i\lambda(\xi')t}, \tag{38.34}$$

and the functions $\hat{F}_j$, defined in (38.22), are given by

$$\hat{F}_j(\xi', t) = \{D_\tau b_j(\xi', \lambda(\xi'))\}^{-1} te^{i\lambda(\xi')t}, \tag{38.35}$$

where $D_\tau = -i\,\partial/\partial\tau$.

38.5. Consider the boundary problem (38.9)–(38.10) taking $p(D) = \Delta^2$ (Δ is the Laplace operator in $\mathbf{R}^n$), $b_1(D) = \Delta$, $b_2(D) = \Delta D_{x^n}$ ($D_{x^j} = -i\,\partial/\partial x^j$). In this case, compute $\hat{E}_\pm(\xi', t)$ and $\hat{F}_j(\xi', t)$, $j = 1, 2$. (Cf. Exercises 37.8 and 37.9.)

38.6. Prove that conditions (38.25) and (38.26) are equivalent.

38.7. Let Δ be the Laplace operator in $\mathbf{R}^2$, where we shall use polar coordinates r, θ. Let Ω denote the open unit disk $r < 1$, Γ the unit circumference $r = 1$. Describe all first-order operators $B(D_\theta, D_r)$ such that

$$\Delta u = f \in L^2(\Omega) \quad \text{in } \Omega, \tag{38.36}$$

$$B(D_\theta, D_r)u = g \in H^{1/2}(\Gamma) \quad \text{on } \Gamma, \tag{38.37}$$

is a Lopatinski boundary value problem.

38.8. Let $P(x, D)$ be a strongly elliptic operator on $\bar{\Omega}$ of order $m = 2m_0$ [thus $P(x, D)$ satisfies (38.14)]. For $j = 1, \ldots, J = m_0$ set $B_j = (\partial/\partial\nu)^{m_j}$, where $m_1 < m_2 < \cdots < m_J < 2m_0 - 1$. Prove that the boundary operators

$\{B_1, \ldots, B_J\}$ cover $P(x, D)$ (see p. 374; $\partial/\partial\nu$ is the differentiation along the normal to Γ, outwardly oriented).

38.9. Same notation as in Exercise 38.8. For $j = J + 1, \ldots, 2m_0$, set $B_j^\natural = (\partial/\partial\nu)^{m_j}$ with $m_{J+1} < \cdots < m_{2J} < 2m_0 - 1$ such that $\{m_1, \ldots, m_{2J}\}$ is a permutation of the interval $\{0, 1, \ldots, 2m_0 - 1\} \subset \mathbf{Z}_+$. Let $P(x, D)^*$ be the *formal adjoint* of $P(x, D)$ (see Remark 37.2). Prove that there are $2J$ boundary operators $C_1, \ldots, C_J, C_{J+1}^\natural, \ldots, C_{2J}^\natural$, uniquely determined, with the following properties:

(38.38) *the order of* C_j *is* $2m_0 - 1 - m_{J+j}$ *and that of* $C_j^\natural$ *is* $2m_0 - 1 - m_j$,

(38.39) *for all* $u, v \in C^\infty(\bar{\Omega})$,

$$\int_\Omega P(x, D)u\,\bar{v}\,dx - \int_\Omega u\,\overline{P(x, D)^*v}\,dx = \sum_{j=1}^{J} \int_\Gamma B_j u\,\overline{C_{J+j}^\natural v}\,d\sigma - \int_\Gamma B_{J+j} u\,\overline{C_j v}\,d\sigma.$$

The boundary value problem defined by $(P(x, D)^*, C_1, \ldots, C_J)$ is sometimes called the *adjoint* of the problem defined by $(P(x, D), B_1, \ldots, B_J)$. Formula (38.39) is called a *generalized Green formula.*

38.10. Use the same notation as in Exercise 38.9. We take $P(x, D)$ to be a second-order strongly elliptic differential operator on $\bar{\Omega}$, $J = 1$, and $B_1 = I$, the identity. Describe the adjoint boundary problem and compare with the problem in §37.3.

38.11. Use the same notation as in Exercise 38.9. Let $P(x, D)$ be an arbitrary strongly elliptic operator of order $m = 2m_0$ on $\bar{\Omega}$; take $J = m_0$ and $B_j = (\partial/\partial\nu)^{j-1}$ ($\partial/\partial\nu$ is the differentiation along the normal to Γ, outwardly oriented). Describe the adjoint problem.

CHAPTER IV

Mixed Problems and Evolution Equations

39

Functions and Distributions Valued in Banach Spaces

As we shall see in the next section, mixed problems combine certain features of the Cauchy problem (Chapter II) with features borrowed from the boundary value problems (Chapter III). As in the Cauchy problem there are $n + 1$ independent variables: n "space" variables $x = (x^1, \ldots, x^n)$, and one "time" variable t. We shall deal with functions and distributions defined in a "cylinder" $\Omega \times]0, T[$, where Ω is an open subset of $\mathbf{R}^n$ (where x varies) and T a number greater than zero (thus t ranges in the interval $]0, T[$—or in any interval of the real line). It is highly convenient to regard these functions of (x, t) as functions of t, valued in a function space in the variable x: Most of the time, the latter will be one of the Sobolev spaces, $H^1(\Omega)$, $H_0^1(\Omega)$, $H^{-1}(\Omega)$, ... Such an approach leads to a considerable streamlining of the arguments, and also, as should be expected, to fairly general results, which can then be applied to a variety of cases—often beyond those used as a starting point. But it has an obvious drawback; that of requiring the reader to be familiar with functions and distributions valued in infinite-dimensional linear spaces.

The purpose of this section is to provide the basic facts about such functions and distributions. In view of our needs in the remainder of the chapter, the basic facts are straightforward generalizations of the scalar case. They hardly require special mention, and to a large extent our intentions here are more psychological than mathematical: We hope to convince the reader that there is not much difference between distributions valued in a (complex) Hilbert space and scalar ones. Essentially the same rules apply to both kinds.

39.1 Continuity and Differentiability of Functions Valued in a Banach Space

Let J denote an open interval $T_0 < t < T_1$ in the real line. In most of what follows it could be replaced by an open subset of a d-dimensional Euclidean space $\mathbf{R}^d$ ($d \geqq 1$) or by a smooth manifold. We shall deal first with functions defined in J and valued in a complex Banach space $\mathbf{E}$. Since $\mathbf{E}$ is a topological space it is obvious what we mean by such a function $\mathbf{f}$ being continuous at some point t of J, or in some subset of J. To say that $\mathbf{f}$ is differentiable at t is equally obvious: It means that the difference quotient

$$\delta_h \mathbf{f}(t) = h^{-1}\{\mathbf{f}(t+h) - \mathbf{f}(t)\} \tag{39.1}$$

converges in $\mathbf{E}$ as $h \neq 0$ converges to 0 (h real and small). Note that this definition fully exploits the fact that $\mathbf{E}$ is a topological linear space: In addition to convergence, we have also used vector subtraction and scalar multiplication of vectors. The limit of $\delta_h \mathbf{f}(t)$ is the derivative of $\mathbf{f}$ at t, $\mathbf{f}'(t)$ [or $(d\mathbf{f}/dt)(t)$ or $\dot{\mathbf{f}}(t)$]. If it exists regardless of what the point t of J is, it defines a function $\mathbf{f}'$ in J. If the latter function is continuous, we say that $\mathbf{f}$ is once continously differentiable, or C^1. More generally, for any k, $1 \leqq k \leqq +\infty$, we say that $\mathbf{f}$ is C^k, or k-times continuously differentiable, if $\mathbf{f}'$ is $C^{k'}$ for every $k' < k$.

We say that $\mathbf{f}$ is *analytic* in J if, whatever $t_0 \in J$, the Taylor series of $\mathbf{f}$ at t_0,

$$\sum_{k=0}^{+\infty} \frac{1}{k!} \mathbf{f}^{(k)}(t_0)(t - t_0)^k, \tag{39.2}$$

converges in $\mathbf{E}$ uniformly in some compact neighborhood of t_0 in J. This implies that $\mathbf{f}$ can be extended as a holomorphic function to the complex values of t in some open neighborhood U of J in the complex plane $\mathbf{C}$. The fact that $\mathbf{f}$ is holomorphic in U means that, whatever $t \in U$, the quotient (39.1) converges in $\mathbf{E}$ when $h \neq 0$ converges to 0 *in the complex plane*.

It should be noted that one can equip $\mathbf{E}$ with a topology different (at least when $\dim \mathbf{E} = +\infty$) from the one defined by the norm, still compatible with its linear structure, and yet relevant to many a situation, for instance, with the weak topology $\sigma(\mathbf{E}, \mathbf{E}^*)$, or with the topology of uniform convergence on the compact subsets of the dual $\mathbf{E}^*$ of $\mathbf{E}$. This leads to different classes of continuous, continuously differentiable, analytic functions valued in $\mathbf{E}$: Such a function $\mathbf{f}$ will be, for instance, weakly (or scalarly) continuous (resp. continuously differentiable, resp. analytic) if the same is true of every scalar function

$$t \mapsto \langle \mathbf{x}^*, \mathbf{f}(t) \rangle, \qquad \mathbf{x}^* \in \mathbf{E}^* \tag{39.3}$$

(the bracket expresses the duality between **E** and its topological dual, **E***). We shall not pursue here the question of the relations between the various classes of functions defined by using different linear (locally convex) topologies on **E** (see Exercises 39.2 and 39.3).

39.2 Integration of Functions Valued in a Banach Space

The definition of the *Riemann integral* of a continuous function $\mathbf{f} : J \to \mathbf{E}$ is immediate:

$$\int_{T_0}^{T_1} \mathbf{f}(t)\, dt \tag{39.4}$$

is the limit of the Riemann sums

$$\sum_{T_0 < t_1 < \cdots < t_k < T_1} \mathbf{f}(t_j')(t_{j+1} - t_j). \tag{39.5}$$

The fact that the Riemann sums make up a convergent net (when **f** is continuous) follows directly from the fact that **E** is *complete*. The operation (39.4) extends, as *Lebesgue integration*, to functions $\mathbf{f} \in L^1(J; \mathbf{E})$.

The Lebesgue spaces $L^p(J; \mathbf{E})$ $(1 \leqq p < +\infty)$ are not difficult to define. It suffices to complete the linear space of **E**-valued continuous functions with compact support, $C_c^0(J; \mathbf{E})$,for the natural norm

$$\left(\int_J \|\mathbf{f}(t)\|_{\mathbf{E}}^p\, dt\right)^{1/p} \qquad (= \|\mathbf{f}\|_{L^p}). \tag{39.6}$$

The trouble with this definition is that the elements of $L^p(J; \mathbf{E})$ are "ideal objects." To recognize them as classes of functions modulo the standard equivalence relation of being equal almost everywhere (a.e.) requires the introductions of the standard concepts of Lebesgue theory, and first of all of measurable functions. For the sake of brevity, we shall follow the Lusin approach:

Definition 39.1. The function $\mathbf{f} : J \to \mathbf{E}$ *is measurable if, given any compact set* $K \subset J$ *and any number* $\varepsilon > 0$, *there is a compact subset* K' *of* K *such that the measure of* $K \backslash K'$ *is less than* ε *and that the restriction of* **f** *to* K' *is continuous.*

We may then introduce the spaces $\mathscr{L}^p(J; \mathbf{E})$ consisting of the functions $f : J \to \mathbf{E}$ having the following two properties: (i) **f** is measurable; (ii) $\|\mathbf{f}\|_{\mathbf{E}}$ belongs to the "scalar" space $L^p(J) = L^p(J; \mathbf{C})$. This definition is also valid for $p = +\infty$. Note that if **f** is measurable (with values in **E**), its norm $\|\mathbf{f}\|_{\mathbf{E}}$ is also measurable (with values in $\mathbf{R}_+$). Then let $\mathscr{N}^p$ be the linear subspace of

$\mathscr{L}^p(J; \mathbf{E})$ consisting of the functions $\mathbf{f}$ which vanish almost everywhere or, which amounts to the same, such that

$$\int_J \|\mathbf{f}(t)\|_{\mathbf{E}}^p \, dt = 0. \tag{39.7}$$

The completion $L^p(J; \mathbf{E})$ of $C_c^0(J; \mathbf{E})$ for the norm (39.6) can be identified with the quotient space $\mathscr{L}^p(J; \mathbf{E})/\mathscr{N}^p$—also "normwise," if we equip the latter with the quotient norm derived from (39.6) [which can be regarded as a seminorm on $\mathscr{L}^p(J; \mathbf{E})$]. The quotient Banach space $\mathscr{L}^\infty(J; \mathbf{E})/\mathscr{N}^\infty$ is the space $L^\infty(J; \mathbf{E})$.

The reader should beware that, although every $L^p(J; \mathbf{E})$ $(1 \leqq p \leqq +\infty)$ is a Banach space, in general, that is, *unless* **E** *itself is a Hilbert space*, $L^2(J; \mathbf{E})$ is *not* a Hilbert space, and unless **E** itself is reflexive, $L^p(J; \mathbf{E})$ $(1 < p < +\infty)$ is not reflexive. It is true, however, that if $\mathbf{E}^*$ is the dual of **E**, equipped with the dual Banach space structure, the dual of $L^p(J; \mathbf{E})$ for $1 \leqq p < +\infty$ is canonically identifiable to $L^q(J; \mathbf{E}^*)$, $q = p/(p-1)$. The duality is expressed by

$$\mathbf{f} \mapsto \langle \mathbf{f}, \mathbf{g} \rangle = \int_J \langle \mathbf{f}(t), \mathbf{g}(t) \rangle \, dt, \qquad \mathbf{g} \in L^q(J; \mathbf{E}^*). \tag{39.8}$$

If **E** is a Hilbert space, $L^2(J; \mathbf{E})$ is equipped with the canonical Hilbert space structure defined by the inner (Hermitian) product

$$(\mathbf{f}, \mathbf{g})_{L^2(J;\mathbf{E})} = \int_J (\mathbf{f}(t), \mathbf{g}(t))_{\mathbf{E}} \, dt. \tag{39.9}$$

In relation with the latter remark, we might also observe that, when **E** is a Hilbert space, the *Fourier transformation* (with respect to the variable t)

$$\hat{\mathbf{f}}(\tau) = \int_{-\infty}^{+\infty} e^{-i\tau t} \mathbf{f}(t) \, dt \tag{39.10}$$

defines an isomorphism (for the Hilbert space structure) between $L^2(\mathbf{R}_t^1; \mathbf{E})$ and $L^2(\mathbf{R}_\tau^1; \mathbf{E})$. This is no longer true, of course, when **E** ceases to be a Hilbert space. When **E** is merely a Banach space, and $L^2(\mathbf{R}^1; \mathbf{E})$ carries the canonical Banach space structure, the Fourier transformation is an isomorphism for the locally convex structures underlying the Banach space structures, but not for the latter: More specifically, the Plancherel formula is no longer true (Exercise 39.6).

39.3 Distributions Valued in Banach Spaces

A scalar distribution T in J is a continuous linear map $C_c^\infty(J) \to \mathbf{C}$. It is therefore natural to define an **E**-valued distribution as a continuous linear map $C_c^\infty(J) \to \mathbf{E}$. The space of such distributions will be denoted by $\mathscr{D}'(J; \mathbf{E})$.

It is equipped with the topology of uniform convergence (of mappings valued in $\mathbf{E}$) on the bounded subsets of $C_c^\infty(J)$. Any locally L^1 function valued in $\mathbf{E}$, $\mathbf{f}$, defines an $\mathbf{E}$-valued distribution by the formula

$$\phi \mapsto \int \phi(t)\mathbf{f}(t)\,dt, \qquad \phi \in C_c^\infty(J). \tag{39.11}$$

As in the scalar case (39.11) defines an *injection* of $L^1_{\text{loc}}(J; \mathbf{E})$ into $\mathscr{D}'(J; \mathbf{E})$ (the injection is continuous for the natural topologies). Any element of $L^1_{\text{loc}}(J; \mathbf{E})$, viewed as a distribution, is called a *function*. A distribution $\mathbf{T}$ in J with values in $\mathbf{E}$ is called a *Radon measure* in J, valued in $\mathbf{E}$, if it can be extended as a continuous linear map $C_c^0(J) \to \mathbf{E}$. Of course, functions are particular cases of Radon measures: $L^1_{\text{loc}}(J; \mathbf{E}) \subset \mathscr{M}(J; \mathbf{E})$.

The standard operations on distributions are defined as in the scalar case, for instance, differentiation of distributions:

$$\partial_t \mathbf{T}(\phi) = -\mathbf{T}(\partial_t \phi), \qquad \partial_t = \frac{d}{dt}, \tag{39.12}$$

or multiplication by a scalar C^∞ function α:

$$\alpha\mathbf{T}(\phi) = \mathbf{T}(\alpha\phi). \tag{39.13}$$

Binary operations such as *tensor product* and *convolution* are more delicate, since there is not necessarily a *pairing* between the values of distributions considered here. We return to this important question at the end of this section. Of course, there is no difficulty in defining the tensor product between distributions, all of which, except possibly one, are scalar. We leave it to the reader.

The *local structure theorem* is valid for $\mathbf{E}$-valued distributions: This is because $\mathbf{E}$ is normed. In general, when $\mathbf{E}$ is a locally convex space, e.g., a Fréchet space but not a Banach space, the theorem is not valid. Thus $\mathbf{E}$ being a Banach space, $\mathbf{T}$ an arbitrary $\mathbf{E}$-valued distribution in J, and J' an arbitrary relatively compact open subset of J, we may write

$$\mathbf{T} = \sum_{k=0}^{M} \partial_t^k \mathbf{f}_k \quad \text{in } J', \tag{39.14}$$

where the $\mathbf{f}_k$ can be taken to be Radon measures, or continuous functions, or L^2 functions, in J, with values in $\mathbf{E}$. The proof is similar to the one in the scalar case.

We shall now define the Fourier transform of an $\mathbf{E}$-valued tempered distribution. First of all, let us introduce the space $\mathscr{S}(\mathbf{E})$ of C^∞ functions in $\mathbf{R}^1$, valued in $\mathbf{E}$, rapidly decaying at infinity, that is, such that

$$\sup_{t \in \mathbf{R}^1} \left\{(1 + t^2)^k \sum_{j=0}^{M} \|\partial_t^j \mathbf{u}(t)\|_{\mathbf{E}}\right\} < +\infty, \qquad k, M = 0, 1, \ldots. \tag{39.15}$$

The topology on $\mathscr{S}(\mathbf{E})$ is defined by the norms in (39.15). It is easily seen (as in the scalar case) that the Fourier transformation (39.10) defines an isomorphism of $\mathscr{S}(\mathbf{E})$ onto itself. Let us then define $\mathscr{S}'(\mathbf{E})$ as the space of continuous linear mappings $\mathscr{S} = \mathscr{S}(\mathbf{C}) \to \mathbf{E}$, identified with a subspace of $\mathscr{D}'(\mathbf{R}^1; \mathbf{E})$, that of the *tempered* E-valued distributions on the real line. The Fourier transform $\mathscr{F}\mathbf{T}$ of an element $\mathbf{T}$ of $\mathscr{S}'(\mathbf{E})$ is defined by the formula

$$(\mathscr{F}\mathbf{T})(\phi) = \mathbf{T}(\mathscr{F}\phi), \qquad \phi \in \mathscr{S}. \tag{39.16}$$

It is seen at once that $\mathscr{F}$, thus defined on $\mathscr{S}'(\mathbf{E})$, extends the Fourier transformation (39.10) defined, say, on $L^2(\mathbf{R}^1; \mathbf{E})$. It is an isomorphism of $\mathscr{S}'(\mathbf{E})$ onto itself. If $\mathbf{E} = \mathbf{F}^*$, the dual of a Banach space $\mathbf{F}$, $\mathscr{F} : \mathscr{S}'(\mathbf{E}) \to \mathscr{S}'(\mathbf{E})$ is the transpose of $\mathscr{F} : \mathscr{S}(\mathbf{F}) \to \mathscr{S}(\mathbf{F})$. Of course, we could not have adopted this as a definition of the Fourier transform, since not every Banach space is the dual of another Banach space.

The definition of the Sobolev spaces on the real line, with values in the Banach space $\mathbf{E}$, follows the usual procedure: If s is any real number, $H^s(\mathbf{E})$ is the space of E-valued tempered distributions $\mathbf{u}$ whose Fourier transform $\hat{\mathbf{u}}(\tau)$ belongs to the space $L^2(\mathbf{R}^1; \mathbf{E})$ with respect to the measure $(1 + \tau^2)^s\, d\tau$. We have continuous injections, with norms $\leqq 1$, $H^s(\mathbf{E}) \subsetneq H^{s'}(\mathbf{E})$ if $s' < s$. The space $\mathscr{S}(\mathbf{E})$ is dense in each $H^s(\mathbf{E})$ and $H^{-s}(\mathbf{E}^*)$ can be canonically identified with the dual of $H^s(\mathbf{E})$. Thus $H^s(\mathbf{E})$ is reflexive if and only if $\mathbf{E}$ is reflexive. When $\mathbf{E}$ is a Hilbert space, one equips $H^s(\mathbf{E})$ with the natural Hilbert space structure derived from that of $\mathbf{E}$. The local Sobolev spaces can be defined in the obvious way: $H^s_{\text{loc}}(J; \mathbf{E})$ is the space of distributions $\mathbf{T}$ defined in J, valued in $\mathbf{E}$, such that $\alpha\mathbf{T} \in H^s(\mathbf{E})$ whatever $\alpha \in C^\infty_c(J)$. As for $H^s_c(J; \mathbf{E})$, it is the subspace of $H^s(\mathbf{E})$ made up of the elements of $H^s(\mathbf{E})$ which have compact support contained in J.

39.4 Binary Operations on E-Valued Distributions

There are many operations bearing on pairs (or N-tuples) of distributions and functions which we have not yet looked at—operations which are vital to the processes of analysis, such as multiplication of a distribution with a (smooth) function, convolution of two distributions, tensor product of two distributions (which, in the latter case, must "depend" on independent variables). The difficulty lies in the fact that we must be given a "natural" way to pair the values, that is, to pair vectors, possibly belonging to different linear spaces.

To handle all these situations in a unified manner, we introduce three Banach spaces $\mathbf{E}$, $\mathbf{F}$, $\mathbf{G}$ and a *continuous bilinear* map $\beta : \mathbf{E} \times \mathbf{F} \to \mathbf{G}$. Let us mention four important examples of such mappings:

Example 39.1. Take $\mathbf{F} = \mathbf{E}^*$, the strong dual of $\mathbf{E}$, and $\mathbf{G} = \mathbf{C}$, the complex field, β the *duality bracket* mapping $(\mathbf{e}, \mathbf{e}^*) \mapsto \langle \mathbf{e}^*, \mathbf{e} \rangle$.

Example 39.2. Let $\mathbf{E} = \mathbf{F} = \mathbf{G}$ be a Banach algebra and β the multiplication in it.

Example 39.3. Let $\mathbf{X}$, $\mathbf{Y}$, $\mathbf{Z}$ be three Banach spaces and $\mathbf{E} = L(\mathbf{X}; \mathbf{Y})$, the space of bounded linear operators $\mathbf{X} \to \mathbf{Y}$, equipped with the operator norm, $\mathbf{F} = L(\mathbf{Y}; \mathbf{Z})$ and $\mathbf{G} = L(\mathbf{X}; \mathbf{Z})$. Take β to be the *composition mapping* $(A, B) \mapsto B \circ A$.

Example 39.4. The preceding three examples are particular cases of the following: Take $\mathbf{F} = L(\mathbf{E}; \mathbf{G})$ and let β be the mapping $(e, A) \mapsto Ae$.

Let us refer to the continuous bilinear mapping $\beta : \mathbf{E} \times \mathbf{F} \to \mathbf{G}$ as a *pairing.*

PROPOSITION 39.1. *If β is a pairing,* $\mathbf{u}$ (*resp.* $\mathbf{v}$) *an element of* $C_c^\infty(J; \mathbf{E})$ [*resp. of* $C^\infty(J; \mathbf{F})$] *,then* $t \mapsto \beta(\mathbf{u}(t), \mathbf{v}(t))$ *belongs to* $C_c^\infty(J; \mathbf{G})$.

This proposition is evident. It follows at once from the closed graph theorem that the bilinear mapping

$$(\mathbf{u}, \mathbf{v}) \mapsto \beta(\mathbf{u}, \mathbf{v}) : C_c^\infty(J; \mathbf{E}) \times C^\infty(J; \mathbf{F}) \to C_c^\infty(J; \mathbf{G}) \tag{39.17}$$

is separately continuous. Consider now the mapping

$$\phi \mapsto \int \beta(\mathbf{u}(t), \mathbf{v}(t))\phi(t)\, dt \tag{39.18}$$

from $C_c^\infty(J)$ into $\mathbf{G}$. It is immediately checked to be continuous and hence to define a distribution $\mathbf{u} \underset{\beta}{\cdot} \mathbf{v}$ on J, with values in $\mathbf{G}$. For fixed $\mathbf{v}$, and using the density of $C_c^\infty (J; \mathbf{E})$ in $\mathscr{D}' (J; \mathbf{E})$, we may extend the mapping $\mathbf{u} \mapsto \mathbf{u} \underset{\beta}{\cdot} \mathbf{v}$ to the whole of $\mathscr{D}'(J; \mathbf{E})$. We have thus defined the *β-multiplication* of $\mathbf{E}$-valued distributions in J with $\mathbf{F}$-valued C^∞ functions (also defined in J). One can check that the β-multiplication is a *continuous* bilinear map

$$\mathscr{D}'(J; \mathbf{E}) \times C^\infty(J; \mathbf{F}) \to \mathscr{D}'(J; \mathbf{G}).$$

The reader may apply this definition in Examples 39.1 to 39.3. He will see, for instance, that if $\mathbf{T}$ is an $\mathbf{E}$-valued distribution and $A(t)$ a C^∞ function of t with values in the Banach space of operators $L(\mathbf{E}; \mathbf{G})$, $A(t)\mathbf{T}$ is well defined as a $\mathbf{G}$-valued distribution.

Let us momentarily denote by x the variable in J, by y the one in another open interval J'. Let $\mathbf{S}$ (resp. $\mathbf{T}$) be an $\mathbf{E}$-valued (resp. $\mathbf{F}$-valued) distribution in J (resp. in J'). In the same manner as in the scalar case we may prove that

$$\phi(x, y) \mapsto \mathbf{T}_y(\phi(x, y))$$

is a continuous linear map of $C_c^\infty(J \times J')$ into $C_c^\infty(J; \mathbf{F})$. We may therefore form the pairing $\beta(\mathbf{S}_x, \mathbf{T}_y(\phi(x, y))) \in \mathscr{D}'(J; \mathbf{G})$. If $\psi(x)$ is any scalar C^∞ function with compact support in J, equal to one in a neighborhood of the x projection of the support of ϕ, we may set

$$(\mathbf{S}_x \underset{\beta}{\otimes} \mathbf{T}_y)(\phi) = \beta(\mathbf{S}_x, \mathbf{T}_y(\phi(x, y)))(\psi). \tag{39.19}$$

It is checked at once that the definition (39.19) does not depend on the choice of the cutoff function ψ. By applying it to decomposable functions $\phi(x, y) = \sum_{j=1}^N \phi_j^1(x)\phi_j^2(y)$ and availing ourselves of Fubini's theorem for distributions, we see also that we could have interchanged the role of x and y in order to arrive to the definition [i.e., first compute $\mathbf{S}_x(\phi(x, y)) \in C_c^\infty(J'; \mathbf{E})$ and form $\beta(\mathbf{S}_x(\phi(x, y)), \mathbf{T}_y)$]. We have thus defined the *β-tensor product* $\mathbf{S}_x \underset{\beta}{\otimes} \mathbf{T}_y$: It defines a separately continuous bilinear map

$$\mathscr{D}'(J; \mathbf{E}) \times \mathscr{D}'(J'; \mathbf{F}) \to \mathscr{D}'(J \times J'; \mathbf{G}).$$

From there we can easily define the *β-convolution* $\mathbf{S} \underset{\beta}{*} \mathbf{T}$ of two distributions on the real line, valued in **E** and in **F**, respectively. We use the formula

$$(\mathbf{S} \underset{\beta}{*} \mathbf{T})(\phi) = (\mathbf{S}_x \underset{\beta}{\otimes} \mathbf{T}_y)(\phi(x + y)), \qquad \phi \in C_c^\infty(\mathbf{R}^1). \tag{39.20}$$

As usual the formula (39.20) does not make sense for arbitrary distributions; the pair **S**, **T** must be "convolvable," i.e., meet certain requirements of support or of decay at infinity. It will make sense, for instance, when either **S** or **T** has compact support, or when both **S** and **T** belong to L^1 (with values in **E** and **F**, respectively).

Much more can be said, but at the elementary level where we have placed ourselves most of it would be a mere repetition of the scalar theory. It is so for most properties of the Fourier transformation (e.g., the Paley–Wiener–Schwartz theorem) and of the convolution, and their mutual relations.

Exercises

(In the following exercises, **E**, **F**, **G** stand for complex Banach spaces and J for an open interval in the real line, where the variable is t.)

39.1. Let m be a nonnegative integer, $A(t)$ a C^m function in J, valued in the Banach space of bounded linear operators of **E** into **F**, $L(\mathbf{E}; \mathbf{F})$. Show that its transpose, ${}^tA(t)$, is a C^m function in J, valued in $L(\mathbf{F}^*; \mathbf{E}^*)$ (the asterisk indicates duals).

39.2. Let m be an integer ≥ 1, $\mathbf{u}(t)$ a mapping of J into **E**. Prove that, if the complex-valued function $\langle \mathbf{x}^*, \mathbf{u}(t) \rangle$ in J is C^m, whatever the continuous linear functional $\mathbf{x}^*$ on $\mathbf{E}$, then the function $\mathbf{u}(t)$ itself is C^{m-1}. [*Hint*: Prove first that if all the functions $\langle \mathbf{x}^*, \mathbf{u}(t)\rangle$, $\mathbf{x}^* \in \mathbf{E}^*$, are Lipschitz continuous in J, the same is true of u itself.]

39.3. Let $\mathbf{H}$ denote a Hilbert space, $\mathbf{e}_1, \ldots, \mathbf{e}_m, \ldots$, a complete orthonormal system in $\mathbf{H}$ (assume $\dim \mathbf{H} = +\infty$). Consider the function $\mathbf{u}(t)$ in the closed interval $0 \leqq t \leqq 1$, defined as follows:

$$\mathbf{u}(0) = 0, \quad \mathbf{u}(t) = (1-\theta)\mathbf{e}_{m+1} + \theta \mathbf{e}_m$$

$$\text{if} \quad t = \theta(1/m) + (1-\theta)\frac{1}{m+1} \qquad (0 \leqq \theta \leqq 1).$$

Prove that, given any $\mathbf{x}^* \in \mathbf{H}^*$, the scalar function $\langle \mathbf{x}^*, \mathbf{u}(t)\rangle$ is continuous in $[0, 1]$, but that $\mathbf{u}$ is not continuous at $t = 0$, as a function valued in the normed space $\mathbf{H}$.

39.4. Let $\mathcal{O}$ be an open subset in the complex plane, where the variable is denoted by z. Let β denote a continuous bilinear map of $\mathbf{E} \times \mathbf{F}$ into $\mathbf{G}$, $\mathbf{u}$ (resp. $\mathbf{v}$) a holomorphic mapping of $\mathcal{O}$ into $\mathbf{E}$ (resp. $\mathbf{F}$). Prove that $\beta(\mathbf{u}, \mathbf{v})$ is a holomorphic mapping of $\mathcal{O}$ into $\mathbf{G}$.

39.5. Let K be a compact interval in the real line, $\mathbf{f}(t)$ a continuous mapping of K into the Banach space $\mathbf{E}$. Prove that the linear functional on the dual $\mathbf{E}^*$ of $\mathbf{E}$,

$$\mathbf{x}^* \mapsto \int_K \langle \mathbf{x}^*, \mathbf{f}(t)\rangle \, dt \tag{39.21}$$

defines a unique element of $\mathbf{E}$, and that this element is equal to the Riemann integral defined in §39.2. Would this have been true if we had not assumed $\mathbf{E}$ to be complete, but only normed?

39.6. Let $\mathbf{E}$ be the two-dimensional space $\mathbf{C}^2$, equipped with the norm

$$\|x\|_{\mathbf{E}} = \sup(|x^1|, |x^2|), \qquad x = (x^1, x^2) \in \mathbf{C}^2.$$

Give an example of an L^2 function $\mathbf{f}(t)$ in the real line, valued in $\mathbf{E}$, with Fourier transform $\hat{\mathbf{f}}(\tau)$ such that

$$\int_{-\infty}^{+\infty} \|\mathbf{f}(t)\|_{\mathbf{E}}^2 \, dt \neq \frac{1}{2\pi} \int_{-\infty}^{+\infty} \|\hat{\mathbf{f}}(\tau)\|_{\mathbf{E}}^2 \, d\tau. \tag{39.22}$$

39.7. Describe the set of pairs $(s, s') \in \mathbf{R}^2$ such that the Dirac measure on the diagonal of $\mathbf{R}^n \times \mathbf{R}^n$, $\delta(x-y)$, as a distribution of the variable x valued in the space of distributions in y, belongs to the space $H^s(\mathbf{R}^n_x; H^{s'}(\mathbf{R}^n_y))$. Derive from the result that the local structure theorem does not hold in $\mathscr{D}'(\mathbf{R}^1_x; C^\infty(\mathbf{R}^1_y))$.

39.8. Let $\mathbf{H}$ be a Hilbert space, A a densely defined, possibly unbounded, self-adjoint linear operator on $\mathbf{H}$. Prove the following assertion: In order that every distribution $\mathbf{u}(t)$, defined in the real line and valued in $\mathbf{H}$, which is a solution of the homogeneous equation

$$\frac{d\mathbf{u}}{dt} = A\mathbf{u}, \tag{39.23}$$

be a C^1 function of t, valued in $\mathbf{H}$, it is necessary and sufficient that A be bounded. Then all solutions of (39.23) are analytic functions of t in the real line, valued in $\mathbf{H}$ (and in fact can be continued to the complex values of t as entire functions in $\mathbf{C}$).

39.9. Let $\mathbf{E}$ be a Banach space, $\mathbf{f}$ a locally integrable function, defined in the real line and valued in $\mathbf{E}$. Prove that there is a subset S of $\mathbf{R}^1$, having Lebesgue measure zero, and a *separable* closed linear subspace $\mathbf{E}_0$ of $\mathbf{E}$ such that $\mathbf{f}(t) \in \mathbf{E}_0$ for all $t \in \mathbf{R}^1 \backslash S$.

39.10. Let $\mathbf{E}$, $\mathbf{F}$ be two Banach spaces, $L(\mathbf{E}; \mathbf{F})$ the Banach space of bounded linear operators of $\mathbf{E}$ into $\mathbf{F}$, $t \mapsto A(t)$ a mapping of $\mathbf{R}^1$ into $L(\mathbf{E}; \mathbf{F})$ such that the following holds:

(39.24) *Given any* $\mathbf{u} \in \mathbf{E}$, *any* $\mathbf{v}^* \in \mathbf{F}^*$ (dual of $\mathbf{F}$), $t \mapsto \langle A(t)\mathbf{u}, \mathbf{v}^* \rangle$ *is a measurable function in* $\mathbf{R}^1$.

Is it then true that, whenever $\mathbf{u}(t)$ is a measurable function in $\mathbf{R}^1$, valued in $\mathbf{E}$, $A(t)\mathbf{u}(t)$ is a measurable function in $\mathbf{R}^1$ valued in $\mathbf{F}$?

39.11. Let $\mathbf{E}$, $\mathbf{F}$, $A(t)$ be as in Exercise 39.10 but suppose now that $A(t) \in L^\infty(0, T; L(\mathbf{E}; \mathbf{F}))$. Prove that $\mathbf{u} \mapsto A\mathbf{u}$ is a bounded linear map of $L^2(0, T; \mathbf{E})$ into $L^2(0, T; \mathbf{F})$.

39.12. Let Ω be a bounded open subset of $\mathbf{R}^n$, having a C^∞ boundary and lying on one side of it. Let m be an integer $\geqq 1$ and let $g_j(\cdot, t)$ $(j = 0, \ldots, m-1)$ be m $C^k(0 \leqq k \leqq +\infty$, k independent of $j)$ functions of t in the closed interval $[0, T]$ $(T > 0)$ valued in the Sobolev spaces $H^{m-j-1/2}(\Gamma)$, respectively. Show that there is a C^k function of t, $0 \leqq t \leqq T$, valued in $H^m(\Omega)$ such that, for each $j = 0, \ldots, m-1$,

$$\gamma_j(g(\cdot, t)) = \left.\frac{\partial^j g}{\partial \nu^j}\right|_\Gamma = g_j(\cdot, t), \qquad 0 \leqq t \leqq T.$$

39.13. Let $\mathbf{E}$, $\mathbf{F}$ be two Banach spaces, $L_s(\mathbf{E}; \mathbf{F})$ the space of bounded linear operators $\mathbf{E} \to \mathbf{F}$ equipped with the topology of pointwise convergence in $\mathbf{E}$. Let X be a compact Hausdorff topological space, $\mathbf{f}$ a continuous mapping of X into $\mathbf{E}$, A a continuous mapping of X into $L_s(\mathbf{E}; \mathbf{F})$. Is it true that $A\mathbf{f}$ is a continuous mapping of X into $\mathbf{F}$?

40

Mixed Problems. Weak Form

Consider a region (i.e., an open set) Ω in space, bounded by a hypersurface $\Gamma = \partial\Omega$. Suppose that Ω is filled with matter, for instance that Ω is a wall, or a metal bar, and that a certain temperature g is maintained on Γ, by means of some heating (or cooling) device. We may wish to study the temperature at the various points of Ω. Of course this will evolve in time, as the heat percolates from the surface to the interior: It is a function $u = u(x, t)$ of space and time. It will depend on its initial values, that is, on the temperature distribution at the time (usually taken as the origin) when we started to heat the outer surface Γ. This distribution of temperature at time $t = 0$ will be denoted by $u_0(x)$. Note that the temperature maintained on Γ, g, need not be homogeneous in space, it may vary from point to point; in other words, it may depend on x. In fact, it may also vary with time, and therefore $g = g(x, t)$. A very simple argument based on the way heat is transferred, from atom to atom, inside the matter which fills Ω, shows that the variation of the temperature u in Ω in space and time obeys (up to a certain degree of approximation) the heat equation

$$\frac{\partial u}{\partial t} = k\,\Delta u, \tag{40.1}$$

where k is the conductivity constant, which depends on the nature of the material which fills the region Ω. Equation (40.1) is valid, provided that this material is *homogeneous* (i.e., its characteristics, reflected in k, do not depend on x nor, for that matter, on t), *isotropic* (propagation of heat is the same in all directions), and that there is *no creation or absorption* of heat, for instance due to chemical reactions, inside Ω. In case the latter phenomenon takes place, Eq. (40.1) should be replaced by the inhomogeneous heat equation

$$\frac{\partial u}{\partial t} = k\,\Delta u + f, \tag{40.2}$$

where $f = f(x, t)$ is a function in Ω describing the creation or the absorption of heat. In dealing with more complicated processes, for instance when homogeneity and isotropy are lacking, one may be forced to replace the operator $-k\Delta$ by a more general *elliptic* (see Definition 23.2) second-order differential operator

$$A = A\left(x, t, \frac{\partial}{\partial x}\right) = -\sum_{i,j=1}^{n} a^{ij}(x, t)\frac{\partial^2}{\partial x^i\,\partial x^j} + \sum_{j=1}^{n} a^j(x, t)\frac{\partial}{\partial x^j} + c(x, t), \tag{40.3}$$

and indeed we shall be looking at the analog of (40.2) where $A(x, t, \partial/\partial x)$ has been substituted for $-k\Delta$. But it is useful to keep in mind the meaning of the problems we shall be studying, in the case of Eq. (40.2) (where, however, possibly after a change of the units, we may assume $k = 1$). Suppose thus that u represents the temperature in the region Ω, determined by the initial temperature u_0 and by the surface temperature g. We shall have

$$\frac{\partial u}{\partial t} = \Delta u + f(x, t) \quad \text{in } \Omega, \tag{40.4}$$

$$u = u_0(x) \quad \text{in } \Omega \qquad \text{when} \quad t = 0, \tag{40.5}$$

$$u = g(x, t) \quad \text{on } \Gamma \qquad \text{for every} \quad t > 0. \tag{40.6}$$

One recognizes in (40.6) the kind of boundary condition encountered in the Dirichlet problem. On the other hand, (40.5) is analogous to the initial condition in the Cauchy problem. In view of this double analogy, Problem (40.4)–(40.5)–(40.6) is often called a *mixed problem.*

Let us point out that the boundary condition adjoined to (40.4)–(40.5) need not be of the Dirichlet type. It might very well be of the Neumann type (cf. Sect. 37): Instead of imposing the temperature on the boundary Γ, we may fix the *heat flow* (or *flux*) through Γ. This amounts to replacing (40.6) by the new condition (cf. §37.1):

$$\frac{\partial u}{\partial \nu} = h(x, t) \quad \text{on } \Gamma \qquad \text{for every} \quad t > 0, \tag{40.7}$$

where $\partial/\partial\nu$ denotes, as usual, the normal derivative at the point x of Γ. One may also impose a "hybrid" kind of boundary condition: Maintain the temperature on some part of the boundary while fixing the heat flux on the remaining portion of it. Let us write

$$\Gamma = \Gamma_0 \cup \Gamma_1, \qquad \Gamma_0 \cap \Gamma_1 = \varnothing,$$

and require (cf. §37.2)

$$u = g(x, t) \quad \text{on } \Gamma_0, \qquad \frac{\partial u}{\partial \nu} = h(x, t) \quad \text{on } \Gamma_1 \qquad \text{for every} \quad t > 0. \tag{40.8}$$

The heat operator $\partial/\partial t - \Delta$ is the archetype of the *parabolic operators*. Mixed problems may, however, be posed for other than parabolic operators —e.g., for hyperbolic ones. If the hyperbolic operator under study happens to be second order, there will be *two* initial conditions, in accordance with the requirements in the Cauchy problem: We will have to preassign the values of u and of its first derivative with respect to the time t, $\partial u/\partial t$, at $t = 0$. Mixed problems for second-order hyperbolic equations (and systems of equations) occur naturally in the theory of sound, for instance to describe the evolution of the air pressure inside a room where noise is produced, or in electromagnetism, e.g., to describe the evolution of the electromagnetic field in some region of space (in the latter example, one would deal with a system of equations, specifically the Maxwell equations), and so on. The simplest examples of mixed problems for hyperbolic equations are of the kind

$$\frac{\partial^2 u}{\partial t^2} = \Delta u + f(x, t) \quad \text{in } \Omega, \tag{40.9}$$

$$u = u_0(x), \qquad \frac{\partial u}{\partial t} = u_1(x) \quad \text{in } \Omega \qquad \text{at} \quad t = 0, \tag{40.10}$$

$$u = g(x, t) \quad \text{on } \Gamma \qquad \text{for every} \quad t > 0. \tag{40.11}$$

In physical applications there would be a factor of the form V^2 (where V represents a velocity) in front of the Laplacian Δ, at the right in (40.9). We have assumed that the units of space and time have been chosen here so as to have $V = 1$. The Dirichlet boundary conditions (40.11) can be replaced by a different kind of boundary condition, e.g., by conditions of the Neumann kind or by a mixture of Dirichlet and Neumann conditions, exactly like in the parabolic case.

We shall focus our attention mainly on mixed problems for parabolic equations, with Dirichlet boundary conditions, of which (40.4)–(40.5)–(40.6) is the prototype. We shall, however, consider the case where $-\Delta$ is replaced by a more general elliptic operator, A, like that given in (40.3). On the other hand, it is convenient to restrict the variation of the time t to some interval $[0, T]$, $T > 0$, although in certain instances we will want to let it vary from $-\infty$ or from 0 to $+\infty$. Thus the problem we propose to study is the following one:

$$\frac{\partial u}{\partial t} + A\left(x, t, \frac{\partial}{\partial x}\right)u = f(x, t) \quad \text{in } \Omega \times]0, T[, \tag{40.12}$$

$$u(x, 0) = u_0(x) \qquad \text{in } \Omega, \tag{40.13}$$

$$u(x, t) = g(x, t) \quad \text{in } \Gamma \times]0, T[. \tag{40.14}$$

We shall state our hypotheses on the data of this problem in more precise terms. In later sections, we also study some of its variants, e.g., when the boundary condition (40.14) is replaced by (40.7) or (40.8), as well as the hyperbolic mixed problems such as (40.9)–(40.10)–(40.11). We mainly follow the variational approach—i.e., deal with the weak problem, and therefore with the Sobolev spaces $H^m(\Omega)$, $H_0^m(\Omega)$, $H^{-m}(\Omega)$, ... (cf. Sects. 22 and 23).

First we must state in a precise manner our hypotheses on the differential operator A. It is convenient to rewrite it in the variational form [cf. (23.6)]

$$A\left(x, t, \frac{\partial}{\partial x}\right) = -\sum_{i,j=1}^{n} \frac{\partial}{\partial x^i} a^{ij}(x, t) \frac{\partial}{\partial x^j} + \sum_{j=1}^{n} b^j(x, t) \frac{\partial}{\partial x^j} + c(x, t). \tag{40.15}$$

We begin with the assumption on the nature of the coefficients:

$$a^{ij}, b^j, c \in L^\infty(\Omega \times]0, T[) \qquad \text{for all} \quad i, j = 1, \ldots, n. \tag{40.16}$$

At this stage let us mention the following result:

LEMMA 40.1. *If (40.16) holds, $A(x, t, \partial/\partial x)$ defines a bounded linear operator*

$$L^2(0, T; H^1(\Omega)) \to L^2(0, T; H^{-1}(\Omega)).$$

Proof. It suffices to prove the assertion when A is replaced by any one of the terms $(\partial/\partial x^i)a^{ij}(x, t)(\partial/\partial x^j)$, $b^j(x, t)\, \partial/\partial x^j$, $c(x, t)$. Since $\partial/\partial x^j$ is a bounded linear operator $L^2(0, T; H^m(\Omega)) \to L^2(0, T; H^{m-1}(\Omega))$, $m = \ldots, -1, 0, 1, \ldots$, it is enough to show that multiplication by an arbitrary element of

$$L^\infty(\Omega \times]0, T[)$$

defines a bounded linear operator of $L^2(0, T; H^0(\Omega)) = L^2(\Omega \times]0, T[)$ into itself, which is well known. Q.E.D.

Now we come to the *ellipticity* assumption about A. It will be stronger than the standard uniform ellipticity as described in (23.13). In particular, if it holds for A it will not hold for $-A$. Observe that the change from A to $-A$ corresponds, roughly speaking, to a time inversion. But the parabolic problems are *not* invariant under time inversion—unlike the hyperbolic ones (generally speaking). This has something to do with the difficulty of sending back (to where it came from) heat which has been diffusing into some matter (or, at a higher level of sophistication, with the difficulty of putting Humpty-Dumpty back together again). At any rate, our condition will be

(40.17) *for some $c_0 > 0$ and (almost) every $(x, t) \in \Omega \times]0, T[$:*

$$c_0|\zeta|^2 \leq \operatorname{Re} \sum_{i,j=1}^{n} a^{ij}(x, t)\zeta_i\bar{\zeta}_j \qquad \text{for all} \quad \zeta \in \mathbf{C}^n. \tag{40.18}$$

Let us mention a simple, but important, situation where (40.17) is satisfied. This is when the coefficients a^{ij} are real and symmetric, i.e., $a^{ij} = a^{ji}$, and the quadratic form

$$\sum_{i,j=1}^{n} a^{ij}(x, t)\zeta_i \bar{\zeta}_j$$

is positive definite, uniformly with respect to (x, t).

Now we come to the interpretation of the boundary conditions. In accordance with the precedent set in the "weak" Dirichlet problem (Sects. 22 and 23) we suppose that $g(x, t)$ is a *trace* on the boundary Γ, depending on t, more precisely that it is a function of t, $0 < t < T$, valued in the space of traces $\mathrm{Tr}^1(\Omega) = H^1(\Omega)/H_0^1(\Omega)$. But, just as in the Dirichlet problem, it is convenient to introduce a representative $\tilde{g} = \tilde{g}(x, t) \in H^1(\Omega)$ of g. For each t, or for almost each t, $0 < t < T$, the trace of $\tilde{g}$, i.e., its canonical image in the quotient space $\mathrm{Tr}^1(\Omega)$, is equal to g. Condition (40.6) will then be restated as follows:

$$(40.19) \qquad u - \tilde{g} \in H_0^1(\Omega) \qquad \text{for almost every} \quad t, \qquad 0 < t < T.$$

But in the present discussion the matter of the regularity of the intervening functions, either data or unknown, cannot be waved aside. We must decide what kind of regularity in t to assign to $\tilde{g}$ or to require from u. From empirical considerations it follows that a "good" function space is the one we now define:

Definition 40.1. We denote by $\mathbf{\Phi}$ *the space of functions* w *such that*

$$(40.20) \qquad w \in L^2(0, T; H^1(\Omega)), \qquad \frac{dw}{dt} \in L^2(0, T; H^{-1}(\Omega)),$$

equipped with its natural norm

$$(40.21) \qquad |||w||| = \left(\int_0^T \{\|w(\cdot, t)\|_1^2 + \|w_t(\cdot, t)\|_{-1}^2\}\, dt\right)^{1/2}.$$

It is seen at once that $\mathbf{\Phi}$ is a Hilbert space and that $C^\infty([0, T]; H^1(\Omega))$ is dense in it. The notation $\mathbf{\Phi}$ might create some confusion if we are dealing with more than one interval $[0, T]$ or more than an open set Ω. More generally we ought to introduce spaces $\mathbf{\Phi}(a, b; \Omega)$. But since the only purpose of making a formal definition of $\mathbf{\Phi}$ is to simplify the statements and the proofs which follow, we shall leave it at that.

Definition 40.2. We denote by $\mathbf{\Phi}_0$ *the subspace of* $\mathbf{\Phi}$ *consisting of the functions* w *such that* $w(\cdot, t) \in H_0^1(\Omega)$ *for almost every* t, $0 < t < T$.

Since $H_0^1(\Omega)$ is closed in $H^1(\Omega)$, so is $\mathbf{\Phi}_0$ in $\mathbf{\Phi}$. We shall always equip $\mathbf{\Phi}_0$ with the Hilbert space structure induced by that of $\mathbf{\Phi}$.

Condition (40.19) now reads

$$u - \tilde{g} \in \mathbf{\Phi}_0 . \tag{40.22}$$

Observe that every function $w \in \mathbf{\Phi}$ can be extended to the *closed* interval $[0, T]$ as an absolutely continuous function valued in $H^{-1}(\Omega)$ and, as such, it has a limit at $+0$ in $H^{-1}(\Omega)$. We could interpret the initial condition (40.13) in that light. But there are advantages in seeking a stronger kind of continuity —continuity not in $H^{-1}(\Omega)$ but in $H^0(\Omega) = L^2(\Omega)$. As a matter of fact, this is automatically achieved if we substitute $\mathbf{\Phi}_0$ for $\mathbf{\Phi}$ as we now show. Let us provide the space $C^0([0, T]; L^2(\Omega))$ with the "maximum norm"

$$\sup_{0<t<T} \|u(\cdot, t)\|_0 = \sup_{0<t<T} \left\{\int_\Omega |u(x, t)|^2 \, dx\right\}^{1/2}.$$

LEMMA 40.2. *The natural injection*

$$C^\infty([0, T]; H_0^1(\Omega)) \to C^0([0, T]; L^2(\Omega)) \tag{40.23}$$

can be extended as a continuous injection $\mathbf{\Phi}_0 \to C^0([0, T]; L^2(\Omega))$.

Proof. Let $u \in \mathbf{\Phi}_0$; we define a function on the symmetric interval $]-T, T[$ by setting $\tilde{u}(t) = u(t)$ for $t > 0$, $\tilde{u}(t) = u(-t)$ for $t < 0$. It is clear that $u \mapsto \tilde{u}$ is a continuous injection of $\mathbf{\Phi}_0 = \mathbf{\Phi}_0(0, T)$ into $\mathbf{\Phi}_0(-T, T)$. Let now $\alpha \in C^\infty(\mathbf{R})$, $\alpha(t) = 0$ for $t < -T$, $\alpha(t) = 1$ for $t > 0$. Let $u \in C^\infty([0, T]; H_0^1(\Omega))$. We have

$$\begin{aligned} \|(\alpha\tilde{u})(\cdot, t)\|_0^2 &= 2 \operatorname{Re} \int_{-T}^t \langle(\alpha\tilde{u})(\cdot, s), \overline{(\alpha\tilde{u})_t(\cdot, s)}\rangle \, ds \\ &\leqq \int_{-T}^T \{\|(\alpha\tilde{u})(\cdot, s)\|_1^2 + \|(\alpha\tilde{u})_t(\cdot, s)\|_{-1}^2\} \, ds \\ &\leqq C(\alpha) \int_{-T}^T \{\|\tilde{u}(\cdot, s)\|_1^2 + \|\tilde{u}_t(\cdot, s)\|_{-1}^2\} \, ds \\ &\leqq 2C(\alpha) |||u||| \end{aligned} \tag{40.24}$$

(for the mapping $u \mapsto \tilde{u}$ has norm 2). Since the restriction of $\alpha\tilde{u}$ to $[0, T]$ is equal to u, we see at once that the natural injection (40.23) is continuous when the first space carries the norm $||| \;\; |||$ induced by $\mathbf{\Phi}_0$ [defined in (40.21)], hence has a unique continuous extension to $\mathbf{\Phi}_0$ [where the space $C^\infty([0, T]; \mathrm{H}_0^1(\Omega))$ is dense]. Q.E.D.

Thus the elements of $\mathbf{\Phi}_0$ can be regarded as continuous functions in the closed interval $[0, T]$, valued in $L^2(\Omega)$. If the same were true of the elements of $\mathbf{\Phi}$, we would have a very convenient interpretation of the initial condition (40.13). Actually, this is the case when Ω is bounded and has a C^1 boundary (and lies on one side of it; Exercise 40.4). But in order not to lose generality, we shall not restrict our choice of the open set Ω. Rather, we are going to limit (very mildly) the choice of the function g entering our boundary condition: We are going to require that $\tilde{g}(\cdot, t)$ converge in $L^2(\Omega)$ as $t \to +0$.

Let us state the existence and uniqueness theorem:

THEOREM 40.1. *Suppose that the coefficients of* $A(x, t, \partial/\partial x)$, *in the variational form* (40.15), a^{ij}, b^j, $c^i (1 \leq i, j \leq n)$ *all belong to* $L^\infty(\Omega \times]0, T[)$ *and that* $A(x, t, \partial/\partial x)$ *satisfies the ellipticity assumption* (40.17).

Then, to every $f \in L^2(0, T; H^{-1}(\Omega))$, *every* $u_0 \in L^2(\Omega)$, *and every* $\tilde{g} \in L^2(0, T; H^1(\Omega))$ *such that*

(40.25)
$$\frac{\partial \tilde{g}}{\partial t} \in L^2(0, T; H^{-1}(\Omega)),$$

(40.26) $\tilde{g}(\cdot, t)$ *converges in* $L^2(\Omega)$ *to a function* $\tilde{g}(\cdot, 0)$ *as* $t > 0$ *goes to zero,*

there is a unique function $u \in L^2(0, T; H^1(\Omega))$ *having the following properties:*

(40.27)
$$\frac{\partial u}{\partial t} + A(x, t, \partial/\partial x)u = f \quad \text{in } \Omega \times]0, T[,$$

(40.28) $u(\cdot, t) - \tilde{g}(\cdot, t) \in H_0^1(\Omega)$ *for almost every* t, $0 < t < T$,

(40.29) $u(\cdot, t)$ *converges in* $L^2(\Omega)$ *to* u_0 *as* $t \to +0$.

Remark 40.1. If $u \in L^2(0, T; H^1(\Omega))$ satisfies (40.27), u belongs to $\mathbf{\Phi}$ (Definition 40.1). If, moreover, (40.28) holds, then $u - \tilde{g} \in \mathbf{\Phi}_0$ and, therefore, by Lemma 40.2, $u - \tilde{g}$ is a continuous mapping of the closed interval $[0, T]$ into $L^2(\Omega)$. Then, by (40.26), we see that $u(\cdot, t)$ must converge in $L^2(\Omega)$ as $t \to +0$, to a limit which we shall denote by $u(\cdot, 0)$. Condition (40.29) simply means that

(40.30)
$$u(\cdot, 0) = u_0.$$

Remark 40.2. We may apply Theorem 40.1 in the case where $f \equiv 0$, $\tilde{g} \equiv 0$. By virtue of Lemma 40.2 we obtain the following result:

THEOREM 40.2. *The mapping* $u \mapsto u(\cdot, 0)$ *from* $\mathbf{\Phi}_0$ *into* $L^2(\Omega)$ *is surjective.*

Such results as Theorem 40.2 can be generalized to Hilbert spaces (or Banach spaces) other than $H_0^1(\Omega)$, $L^2(\Omega)$, $H^{-1}(\Omega)$. They are very simple particular cases of the *theory of interpolation* (and of the *trace* method of interpolation; cf. Exercise 25.1). Theorem 40.2 implies that $L^2(\Omega) = H^0(\Omega)$ is an interpolation space between $H_0^1(\Omega)$ and $H^{-1}(\Omega)$.

Remark 40.3. The validity of Theorem 40.2 explains why there is *no compatibility condition*, in Theorem 40.1, linking u_0 and $\tilde{g}$. In view of (40.28)–(40.29) one could have expected that $u_0 - \tilde{g}(\cdot, 0)$ should belong to $H_0^1(\Omega)$. But notice that such a requirement would be meaningless, since u_0 belongs to $L^2(\Omega)$. More important, the value of $\tilde{g}$ at $t = 0$ cannot be relevant to the statement of Theorem 40.1: Indeed Theorem 40.2 tells us there always is a function h in $\mathbf{\Phi}_0$ such that $h(\cdot, 0) = \tilde{g}(\cdot, 0)$. The requirement (40.28) on u would not be modified if we substituted $\tilde{g} - h$ for $\tilde{g}$.

Remark 40.4. Theorem 40.1 may of course be applied to $\Omega = \mathbf{R}^n$. Then, the boundary condition (40.28) becomes void, as $H^1(\mathbf{R}^n) = H_0^1(\mathbf{R}^n)$. In this case, Theorem 40.1 asserts that, given any $f \in L^2(0, T; H^{-1}(\mathbf{R}^n))$, there is a unique solution $u \in L^2(0, T; H^1(\mathbf{R}^n))$ of the equation $u_t + Au = f$ with arbitrarily preassigned initial value $u(\cdot, 0)$ in $L^2(\mathbf{R}^n)$. This is no longer a *mixed problem*—it is a *Cauchy problem* (see Chapter II).

Remark 40.5. Usually, an existence and uniqueness statement like Theorem 40.1 is complemented by a statement of continuous dependence on the data. In the present situation, this would take the form that *the mapping*

$$(f, u_0, \tilde{g}) \mapsto u,$$

from $L^2(0, T; H^{-1}(\Omega)) \times L^2(\Omega) \times \mathbf{\Phi}$ *into* $L^2(0, T; H^1(\Omega))$ *is continuous.* As a matter of fact, assertions like this one usually follow at once from the existence and uniqueness of the solution, via the closed graph theorem. But in our case, we have the additional condition (40.26) which perturbs the situation somewhat. Thus $\tilde{g}$ and u belong to a linear subspace of $\mathbf{\Phi}$ (Definition 40.1) which, at first sight, is not closed. Therefore we cannot apply the closed graph theorem. Nevertheless continuity holds. This follows from an estimate derived at the same time as the existence of the solution [Estimate (41.26)].

Exercises

40.1. Let $\mathbf{V}$ be a Hilbert space, $\overline{\mathbf{V}}'$ its antidual, $\mathbf{H}$ a third Hilbert space such that we have continuous injections, with dense images, $\mathbf{V} \subsetneq \mathbf{H} \subsetneq \overline{\mathbf{V}}'$. Furthermore we assume that the injection $\mathbf{H} \subsetneq \overline{\mathbf{V}}'$ is the "adjoint" of the injection $\mathbf{V} \subsetneq \mathbf{H}$. Let T be any number greater than zero and call $\mathbf{\Phi}_0(0, T; \mathbf{V})$ the space of $\mathbf{V}$-valued functions $\mathbf{u}(t)$ in $[0, T]$ such that

$$\text{(40.31)} \qquad \mathbf{u} \in L^2(0, T; \mathbf{V}), \qquad \frac{d\mathbf{u}}{dt} \in L^2(0, T; \overline{\mathbf{V}}'),$$

equipped with the Hilbert norm

$$\text{(40.32)} \qquad |||\mathbf{u}||| = \left(\int_0^T \left\{ \|\mathbf{u}(t)\|_{\mathbf{V}}^2 + \left\| \frac{d\mathbf{u}}{dt}(t) \right\|_{\mathbf{V}'}^2 \right\} dt \right)^{1/2}.$$

Prove that $C^\infty([0, T]; \mathbf{V})$ is dense in $\mathbf{\Phi}_0(0, T; \mathbf{V})$ and that the natural injection of $C^\infty([0, T]; \mathbf{V})$ into $C^0([0, T]: \mathbf{H})$ can be extended as a continuous injection of $\mathbf{\Phi}_0(0, T; \mathbf{V})$ into $C^0([0, T]: \mathbf{H})$.

40.2. For almost every t, $0 \leqq t \leqq T$, let $A(t)$ be a bounded linear operator of $\mathbf{V}$ into $\overline{\mathbf{V}}'$ (see Exercise 40.1), such that, whatever $\mathbf{v}, \mathbf{w} \in \mathbf{V}$,

(40.33) *the function* $t \mapsto \langle A(t)\mathbf{v}, \mathbf{w}\rangle^-$ *belongs to* $L^\infty(0, T)$.

Prove that $\mathbf{v}(t) \mapsto A(t)\mathbf{v}(t)$ defines a bounded linear map of $L^2(0, T; \mathbf{V})$ into $L^2(0, T; \mathbf{V}')$. ($\langle\ ,\ \rangle^-$ is the bracket of the antiduality between $\mathbf{V}$ and $\overline{\mathbf{V}}'$.)

40.3. Consider an operator

$$A(x, t, \partial/\partial x) = \sum_{|p|, |q| \leqq m} (-1)^{|p|}\left(\frac{\partial}{\partial x}\right)^p a_{p,q}(x, t)\left(\frac{\partial}{\partial x}\right)^q, \qquad m \geqq 1, \tag{40.34}$$

whose coefficients belong to $L^\infty(\Omega \times [0, T])$. State precisely the hypotheses and the conclusion in the theorem analogous to Theorem 40.1 where $A(x, t, \partial/\partial x)$ is chosen according to (40.34) (state with special care what the boundary conditions and the hypotheses on the boundary data should be).

40.4. Let T be a strictly positive number and $u(x, t)$ a function in $\mathbf{R}^1_+ \times]0, T[$ such that

$$u \in L^2(0, T; H^1(\mathbf{R}^1_+)), \qquad \frac{du}{dt} \in L^2(0, T; H^{-1}(\mathbf{R}^1_+)).$$

Call $\tilde{u}(x, t)$ the function in $\mathbf{R}^1 \times]0, T[$ equal to $u(x, t)$ for $x > 0$ and to $u(-x, t)$ for $x < 0$. Show that we have

$$\tilde{u} \in L^2(0, T; H^1(\mathbf{R}^1)), \qquad \frac{d\tilde{u}}{dt} \in L^2(0, T; H^{-1}(\mathbf{R}^1)).$$

Indicate how we can derive from this the following result:

LEMMA 40.3. *Let Ω be a bounded open subset of $\mathbf{R}^n$ whose boundary is a C^1 hypersurface (and which lies on one side of it). Let Ω_0 be an open subset of $\mathbf{R}^n$ containing $\overline{\Omega}$. There is a continuous linear mapping $\varepsilon_\Omega : H^1(\Omega) \to H^1_0(\Omega_0)$ such that $r_\Omega \varepsilon_\Omega =$ Identity in $H^1(\Omega)$* (r_Ω is the mapping "restriction to Ω"), *and furthermore such that ε_Ω extends as a continuous linear mapping of $\mathbf{\Phi}(0, T; \Omega)$ into $\mathbf{\Phi}_0(0, T; \Omega_0)$* (Definitions 40.1 and 40.2).

40.5. Let Ω be as in Lemma 40.3. Show that the space $\mathbf{\Phi}$ (Definition 40.1) can be naturally, and continuously, injected into $C^0([0, T]; L^2(\Omega))$.

40.6. Let Ω be as in Lemma 40.3 but suppose now that its boundary is a C^{m+1} hypersurface ($m \geqq 1$ arbitrary). Consider the space $\mathbf{\Phi}^m(0, T; \Omega)$ of functions $u(x, t)$ in $\Omega \times]0, T[$ such that

$$u \in L^2(0, T; H^{m+1}(\Omega)), \qquad \frac{du}{dt} \in L^2(0, T; H^{m-1}(\Omega)).$$

Prove that $\mathbf{\Phi}^m(0, T; \Omega)$ can be naturally injected into $C^0([0, T]; H^m(\Omega))$ and that the injection is continuous when we equip $\mathbf{\Phi}^m(0, T; \Omega)$ with the obvious norm. [*Hint*: Use Theorem 26.A.3 and Remark 26.A.1.)

41

Energy Inequalities. Proof of Theorem 40.I: Existence and Uniqueness of the Weak Solution to the Parabolic Mixed Problem

For $\tau > 0$ we define the sesquilinear form on $H^1(\Omega) \times H^1(\Omega)$:

$$a_\tau(t; u, v) = \tau \int_\Omega u\bar{v}\, dx + \sum_{i,j=1}^{n} \int_\Omega a^{ij}(x, t) \frac{\partial u}{\partial x^j} \frac{\partial \bar{v}}{\partial x^i}\, dx$$

$$+ \sum_{j=1}^{n} \int_\Omega b^j(x, t) \frac{\partial u}{\partial x^j} \bar{v}\, dx + \int_\Omega c(x, t) u\bar{v}\, dx.$$

It is the analog, in the present situation, of the functional (23.7). Our hypotheses (40.16) and (40.17) imply at once

(41.1) *There is* $\tau_0 > 0$ *such that, for all* $\tau > \tau_0$, *all* $u \in H^1(\Omega)$ *and almost all* t, $0 < t < T$,

(41.2) $$c_0 \|u\|_1^2 \leqq 2 \operatorname{Re} a_\tau(t; u, u).$$

We recall that $\|u\|_1^2$ is the norm of u in $H^1(\Omega)$. The constant c_0 is the same as in (40.17).

In (41.2) we may take u dependent on t. As a matter of fact, we derive from (41.2), for arbitrary $u \in L^2(0, T; H^1(\Omega))$,

(41.3) $$c_0 \int_0^T \|u\|_1^2\, dt \leqq 2 \operatorname{Re} \int_0^T a_\tau(t; u, u)\, dt.$$

Now we define two sesquilinear forms on $\mathbf{\Phi}_0 \times \mathbf{\Phi}_0$ (Definition 40.2):

$$\mathfrak{A}_\tau(u, v) = \int_0^T \{-\langle u, \bar{v}_t\rangle + a_\tau(t; u, v)\}\, dt, \tag{41.4}$$

$$\mathfrak{A}'_\tau(u, v) = \int_0^T \{\langle u_t, \bar{v}\rangle + a_\tau(t; u, v)\}\, dt. \tag{41.5}$$

We observe that (for $u \in \mathbf{\Phi}_0$)

$$2 \operatorname{Re} \int_0^T \langle u, \bar{u}_t\rangle\, dt = \int_\Omega (|u(x, T)|^2 - |u(x, 0)|^2)\, dx \tag{41.6}$$

(cf. Lemma 40.2).

If we combine (41.3) with (41.6), we obtain the following important "energy inequalities":

LEMMA 41.1. *If (40.16) and (40.17) hold, then we have, for all $\tau > \tau_0$ and all $u \in \mathbf{\Phi}_0$* (Definition 40.2)

$$c_0 \int_0^T \|u(\cdot, t)\|_1^2\, dt + \int_\Omega |u(x, 0)|^2\, dx \leqq 2 \operatorname{Re} \mathfrak{A}_\tau(u, u) + \int_\Omega |u(x, T)|^2\, dx, \tag{41.7}$$

$$c_0 \int_0^T \|u(\cdot, t)\|_1^2\, dt + \int_\Omega |u(x, T)|^2\, dx \leqq 2 \operatorname{Re} \mathfrak{A}'_\tau(u, u) + \int_\Omega |u(x, 0)|^2\, dx. \tag{41.8}$$

The proof of Theorem 40.1 is based on the exploitation of these inequalities: (41.7) will lead to the existence of the solution, (41.8) to its uniqueness. In some way the parameter $\tau > 0$ must be introduced. This is done by transforming the unknown u. We set

$$U = e^{-\tau t}(u - \tilde{g}).$$

It is clear that $U \in \mathbf{\Phi}_0$ and that

$$U_t + (\tau + A)U = F \tag{41.9}$$

with

$$F = e^{-\tau t}\left(f - \frac{\partial \tilde{g}}{\partial t} - A\tilde{g}\right) \in L^2(0, T; H^{-1}(\Omega)). \tag{41.10}$$

The initial condition (40.28) now reads

$$U(\cdot, 0) = U_0 \tag{41.11}$$

where

$$U_0 = u_0 - \tilde{g}(\cdot, 0). \tag{41.12}$$

I Existence of the Solution

We shall apply the following generalization (due to J. L. Lions) of the Lax–Milgram lemma (Lemma 23.1):

LEMMA 41.2. *Let* $\mathbf{E}$ *be a Hilbert space,* $\mathfrak{h}$ *a linear subspace of* $\mathbf{E}$, $\mathfrak{A}(\mathbf{w}, \mathbf{h})$ *a sesquilinear functional on* $\mathbf{E} \times \mathfrak{h}$ *having the following properties*:

(41.13) *for each fixed* $\mathbf{h} \in \mathfrak{h}$, $\mathbf{w} \mapsto \mathfrak{A}(\mathbf{w}, \mathbf{h})$ *is a continuous linear functional on* $\mathbf{E}$;

(41.14) *there is* $c_0' > 0$ *such that, for every* $\mathbf{h} \in \mathfrak{h}$,

$$c_0' \|\mathbf{h}\|_{\mathbf{E}}^2 \leqq |\mathfrak{A}(\mathbf{h}, \mathbf{h})|.$$

Conclusion. There is a bounded linear map G *of the antidual* $\overline{\mathbf{E}}'$ *of* $\mathbf{E}$ *into* $\mathbf{E}$, *with norm* $\leqq c_0'^{-1}$, *such that, for every continuous antilinear functional* λ *on* $\mathbf{E}$,

$$\mathfrak{A}(G\lambda, \mathbf{h}) = \lambda(\mathbf{h}) \qquad \textit{for every} \quad \mathbf{h} \in \mathfrak{h}. \tag{41.15}$$

Proof. By (41.13) to every $\mathbf{h} \in \mathfrak{h}$ there is an element $R\mathbf{h}$ of $\mathbf{E}$ such that $\mathfrak{A}(\mathbf{w}, \mathbf{h}) = (\mathbf{w}, R\mathbf{h})_{\mathbf{E}}$ for all $\mathbf{w} \in \mathbf{E}$. By (41.14) we have

$$\|\mathbf{h}\|_{\mathbf{E}} \leqq c_0'^{-1} \|R\mathbf{h}\|_{\mathbf{E}}, \qquad \forall \mathbf{h} \in \mathfrak{h}. \tag{41.16}$$

Thus R defines an injective linear map $\mathfrak{h} \to \mathbf{E}$ (in general, not continuous). Also (41.16) shows that the mapping $R\mathbf{h} \mapsto \mathbf{h}$, from $R\mathfrak{h}$ equipped with the norm of $\mathbf{E}$, into $\mathbf{E}$, is continuous. It extends by continuity to the closure of $R\mathfrak{h}$ in $\mathbf{E}$ and, by setting it equal to zero on the orthogonal of $R\mathfrak{h}$, to the whole of $\mathbf{E}$. This extension, which we denote by G_1^*, has a norm $\leqq c_0'^{-1}$. Its adjoint, which we denote by G_1, is a bounded linear operator $\mathbf{E} \to \mathbf{E}$ with the same norm. We have, for all $\mathbf{w} \in \mathbf{E}$ and all $\mathbf{h} \in \mathfrak{h}$,

$$\mathfrak{A}(G_1 \mathbf{w}, \mathbf{h}) = (G_1 \mathbf{w}, R\mathbf{h})_E = (\mathbf{w}, G_1^* R\mathbf{h})_E = (\mathbf{w}, \mathbf{h})_E.$$

Now let J denote the canonical isomorphism of $\overline{\mathbf{E}}'$ onto $\mathbf{E}$. We set $G\lambda = G_1 J\lambda$ for all $\lambda \in \overline{\mathbf{E}}'$. Q.E.D.

We shall apply Lemma 41.2 with the following choice of $\mathbf{E}$: It will be the space of pairs (v, v_0), $v \in L^2(0, T; H_0^1(\Omega))$, $v_0 \in L^2(\Omega)$. The norm on $\mathbf{E}$ will be the natural one:

$$\|(v, v_0)\|_{\mathbf{E}} = \left\{ \int_0^T \|v(\cdot, t)\|_1^2 \, dt + \int_\Omega |v_0(x)|^2 \, dx \right\}^{1/2}.$$

As for $\mathfrak{h}$ it will be the subspace of $\mathbf{E}$ consisting of the pairs (v, v_0) such that $v \in \mathbf{\Phi}_0$, $v_0 =$ limit of $v(\cdot, t)$ as $t \to +0$ in $L^2(\Omega)$, and furthermore such that

(41.17) *the limit in* $L^2(\Omega)$ *of* $v(\cdot, t)$ *as* $t < T$ *tends to* T *is zero.*

(We have tacitly applied Lemma 40.2.)

In our situation, the form $\mathfrak{A}$ of Lemma 41.2 will be the form

$$((w, w_0), (h, h_0)) \mapsto \mathfrak{A}_\tau(w, h), \qquad \mathfrak{A}_\tau \text{ given by (41.4).}\dagger$$

The energy estimate (41.7) shows, in view of (41.17), that the hypothesis (41.14) of Lemma 41.2 is verified; as for (41.13) it is verified in view of the fact that $\mathfrak{h} \subset \mathbf{\Phi}_0$. We may therefore take advantage of the conclusion. The continuous functional λ on $\mathbf{E}$ will be

$$(v, v_0) \mapsto \int_0^T \langle F, \bar{v} \rangle \, dt + \int_\Omega U_0(x) \bar{v}_0(x) \, dx, \tag{41.18}$$

where F and U_0 are the functions in (41.9) and (41.11), respectively, and $\langle \, , \rangle$ is the bracket of the duality between $H^{-1}(\Omega)$ and $H_0^1(\Omega)$. We reach the conclusion that there is $(V, V_0) \in \mathbf{E}$ such that

$$\mathfrak{A}_\tau(V, h) = \int_0^T \langle F, \bar{h} \rangle \, dt + \int_\Omega U_0(x) \bar{h}(x, 0) \, dx, \qquad \forall h \in \mathfrak{h}. \tag{41.19}$$

Let us first choose h in $C_c^\infty(]0, T[; H_0^1(\Omega))$. In this case, Eq. (41.19) reads

$$\int_0^T \langle V_t + (\tau + A)V, \bar{h} \rangle \, dt = \int_0^T \langle F, \bar{h} \rangle \, dt, \tag{41.20}$$

which implies

$$V_t + (\tau + A)V = F \tag{41.21}$$

in the sense of distributions in $]0, T[$ with values in $H^{-1}(\Omega)$, or, also, in the sense of distributions in $\Omega \times]0, T[$. Equation (41.21) implies at once that $V_t \in L^2(0, T; H^{-1}(\Omega))$, hence $V \in \mathbf{\Phi}_0$. In particular, V may be viewed as a continuous function in the closed interval $[0, T]$ with values in $L^2(\Omega)$. If then we return to (41.19), we see by integration by parts that

$$\mathfrak{A}_\tau(V, h) = \langle V, \bar{h} \rangle \Big|_{t=0} + \int_0^T \langle V_t + (\tau + A)V, \bar{h} \rangle \, dt$$

and therefore, by (41.21), that

$$\langle V, \bar{h} \rangle \Big|_{t=0} = \int_\Omega V(x, 0) \bar{h}(x, 0) \, dx = \int_\Omega U_0(x) \bar{h}(x, 0) \, dx. \tag{41.22}$$

Clearly we may take $h \in \mathfrak{h}$ such that $h(x, 0)$ is any given element of $H_0^1(\Omega)$. Thus (41.22) implies $V(x, 0) = U_0(x)$ a.e. in Ω. We see that V is a solution of (41.9)–(41.11).

† It is clear that, whereas $\mathfrak{A}'_\tau(w, h)$ can be extended to $\mathbf{\Phi}_0 \times L^2(0, T; H_0^1(\Omega))$, $\mathfrak{A}_\tau(w, h)$ can be extended as a sesquilinear functional on $L^2(0, T; H_0^1(\Omega)) \times \mathbf{\Phi}_0$.

II Uniqueness of the Solution

By subtraction, it is a matter of proving that if $W \in \mathbf{\Phi}_0$ satisfies

$$W_t + (\tau + A)W = 0, \tag{41.23}$$

$$W(\cdot, 0) = 0, \tag{41.24}$$

we must have $W \equiv 0$. By (41.5) and (41.23) we have

$$\mathfrak{A}'_\tau(W, W) = \int_0^T \langle W_t + (\tau + A)W, \overline{W} \rangle \, dt = 0,$$

and hence our contention by (41.8) [in view of (41.24)].

III Estimation of the Solution

Let us return briefly to the argument of part I (existence of the solution) of the present section, in particular, to our choice of the spaces $\mathbf{E}$, $\mathfrak{h}$, of the form $\mathfrak{A}$, etc., when we have applied Lemma 41.2. We observe that the norm, in $\overline{\mathbf{E}}'$, of the antilinear functional λ, given by (41.18), is

$$\|\lambda\| = \left\{\int_0^T \|F(\cdot, t)\|^2_{H^{-1}(\Omega)} \, dt + \int_\Omega |U_0(x)|^2 \, dx\right\}^{1/2}.$$

If then we apply the conclusion of Lemma 41.2 and, in particular, the estimate on the norm of the operator G, we find that the norm, in $\mathbf{E}$, of the solution U of (41.9)–(41.11), is $\leqq 2c_0^{-1}\|\lambda\| = C_0\|\lambda\|$. In other words,

$$\int_0^T \|U(\cdot, t)\|^2_{H^1(\Omega)} \, dt \leqq C_0^2\left\{\int_0^T \|F(\cdot, t)\|^2_{H^{-1}(\Omega)} \, dt + \|U_0\|^2_{L^2(\Omega)}\right\}. \tag{41.25}$$

At this point we return to the original unknown u, and express U in terms of u and $\tilde{g}$. We derive from (41.25) the following estimate:

$$\begin{aligned} &\int_0^T \|u(\cdot, t) - \tilde{g}(\cdot, t)\|^2_{H^1(\Omega)} \, dt \\ &\leqq K^2\left\{\int_\Omega |u_0(x) - \tilde{g}(x, 0)|^2 \, dx + \int_0^T \|f(\cdot, t)\|^2_{H^{-1}(\Omega)} \, dt \right. \\ &\quad \left. + \int_0^T \{\|\tilde{g}(\cdot, t)\|^2_{H^1(\Omega)} + \|\tilde{g}_t(\cdot, t)\|^2_{H^{-1}(\Omega)}\} \, dt\right\}. \end{aligned} \tag{41.26}$$

This is the estimate announced in Remark 40.5. The constant K depends on Ω, $A(x, t, \partial/\partial x)$, and also on the number $T > 0$.

Exercises

41.1. Let $\mathbf{V}$, $\mathbf{H}$ be as in Exercise 40.1, and the operator-valued function $A(t)$ as in Exercise 40.2. Let us assume now that there are two constants $\lambda_0 \in \mathbf{R}$, $c_0 > 0$ such that, *for almost every* $t \in [0, T]$,

$$\operatorname{Re}\langle A(t)\mathbf{v}, \mathbf{v}\rangle^{-} + \lambda_0 \|\mathbf{v}\|_{\mathbf{H}}^2 \geqq c_0 \|v\|_{\mathbf{V}}^2, \qquad \forall \mathbf{v} \in \mathbf{V}. \tag{41.27}$$

Duplicate the argument in Sect. 41 so as to prove the following theorem:

THEOREM 41.1. *Under the preceding hypotheses, given any function* $\mathbf{f}(t) \in L^2(0, \mathbf{T}; \overline{\mathbf{V}}')$ *and any element* $\mathbf{u}_0 \in \mathbf{H}$, *there is a unique function* $\mathbf{u} \in L^2(0, T; \mathbf{V})$ *such that*

$$\frac{d\mathbf{u}}{dt} + A(t)\mathbf{u} = \mathbf{f} \quad \textit{in }]0, T[; \tag{41.28}$$

$$\mathbf{u}(0) = \mathbf{u}_0 . \tag{41.29}$$

[By the result in Exercise 40.2, we derive $d\mathbf{u}/dt \in L^2(0, T; \overline{\mathbf{V}}')$ from Eq. (41.28) and by the one in Exercise 40.1 we see that $\mathbf{u}$ can be regarded as a continuous function, in the closed interval $[0, T]$, valued in $\mathbf{H}$, from which we obtain the meaning of (41.29).]

41.2. Write down explicitly and check the energy estimates (41.7) and (41.8) in the case where the operator $A(x, t, \partial/\partial x)$ in Theorem 40.1 is minus the Laplace operator in the space variables x.

41.3. Let $\mathbf{V}$ and $A(t)$ be as in Exercise 40.2. Let $\mathbf{v} \in L^2(0, T; \mathbf{V})$ be a solution of the equation

$$\frac{d\mathbf{v}}{dt} + A(t)\mathbf{v} = 0 \quad \text{in }]0, T[. \tag{41.30}$$

Show that there is a constant $C > 0$ such that

$$\left\| \frac{d\mathbf{v}}{dt} \right\|_{\overline{\mathbf{V}}'} \leqq C \|\mathbf{v}\|_{\mathbf{V}'} . \tag{41.31}$$

Apply the method in Exercise 20.2 to prove that if $\mathbf{v}(0) = 0$ [$\mathbf{v}(t)$ is an absolutely continuous function in the closed interval $0 \leqq t \leqq T$ valued in $\overline{\mathbf{V}}'$], then necessarily $\mathbf{v}(t) = 0$ for all $t < T$.

41.4. Let Ω be a bounded open subset of $\mathbf{R}^n$ whose boundary Γ is a C^∞ hypersurface and which lies on one side of it. Let $\Gamma = \Gamma_0 \cup \Gamma_1$, $\Gamma_0 \cap \Gamma_1 = \varnothing$ and let $g(x, t)$ be a C^1 function of $t \in \mathbf{R}^1$, vanishing for $t > 0$, valued in

$H^2(\Omega)$. Let $u_0(x) \in L^2(\Omega)$. Show how to apply Theorem 41.1 to prove that the mixed problem:

$$u_t = \Delta u \quad \text{in } \Omega \times]0, T[, \tag{41.32}$$

$$u(x, 0) = u_0(x) \quad \text{in } \Omega, \tag{41.33}$$

$$u - g = 0 \quad \text{on } \Gamma_0, \qquad \frac{\partial}{\partial \nu}(u - g) = 0 \quad \text{on } \Gamma_1, \qquad \forall t > 0, \tag{41.34}$$

has one and only one solution $u \in L^2(0, T; \mathbf{V})$ for a suitable choice of $\mathbf{V}$, $H_0^1(\Omega) \subset \mathbf{V} \subset H^1(\Omega)$, and under suitable hypotheses on u_0 and g (state the latter hypotheses precisely).

41.5. Let Ω and $\Gamma = \partial\Omega$ be as in Exercise 41.4. Let

$$P(x, t, D) = \sum_{|\alpha| \leqq 2m} a_\alpha(x, t) D^\alpha$$

be a uniformly strongly elliptic operator in Ω (see Sect. 36; $m \geqq 1$). Make the assumption that all partial derivatives with respect to x of order $\leqq m$, of every coefficient a_α, belong to $L^\infty(\Omega \times [0, T])$. Let $u_0 \in L^2(\Omega)$, $g_j(\cdot, t)$ be a once continuously differentiable function of t, $0 \leqq t \leqq T$, with values in $H^{m-j-1/2}(\Gamma)$ for each $j = 0, \ldots, m-1$. Show how to apply Theorem 41.1 to derive that, under suitable additional conditions linking u_0 and the g_j, which the student must precisely state (cf. Exercise 40.3), the mixed problem

$$\frac{du}{dt} + P(x, t, D_x)u = 0 \quad \text{in } \Omega \times]0, T[, \tag{41.35}$$

$$u(x, 0) = u_0(x) \quad \text{in } \Omega, \tag{41.36}$$

$$\frac{\partial^j}{\partial \nu^j} u|_\Gamma = g_j, \qquad j = 0, \ldots, m-1, \tag{41.37}$$

has a unique solution $u \in \mathbf{\Phi}^m(0, T; \Omega)$ (see Exercise 40.6).

42

Regularity of the Weak Solution with Respect to the Time Variable

We have established the existence and uniqueness of the solution u to the mixed problem (40.27)–(40.28)–(40.29) and we should now examine its regularity, under appropriate regularity assumptions on the data. But if we compare the present situation to that of boundary value problems, and regard a mixed problem as a boundary value problem, of a peculiar kind, relative to a truncated cylinder in a space $\mathbf{R}^{n+1}$, namely $\Omega \times]0, T[$, we cannot help but notice the fact that the region in question can never be smooth: It must always have corners at the boundary of the set Ω lying in the hyperplane $t = 0$ (unless Ω is the whole space $\mathbf{R}^n$, and then there is no boundary at all—but then our problem is a global Cauchy problem!). Indeed, such corners introduce singularities, even in the simplest of situations (see Exercise 44.2). One may then be willing to limit the investigation of the regularity of the solution near the portions of the boundary which are sufficiently smooth, and indeed this can be carried out by methods not unlike those used in the elliptic boundary problems (Sect. 27), and which are, of course, rather complicated. Actually, this comment would also apply to the study of the regularity with respect to the space variables only. Under suitable assumptions, the latter can be derived from analogous results for elliptic equations, of the kind obtained for $\lambda - \Delta$ in Sect. 27 (but, needless to say, more general). When the coefficients of the operator under study are independent of t, one can go rather far in this direction and at not too great a cost, for instance by the continuous semigroups method (Sect. 45) or by the method of the Laplace transform (Sects. 43 and 44). Indications of what can be achieved in this way are given in Exercise 44.4.

In this section, we merely discuss the regularity with respect to the time variable t. The reason we are able to do so is that such a discussion could as well apply to abstract evolution equations (where of course the peculiarities of the domain or of the data are not there to create complications). However, a discussion of the regularity with respect to the space variables must perforce take these peculiairities into consideration, and would necessarily greatly increase the technicality of the text. At the relatively elementary level where this book would like to remain, the indications contained in Exercise 44.4 should suffice.

Thus, we must assume that the data are suitably regular with respect to t: This concerns the coefficients of the differential operator $A(x, t, \partial/\partial x)$, the right-hand side $f(x, t)$, and the function $\tilde{g}(x, t)$ which determines the boundary value of the solution. It does not concern the open set Ω (in particular, its boundary $\Gamma = \partial\Omega$) nor the initial datum u_0—although, as we shall see, stronger regularity properties on u, f, $\tilde{g}$ with respect to t will require more stringent regularity conditions *with respect to* x, on the part of u_0 [cf. (42.19)].

If we wish to get a clear idea of the kind of results to expect, and of the methods to use in order to prove them, it is perhaps advisable to study thoroughly the case of *first-order* differentiability. This is what we are now going to do. Let us make the following assumptions (in addition to those already made in Theorem 40.1):

(42.1) *The coefficients* $a^{ij}(x, t)$, $b^j(x, t)$, $c(x, t)$ *of* $A(x, t, \partial/\partial x)$ *have first-order* t *derivatives* (in the distribution sense)—*all of which belong to* $L^\infty(\Omega \times]0, T[)$.

Let us denote by $A_t = A_t(x, t, \partial/\partial x)$ the differential operator obtained by differentiating with respect to t all the coefficients of A:

$$A_t v = (\partial/\partial t)(Av) - Av_t.$$

Note that (42.1) is equivalent to the following: The coefficients a^{ij}, b^j, c are uniformly Lipschitz continuous functions of t, in $[0, T]$, with values in $L^\infty(\Omega)$.

Next we introduce the hypotheses that

(42.2) $f_t \in L^2(0, T; H^{-1}(\Omega))$,

(42.3) $\tilde{g}_t \in L^2(0, T; H^1(\Omega))$, $\quad \tilde{g}_{tt} \in L^2(0, T; H^{-1}(\Omega))$ $\quad$ (i.e., $\tilde{g}_t \in \mathbf{\Phi}$),

(42.4) $\tilde{g}_t(\cdot, t)$ *converges in* $L^2(\Omega)$ *to a function* $\tilde{g}_t(\cdot, 0)$ *as* $t \to +0$.

Needless to say, we are assuming that $\tilde{g} \in L^2(0, T; H^1(\Omega))$; therefore, under the present hypotheses, $\tilde{g}$ is an absolutely continuous function, in the closed interval $[0, T]$, with values in $H^1(\Omega)$, and the initial value $\tilde{g}(\cdot, 0)$ in (40.26) must belong to $H^1(\Omega)$.

We might ask whether, under hypotheses (42.1) to (42.4), we may conclude that the solution u of (40.27)–(40.28)–(40.29) has the following properties:

(42.5) $$u_t \in \mathbf{\Phi},$$

(42.6) $$u_t(\cdot, t) \text{ converges in } L^2(\Omega) \text{ as } t \to +0.$$

Condition (42.5) implies that u, like $\tilde{g}$, is absolutely continuous in $[0, T]$ with values in $H^1(\Omega)$. Therefore, its initial value u_0 must lie in $H^1(\Omega)$, and not merely in $L^2(\Omega)$. Furthermore, condition (40.28) would imply that $u - \tilde{g} \in C^0([0, T]; H_0^1(\Omega))$, and consequently, we should have $u_0 - \tilde{g}(\cdot, 0) \in H_0^1(\Omega)$. From now on we assume that this is so. Note that, Au, like f, is an absolutely continuous function in $[0, T]$ valued in $H^{-1}(\Omega)$. As a consequence of this, the initial value $u_t(\cdot, 0)$ is completely determined by Eq. (40.27):

(42.7) $$u_t(\cdot, 0) = u_1 = f(\cdot, 0) - A(x, 0, \partial/\partial x)u_0 .$$

By (42.6) we must have $u_1 \in L^2(\Omega)$. Generally speaking, we see that our regularity requirements with respect to t have led to increased regularity with respect to x [and also, of course, with respect to (x, t)]. If we differentiate (40.27) with respect to t, we obtain

(42.8) $$u_{tt} + A(x, t, \partial/\partial x)u_t = f_t - A_t(x, t, \partial/\partial x)u.$$

This suggests that we look at the following mixed problem:

(42.9) $$v_t + A(x, t, \partial/\partial x)v = f_t - A_t u \quad \text{in } \Omega \times]0, T[,$$

(42.10) $$v(\cdot, t) - \tilde{g}_t(\cdot, t) \in H_0^1(\Omega) \qquad \text{for almost every} \quad t, \quad 0 < t < T,$$

(42.11) $$v(\cdot, t) \text{ converges in } L^2(\Omega) \text{ to } u_1 \text{ as } t \to +0.$$

By Theorem 40.1 we know that there is a unique solution $v \in \Phi$ to this problem. We are going to show that $v = u_t$. This will prove (42.5) and (42.6). In order to prove this, we shall prove that

$$w(\cdot, t) = u_0 + \int_0^t v(\cdot, t')\, dt'$$

is equal to u. Observe that $w \in C^0([0, T]; H^1(\Omega))$ and that $w_t \in \Phi$. We have

$$w_t = v = u_1 - \int_0^t Av\, dt + f - f(\cdot, 0) - \int_0^t A_t u\, dt$$

$$= Aw + \int_0^t A_t(w - u)\, dt + f,$$

in view of the expression of u_1 [in (42.7)] and of the fact that, by integration by parts,

$$\int_0^t Av\, dt = \int_0^t Aw_t\, dt = Aw - A(x, 0, \partial/\partial x)u_0 - \int_0^t A_t w\, dt.$$

By subtraction from (40.27) we obtain

(42.12) $$(u - w)_t + A(u - w) = \int_0^t A_t(u - w)\, dt.$$

On the other hand,

$$w(\cdot, t) - \tilde{g}(\cdot, t) = u_0 - \tilde{g}(\cdot, 0) + \int_0^t [v(\cdot, t) - \tilde{g}_t(\cdot, t)]\, dt,$$

from which, by (42.10),

$$w(\cdot, t) - \tilde{g}(\cdot, t) - [u_0 - \tilde{g}(\cdot, 0)] \in H_0^1(\Omega) \quad \text{a.e. in } [0, T].$$

By the remark following conditions (42.5)–(42.6) we know that $u_0 - \tilde{g}(\cdot, 0) \in H_0^1(\Omega)$. Thus $w(\cdot, t) - \tilde{g}(\cdot, t) \in H_0^1(\Omega)$ a.e. in $[0, T]$, and therefore, by (40.28),

(42.13) $\quad u(\cdot, t) - w(\cdot, t) \in H_0^1(\Omega) \qquad$ for almost every $\ t, \quad 0 < t < T.$

Finally,

(42.14) $\quad u(\cdot, t) - w(\cdot, t)$ converges to 0 in $L^2(\Omega)$ as $t \to +0$.

As a matter of fact, by (42.13), $u - w \in \mathbf{\Phi}_0$ hence $u - w \in C^0([0, T]; L^2(\Omega))$ (Lemma 40.2). We therefore have the right to rephrase condition (42.14) in the form:

$$(u - w)(\cdot, 0) = 0. \tag{42.15}$$

Consider now the following problem:

$$h \in \mathbf{\Phi}_0, \tag{42.16}$$

$$h_t + Ah = f_1 \in L^2(0, T; H^{-1}(\Omega)), \tag{42.17}$$

$$h(\cdot, 0) = 0. \tag{42.18}$$

By Theorem 40.1 we know that h is uniquely determined. Moreover, the estimate (41.26) simplifies considerably in the present situation. It reads, here,

$$\int_0^T \|h(\cdot, t)\|_{H^1(\Omega)}^2\, dt \leqq K^2 \int_0^T \|f_1(\cdot, t)\|_{H^{-1}(\Omega)}^2\, dt. \tag{42.19}$$

Let us momentarily fix T and let T' be any number, $0 < T' \leqq T$. Let $f_1^\natural$ be any element of $L^2(0, T'; H^{-1}(\Omega))$ and let $h^\natural$ denote the solution of (42.16)–(42.17)–(42.18) but with T' substituted for T and $f_1^\natural$ for f_1. We may extend $f_1^\natural$ to $[0, T]$ by setting the extension f_1 to be zero in $]T', T]$, and then let h denote the solution of the same problem (but now with T and f_1). By the uniqueness of the solution, we must have $h = h^\natural$ in $[0, T'[$. By (42.19) we obtain

$$\begin{aligned} \int_0^{T'} \|h^\natural(\cdot, t)\|_{H^1(\Omega)}^2\, dt &\leqq K^2 \int_0^T \|f_1(\cdot, t)\|_{H^{-1}(\Omega)}^2\, dt \\ &= K^2 \int_0^{T'} \|f_1^\natural(\cdot, t)\|_{H^{-1}(\Omega)}^2\, dt. \end{aligned} \tag{42.20}$$

This shows that the constant K^2 can be chosen independently of T in any finite interval $[0, T_0]$.

We choose now $h = u - v$. Then (42.16)–(42.17)–(42.18) are satisfied with

$$f_1 = \int_0^t A_t h \, dt \qquad \text{[cf. (42.12)]}.$$

By virtue of our hypothesis (42.1) on the coefficients of $A(x, t, \partial/\partial x)$ and of Lemma 40.1, we have

$$\|f_1(\cdot, t)\|^2_{H^{-1}(\Omega)} \leqq K_1^2 \left\{ \int_0^t \|h(\cdot, t')\|_{H^1(\Omega)} \, dt' \right\}^2. \tag{42.21}$$

Now let T' be the largest number, $0 \leqq T' \leqq T$, such that $h \equiv 0$ in $[0, T']$; clearly, T' exists (a priori, it could be zero). From (42.21) we derive that $f_1 \equiv 0$ for $0 \leqq t \leqq T'$, hence, by (42.19) and (42.21),

$$\int_{T'}^T \|h(\cdot, t)\|^2_{H^1(\Omega)} \, dt \leqq K^2 K_1^2 \int_{T'}^T \left\{ \int_{T'}^t \|h(\cdot, t')\|_{H^1(\Omega)} \, dt' \right\}^2 dt$$

$$\leqq K^2 K_1^2 \int_{T'}^T (t - T') \, dt \int_{T'}^T \|h(\cdot, t')\|^2_{H^1(O)} \, dt'.$$

Suppose we had $T' < T$. We would derive from the preceding inequalities

$$1 \leqq K^2 K_1^2 (T - T')^2/2. \tag{42.22}$$

But by our earlier remark on the fact that K can be taken to be independent of $T \leqq T_0$ (and that the same is evidently true of K_1) we may now make T converge to $T' < T$; (42.22) is absurd! We have proved that we must have $T' = T$, i.e., h vanishes identically, that is, $u = w$ in $[0, T]$, and therefore $v = u_t$.

In conclusion, we have proved the following result, *when* $m = 1$:

THEOREM 42.1. *In addition to all the hypotheses in Theorem 40.1 we make the following ones*:

(42.23) *all the derivatives of order* $\leqq m$ *with respect to* t, *of the coefficients* a^{ij}, b^j, c $(1 \leqq i, j \leqq n)$ *of* $A(x, t, \partial/\partial x)$ *belong to* $L^\infty(\Omega \times]0, T[)$,

(42.24) $f^{(j)} \in L^2(0, T; H^{-1}(\Omega))$, $\quad 0 \leqq j \leqq m$,

(42.25) $\tilde{g}^{(j)} \in L^2(0, T; H^1(\Omega))$, $\quad 0 \leqq j \leqq m$, $\quad \tilde{g}^{(m+1)} \in L^2(0, T; H^{-1}(\Omega))$,

(42.26) $\tilde{g}^{(m)}(\cdot, t)$ *converges in* $L^2(\Omega)$ *as* $t \to +0$.

Let then u denote the solution of problem (40.27)–(40.28)–(40.29), whose existence and uniqueness are asserted in Theorem 40.1. Assume that the following is true:

$$u^{(j)}(\cdot, 0) \in L^2(\Omega), \qquad 0 \leqq j \leqq m, \tag{42.27}$$

$$u^{(j)}(\cdot, 0) - \tilde{g}^{(j)}(\cdot, 0) \in H_0^1(\Omega), \qquad 0 \leqq j \leqq m - 1. \tag{42.28}$$

Then the following is true:

(42.29) $u^{(j)} \in L^2(0, T; H^1(\Omega))$, $\quad 0 \leqq j \leqq m$, $\quad u^{(m+1)} \in L^2(0, T; H^{-1}(\Omega))$,

(42.30) $u^{(j)}(\cdot, t) - \tilde{g}^{(j)}(\cdot, t) \in H_0^1(\Omega)$ *for almost every* $t, 0 < t < T$, *and for every* $j = 0, \ldots, m-1$,

(42.31) $u^{(m)}(\cdot, t)$ *converges in* $L^2(\Omega)$ *as* $t \to +0$.

The proof of Theorem 42.1 for general $m > 0$ is straightforward: by induction on m and by applying Theorem 42.1 when $m = 1$ (which we have proved) to $u^{(m-1)}$. We leave the details to the student.

A remark is in order concerning the conditions (42.27) and (42.28): The reader should not get the impression that they bear on the derivatives of the solution u, rather than on the data [the conditions (42.23) to (42.26) obviously bear on the data]. As a matter of fact, (42.27)–(42.28) bear on the initial datum u_0 and on the derivatives $f^{(j)}(\cdot, 0)$ of f at $t = 0$. Indeed, from the differential equation (40.27) we derive that u_t, and more generally all the $u^{(j)}$, $0 \leqq j \leqq m$, are continuous functions in the closed interval $[0, T]$ with values in $H^{-1}(\Omega)$ [we are tacitly using the hypotheses (42.23) and (42.24)], therefore $u^{(j)}(\cdot, 0)$ is well defined, for $0 \leqq j \leqq m$, as an element of $H^{-1}(\Omega)$. Furthermore, this element can be expressed as a *linear* function of u_0 and of the $f^{(j')}(\cdot, 0)$ for $j' < j$—although expressing it can be cumbersome (especially if j is very large). For instance,

$$\begin{aligned} u_t(\cdot, 0) &= -A(x, 0, \partial/\partial x)u_0 + f(\cdot, 0), \\ u_{tt}(\cdot, 0) &= -A(x, 0, \partial/\partial x)u_t(\cdot, 0) - A_t(x, 0, \partial/\partial x)u_0 + f_t(\cdot, 0) \\ &= A(x, 0, \partial/\partial x)^2 u_0 - A(x, 0, \partial/\partial x)f(\cdot, 0) \\ &\quad - A_t(x, 0, \partial/\partial x)u_0 + f_t(\cdot, 0), \end{aligned}$$

and so on.

A particular case where the statement simplifies considerably—insofar as the conditions (42.27) and (42.28) are automatically satisfied—is that where

(42.32) $$f^{(j)}(\cdot, 0) = 0, \quad 0 \leqq j \leqq m-1,$$

(42.33) $$\tilde{g}^{(j)}(\cdot, 0) = 0, \quad 0 \leqq j \leqq m-1,$$

(42.34) $$u_0 = 0.$$

Indeed, in this case, all the $u^{(j)}(\cdot, 0)$, $0 \leqq j \leqq m$, vanish. Note that if we combine (42.24) with (42.32), we see that we can extend f to $]-\infty, T[$ by setting $f = 0$ for $t < 0$, and that, thus extended,

(42.35) $$f^{(j)} \in L^2(-\infty, T; H^{-1}(\Omega)), \quad 0 \leqq j \leqq m; \qquad f \equiv 0 \quad \text{for} \quad t < 0,$$

similarly,

(42.36) $$\tilde{g}^{(j)} \in L^2(-\infty, T; H^1(\Omega)), \quad 0 \leqq j \leqq m; \qquad \tilde{g} \equiv 0 \quad \text{for} \quad t < 0.$$

We obtain the following

COROLLARY 42.1. *Suppose, in addition to the hypotheses in Theorem 40.1 and to the hypotheses (42.23) to (42.26), that (42.32)–(42.33)–(42.34) hold. Then we have* $u^{(j)}(\cdot, 0) = 0$ *for every* $j = 0, \ldots, m$.

In other words, if we extend the solution u by zero to $]-\infty, 0]$, we see that

(42.37) $$u^{(j)} \in L^2(-\infty, T; H^1(\Omega)), \qquad 0 \leqq j \leqq m;$$
$$u^{(m+1)} \in L^2(-\infty, T; H^{-1}(\Omega)), \qquad u \equiv 0 \quad \text{for} \quad t < 0.$$

Exercises

42.1. Let $\mathbf{V}$, $\mathbf{H}$ be as in Exercise 40.1 and let A be a bounded linear operator $\mathbf{V} \to \overline{\mathbf{V}}'$ such that

(42.38) $$\operatorname{Re}\langle A\mathbf{v}, \mathbf{v}\rangle^- + \lambda_0 \|\mathbf{v}\|_{\mathbf{H}}^2 \geqq c_0 \|\mathbf{v}\|_{\mathbf{V}}^2, \qquad \forall \mathbf{v} \in \mathbf{V},$$

for suitable constants λ_0, c_0 [cf. (41.27)]. Let $\mathbf{u}(t) \in L^2(0, T; \mathbf{V})$ satisfy

(42.39) $$\mathbf{u}_t + A\mathbf{u} = 0, \quad 0 < t < T; \qquad \mathbf{u}(0) = \mathbf{u}_0 \in \mathbf{H}.$$

Prove that we have

(42.40) $$\mathbf{u}^{(j)} \in L^2(0, T; \mathbf{V}) \quad (0 \leqq j \leqq m); \qquad \mathbf{u}^{(m+1)} \in L^2(0, T; \overline{\mathbf{V}}'),$$

if and only if

(42.41) $$A^j \mathbf{u}_0 \in \mathbf{H}, \qquad j = 0, \ldots, m.$$

42.2. Use the same hypotheses (and notation) as in Exercise 41.1. Let K be the canonical linear isometry of $\mathbf{V}$ onto $\overline{\mathbf{V}}'$. We may regard K as a positive operator, generally unbounded, in $\mathbf{H}$. Let us assume that we have

(42.42) $$A^{(j)} \in L^\infty(0, T; L(\mathbf{V}; \overline{\mathbf{V}}')), \qquad 0 \leqq j \leqq m;$$

(42.43) $$\mathbf{f}^{(j)} \in L^2(0, T; \overline{\mathbf{V}}'), \qquad 0 \leqq j \leqq m.$$

Let $\mathbf{u}(t) \in L^2(0, T; \mathbf{V})$ be the solution of (41.28)–(41.29) (see Exercise 41.1 and Theorem 41.1). Prove that (42.42)–(42.43) implies that $\mathbf{u}$ is a C^m function, in the closed interval $[0, T]$, valued in $\overline{\mathbf{V}}'$.

Show that the following properties of $\mathbf{f}$ and $\mathbf{u}_0$,

(42.44) $$K^{m-1-j}\mathbf{f}^{(j)}(0) \in \mathbf{H}, \quad j = 0, \ldots, m-1; \qquad K^m \mathbf{u}_0 \in \mathbf{H},$$

imply that

(42.45) $$\mathbf{u}^{(j)}(0) \in \mathbf{V}, \quad j = 0, \ldots, m-1; \qquad \mathbf{u}^{(m)}(0) \in \mathbf{H}.$$

42.3. Using the result of the preceding exercise and adapting the reasoning in Sect. 42, prove the following:

THEOREM 42.2. *In addition to the hypotheses in Theorem 41.1, we suppose that (42.42), (42.43), and (42.44) hold.*

Then the solution $\mathbf{u}(t)$ *of (41.28)–(41.29) has the following property*:

(42.46) $\mathbf{u}^{(j)} \in L^2(0, T; \mathbf{V}), \quad j = 0, \ldots, m; \qquad \mathbf{u}^{(m+1)} \in L^2(0, T; \mathbf{V}')$.

Derive from (42.46) that $\mathbf{u}$ is a C^{m-1} (resp. a C^m) function in the closed interval $[0, T]$, valued in $\mathbf{V}$ (resp. in $\mathbf{H}$).

42.4. Apply Theorem 42.2 in the case where $\mathbf{V} = H^1(\mathbf{R}^n)$, $\mathbf{H} = L^2(\mathbf{R}^n)$, and $-A(t) = \Delta$, the Laplace operator in the space variables (i.e., in $\mathbf{R}^n$). Prove that, in this case, we have

(42.47) $$\mathbf{u}^{(j)}(0) \in H^{2(m-j)}(\mathbf{R}^n), \qquad j = 0, \ldots, m.$$

Can anything be asserted about $\mathbf{u}^{(m+1)}(0)$?

43

The Laplace Transform

Treatment of mixed problems such as the ones we have been studying in the preceding sections is easier when the coefficients of the differential operator $A(x, t, \partial/\partial x)$ do not depend on the time variable t. In this case, two fairly general methods, in addition to the abstract method of Sects. 40 and 41, are available to us: the Laplace transform method and the theory of semigroups. These methods lead to rather precise results, and have a wide range of application, beyond the particular situation in which we are going to use them. We shall begin with the Laplace transform. We shall briefly recall its definition, its main properties—with a twist in the direction of functions, and distributions!, valued in a Banach space **E** (where the norm will be denoted by $\| \ \|_{\mathbf{E}}$).

We deal with functions and distributions defined in the real line **R**. Traditionally, the *Laplace transform* of a function $f(t)$ is defined (formally) by

$$\mathscr{L}f(\sigma + i\tau) = \int_0^{+\infty} e^{-(\sigma+i\tau)t} f(t)\, dt, \tag{43.1}$$

where $\sigma + i\tau$ is a complex variable (often denoted by p). That there must be some relation between the Fourier and the Laplace transform is evident. We know that the former is closely linked with the underlying additive group structure of **R**—therefore, the latter should partake of this link. But the positive half-line $\mathbf{R}_+ = [0, +\infty[$ is not an additive group, and the first thing we do is to extend the integration to the whole real line—after extending f by zero to $]-\infty, 0[$. The formal definition (43.1) can now be rewritten

$$\mathscr{L}f(\sigma + i\tau) = \int_{-\infty}^{+\infty} e^{-i\tau t}(e^{-\sigma t} f(t))\, dt, \tag{43.2}$$

where, we repeat, $f(t) = 0$ for $t < 0$. Formula (43.2) underlines the fact that $\mathscr{L}f(\sigma + i\tau)$ is the *Fourier transform* of $e^{-\sigma t}f(t)$, where the real number σ plays the role of a parameter. Let us agree to go on writing t for the variable on $\mathbf{R}^1$, τ for the one in the dual of $\mathbf{R}^1$, that is, on the Fourier transform side. Then, still formally, we shall set for any distribution T on the real line, $T = 0$ in $]-\infty, 0[$, i.e., $T \in \mathscr{D}'_+$,

$$\mathscr{L}T(\sigma + i\tau) = \mathscr{F}(e^{-\sigma t}T). \tag{43.3}$$

Of course, the right-hand side of (43.3) will make no sense—unless $e^{-\sigma t}T$ is "Fourier-transformable," which is to say, *tempered*. Suppose for a moment that T is a continuous function f. If for some σ_0 real, $e^{-\sigma_0 t}f$ is tempered, then, because of the fact that $f(t) = 0$ for $t < 0$, $e^{-\sigma t}f$, $\sigma > \sigma_0$, is not only tempered: In fact, it decays at infinity faster than some exponential $e^{-\varepsilon t}$. For such σ the integral at the right in (43.2) makes sense and we have the right to differentiate with respect to σ and to τ under the integral sign. If we apply the differential operator $\partial/\partial\sigma + i\ \partial/\partial\tau$ to that right-hand side, we obtain zero, which means that $\mathscr{L}f(\sigma + i\tau)$ is a holomorphic function of $\sigma + i\tau$ in the "vertical" half-plane $\sigma > \sigma_0$. The same observation applies to distributions:

PROPOSITION 43.1. *Suppose that $T \in \mathscr{D}'_+$ and that $e^{-\sigma_0 t}\ T$ is tempered (σ_0 real). Then $\mathscr{L}T(\sigma + i\tau) = \mathscr{F}(e^{-\sigma t}T)$ is a holomorphic function of $\sigma + i\tau$ in the half-plane $\sigma > \sigma_0$.*

The preceding statement remains valid if the distribution T is valued in the Banach space **E**. The only thing which must be clarified is the meaning of "tempered." If we denote, as usual, by $\mathscr{S}$ the space of C^∞ functions on the real line whose derivatives of all order decay at infinity faster than any power of $|t|^{-1}$, the space $\mathscr{S}'(\mathbf{E})$ of tempered **E**-valued distributions is, by definition, the space of continuous linear mappings of $\mathscr{S}$ into **E** (these mappings are regarded as **E**-valued distributions by restricting them from $\mathscr{S}$ to the dense subspace C_c^∞). One can then prove that *a distribution* **E**, *defined on the real line, and valued in* **E**, *is tempered if and only if it can be written*

$$\mathbf{T} = \sum_{j=0}^{m} \left(\frac{d}{dt}\right)^j \mathbf{f}_j$$

where the $\mathbf{f}_j$ are continuous functions $\mathbf{R} \to \mathbf{E}$ with the property that, for some integer $k \geqq 0$, the nonnegative functions

$$(1 + |t|)^{-k}\|\mathbf{f}_j(t)\|_{\mathbf{E}}, \qquad j = 0, \ldots, m,$$

are bounded on **R** (both m and k depend on **T**). Thus the properties of tempered scalar distributions extend without modification to tempered distributions valued in Banach spaces.

We shall say that $\mathbf{T} \in \mathscr{D}'_+(\mathbf{E})$ is *Laplace-transformable* if $e^{-\sigma_0 t}\mathbf{T}$ is tempered for some real number σ_0.

The *inversion formula* for the Laplace transform follows at once from the inversion (or reciprocity) formula for the Fourier transform and from (43.3):

$$T = \mathscr{F}^{-1}(e^{\sigma t}\mathscr{L}T(\sigma + i\tau)). \tag{43.4}$$

The reader should keep in mind that $\mathscr{F}^{-1}$ transforms distributions in the variable τ into distributions in t; σ continues to play the role of a parameter. In a formal sense (which could be made rigorous), (43.4) can be rewritten as

$$T = (2\pi)^{-1} \int_{-\infty}^{+\infty} e^{(\sigma + i\tau)t}\mathscr{L}T(\sigma + i\tau)\, d\tau,$$

that is,

$$T = \frac{1}{2\pi i} \int_{\gamma} e^{pt}\mathscr{L}T(p)\, dp, \tag{43.5}$$

where γ is a vertical line $\{p;\ \mathrm{Re}\, p = \sigma > \sigma_0\}$. Formula (43.5) is classical. When dealing with distributions, it should be suitably interpreted.

The advantages of the Laplace transform are akin to those of the Fourier transform: They lead to a symbolic calculus which, more precisely, means that they transform convolution into multiplication (when these make sense). It is well known that $\mathscr{D}'_+$ *is a commutative convolution algebra.* Note that if f and g are locally integrable functions, both vanishing for $t < 0$, their convolution can be defined, regardless of their rates of growth at infinity. Indeed,

$$(f * g)(t) = \int_{-\infty}^{+\infty} f(t-s)g(s)\, ds = \int_0^t f(t-s)g(s)\, ds. \tag{43.6}$$

The same circumstances makes the definition work for two distributions S, T vanishing for $t < 0$. We note that if $\varphi \in C_c^\infty$,

$$(\check{T} * \varphi)(t) = \langle T_s, \varphi(s + t)\rangle$$

is a C^∞ function of t, vanishing identically for $t > t_0$ [assuming that $\varphi(t) = 0$ for $t > t_0$]. Since $S = 0$ when $t < 0$, the intersection of the supports of S and $\check{T} * \varphi$ is compact, and we may form

$$\langle S, \check{T} * \varphi\rangle = \langle S_t, \langle T_s, \varphi(s + t)\rangle\rangle,$$

which, by the Fubini theorem for distributions, is equal to

$$\langle T_t, \langle S_s, \varphi(s + t)\rangle\rangle = \langle T, \check{S} * \varphi\rangle.$$

By definition, $S * T$ is the distribution $\varphi \mapsto \langle S, \check{T} * \varphi\rangle$. This extends (43.6) at once.

The commutativity of the convolution in $\mathscr{D}'_+$ is evident; there is a unit element, δ. Furthermore, $\mathscr{D}'_+$ *is an integral domain* (this follows at once from Theorem 43.2, the *theorem of supports*; for a proof, see Exercises 43.4 to 43.6).

PROPOSITION 43.2. *Let $S, T \in \mathscr{D}'_+$. Suppose that both $e^{-\sigma_0 t}S$ and $e^{-\sigma_0 t}T$ are tempered. Then, in the half-plane $\sigma > \sigma_0$,*

$$\mathscr{L}(S * T) = (\mathscr{L}S)(\mathscr{L}T). \tag{43.7}$$

Proof. If $\sigma > \sigma_0$, not only are $e^{-\sigma t}S$, $e^{-\sigma t}T$ tempered: They are rapidly decaying at infinity. Their convolution is therefore well defined and rapidly decaying at infinity. We have

$$e^{-\sigma t}(S * T) = (e^{-\sigma t}S) * (e^{-\sigma t}T), \tag{43.8}$$

hence $\mathscr{L}(S * T)$ is a holomorphic function of $\sigma + i\tau$ in the half-plane $\sigma > \sigma_0$. If we apply the Fourier transformation to both sides of (43.8) and use the fact that, under the present circumstances, $\mathscr{F}(U * V) = \mathscr{F}U\mathscr{F}V$, we immediately obtain (43.7). Q.E.D.

When dealing with **E**-valued distributions, the convolution makes no sense, even when the distributions vanish for $t < 0$, unless **E** is a Banach algebra (cf., Example 39.2). We need some kind of *pairing* (or coupling) between the value spaces. On this subject we refer to §39.4. The pairing which, from our viewpoint, will be the most valuable is the one between elements of **E** and elements of $L(\mathbf{E}; \mathbf{F})$, that is, bounded linear operators of **E** into another Banach space **F**. As usual, $L(\mathbf{E}; \mathbf{F})$ will be equipped with the operator norm, which we denote simply by $\| \ \|$ (see Example 39.4). When $\mathbf{S} \in \mathscr{D}'_+(L(\mathbf{E}; \mathbf{F}))$ and $\mathbf{T} \in \mathscr{D}'_+(\mathbf{E})$, the convolution $\mathbf{S} * \mathbf{T}$ is defined in the obvious manner:

$$(\mathbf{S} * \mathbf{T})(\varphi) = \mathbf{S}(\check{\mathbf{T}} * \varphi), \qquad \varphi \in C_c^\infty. \tag{43.9}$$

For general spaces **E** and **F** it makes no sense to ask whether such a convolution is commutative: Operators act on vectors, but the latter do not "act" on the former. Nor does it make sense to seek a unit element. When $\mathbf{E} = \mathbf{F}$, the situation is a bit more standard. Then $\mathscr{D}'_+(L(\mathbf{E}; \mathbf{E}))$ is a convolution algebra, never a commutative one unless $\dim \mathbf{E} = 1$. But it has a unit element, $I_\mathbf{E}\,\delta$, where $I_\mathbf{E}$ is the identity mapping of **E** and δ the Dirac measure on the real line (at the origin). Then, for convolution, $\mathscr{D}'_+(\mathbf{E})$ is a $\mathscr{D}'_+(L(\mathbf{E}; \mathbf{E}))$-(left)-module.

It suffices to state that formula (43.7) remains valid under the hypotheses of Proposition 43.2, when S is valued in $L(\mathbf{E}; \mathbf{F})$ and T in **E**.

The reader may wonder why the Laplace transform is now more important to us than the Fourier transform. The reason is that, if we content ourselves with dealing with distributions which vanish for $t < 0$, then we can apply the Laplace transform to far more distributions than we can the Fourier transform—the Laplace transform makes sense for distributions which grow exponentially at $+\infty$, whereas only the tempered ones are Fourier-transformable. This is quite important if we wish to apply these transformations to differential equations: For instance, the solutions of an ordinary differential equation on the real line are linear combinations of exponential monomials of the form $t^k e^{\zeta t}$, which, after multiplication by the Heaviside function $H(t) = 0$ for $t < 0$, $H(t) = 1$ for $t > 0$, are always Laplace-transformable, whereas they are not tempered, unless Re $\zeta \leqq 0$.

The Paley–Wiener–Schwartz theorem, which characterizes the Fourier transforms of the distributions with compact support, has an analog in the theory of the Laplace transformation. The analog characterizes the Laplace transforms of the distributions with support in $[0, +\infty]$:

THEOREM 43.1. *Let* $\mathbf{h}(p)$ *denote a holomorphic function in the half-plane* Re $p > \sigma_0$, *valued in the Banach space* $\mathbf{E}$. *The two following conditions are equivalent*:

(43.10) *there is a distribution* $\mathbf{T} \in \mathcal{D}'_+(\mathbf{E})$ *whose Laplace transform is equal to* $\mathbf{h}(p)$;

(43.11) *there is* σ_1 *real*, $\sigma_0 \leqq \sigma_1 < +\infty$, *a constant* $C > 0$, *and an integer* $k \geqq 0$ *such that, for all complex numbers* p, Re $p > \sigma_1$,

$$\|\mathbf{h}(p)\|_{\mathbf{E}} \leqq C(1 + |p|)^k. \tag{43.12}$$

Proof. We shall consider only scalar functions and distributions; the proof is identical when they are valued in $\mathbf{E}$.

Let us first prove that (43.10) implies (43.11). In (43.10) it is tacitly understood that $\mathbf{T} = T$ is Laplace-transformable. Let therefore σ_0' be a real number such that $T_t \exp(-\sigma_0' t)$ is tempered, which means that

$$e^{-\sigma_0' t} T_t = \sum_{j=0}^{k} (d/dt)^j f_j,$$

where the f_j are continuous functions on the real line, with the property that, for some integer $M \geqq 0$,

$$|f_j(t)| \leqq M(1 + |t|)^M \qquad \text{for all} \quad t \in \mathbf{R}, \quad j = 0, \ldots, k. \tag{43.13}$$

Let now $\sigma_1 > \sup(\sigma_0, \sigma_0')$ and take $\sigma = \text{Re}\, p \geqq \sigma_1$. We have

$$e^{-\sigma t} T_t = \sum_{j=0}^{k} \exp[-(\sigma - \sigma_0')t](d/dt)^j f_j$$

$$= \sum_{j=0}^{k} \left(\frac{d}{dt} + \sigma - \sigma_0'\right)^j g_j$$

where $g_j = \exp[-(\sigma - \sigma_0')t] f_j$. Setting $\varepsilon = \sigma_1 - \sigma_0'$ and availing ourselves of (43.13), we see that

$$|g_j(t)| \leqq M(1 + |t|)^M e^{-\varepsilon t}, \qquad \forall t \in \mathbf{R}, \quad 0 \leqq j \leqq k.$$

Consequently, the Fourier transforms $\hat{g}_j$ of the g_j are *bounded* C^∞ functions of τ. We obtain that

$$h(p) = \mathscr{L}T(p) = \mathscr{F}(e^{-\sigma t}T_t) = \sum_{j=0}^{k} (p - \sigma_0')^j \hat{g}_j(\tau),$$

from which (43.12) follows at once.

Next we prove that (43.11) implies (43.10). We may assume that σ_1 is greater than zero. We write, as usual, $p = \sigma + i\tau$, and we consider, for fixed $\sigma > \sigma_1$, the function $w(p) = p^{-k-2}h(p)$. It is an integrable function of τ on the real line. Let us set

$$f(t) = \frac{1}{2\pi}\int_{-\infty}^{+\infty} e^{(\sigma+i\tau)t} w(\sigma + i\tau)\, d\tau = \frac{1}{2\pi i}\int_{\gamma_\sigma} e^{pt} w(p)\, dp,$$

where γ_σ denotes the (oriented) vertical line Re $p = \sigma$. In view of the decay of $w(\sigma + i\tau)$ as $\tau \to \pm\infty$ [which follows from (43.12)] we may apply the Cauchy integral theorem and conclude, as expected, that the integral defining $f(t)$ is independent of $\sigma > \sigma_1$. As a matter of fact, we have

$$|f(t)| \leqq C_0 e^{\sigma t} \int_{-\infty}^{+\infty} (1 + |\tau|)^{-2}\, d\tau, \tag{43.14}$$

where C_0 is a positive constant, independent of $\sigma > \sigma_1$. Take $t < 0$ fixed and make σ go to $+\infty$: The right-hand side goes to zero, hence the left-hand side, which is independent of σ, must be zero. This shows that $f(t) = 0$ for $t < 0$.

The inequality (43.14) also proves that $e^{-\sigma t}f$ is rapidly decaying at $+\infty$ for any $\sigma > \sigma_1$. We set

$$T = (d/dt)^{k+2} f. \tag{43.15}$$

It is clear that $T \in \mathscr{D}'_+$ and that it is Laplace-transformable. By our definition of f, the Laplace transform of T is identical to h.

Exercises

43.1. Let T be an arbitrary number greater than zero, u a distribution belonging to $\mathscr{D}'_+$. Prove that there is a continuous function f in the real line $\mathbf{R}^1$, vanishing for $t < 0$, and an integer $m \geqq 0$, depending on u, such that

$$u = (d/dt)^m f \quad \text{in }]-\infty, T[. \tag{43.16}$$

43.2. Prove that any distribution $u \in \mathscr{D}'_+$ can be written as a series

$$u = \sum_{j=0}^{+\infty} (d/dt)^{m_j} f_j, \tag{43.17}$$

where the f_j are continuous functions in the whole real line, the m_j form a nondecreasing sequence of integers $\geqq 0$, and for each $j = 0, 1, \ldots, f_j(t) = 0$ unless $j - 1 \leqq t < j + 1$.

43.3. Suppose that the distribution $u \in \mathscr{D}'_+$ is Laplace-transformable, i.e., $e^{-\sigma t}u$ is tempered for some real number σ. Prove that there is a continuous function f in $\mathbf{R}^1$, $f(t) = 0$ for $t < 0$, also Laplace-transformable, and an integer m such that $u = (d/dt)^m f$ in the whole real line.

43.4. Derive from Theorem 43.1 that a distribution $u \in \mathscr{D}'_+$, which is Laplace-transformable, vanishes in the open half-line $t < t_0$ if and only if there are constants C, $k > 0$, and $\sigma_1 \in \mathbf{R}$ such that

$$|\mathscr{L}u(p)| \leqq C(1 + |p|^k) \exp[-t_0(\operatorname{Re} p)], \qquad p \in \mathbf{C},\ \operatorname{Re} p > \sigma_1. \tag{43.18}$$

Derive from this that, given *any* distribution $w \in \mathscr{D}'_+$, if $w * w = 0$ for $t < T$, we must necessarily have $w = 0$ for $t < T/2$.

43.5. Call a the largest real number having the following property:

(43.19) *Given any $T > 0$, any two distributions $u, v \in \mathscr{D}'_+$,*

$$u * v = 0 \quad \textit{for} \quad t < T \Rightarrow u * (tv) = 0 \quad \textit{for} \quad t < aT. \tag{43.20}$$

Prove that $0 \leqq a \leqq 1$. Show that, if $u * v = 0$ for $t < T$, we also have

$$(tu) * v + u * (tv) = 0, \qquad \text{for } t < T. \tag{43.21}$$

Derive from (43.19) and (43.21) that then

$$(tu) * (tv) = 0 \qquad \text{for} \quad t < a^2 T, \tag{43.22}$$

$$[u * (tv)] * [u * (tv)] = 0 \qquad \text{for} \quad t < (1 + a^2)T. \tag{43.23}$$

[*Hint*: Convolve (43.20) with $u * (tv)$.] By applying the conclusion of Exercise 43.4 in conjunction with (43.23), derive that $a = 1$.

43.6. From the conclusion in Exercise 43.5, conclude that if $u, v \in \mathscr{D}'_+$ are such that $u * v = 0$ if $t < T$, then necessarily

$$u * (e^{pt}v) = 0 \qquad \text{for all} \quad t < T, \quad p \in \mathbf{C}. \tag{43.24}$$

Show that this implies, when u and v are *continuous* functions, that

$$u(t - s)v(s) = 0 \qquad \text{for all} \quad s \in \mathbf{R}, \quad t < T, \tag{43.25}$$

and that if there is a sequence of numbers $s_j > 0$, converging to 0, such that $v(s_j) \neq 0$, then necessarily $u(t) = 0$ for $t < T$. From this, either by regularization or by the conclusion of Exercise 43.2, derive the *theorem of supports* in one variable:

THEOREM 43.2. *Let u, v be distributions belonging to $\mathcal{D}'_+$. Assume that the support of v contains the origin. Then, if the convolution $u * v$ vanishes for $t < T$, u itself must vanish for $t < T (T > 0)$.*

43.7. In what follows, $H(t)$ denotes Heaviside's function, equal to one for $t > 0$ and to zero for $t < 0$, and $\Gamma(\alpha) = \int_0^{+\infty} e^{-t} t^{\alpha-1}\, dt$ (Re $\alpha > 0$) is the Euler gamma function. Compute the Laplace transforms of the following distributions:

(i) $\delta^{(j)}$, $j = 0, 1, \ldots$ (δ is the Dirac measure);
(ii) $H(t)t^{\alpha}/\Gamma(\alpha + 1)$, Re $\alpha > -1$;
(iii) $H(t)\cos t$, $H(t)\sin t$;
(iv) $H(t)e^{\zeta t}t^{\alpha}/\Gamma(\alpha + 1)$, $\zeta \in \mathbf{C}$, Re $\alpha > -1$.

43.8. Let $f(p)$, $g(p)$ be two polynomials in one variable, with complex coefficients, leading coefficient one, and no common factors. Prove that there is a unique distribution $u \in \mathcal{D}'_+$ whose Laplace transform is equal to the rational function $f(p)/g(p)$ and compute it.

43.9. Show that if α is a real number $\geqq 0$, the function $e^{-\alpha\sqrt{p}} p^{-1/2}$ is the Laplace transform of $[H(t)/\sqrt{\pi t}] \exp(-\alpha^2/4t)$.

44

Application of the Laplace Transform to the Solution of Parabolic Mixed Problems

We return now to the differential operator $A(x, t, \partial/\partial x)$ of Sect. 40 and following, but we assume throughout the present section that its coefficients are independent of t. We denote the operator by $A(x, \partial/\partial x)$ or simply by A:

$$A = A(x, \partial/\partial x) = -\sum_{j,k=1}^{n} \frac{\partial}{\partial x^k} a^{jk}(x) \frac{\partial}{\partial x^j} + \sum_{j=1}^{n} b^j(x) \frac{\partial}{\partial x^j} + c(x),$$

with the following assumptions:

(44.1) a^{jk} $(1 \leqq j, k \leqq n)$, b^j $(1 \leqq j \leqq n)$, *c belong to* $L^\infty(\Omega)$;

(44.2) *there is a constant* $c_0 > 0$ *such that, for almost all x in Ω and for all* $\zeta \in \mathbf{C}^n$,

$$\operatorname{Re} \sum_{j,k=1}^{n} a^{jk}(x)\zeta_j\bar{\zeta}_k \geqq c_0 |\zeta|^2.$$

As in Sect. 41 we associate with A a sesquilinear form on $H^1(\Omega) \times H^1(\Omega)$, except that we want now to have the parameter τ, which was real, replaced by the complex parameter p:

$$a_p(u, v) = p \int_\Omega u\bar{v}\, dx + \sum_{j,k=1}^{n} \int_\Omega a^{jk}(x) \frac{\partial u}{\partial x^j} \frac{\partial \bar{v}}{\partial x^k}\, dx$$
$$+ \sum_{j=1}^{n} \int_\Omega b^j(x) \frac{\partial u}{\partial x^j} \bar{v}\, dx + \int_\Omega c(x) u\bar{v}\, dx.$$

The analog of (41.1) is valid here:

(44.3) *There is σ_0 real such that, for all $p \in \mathbf{C}$,* $\operatorname{Re} p > \sigma_0$, *for all $u \in H^1(\Omega)$,*

(44.4) $$c_0 \|u\|^2_{H^1(\Omega)} \leqq 2 \operatorname{Re} a_p(u, u).$$

From the Lax–Milgram theorem (Lemma 23.1) we derive that $p + A$ is an isomorphism of $H_0^1(\Omega)$ onto $H^{-1}(\Omega)$ [we recall that if $u, v \in H_0^1(\Omega)$,

$$(44.5) \qquad a_p(u, v) = \langle (p + A(x, \partial/\partial x))u, \bar{v} \rangle.]$$

Let $\tilde{G}(p)$ denote the inverse of $(p + A) : H_0^1(\Omega) \to H^{-1}(\Omega)$. Of course, $\tilde{G}(p)$ is an isomorphism of $H^{-1}(\Omega)$ onto $H_0^1(\Omega)$ [$\tilde{G}(p)$ is the Green operator of $p - A$ for the weak Dirichlet problem in Ω]. From (44.4) we derive

(44.6) *The operator norm of* $\tilde{G}(p)$ *is* $\leqq 2/c_0$ *for all complex numbers* p *such that* $\operatorname{Re} p > \sigma_0$.

We note that $(p + A)$ is obviously a holomorphic function of p, $\operatorname{Re} p > \sigma_0$, valued in $L(H_0^1(\Omega); H^{-1}(\Omega))$. We shall apply the following result:

LEMMA 44.1. *Let $\mathcal{O}$ be an open subset of the complex plane, $T(p)$ a holomorphic function of p in $\mathcal{O}$ valued in* $L(\mathbf{E}; \mathbf{F})$ ($\mathbf{E}$, $\mathbf{F}$ are two Banach spaces) *such that, for each $p \in \mathcal{O}$, $T(p)$ is an isomorphism of* $\mathbf{E}$ *onto* $\mathbf{F}$. *Then $T(p)^{-1}$ is a holomorphic function of p in $\mathcal{O}$, valued in* $L(\mathbf{F}; \mathbf{E})$.

Proof. We begin by showing that $T^{-1}(p)$ is a continuous function of p in $\mathcal{O}$ with values in $L(\mathbf{F}; \mathbf{E})$.

Let p_0 denote an arbitrary point in $\mathcal{O}$. Since $T(p_0)$ is an isomorphism, there is a constant $C_0 > 0$ such that

$$\|\mathbf{e}\|_{\mathbf{E}} \leqq C_0 \|T(p_0)\mathbf{e}\|_{\mathbf{F}} \qquad \text{for all} \quad \mathbf{e} \in \mathbf{E}.$$

Therefore, for $p \in \mathcal{O}$ sufficiently close to p_0,

$$\begin{aligned} \|\mathbf{e}\|_{\mathbf{E}} &\leqq C_0 \|T(p)\mathbf{e}\|_{\mathbf{F}} + C_0 \|T(p) - T(p_0)\| \; \|\mathbf{e}\|_{\mathbf{E}} \\ &\leqq C_0 \|T(p)\mathbf{e}\|_{\mathbf{F}} + \tfrac{1}{2}\|\mathbf{e}\|_{\mathbf{E}}. \end{aligned}$$

This shows that the operator norm of $T(p)^{-1}$ is locally bounded in $\mathcal{O}$. But

$$T^{-1}(p) - T^{-1}(p_0) = T^{-1}(p_0)\{T(p_0) - T(p)\}T^{-1}(p),$$

which shows that, for p close enough to p_0,

$$\|T^{-1}(p) - T^{-1}(p_0)\| \leqq 2C_0^2 \|T(p_0) - T(p)\|.$$

Thus $T^{-1}(p)$ is a continuous function of p in $\mathcal{O}$. We have $T(p)T^{-1}(p) = I_{\mathbf{F}}$, the identity mapping of $\mathbf{F}$. We apply the Cauchy–Riemann operator $\partial/\partial\bar{p}$ to both sides, and obtain $T(p)(\partial/\partial\bar{p})T^{-1}(p) = 0$, from which [after multiplication by $T^{-1}(p)$ on the left], $(\partial/\partial\bar{p})T^{-1}(p) = 0$, which is what we wanted.

Q.E.D.

We are now in a position to apply Theorem 43.1 : $\tilde{G}(p)$ is a holomorphic function of p in the half-plane $\operatorname{Re} p > \sigma_0$, valued in the Banach space

$L(H^{-1}(\Omega); H_0^1(\Omega))$ whose operator norm is bounded independently of p. Hence it is the Laplace transform of a distribution $G_t \in \mathscr{D}'_+(L(H^{-1}(\Omega); H_0^1(\Omega))$, which satisfies the following two equations:

$$(\partial/\partial t + A(x, \partial/\partial x))G_t = I\,\delta_t, \tag{44.7}$$

$$G_t(\partial/\partial t + A(x, \partial/\partial x)) = I\,\delta_t, \tag{44.8}$$

where I denotes the identity mapping of $H^{-1}(\Omega)$ in (44.7) and the one of $H_0^1(\Omega)$ in (44.8) [δ_t denotes the Dirac measure in the t variable, and $I\,\delta_t$ is the operator-valued distribution which to any test function φ assigns the value $\varphi(0)I$]. Equations (44.7) and (44.8) are the Laplace transforms of the inversion equations $(p + A)\tilde{G}(p) = I$, $\tilde{G}(p)(p + A) = I$.

We consider the mixed problem

$$\frac{\partial u}{\partial t} + A(x, \partial/\partial x)u = f \quad \text{in } \Omega \times]0, T[, \tag{44.9}$$

$$u - \tilde{g} \in H_0^1(\Omega) \quad \text{almost everywhere in } [0, T], \tag{44.10}$$

$$u(\cdot, 0) = u_0, \tag{44.11}$$

under the assumptions of the previous sections:

$$f \in L^2(0, T; H^{-1}(\Omega)), \qquad u_0 \in L^2(\Omega), \tag{44.12}$$

$$\tilde{g} \in L^2(0, T; H^1(\Omega)), \qquad \frac{d\tilde{g}}{dt} \in L^2(0, T; H^{-1}(\Omega)), \tag{44.13}$$

(44.14) *as* $t > 0$ *goes to* 0, $\tilde{g}(\cdot, t)$ *converges in* $L^2(\Omega)$ [to $\tilde{g}(\cdot, 0)$]

Of course, we know by Theorem 40.1 that there is a unique solution u endowed with good properties. But the question here is how to express this solution in terms of the data, by means of the operator G_t.

In order to do this, we begin by substituting $U = u - \tilde{g}$ for u and by setting

$$F = f - \left\{\frac{\partial \tilde{g}}{\partial t} + A(x, \partial/\partial x)\tilde{g}\right\} \qquad [\in L^2(0, T; H^{-1}(\Omega))]. \tag{44.15}$$

Our problem is now transformed into the following one:

$$\frac{\partial U}{\partial t} + A(x, \partial/\partial x)U = F \quad \text{in } \Omega \times]0, T[, \tag{44.16}$$

$$U \in L^2(0, T; H_0^1(\Omega)), \tag{44.17}$$

$$U(\cdot, t) \text{ converges to } U_0 \text{ in } L^2(\Omega) \text{ as } t \to +0. \tag{44.18}$$

We know solution U exists, is unique, and belongs to $C_0([0, T], L^2(\Omega))$ (Lemma 40.2). We wish to express U in terms of F and U_0 by means of G_t.

First of all, we extend F to $-\infty < t < +\infty$ by setting it equal to zero outside of $[0, T]$. We may now regard F as an element of $\mathscr{D}'_+(H^{-1}(\Omega))$, with support contained in the closed interval $[0, T]$.

Next we define $V = U$ in $[0, T]$, $V = 0$ for $t < 0$ and $t > T$. We have, by the standard formula for distribution derivatives of continuous functions with jumps [here the functions are valued in $H^{-1}(\Omega)$],

$$\frac{dV}{dt} = \left\{\frac{dU}{dt}\right\} + U_0\delta_0 - U(\cdot, T)\delta_T$$

where $\{dU/dt\}$ denotes the element of $L^2(-\infty, +\infty; H^{-1}(\Omega))$, equal to dU/dt in $[0, T]$ and to zero outside of $[0, T]$; δ_0 (resp. δ_T) is the Dirac measure at $t = 0$ (resp. at $t = T$). Finally we obtain

$$\frac{dV}{dt} + A\left(x, \frac{\partial}{\partial x}\right)V = F + U_0\delta_0 - U(\cdot, T)\delta_T \quad \text{in } \Omega \times \mathbf{R}. \tag{44.19}$$

This yields at once

$$V = G * F + GU_0 - (GU(\cdot, T)) * \delta_T \qquad (\text{in } \Omega \times \mathbf{R}), \tag{44.20}$$

where the convolutions are performed with respect to the variable t while G acts as an operator on the values of F, U_0, $U(\cdot, T)$. These values belong to $H^{-1}(\Omega)$, and the result of the action of G is to map them into $H_0^1(\Omega)$. Observe that the support of $(GU(\cdot, T)) * \delta_T = G_{t-T}\, U(\cdot, T)$ is contained in $[T, +\infty[$. Consequently, if we restrict all functions and distributions to the open interval $]0, T[$, we get $V = GU_0 + G * F$, that is,

$$U = GU_0 + G * F \quad \text{in } \Omega \times]0, T[. \tag{44.21}$$

This is the natural generalization of the standard formula for ordinary differential equations [see (11.21)]. In practice, in order to express, or to compute, the solution u of (44.9)–(44.10)–(44.11), it is best to use (44.21). But if one wants to have a general formula for u, in terms of u_0, $\tilde{g}$, and f, it is possible to derive one from (44.21)—provided that one proceeds with care.

Let us write $\tilde{g}_0$ instead of $\tilde{g}(\cdot, 0)$. From (44.21) we draw

$$u = \tilde{g} + G(u_0 - \tilde{g}_0) + G * f - G * \left\{\frac{\partial \tilde{g}}{\partial t} + A\left(x, \frac{\partial}{\partial x}\right)\tilde{g}\right\}.$$

Let us apply once again the formula for the distribution derivative of a continuous function having a "finite" jump at the origin. We obtain

$$H(t)\left\{\frac{\partial \tilde{g}}{\partial t} + A\left(x, \frac{\partial}{\partial x}\right)\tilde{g}\right\} = \left\{\frac{\partial}{\partial t} + A\left(x, \frac{\partial}{\partial x}\right)\right\}H(t)\tilde{g} - \delta_0\tilde{g}_0,$$

whence

$$u = Gu_0 + G * f + M * \tilde{g}, \tag{44.21'}$$

where

$$M * \tilde{g} = \tilde{g} - G * \left\{\frac{\partial}{\partial t} + A\left(x, \frac{\partial}{\partial x}\right)\right\} H(t)\tilde{g}.$$

It would be a mistake to apply formula (44.8) and conclude from it that $M * \tilde{g}$ vanishes identically. Indeed, the two sides in (44.8) operate on $H_0^1(\Omega)$. But the values of $\tilde{g}$ do not lie in the latter space, outside of the trivial case where we could have taken $\tilde{g} \equiv 0$ to begin with. What can be seen right away is that, if $\tilde{g}_1$ is another function, valued in $H^1(\Omega)$, like $\tilde{g}$, and furthermore such that $\tilde{g} - \tilde{g}_1 \in H_0^1(\Omega)$ for almost every $t > 0$, then necessarily

$$M * \tilde{g} = M * \tilde{g}_1.$$

This means that $M *$ acts on functions g of t (as a convolution with respect to t) valued in the quotient space $H^1(\Omega)/H_0^1(\Omega)$, that is, the space of *traces* of functions belonging to $H^1(\Omega)$ on the boundary $\partial\Omega$ of Ω (Sect. 26). Under suitable regularity assumptions on Ω and on $A(x, \partial/\partial x)$, $M * g$ is a convolution operator on t, acting as an operator from $H^{1/2}(\partial\Omega)$ to $H^1(\Omega)$:

$$(M * g)(x, t) = \int_0^t \int_{\partial\Omega} M(x, y; t - s) g(y, s)\, d\sigma_y\, ds, \qquad x \in \Omega, \quad t > 0, \tag{44.21''}$$

where $d\sigma_y$ is the area measure on $\partial\Omega$, and $g(y, t)$ is a suitably regular function of $t > 0$, valued in $H^{1/2}(\partial\Omega)$. Of course, $M * g$ is then the solution of the mixed problem

$$\frac{\partial u}{\partial t} + A\left(x, \frac{\partial}{\partial x}\right) u = 0 \quad \text{in } \Omega \times]0, T[,$$

$$u(x, 0) = 0, \quad x \in \Omega; \qquad u(y, t) = g(y, t), \quad y \in \partial\Omega, \quad t > 0.$$

We shall now discuss some examples. They concern the *heat* operator

$$L = \frac{\partial}{\partial t} - \Delta, \qquad \Delta = \left(\frac{\partial}{\partial x^1}\right)^2 + \cdots + \left(\frac{\partial}{\partial x^n}\right)^2.$$

Example 44.1 The Fundamental Solution of the Heat Equation ($\Omega = \mathbf{R}^n$)

The problem under study is the Cauchy problem

$$\frac{\partial u}{\partial t} - \Delta u = f \in L^2(0, T; H^{-1}(\mathbf{R}^n)), \tag{44.22}$$

$$u|_{t=0} = u_0 \in L^2(\mathbf{R}^n), \tag{44.23}$$

where u denotes the unique solution belonging to $L^2(0, T; H^1(\mathbf{R}^n))$. In this case, since there are no boundary conditions, we have $u = U$, and therefore [cf. (44.21)]

(44.24) $$u = Gu_0 + G * f \quad \text{in } \mathbf{R}^n \times]0, T[.$$

We wish to compute the operator G explicitly. Its Laplace transform $\tilde{G}(p)$ is the inverse of $p - \Delta : H^1(\mathbf{R}^n) \to H^{-1}(\mathbf{R}^n)$. By Fourier transformation, we see at once that

(44.25) $$\tilde{G}(p)\varphi(x) = (2\pi)^{-n} \int e^{ix\cdot\xi}(p + |\xi|^2)^{-1}\hat{\varphi}(\xi)\, d\xi, \qquad \varphi \in C_c^\infty(\mathbf{R}^n).$$

Here $\hat{\varphi}$ denotes the Fourier transform of φ:

$$\hat{\varphi}(\xi) = \int e^{-ix\cdot\xi}\varphi(x)\, dx.$$

We may apply the inversion formula (43.5) for the Laplace transform

$$G_t \varphi = (2\pi)^{-n} \int e^{ix\cdot\xi} \left(\frac{1}{2\pi i} \int_{\gamma_\sigma} e^{pt} \frac{dp}{p + |\xi|^2}\right) \hat{\varphi}(\xi)\, d\xi,$$

where γ_σ is a vertical line $\operatorname{Re} p = \sigma > 0$. At first sight, the exchange in the order of integrations might not seem legitimate, since the integrand in the integral with respect to p is not absolutely integrable (over γ_σ). But in fact we are applying the Fubini theorem for distributions and the reader may imagine that there is tacitly a convergence factor of the kind $\chi(\operatorname{Im} p)$, with $\chi \in C_c^\infty(\mathbf{R}^1)$. converging appropriately to 1. Let us compute

$$K(\xi, t) = \frac{1}{2\pi i} \int_{\gamma_\sigma} e^{pt} \frac{dp}{p + |\xi|^2}.$$

The correct way to handle this is to regard $K(\xi, t)e^{-\sigma t}$ as the inverse Fourier transform of the function of τ,

$$(\sigma + i\tau + |\xi|^2)^{-1}$$

which is clearly square-integrable over γ_σ. We know that $K \equiv 0$ for $t < 0$. In fact (cf. Exercise 43.3, part iv), we know that

(44.26) $$K(\xi, t) = H(t) \exp(-|\xi|^2 t).$$

Observe that

$$\frac{\partial K}{\partial t} + |\xi|^2 K = \delta_t;$$

in other words, K is the fundamental solution of the Fourier transform of the differential operator $\partial/\partial t - \Delta$ with respect to x. Since

$$G_t \varphi = (2\pi)^{-n} \int e^{ix\cdot\xi} K(\xi, t)\hat{\varphi}(\xi)\, d\xi,$$

which shows that $K(\xi, t)$ is the Fourier transform of G with respect to x, we immediately obtain

$$G = (2\sqrt{\pi t})^{-n} H(t) \exp\left(-\frac{|x|^2}{4t}\right), \tag{44.27}$$

which is the standard fundamental solution of the heat equation (see Sect. 6, §6.1).

Example 44.2 Temperature Distribution inside a Metal Bar

This example concerns a mixed problem where the number of space variables is *one* and the open set Ω is a finite interval, say $]a, b[$. It is "concretely" visualized as the problem of determining the distribution and variation of the temperature inside a metal bar, or a homogeneous wall, whose thickness is finite (value: $L = b - a$) but whose other dimensions, height and width, are infinite, and whose two faces (corresponding to $x = a$ and to $x = b$) are heated according to a given program (represented, for us, by the function $\tilde{g}$). The precise statement of the problem is

$$\frac{\partial u}{\partial t} - \frac{\partial^2 u}{\partial x^2} = 0, \qquad a < x < b, \qquad 0 < t < T, \tag{44.28}$$

$$u(a, t) = g_a(t), \qquad u(b, t) = g_b(t), \qquad 0 < t < T, \tag{44.29}$$

$$u(x, 0) = u_0(x), \qquad a < x < b. \tag{44.30}$$

In agreement with the general case, we assume $u_0 \in L^2(a, b)$ and that the derivatives of g_a and g_b (together with these functions themselves) belong to $L^2(0, T)$. The choice of the function $\tilde{g}$ is particularly simple, in the present situation:

$$\tilde{g}(x, t) = (b - a)^{-1}\{g_a(t)(b - x) + g_b(t)(x - a)\}.$$

We may perform the change of function $U = u - \tilde{g}$ and set

$$F = -\frac{\partial \tilde{g}}{\partial t} = -(b - a)^{-1}[g_a'(t)(b - x) + g_b'(t)(x - a)].$$

If we set, in addition to the above,

$$U_0 = u_0 - \tilde{g}(\cdot, 0),$$

we transform the problem (44.28)–(44.29)–(44.30) into the problem

(44.31) $$\frac{\partial U}{\partial t} - \frac{\partial^2 U}{\partial x^2} = F, \qquad a < x < b, \quad 0 < t < T,$$

(44.32) $U = 0$ when $x = a$ and when $x = b$, whenever $0 < t < T$,

(44.33) $$U = U_0 \qquad \text{at} \quad t = 0.$$

We wish to compute the operator G in this case. Its Laplace transform $\tilde{G}(p)$ is the Green function of the operator

$$p - \frac{d^2}{dx^2}$$

in the interval $]a, b[$. We shall always assume $\operatorname{Re} p \geqq 0$. We have

(44.34) $$\tilde{G}(p)\varphi(x) = \int_a^b \tilde{G}(x, x'; p)\varphi(x)\, dx, \qquad \varphi \in C^\infty([a, b]),$$

with (see Exercise 29.12)

(44.35) $$\tilde{G}(x, x'; p) = \tilde{G}_\infty(x - x'; p) + \tilde{h}(x, x'; p),$$

where

(44.36) $$\tilde{G}_\infty(y; p) = \frac{1}{2\sqrt{p}}\{H(y)e^{-y\sqrt{p}} + H(-y)e^{y\sqrt{p}}\},$$

and (writing $L = b - a$)

(44.37)
$$\tilde{h}(x, x'; p) = -\frac{1}{2\sqrt{p}}(1 - e^{-2L\sqrt{p}})^{-1}\left\{\exp\left[-2\sqrt{p}\left(\frac{x + x'}{2} - a\right)\right]\right.$$
$$- \exp\left[-2\sqrt{p}\left(L + \frac{x - x'}{2}\right)\right] + \exp\left[-2\sqrt{p}\left(b - \frac{x + x'}{2}\right)\right]$$
$$\left.- \exp\left[-2\sqrt{p}\left(L - \frac{x - x'}{2}\right)\right]\right\}.$$

The reader will easily verify that $\tilde{G}_\infty$ and h are bounded functions of p in any half-plane $\operatorname{Re} p \geqq \sigma_0 > 0$. Let us also recall that $\tilde{G}_\infty(y; p)$ is a fundamental solution of $p - (d/dy)^2$ and that $\tilde{h}(x, x'; p)$ is the solution of the homogeneous equation

$$\{p - (d/dx)^2\}\tilde{h} = 0$$

such that

$$\tilde{h}(a, x'; p) + \tilde{G}_\infty(a - x'; p) = \tilde{h}(b, x'; p) + \tilde{G}_\infty(b - x'; p) = 0$$

for any x', p. As a consequence of this, for all x', $a < x' < b$, p, $\operatorname{Re} p > 0$,

$$\left(p - \frac{\partial^2}{\partial x^2}\right)\tilde{G}(x, x'; p) = \delta(x - x'), \tag{44.38}$$

$$\tilde{G}(a, x'; p) = \tilde{G}(b, x'; p) = 0. \tag{44.39}$$

Of course, throughout this argument, $\sqrt{p}$ denotes the branch of the square-root function which is greater than zero for p real and greater than zero.

Observe that $\tilde{G}(x, x'; p)$ can be represented as an infinite series of terms of the form

$$c\,\frac{1}{\sqrt{p}}\,e^{-\alpha\sqrt{p}}, \qquad \alpha \geqq 0. \tag{44.40}$$

It can be easily shown (or else found in tables of Laplace transforms) that $p^{-1/2}e^{-\alpha\sqrt{p}}(\alpha \geqq 0)$ is the Laplace transform of $H(t)(\pi t)^{-1/2}\exp(-\alpha^2/4t)$ (see Exercise 43.9). If we take this fact into account, we derive from (44.36) that $\tilde{G}_\infty$ is the Laplace transform of the standard fundamental solution of the heat equation [cf. (44.27)]:

$$G_\infty(y, t) = \frac{H(t)}{2\sqrt{\pi t}}\exp\left(\frac{-y^2}{4t}\right). \tag{44.41}$$

The expression of the inverse Laplace transform of $\tilde{h}(x, x'; p)$ is more complicated:

$$\begin{aligned} h(x, x'; t) = \frac{H(t)}{2\sqrt{\pi t}}\sum_{j=1}^{+\infty}\Bigg\{&\exp\left[-\frac{1}{t}\left(jL + \frac{x - x'}{2}\right)^2\right] \\ &+ \exp\left[-\frac{1}{t}\left(jL - \frac{x - x'}{2}\right)^2\right] \\ &- \exp\left[-\frac{1}{t}\left(jL + \frac{x + x'}{2} - b\right)^2\right] \\ &- \exp\left[-\frac{1}{t}\left(jL + a - \frac{x + x'}{2}\right)^2\right]\Bigg\}. \end{aligned} \tag{44.42}$$

In order to derive the expression of the solution u to the problem (44.28)–(44.29)–(44.30) it suffices to apply formula (44.21).

The preceding formulas remain valid when a tends to $-\infty$ and/or b tends to $+\infty$. For instance, let us consider the case $a = 0$, $b = +\infty$. Then,

$$G(x, x'; t) = G_\infty(x - x', t) - G_\infty(x + x', t). \tag{44.43}$$

Let us introduce the function

$$E_1(x, t) = (\pi t)^{-1/2} \int_0^x \exp(-y^2/4t)\, dy \qquad (x \in \mathbf{R}, \quad t \geqq 0). \tag{44.44}$$

Application of the representation formula (44.21) leads straightforwardly to the following expression:

$$u(x, t) = \frac{1}{2} \int_{-x}^{x} \frac{\partial E_1}{\partial y}(y, t) u_0(x - y)\, dy - \int_0^t \frac{\partial E_1}{\partial t}(x, t - s) g(s)\, ds \tag{44.45}$$

for the solution of the mixed problem:

$$\frac{\partial u}{\partial t} = \frac{\partial^2 u}{\partial x^2}, \qquad \forall x > 0, \quad \forall t > 0; \tag{44.46}$$

$$u(x, 0) = u_0(x), \quad \forall x > 0; \qquad u(0, t) = g(t), \quad \forall t > 0. \tag{44.47}$$

Exercises

44.1. Verify that, in the case where u_0 and g are continuous functions in the closed half-line $[0, +\infty[$, the function $u(x, t)$ given by (44.45) is a continuous function of (x, t) for $x > 0, t > 0$, and that the first term on the right-hand side converges to $u_0(x)$ when $t \to +0$, whereas the second term converges to $g(t)$ when $x \to +0$.

44.2. Consider the mixed problem (44.46)–(44.47) where one takes $u_0 \equiv 0$, $g \equiv 1$. Show that the solution u is *not* continuous at the "corner" $(0, 0)$ in the closed quadrant $\{(x, t) \in \mathbf{R}^2;\ x \geqq 0, t \geqq 0\}$.

44.3. Consider the mixed problem (44.46)–(44.47) where one takes $u_0 \equiv 0$. What are the implications on the boundary function $g(t)$ in (44.47), of the hypotheses (44.13) (where $\Omega = \mathbf{R}^1_+$)? Verify that, if (44.13) holds, then, as $t \to +0$,

$$\int_0^{+\infty} |u(x, t)|^2\, dx \to 0.$$

44.4 Let m be an integer $\geqq 1$, Ω a bounded open subset of $\mathbf{R}^n$ with C^m boundary (and lying on one side of it). Consider the problem:

$$\frac{\partial u}{\partial t} - \Delta u = f \quad \text{in } \Omega \times]0, +\infty[, \tag{44.48}$$

$$u|_{t=0} = u_0 \quad \text{in } \Omega, \tag{44.49}$$

$$(u - \tilde{g})(\cdot, t) \in H_0^1(\Omega) \qquad \text{for almost every} \quad t > 0. \tag{44.50}$$

The following hypotheses are made, concerning the data:

$$f \in L^2(0, +\infty; H^{m-2}(\Omega)), \qquad u_0 \in H^{m-1}(\Omega), \tag{44.51}$$

$$\tilde{g} \in L^2(0, +\infty; H^m(\Omega)), \qquad \frac{d\tilde{g}}{dt} \in L^2(0, +\infty; H^{m-2}(\Omega)). \tag{44.52}$$

By using Laplace transformation with respect to t, and applying Lemma 40.3 (Exercise 40.6), prove that the solution u of (44.48)–(44.49)–(44.50) has the following regularity properties with respect to the variables x:

$$u \in L^2(0, +\infty; H^m(\Omega)), \qquad \frac{du}{dt} \in L^2(0, +\infty; H^{m-2}(\Omega)), \tag{44.53}$$

and consequently,

$$u \in C^0([0, +\infty[; H^{m-1}(\Omega)). \tag{44.54}$$

44.5. Let $\mathbf{V} \subset \mathbf{H}$ be two Hilbert spaces. The injection is continuous and has dense image; its "antitranspose" is a continuous injection of $\mathbf{H}$ into the antidual $\overline{\mathbf{V}}'$ of $\mathbf{V}$. Consider a *positive* isomorphism A of $\mathbf{V}$ onto $\overline{\mathbf{V}}'$ (thus $\langle A\mathbf{v}, \bar{\mathbf{v}}\rangle^- \geqq c_0 \|\mathbf{v}\|_{\mathbf{v}}^2$ for every $v \in \mathbf{V}$; $c_0 > 0$). Show that one can define the positive square root $A^{1/2}$ of A and that it is an isomorphism of $\mathbf{V}$ onto $\mathbf{H}$.

Apply the Laplace transformation to the abstract Cauchy problem

$$\frac{d^2\mathbf{u}}{dt^2} + A\mathbf{u} = \mathbf{f}, \qquad \forall t > 0; \tag{44.55}$$

$$\mathbf{u}\Big|_{t=0} = \mathbf{u}_0 \in \mathbf{V}, \qquad \frac{d\mathbf{u}}{dt}\Big|_{t=0} = \mathbf{u}_1 \in \mathbf{H}, \tag{44.56}$$

assuming also that $\mathbf{f}(t)$ is a continuous function of $t \geqq 0$ valued in $\mathbf{H}$. Derive from the formula for the Laplace transforms that in a sense to be made precise, we may write,

(44.57)

$$\mathbf{u}(t) = \cos(tA^{1/2})\mathbf{u}_0 + \sin(tA^{1/2})A^{-1/2}\mathbf{u}_1 + \int_0^t \sin(sA^{1/2})A^{-1/2}\mathbf{f}(t-s)\,ds,$$

for $t \geqq 0$. Compare with (13.10).

44.6. Let Ω be an open subset of $\mathbf{R}^n$. Apply the results of Exercise 44.5 to the case where $\mathbf{V} = H_0^1(\Omega)$, $\mathbf{H} = L^2(\Omega)$, and A is defined by a strongly elliptic differential operator on Ω, $A(x, \partial/\partial x)$. We recall that the sesquilinear form on $H_0^1(\Omega) \times H_0^1(\Omega)$ associated with $A(x, \partial/\partial x)$ must be Hermitian and coercive

(Definition 23.1). Consider then the mixed problem, with Dirichlet-type boundary data:

$$\frac{\partial^2 u}{\partial t^2} + A(x, \partial/\partial x)u = f \quad \text{in } \Omega \times]0, T[\quad (T > 0), \tag{44.58}$$

$$u\bigg|_{t=0} = u_0, \quad \frac{\partial u}{\partial t}\bigg|_{t=0} = u_1 \quad \text{in } \Omega, \tag{44.59}$$

$$u(\cdot, t) - \tilde{g}(\cdot, t) \in H_0^1(\Omega) \quad \text{for a.e. } t \in]0, T[. \tag{44.60}$$

Make the following assumptions about the data:

$$f \in C^0([0, T]; H^0(\Omega)), \quad u_0 \in H_0^1(\Omega), \quad u_1 \in H^0(\Omega), \tag{44.61}$$

$$\tilde{g} \in C^j([0, T]; H^{2-j}(\Omega)), \quad j = 0, 1, 2. \tag{44.62}$$

Prove that

$$u = \frac{\partial G}{\partial t} u_0 + G u_1 + G * f + M * \tilde{g}, \tag{44.63}$$

where G is the inverse Laplace transform of $(p^2 + A)^{-1}$ [regarded, say, as an isomorphism of $L^2(\Omega)$ onto $H_0^1(\Omega) \cap H^2(\Omega)$ under suitable assumptions on Ω], and

$$M * \tilde{g} = \tilde{g} - G * \left(\frac{\partial^2 \tilde{g}}{\partial t^2} + A\tilde{g}\right) \tag{44.64}$$

(the convolutions must be understood with respect to the time t).

44.7. Take Ω to be a bounded interval $]a, b[$ (with length $L > 0$) in the real line and $A = -d^2/dx^2$. Show that the distribution denoted by G in Exercise 44.6 is given, in the present case, by

$$\begin{aligned} G(x, x'; t) = \tfrac{1}{2}\{&K(x - x', t) + K(x' - x, t) - K(x + x', t + 2a) \\ &- K(x + x', t - 2b)\}, \end{aligned} \tag{44.65}$$

where

$$K(y, t) = H(y)H(t - y) + \sum_{j=1}^{+\infty} H(t - y - 2jL) \tag{44.66}$$

[$H(\cdot)$ is Heaviside's function].

45

Rudiments of Continuous Semigroup Theory

We continue to deal with the case where the differential operator $A = A(x, \partial/\partial x)$ is independent of the time variable t. Until now we have regarded A as a bounded linear operator $H_0^1(\Omega) \to H^{-1}(\Omega)$; we have exploited the fact that, for $\tau > 0$ large enough, $\tau + A$ is an isomorphism of the former space onto the latter. But in availing ourselves of the continuity, or boundedness, of A, we have had to pay a price. This has been already alluded to in the preceding section: If A had been a bounded linear operator of a Hilbert space $\mathbf{H}$ *into itself*, we would have been allowed to use the exponential function e^{-tA} in the solution of the initial value problem:

$$\mathbf{u}_t + A\mathbf{u} = f, \qquad 0 < t < T, \tag{45.1}$$

$$\mathbf{u}(0) = \mathbf{u}_0, \tag{45.2}$$

[cf. (44.21) and following remarks]. We are prevented from following this path by the fact that A is an operator from one Hilbert space into a different one. In this case, its exponential makes no sense. But we could regard A as an *unbounded* linear operator in one and the same space, $\mathbf{H} = L^2(\Omega)$, and try to form the exponential e^{-tA}. It is not unreasonable to expect that the latter will turn out to be a *bounded* linear mapping of $\mathbf{H}$ into itself and even, in some instances, a linear automorphism of $\mathbf{H}$.

This approach would have several advantages. For one, we would operate within $L^2(\Omega)$ and make better use of our information, in the case where $f \in L^2(0, T; H^0(\Omega))$—and not merely $f \in L^2(0, T; H^{-1}(\Omega))$. More importantly, we might be able to handle initial value problems for differential operators A which do not fulfill the strong ellipticity condition (44.2). That such expectations are not excessive is shown in the following two examples (both are Cauchy problems).

Example 45.1 Cauchy Problem for the Heat Equation

This is the same as Example 44.1. We shall use the conclusions reached there. In this case, $-A = \Delta$, the Laplace operator in $\mathbf{R}^n$. There is only one possible choice, here, for the operator e^{-tA}, and this is the convolution, with respect to x, with the fundamental solution of the heat equation

$$E(x, t) = \frac{H(t)}{(2\sqrt{\pi t})^n} \exp\left(-\frac{|x|^2}{4t}\right).$$

For $t > 0$, we have

$$\int_{-\infty}^{+\infty} E(x, t)\, dx = 1,$$

and when $t \to +0$, $E(\cdot, f)$ converges to the Dirac distribution $\delta(x)$. It follows from this that convolution with E (with respect to x) defines a bounded linear operator of $L^2(\mathbf{R}^n)$ into itself; when $t \to +0$, this operator converges to the identity.

Example 45.2 Cauchy Problem for the Schrödinger Equation

We look at problem (45.1)–(45.2) when $A = -i\Delta$ (see Sect. 6, Example 6.2). Note that our ellipticity hypothesis (44.2) is not satisfied here. However, it is easy to solve the problem, if $f \in L^2(0, T; L^2(\mathbf{R}^n))$ and $u \in L^2(\mathbf{R}^n)$, simply by performing a Fourier transformation with respect to x. By the argument used in Sect. 6 (*loc. cit.*) we find that e^{-tA} makes sense: It is the convolution in the x variables with the fundamental solution of the Schrödinger operator

$$E(x, t) = \exp\left[-i(n-2)\frac{\pi}{4}\right] \frac{H(t)}{(2\sqrt{\pi t})^n} \exp\left(-\frac{|x|^2}{4it}\right).$$

In order to see that, for each t, convolution with $E(\cdot, t)$ defines a bounded linear operator on $L^2(\mathbf{R}^n)$, it is convenient to return to its Fourier transform (in x):

$$\tilde{E}(\xi, t) = iH(t) \exp(-it|\xi|^2) \tag{45.3}$$

(cf. Example 6.2) and show that multiplication with $\tilde{E}(\cdot, t)$ defines a bounded linear operator on $L^2(\mathbf{R}_n)$. In fact, for $t > 0$, multiplication with $\tilde{E}(\cdot, t)$ defines a *unitary* operator on $L^2(\mathbf{R}_n)$ and therefore, $E \underset{(x)}{*}$ is a unitary operator on $L^2(\mathbf{R}^n)$.

In both preceding examples, we have

$$[E(\cdot, t) *] \circ [E(\cdot, t') *] = E(\cdot, t + t'), \qquad 0 < t, t', \tag{45.4}$$

$$E(\cdot, 0) * = \text{Identity}, \tag{45.5}$$

which are properties one should expect of an exponential e^{-tA}. These properties are generalized in the definition of a continuous semigroup†:

Definition 45.1. Let **E** *be a Banach space. A mapping* $t \mapsto \mathcal{T}_t$ *of* $\mathbf{R}_+$ *into* $L(\mathbf{E}; \mathbf{E})$ *is called a continuous semigroup if it has the following three properties*:

$$\mathcal{T}_s \mathcal{T}_t = \mathcal{T}_{s+t} \qquad (s, t \geqq 0), \tag{45.6}$$

$$\mathcal{T}_0 = I, \text{ the identity of } \mathbf{E}, \tag{45.7}$$

(45.8) *the mapping* $t \mapsto \mathcal{T}_t$ *of* $\mathbf{R}_+$ *into* $L(\mathbf{E}; \mathbf{E})$ *is continuous when the latter space carries the topology of pointwise convergence in* **E**.

The latter property, (45.8), means that, given any $t_0 \in \mathbf{R}_+$ and any $\mathbf{e} \in \mathbf{E}$, $\|(\mathcal{T}_t - \mathcal{T}_{t_0})\mathbf{e}\|_{\mathbf{E}} \to 0$ as $t \to t_0$. The student might ask why we have not adopted a stronger kind of convergence, specifically the convergence in the sense of the operator norm. Had we adopted this kind of convergence, we would have excluded the very important examples 45.1 and 45.2. For instance, when $t > 0$ converges to zero, it is not true that $\tilde{E}(\xi, t)$, given by (45.3) as a multiplication operator on L^2_ξ, converges to the identity in norm: given $t > 0$, we may choose $|\xi| = (\pi/t)^{1/2}$ and thus check that the norm of $\tilde{E}(\xi, t) - \tilde{E}(\xi, 0)$ in L^∞_ξ is equal to 2. As a matter of fact, the only semigroups of operators, continuous for the operator norm, are those of the form $\exp(-tA)$ with *A bounded* (Exercise 45.1).

In the applications of interest to us, we always start with a given differential operator $A = A(x, \partial/\partial x)$, which we regard as an unbounded operator in some Banach space **E**; most of the time $\mathbf{E} = L^2(\Omega)$. We wish to express the solution **u** of the initial value problem (45.1)–(45.2) in terms of the data **f**, $\mathbf{u}_0$ by means of the exponential function e^{-tA}. The latter is not well defined and, under favorable circumstances, its role will be played by a continuous semigroup $\{\mathcal{T}_t\}$ associated with A. If A were bounded, $\mathcal{T}_t$ would be equal to e^{-tA}. Observe that, in such a case, we could recover A from $\mathcal{T}_t$ by the formula

$$A = \lim_{t \to +0} \frac{1}{t}(I - e^{-tA}),$$

† We are relating a semigroup $\mathcal{T}_t$ to the "inverse exponential" e^{-tA} for the sake of consistency in the remainder of the book where, most of the time, A denotes a strongly elliptic differential operator, and hence defines a *positive* linear operator. In accordance with this the infinitesimal generator (see Definition 45.2) will be $-A$, and not A.

where the limit might be taken in the sense of the operator norm. There is an analogous formula when A is unbounded, but the limit must then be taken in the sense of pointwise convergence of operators in $\mathbf{E}$:

Definition 45.2. We shall denote by $\mathscr{D}(A)$ the linear subspace of $\mathbf{E}$ consisting of those vectors $\mathbf{e} \in \mathbf{E}$ such that $h^{-1}(\mathbf{e} - \mathscr{T}_h \mathbf{e})$ converges in $\mathbf{E}$ as $h > 0$ goes to zero. For every $\mathbf{e} \in \mathscr{D}(A)$ we set

$$A\mathbf{e} = \lim_{h \to +0} \frac{1}{h} (\mathbf{e} - \mathscr{T}_h \mathbf{e}).$$

The linear operator $-A : \mathscr{D}(A) \to \mathbf{E}$ is called the infinitesimal generator of the semigroup $\{\mathscr{T}_t\}$.†

A priori $\mathscr{D}(A)$ could be thought of as small, even as consisting only of the origin in $\mathbf{E}$. In fact, we are going to see soon that $\mathscr{D}(A)$ is dense in $\mathbf{E}$. Let us observe right now that every $\mathscr{T}_t$ commutes with A, since $\mathscr{T}_t$ commutes with $h^{-1}(I - \mathscr{T}_h)$, $h > 0$, and thus $\mathscr{T}_t \mathscr{D}(A) \subset \mathscr{D}(A)$.

We shall derive a certain number of properties of the infinitesimal generator A [among these, the fact that $\mathscr{D}(A)$ is dense] which will enable us to characterize all the linear operators (defined on dense linear subspaces of E) which generate continuous semigroups. It helps to keep in touch with the case where A is bounded, i.e., where $\mathscr{T}_t$ is the true exponential e^{-tA}. Note that, in this case, we have

$$\frac{d}{dt}\mathscr{T}_t + A\mathscr{T}_t = 0, \qquad t > 0, \tag{45.9}$$

$$\mathscr{T}_0 = I. \tag{45.10}$$

We have the right to form the Laplace transform of $\mathscr{T}_t$, since $e^{-\|A\|t}\mathscr{T}_t$ is bounded on $\mathbf{R}_+$. More correctly we introduce the Laplace transform $R(p)$ of the operator-valued function $H(t)\mathscr{T}_t$. By (45.9) and (45.10) we have

$$\left(\frac{d}{dt} + A\right)\{H(t)\mathscr{T}_t\} = I\delta, \tag{45.11}$$

where δ is the Dirac distribution. A Laplace transformation yields

$$(p + A)R(p) = I, \tag{45.12}$$

and since $R(p)$ obviously commutes with A, we see that

$$R(p) = (pI + A)^{-1}:$$

$R(p)$ is the *resolvent* of $-A$ [we know that $R(p)$ exists, and is a bounded linear operator on $\mathbf{E}$, for $|p| > \|A\|$].

† See footnote to page 438.

When A is not bounded, although we cannot define e^{-tA} directly, we may define the Laplace transform $R(p)$ of $H(t)\mathcal{T}_t$ (we assume, until further notice, that $-A$ is the infinitesimal generator of the semigroup $\mathcal{T}_t$). This is due to the following lemma.

LEMMA 45.1. *Let $\{\mathcal{T}_t\}$ be a continuous semigroup on* **E**. *There are constants M, $B > 0$ such that $\|\mathcal{T}_t\| \leqq Me^{Bt}$ for all $t \geqq 0$.*

As usual, we have denoted the operator norm by $\| \ \|$.

Proof. Let a be any strictly positive number and let $\mathcal{B}_a$ denote the image of the closed interval $[0, a]$ under the mapping $t \mapsto \mathcal{T}_t$. By (45.8) $\mathcal{B}_a$ is compact for the topology of pointwise convergence on $L(\mathbf{E}; \mathbf{E})$, hence [TVS, D&K, Theorem 33.1] it is bounded for the operator norm. Let us set $M_a = \sup_{0 \leqq t \leqq a} \|\mathcal{T}_t\|$ (note that $M_a \geqq 1$). Given an arbitrary positive number t, let m be the largest integer such that $ma \leqq t$. By (45.6) we have

$$\|\mathcal{T}_t\| = \|\mathcal{T}_{t-ma}\mathcal{T}_{ma}\| = \|\mathcal{T}_{t-ma}\mathcal{T}_a^m\| \leqq M_a^{m+1} \leqq M_a e^{Bma} \leqq M_a e^{Bt},$$

where $B = (1/a) \log M_a$. Q.E.D.

We now have the right to define, for $\operatorname{Re} p > B$, the constant in Lemma 45.1,

$$R(p) = \int_0^{+\infty} e^{-pt}\mathcal{T}_t \, dt.$$

We know that $R(p)$ is a holomorphic function of p, $\operatorname{Re} p > B$, valued in $L(\mathbf{E}; \mathbf{E})$. Since every $\mathcal{T}_t$ commutes with A so does $R(p)$. As a matter of fact we have

PROPOSITION 45.1. *If* $\operatorname{Re} p > B$ (cf. Lemma 45.1), *the range of $R(p)$ is contained in $\mathcal{D}(A)$ and we have*

$$(pI + A)R(p) = R(p)(pI + A) = I. \tag{45.13}$$

Proof. Let $\mathbf{e} \in \mathbf{E}$, $h > 0$ be arbitrary. We have

$$h^{-1}(\mathcal{T}_h - I)R(p)\mathbf{e} = h^{-1} \int_0^{+\infty} e^{-pt}(\mathcal{T}_{t+h} - \mathcal{T}_t)\mathbf{e} \, dt$$

$$= \frac{1}{h}(e^{ph} - 1) \int_h^{+\infty} e^{-pt}\mathcal{T}_t \mathbf{e} \, dt - \frac{1}{h} \int_0^h e^{-pt}\mathcal{T}_t \mathbf{e} \, dt.$$

When $h \to +0$ the first term converges to $pR(p)\mathbf{e}$ and the second one to $-\mathcal{T}_0\mathbf{e} = -\mathbf{e}$. Q.E.D.

PROPOSITION 45.2. *Suppose the complex variable p converges to ∞ in a sector $|\operatorname{Im} p| < C \operatorname{Re} p$ $(C > 0)$. Then, for every $\mathbf{e} \in \mathbf{E}$, $pR(p)\mathbf{e}$ converges to* **e** *in* **E**.

Proof. We have (writing $p = \sigma + i\tau$ with σ, τ real)

$$pR(p)\mathbf{e} - \mathbf{e} = p\int_0^{+\infty} e^{-pt}(\mathscr{T}_t\mathbf{e} - \mathbf{e})\, dt$$

$$= \frac{p}{\sigma}\int_0^{+\infty} e^{-(p/\sigma)t}(\mathscr{T}_{t/\sigma}\mathbf{e} - \mathbf{e})\, dt$$

from which

$$\|pR(p)\mathbf{e} - \mathbf{e}\|_{\mathbf{E}} \leqq (1 + C)\int_0^{+\infty} e^{-t}\|\mathscr{T}_{t/\sigma}\mathbf{e} - \mathbf{e}\|_{\mathbf{E}}\, dt \to 0$$

when $\sigma \to +\infty$, by virtue of Lemma 45.1 and by Lebesgue's dominated convergence theorem. Q.E.D.

COROLLARY 45.1. *$\mathscr{D}(A)$ is dense in* $\mathbf{E}$.

Proof. Combine Propositions 45.1 and 45.2.

We have shown that A has a dense domain and that $R(p) = R(p;\ -A) = (pI + A)^{-1}$ for $\operatorname{Re} p > B$.

By inverse Laplace transformation applied to the equation $(pI + A)R(p) = I$ we see that $\mathscr{T}_t$ satisfies Eq. (45.9) in the distribution sense. Actually, it is not difficult to show that if $\mathbf{e} \in \mathscr{D}(A)$, $\mathscr{T}_t\mathbf{e}$ is differentiable at each point $t > 0$ and therefore $(d/dt)(\mathscr{T}_t\mathbf{e}) = -A\mathscr{T}_t\mathbf{e}$ for $t > 0$ in the classical sense.

We are now in a position to characterize all the infinitesimal generators of the continuous semigroups:

THEOREM 45.1. (Hille–Yosida) *Let A be a linear operator with domain $\mathscr{D}(A)$ dense in the Banach space* $\mathbf{E}$. *Suppose that, for some $\lambda_0 > 0$, the resolvent $R(\lambda;\ -A) = (\lambda I + A)^{-1}$ of $-A$ exists and is a bounded linear operator on* $\mathbf{E}$ *for all integer values of $\lambda > \lambda_0$. Then the following two conditions are equivalent:*

(a) *$-A$ is the infinitesimal generator of a continuous semigroup $\{\mathscr{T}_t\}$;*
(b) *there are constants M, $B \geqq 0$ such that, for all $k = 1, 2, \ldots$, and all integers $m > \sup(\lambda_0, B)$,*

$$\left\|\left(I + \frac{1}{m}A\right)^{-k}\right\| \leqq M\left(1 - \frac{B}{m}\right)^{-k}. \tag{45.14}$$

Proof. (1) (a) $\Rightarrow$ (b). If (a) holds, we have seen that $R(\lambda;\ -A) = R(\lambda) = \int_0^{+\infty} e^{-\lambda t}\mathscr{T}_t\, dt$ for $\operatorname{Re}\lambda > B$, the constant in Lemma 45.1. We may differentiate $R(\lambda)$ with respect to λ under the integral sign. We obtain

$$R^{(k)}(\lambda) = \int_0^{+\infty} (-t)^k e^{-\lambda t}\mathscr{T}_t\, dt.$$

On the other hand, since $R(\lambda) = (\lambda I + A)^{-1}$, we can easily check that

$$R^{(k)}(\lambda) = (-1)^k k!\, R(\lambda)^{k+1},$$

whence by Lemma 45.1,

$$\|R(\lambda)^{k+1}\| \leqq M \int_0^{+\infty} \frac{t^k}{k!}\, e^{-(\mathrm{Re}\,\lambda - B)t}\, dt = M(\mathrm{Re}\,\lambda - B)^{-k-1},$$

from which (45.14) is derived, noting that $pR(p) = (I + p^{-1}A)^{-1}$.

(2) (b) ⇒ (a). Let us set $J_m = (I + (1/m)A)^{-1}$, m an integer greater than λ_0. By (45.14) we know that the J_m, $m > \sup(\lambda_0, B)$, form a bounded set of linear operators on $\mathbf{E}$. If $\mathbf{e} \in \mathscr{D}(A)$, $\mathbf{e} - J_m\mathbf{e} = m^{-1}AJ_m\mathbf{e} = m^{-1}J_m A\mathbf{e}$, hence $\|J_m\mathbf{e} - \mathbf{e}\|_{\mathbf{E}} \leqq (C/m)\|A\mathbf{e}\|_{\mathbf{E}} \to 0$ as $m \to +\infty$. Since $\mathscr{D}(A)$ is dense in $\mathbf{E}$, this means that $J_m\mathbf{e} \to \mathbf{e}$, as $m \to +\infty$, for every $\mathbf{e} \in \mathbf{E}$.

Let us now consider

$${}^m\mathscr{T}_t = \exp(-tAJ_m) = \exp(mt(J_m - I)) = e^{-mt}\exp(mtJ_m), \qquad t \geqq 0.$$

Again by (45.14) we have

$$\|\exp(mtJ_m)\| \leqq \sum_{k=0}^{+\infty} \frac{(mt)^k}{k!}\, \|J_m^k\| \leqq M \exp\left\{m\left(1 - \frac{B}{m}\right)^{-1} t\right\}.$$

Observe that $m(1 - B/m)^{-1} - m = (m/(m - B))B$. We obtain

$$\text{(45.15)} \quad \|{}^m\mathscr{T}_t\| \leqq M \exp\left\{\left(1 - \frac{B}{m}\right)^{-1} Bt\right\}, \qquad \forall m > \sup(\lambda_0, B), \quad \forall t \geqq 0.$$

It is evident that all the operators J_m, ${}^n\mathscr{T}_t$ commute with A and that they commute among themselves. Furthermore,

$$\begin{aligned} {}^m\mathscr{T}_t - {}^n\mathscr{T}_t &= \{\exp[-tA(J_m - J_n)] - I\}\exp(-tAJ_n) \\ &= -\exp(-tAJ_n)\int_0^t A(J_m - J_n)\exp[-sA(J_m - J_n)]\, ds \\ &= -\int_0^t {}^m\mathscr{T}_s\, {}^n\mathscr{T}_{t-s}(J_m - J_n)A\, ds. \end{aligned}$$

Let then $\mathbf{e}$ be an arbitrary element of $\mathscr{D}(A)$. We have, by (45.15),

$$\|{}^m\mathscr{T}_t\mathbf{e} - {}^n\mathscr{T}_t\mathbf{e}\|_{\mathbf{E}} \leqq M^2\|(J_m - J_n)A\mathbf{e}\|_{\mathbf{E}} \int_0^t e^{2Bs}\, ds$$

if $m, n > 2B$. We have seen that $J_m A\mathbf{e}$ converges to $A\mathbf{e}$. It follows that the ${}^m\mathscr{T}_t\mathbf{e}$ form a Cauchy sequence in $\mathbf{E}$ and converge to a limit, which we denote by $\mathscr{T}_t\mathbf{e}$. The convergence is uniform with respect to t in any bounded interval $[0, T]$, $T < +\infty$. Once more by (45.15), we see that, when $0 \leqq t \leqq T$, the ${}^m\mathscr{T}_t$ form a bounded set of linear operators, hence (again by [TVS, D&K,

Prop. 32.5]) we conclude that ${}^m\mathscr{T}_t\mathbf{e}$ converges to $\mathscr{T}_t\mathbf{e}$, for any $\mathbf{e} \in \mathbf{E}$, uniformly with respect to t on bounded intervals. From this it follows at once that, given any $\mathbf{e} \in \mathbf{E}$, $t \mapsto \mathscr{T}_t\mathbf{e}$ is a continuous function of t in $\mathbf{R}_+$; that $\mathscr{T}_s\mathscr{T}_t = \mathscr{T}_{s+t}$ and that $\mathscr{T}_0 = I$, for these properties are true when ${}^m\mathscr{T}$ is substituted for $\mathscr{T}$. In passing, note that we have, by virtue of (45.15),

$$\|\mathscr{T}_t\| \leqq Me^{Bt} \qquad \text{for all} \quad t \geqq 0. \tag{45.16}$$

What remains to be shown is that $-A$ is the infinitesimal generator of the semigroup $\{\mathscr{T}_t\}$. Let $-A'$ be the infinitesimal generator of $\{\mathscr{T}_t\}$. In view of (45.16) we know that, for Re λ sufficiently large, the resolvent $R(\lambda; -A')$ is equal to

$$\int_0^{+\infty} e^{-\lambda t}\mathscr{T}_t\,dt = \lim_{m\to+\infty}\int_0^{+\infty} \exp(-\lambda t)\exp(-tAJ_m)\,dt = \lim_{m\to+\infty}(\lambda I + AJ_m)^{-1}$$
$$= R(\lambda; -A),$$

where the limits are taken in the sense of the pointwise convergence in $\mathbf{E}$. In other words, for Re λ large enough, $\lambda I + A$ and $\lambda I + A'$, which are bijections of $\mathscr{D}(A)$ and $\mathscr{D}(A')$, respectively, onto $\mathbf{E}$, have the same inverse, hence are equal, which is only possible if $A = A'$.

The proof of Theorem 45.1 is now complete.

We have seen that every continuous semigroup $\{\mathscr{T}_t\}$ on $\mathbf{E}$ has an infinitesimal generator $-A$ with dense domain $\mathscr{D}(A)$, whose resolvent $R(\lambda; -A)$ exists and is a bounded linear operator for Re λ larger than some constant $B \geqq 0$, and furthermore has property (b) of Theorem 45.1. Conversely, to any such operator $-A$ corresponds a continuous semigroup $\{\mathscr{T}_t\}$ of which $-A$ is the infinitesimal generator. In fact, the *semigroup* $\{\mathscr{T}_t\}$ *is unique*. This can be shown in several ways. One way is to recall that the Laplace transform is injective [on the Laplace-transformable distributions belonging to $\mathscr{D}'_+(L(\mathbf{E}; \mathbf{E}))$], hence the resolvent $R(\lambda; -A)$ determines $\mathscr{T}_t$ (by inverse Laplace transformation).

As a matter of fact, the proof of the implication (b) ⇒ (a) in Theorem 45.1 has yielded a representation formula from the semigroup $\mathscr{T}_t$ starting from the infinitesimal generator $-A$: it is the expression of $\mathscr{T}_t$ as the *pointwise limit* of the ${}^m\mathscr{T}_t$:

$$\mathscr{T}_t = \lim_{m\to+\infty} \exp\left\{-t\left(I + \frac{1}{m}A\right)^{-1}A\right\}. \tag{45.17}$$

(*Pointwise* means that, given any $\mathbf{e} \in \mathbf{E}$, $\mathscr{T}_t\mathbf{e}$ is the limit in $\mathbf{E}$ of the vectors ${}^m\mathscr{T}_t\mathbf{e}$; we recall that the convergence is uniform with respect to t on every finite interval $0 \leqq t \leqq T < +\infty$.)

Another remark connected with Theorem 45.1 is that *the infinitesimal generator of a continuous semigroup* $\{\mathcal{T}_t\}$, $-A$, *is a closed operator*. This means that the *graph* of $A : \mathcal{D}(A) \to \mathbf{E}$, which is a subset of $\mathbf{E} \times \mathbf{E}$, is closed there, or, equivalently, that if a sequence $\mathbf{e}_j$ in $\mathcal{D}(A)$ converges in $\mathbf{E}$ to an element $\mathbf{e}$ and if $A\mathbf{e}_j$ converges also in $\mathbf{E}$, to an element $\mathbf{f}$, we must have $\mathbf{e} \in \mathcal{D}(A)$ and $A\mathbf{e} = \mathbf{f}$. Indeed, for m large enough, $(I + (1/m)A)^{-1}$ is a continuous linear operator. If we apply it to $(\mathbf{e}_j + (1/m)A\mathbf{e}_j)$, which we know converges to $(\mathbf{e} + (1/m)\mathbf{f}) = \mathbf{g}$, we see that $\mathbf{e}_j$ must converge to $(I + (1/m)A)^{-1}\mathbf{g}$ and since it also converges to $\mathbf{e}$ this means that $\mathbf{e}$ belongs to the range of $(I + (1/m)A)^{-1}$, hence to $\mathcal{D}(A)$ by Proposition 45.1, and that we have

$$\mathbf{g} = \mathbf{e} + \frac{1}{m}\mathbf{f} = \left(I + \frac{1}{m}A\right)\mathbf{e}.$$

This implies at once $\mathbf{f} = A\mathbf{e}$.

Application to Parabolic Mixed Problems

We are now going to indicate briefly how one can apply the theory of continuous semigroups to the solution of certain parabolic mixed problems. In order to simplify the exposition, we shall put some restrictions on the problem under study. We shall deal with a *bounded* open subset Ω of $\mathbf{R}^n$. We consider a second-order linear partial differential operator in Ω, of the kind considered in Sect. 44:

$$A = A\left(x, \frac{\partial}{\partial x}\right) = -\sum_{j,k=1}^{n} \frac{\partial}{\partial x^k} a^{jk}(x) \frac{\partial}{\partial x^j} + \sum_{j=1}^{n} b^j(x) \frac{\partial}{\partial x^j} + c(x)$$

(although, under the present circumstances, the fact that A is written in the variational form is of no importance: our hypotheses will be such that we could commute $\partial/\partial x^k$ and $a^{jk}(x)$ in the double sum above). We assume that A satisfies the uniform *strong ellipticity* assumption (44.2):

$$\text{(45.18)} \qquad \operatorname{Re} \sum_{j,k=1}^{n} a^{jk}(x)\zeta_j\bar{\zeta}_k \geqq c|\zeta|^2, \qquad \text{for all} \quad x \in \Omega,\ \zeta \in \mathbf{C}^n.$$

In addition to these hypotheses we shall make all the necessary assumptions on the regularity of the boundary $\partial\Omega$ of Ω and on the coefficients $a^{jk}(x)$, $b^j(x)$, $c(x)$ to ensure the validity of the following property:

$$\text{(45.19)} \qquad \forall u \in H_0^1(\Omega), \qquad Au \in L^2(\Omega) \quad \Rightarrow \quad u \in H^2(\Omega).$$

When $-A$ is the Laplacian, sufficient conditions for (45.19) to hold are provided in Sect. 27 (see Theorem 27.2) and, as indicated there, the same result extends, by identical methods, to more general differential operators. We know [cf. (44.3) and following remarks] that there is a real number λ_0 such that, if $\lambda > \lambda_0$, $\lambda + A$ is an isomorphism of $H_0^1(\Omega)$ onto $H^{-1}(\Omega)$, which, combined with (45.19), implies that

(45.20) *if* $\lambda > \lambda_0$, $\lambda + A$ *is an isomorphism* (in an obvious sense) *of* $H_0^1(\Omega) \cap H^2(\Omega)$ *onto* $L^2(\Omega)$.

Let us denote by $(\lambda + A)^{-1}$ the inverse mapping, regarded, however, as a bounded linear operator of $L^2(\Omega)$ *into itself*. Its range will be $H_0^1(\Omega) \cap H^2(\Omega)$, which we regard as the *domain* $\mathscr{D}(A)$ of A. Still for $\lambda > \lambda_0$, let us consider an arbitrary function $u \in H_0^1(\Omega) \cap H^2(\Omega)$ and

$$\|(\lambda + A)u\|_0^2 = (\lambda - \lambda_0)^2\|u\|_0^2 + 2(\lambda - \lambda_0)\,\mathrm{Re}(u, (\lambda_0 + A)u)_0 + \|(\lambda_0 + A)u\|_0^2$$

where $(\ ,\ )_0$ and $\|\ \|_0$ denote the inner product and the norm in $L^2(\Omega)$. Again by (44.3) we obtain

$$(\lambda - \lambda_0)^2\|u\|_0^2 + \|(\lambda_0 + A)u\|_0^2 \leqq \|(\lambda + A)u\|_0^2. \tag{45.21}$$

By (45.20) we have the right to substitute $(\lambda + A)^{-1}f$ for u, with f arbitrary in $L^2(\Omega)$. We derive from (45.21):

$$(\lambda - \lambda_0)\|(\lambda + A)^{-1}f\|_0 \leqq \|f\|_0, \qquad \lambda > \lambda_0, \quad f \in L^2(\Omega). \tag{45.22}$$

By iteration we derive

$$\left\|\left(I + \frac{1}{\lambda}A\right)^{-m} f\right\|_0 \leqq (1 - \lambda_0/\lambda)^{-m}\|f\|_0, \qquad m = 0, 1, \ldots, \tag{45.23}$$
$$\text{for all} \quad \lambda > \lambda_0 \quad \text{and all} \quad f \in L^2(\Omega).$$

If we apply Theorem 45.1, we reach the conclusion that $-A$ is the infinitesimal generator of a continuous semigroup $\{\mathscr{T}_t\}$ on $L^2(\Omega)$. This in turn means that we can now solve the mixed problem

$$u_t + Au = f, \qquad u\Big|_{t=0} = u_0, \tag{45.24}$$

where f and u_0 are valued in $L^2(\Omega)$, by the formula

$$u(t) = \mathscr{T}_t u_0 + \int_0^t \mathscr{T}_{t-s} f(s)\, ds. \tag{45.25}$$

Of course, we must specify our assumptions about the regularity of f with respect to t. These may depend on the situation. But observe that, without additional information, the first term on the right-hand side of (45.25) is merely a continuous function valued in $L^2(\Omega)$—although, for each fixed t, $\mathscr{T}_t u_0$ belongs to the domain of A, i.e., $H_0^1(\Omega) \cap H^2(\Omega)$. Thus, if f is integrable with respect to t (say on $[0, T]$), the second term will be absolutely continuous, in other words will be better (in general) than the first one, and the solution u will be *continuous*.

Exercises

(In the following exercises, **E** will always denote a complex Banach space, with norm $\| \|_{\mathbf{E}}$; the operator norm will be denoted by $\| \|$.)

45.1. Let $\{\mathscr{T}_t\}$ be a semigroup of operators on **E**, continuous for the operator norm [this means that (45.6)–(45.7) hold and that $t \mapsto \mathscr{T}_t$ is a continuous mapping of $[0, +\infty[$ into the Banach space $L(\mathbf{E}; \mathbf{E})$]. Let $R(p)$ be the Laplace transform of $H(t)\mathscr{T}_t$, i.e., $R(p) = (pI + A)^{-1}$, $-A$ being the infinitesimal generator of the semigroup $\{\mathscr{T}_t\}$. Prove that, for p real positive, large enough, $R(p)$ is an invertible map of **E** onto itself. Derive from this fact that A is bounded and $\mathscr{T}_t = e^{-tA}$.

45.2. Let $\{U_t\}$ be a continuous semigroup (Definition 45.1) of *unitary* operators on **E** (assumed to be a Hilbert space). Show that one can define, in a unique manner, U_t for $t < 0$, in such a way that $\{U_t\}_{t \in \mathbf{R}}$ becomes a *continuous group* of unitary operators on **E**. Derive from this that the infinitesimal generator of the semigroup $\{U_t\}_{t \geq 0}$ is an antiself-adjoint operator, densely defined, in **E**.

45.3. Can you give an example of a continuous *group* of operators $\{\mathscr{T}_t\}$ on a Hilbert space **E** whose infinitesimal generator is not bounded (Exercise 45.1) nor antiself-adjoint (Exercise 45.2)?

45.4. Let **H** be a Hilbert space, A a densely defined self-adjoint operator in **H**. Assume that A is positive and has a bounded inverse and consider the abstract Cauchy problem:

$$\text{(45.26)} \qquad \frac{d^2\mathbf{u}}{dt^2} + A\mathbf{u} = 0, \quad t \in \mathbf{R}, \qquad \mathbf{u}\Big|_{t=0} = \mathbf{u}_0 \in \mathbf{H}, \qquad \mathbf{u}_t\Big|_{t=0} = \mathbf{u}_1 \in \mathbf{H}.$$

Show that there is a continuous unitary group U_t in $\mathbf{H} \times \mathbf{H}$ such that the solution $\mathbf{u}$ of (45.26) can be linked to the Cauchy data $\mathbf{u}_0$, $\mathbf{u}_1$ by the formula

$$\text{(45.27)} \qquad \begin{pmatrix} \mathbf{u}(t) \\ \dfrac{d\mathbf{u}}{dt}(t) \end{pmatrix} = U_t \begin{pmatrix} \mathbf{u}_0 \\ \mathbf{u}_1 \end{pmatrix}, \qquad t \in \mathbf{R},$$

and this relation leads to the formula (to which the student is asked to give a precise meaning):

$$\text{(45.28)} \qquad \mathbf{u}(t) = \cos(A^{1/2}t)\mathbf{u}_0 + A^{-1/2}\sin(A^{1/2}t)\mathbf{u}_1.$$

Compare with (13.10) and Exercise 44.5.

45.5. Let $\{\mathscr{T}_t\}$ be a continuous semigroup on **E** such that there is a constant $C > 0$ for which the following is true:

(45.29) $\forall \mathbf{e} \in \mathbf{E}$, *the function of* $t \geqq 0$, $\mathscr{T}_t \mathbf{e}$, *can be extended as a continuous function in the sector* $|\text{Im } t| \leqq C \text{ Re } t$ *of the complex plane, holomorphic in the interior* (and valued in **E**).

Prove that there are positive constants M, B such that

$$\|\mathscr{T}_z\| \leqq Me^{B(\text{Re } z)}, \qquad \forall z \in \mathbf{C}, \qquad |\text{Im } z| \leqq C \text{ Re } z. \tag{45.30}$$

Derive from this that the resolvent $R(\lambda)$ of the infinitesimal generator of $\mathscr{T}_t$ is a holomorphic function, valued in the Banach space $L(\mathbf{E}; \mathbf{E})$, in the region

$$\lambda \in \mathbf{C}, \qquad \text{Re } \lambda + C|\text{Im } \lambda| > B. \tag{45.31}$$

45.6. Use the notation of Exercise 45.5. Prove the converse of the result stated there, namely that if the resolvent $R(\lambda)$ of the infinitesimal generator exists and is a holomorphic function of λ, with values in $L(\mathbf{E}; \mathbf{E})$, in the region (45.31), then (45.29) holds, possibly for a smaller value of the constant $C > 0$.

45.7. Let Ω be a bounded open subset of $\mathbf{R}^n$, whose boundary is a smooth hypersurface and which lies on one side of it. Let $P(x, D)$ be a strongly elliptic differential operator in Ω, whose coefficients all belong to $C^\infty(\bar{\Omega})$. Show that $P(x, D)$ defines an unbounded linear operator in $L^2(\Omega)$, with domain $H^{2m}(\Omega) \cap H_0^m(\Omega)$, A, and that the resolvent $R(\lambda)$ of $-A$ is a holomorphic function, valued in $L(L^2(\Omega); L^2(\Omega))$, in a region of the kind (45.31). [The student is permitted to assume the validity for the operator $P(x, D)$ of the analog of Theorem 27.2.]

45.8. Let A be a densely defined self-adjoint operator in a Hilbert space **H**; suppose that A is positive and, in fact, $A \geqq c_0 I$ for some $c_0 > 0$. Show that the continuous semigroup $\{\mathscr{T}_t\}$ generated by $-A$ satisfies the inequality

$$\|\mathscr{T}_t\| \leqq \text{const} \exp(-c_0 t), \qquad \forall t \geqq 0. \tag{45.32}$$

Show that the inverse of A, A^{-1}, which is a bounded linear operator on **E**, verifies

$$A^{-1} = \int_0^{+\infty} \mathscr{T}_t \, dt.$$

45.9. Let **H** and A be as in Exercise 45.8. Let $\mathbf{u}_0$ be an element of **H** and $\mathbf{f}(t)$ a continuous mapping of $[0, +\infty[$ into **H**, converging to $\mathbf{f}_\infty$ in **H** as $t \to +\infty$. Prove that the solution $\mathbf{u}(t)$ of the problem

$$\frac{d\mathbf{u}}{dt} + A\mathbf{u} = \mathbf{f}, \qquad \forall t > 0; \qquad \mathbf{u}(0) = \mathbf{u}_0, \tag{45.33}$$

converges, as $t \to +\infty$, to the solution $\mathbf{u}_\infty$ of the equation

$$A\mathbf{u}_\infty = \mathbf{f}_\infty. \tag{45.34}$$

45.10. Let Ω be a bounded open subset of $\mathbf{R}^n$, whose boundary Γ is a C^∞ hypersurface and which lies on one side of it. Let g be a complex-valued function on Γ, belonging to $H^{3/2}(\Gamma)$, and let $u(x, t)$ be the solution of the mixed problem

$$\frac{\partial u}{\partial t} = \Delta_x u \quad \text{in } \Omega, \qquad \text{for} \quad t > 0; \tag{45.35}$$

$$u\Big|_{t=0} = u_0(x) \in L^2(\Omega), \tag{45.36}$$

$$u = g \quad \text{on } \Gamma. \tag{45.37}$$

Apply the result in Exercise 45.9 to show that, when $t \to +\infty$, $u(x, t)$ converges to the solution $u_\infty(x)$ of the Dirichlet problem:

$$\Delta u_\infty = 0 \quad \text{in } \Omega, \qquad u = g \quad \text{on } \Gamma. \tag{45.38}$$

Could one have predicted the conclusion from physical considerations, interpreting u, u_0, and g as temperatures?

46

Application of Eigenfunction Expansion to Parabolic and to Hyperbolic Mixed Problems

As in Sects. 44 and 45, we continue to deal with a differential operator $A = A(x, \partial/\partial x)$ whose coefficients are L^∞ functions in the open set $\Omega \subset \mathbf{R}^n$, independent of the time variable t:

$$A = -\sum_{j,k=1}^{n} \frac{\partial}{\partial x^j} a^{jk}(x) \frac{\partial}{\partial x^k} + \sum_{j=1}^{n} b^j(x) \frac{\partial}{\partial x^j} + c(x).$$

In this section, however, we shall make use of the eigenvalue expansions described in Sect. 34. We refer throughout to the concepts and notation used in that section. In particular, the operator A will be assumed to be *formally self-adjoint* and *strongly elliptic* [cf. (34.3) through (34.6)]. We shall also make the hypothesis (34.9), namely that the open set Ω is bounded.

We recall that we may find a number κ such that $\kappa + A$, viewed as a continuous linear operator $H_0^1(\Omega) \to H^{-1}(\Omega)$ [which is the dual of $H_0^1(\Omega)$] becomes positive definite and, as a matter of fact, an isomorphism [it suffices to take $\kappa > \lambda_0$; see (34.7)]. Its inverse $G(\kappa)$, when composed on the right with the natural injection of $H_0^1(\Omega)$ into $H^{-1}(\Omega)$, becomes a compact operator on the former space; when composed on the left, it becomes a compact operator on the latter. At any rate, its spectrum is discrete, and consists of a sequence of strictly positive numbers converging to zero. We shall instead look at the eigenvalues of A, λ, related to those of $G(\kappa)$ by the relation

(46.1) $$\lambda = \chi^{-1} - \kappa, \qquad \chi \in \text{spectrum of } G(\kappa).$$

The eigenvalues of A form the sequence (34.11) of real numbers, converging to $+\infty$. Thus, except possibly for a finite number of them, they are all strictly positive.

We shall then make use of the eigenfunctions E_j ($j = 1, 2, \ldots$) of A defined in (34.19). Proposition 34.2 provides us with a new method of solving the mixed problem:

$$\frac{\partial u}{\partial t} + Au = f \quad \text{in } \Omega \times \,]0, T[, \tag{46.2}$$

$$u(\cdot, 0) = u_0, \tag{46.3}$$

where we seek a solution u valued in $H_0^1(\Omega)$ (i.e., assuming zero boundary values; we recall that the situation can always be transformed into this one, by a change of the unknown function). We shall take

$$f \in L^2(0, T; H^{-1}(\Omega)), \qquad u_0 \in L^2(\Omega), \tag{46.4}$$

and seek

$$u \in L^2(0, T; H_0^1(\Omega)) \qquad \text{such that} \quad u_t \in L^2(0, T; H^{-1}(\Omega)). \tag{46.5}$$

Let us apply Proposition 34.2 and write

$$f(x, t) = \sum_{j=1}^{+\infty} f_j(t) E_j, \qquad u_0 = \sum_{j=1}^{+\infty} u_{0j} E_j, \tag{46.6}$$

$$u(x, t) = \sum_{j=1}^{+\infty} u_j(t) E_j. \tag{46.7}$$

For the sake of simplicity let us assume in the remainder of this section that all summations with respect to j range from 1 to $+\infty$, without repeating it each time. Since $AE_j = \lambda_j E_j$, Eqs. (46.2)–(46.3) transform into the following sequence of ordinary differential initial value problems:

$$u_j' + \lambda_j u_j = f_j, \qquad 0 < t < T, \tag{46.8$_j$}$$

$$u_j(0) = u_{0j}, \tag{46.9$_j$}$$

whose unique solution is given by

$$u_j(t) = u_{0j} \exp(-\lambda_j t) + \int_0^t \exp[-\lambda_j(t - s)] f_j(s)\, ds. \tag{46.10}$$

Let us denote by $v_j(t)$ the integral at the right and set

$$v(x, t) = \sum v_j(t) E_j(x). \tag{46.11}$$

We know that

$$\sum |u_{0j}|^2 = \|u_0\|_{L^2(\Omega)} < +\infty. \tag{46.12}$$

On the other hand by virtue of Proposition 34.2

$$(46.13)\qquad \int_0^T \|u(\cdot, t) - v(\cdot, t)\|^2_{H^1(\Omega)}\, dt$$
$$= \sum (\kappa + \lambda_j)|u_{0j}|^2 \int_0^T \exp(-2\lambda_j t)\, dt$$
$$= \sum \frac{\kappa + \lambda_j}{2\lambda_j}(1 - \exp(-2\lambda_j T))|u_{0j}|^2 \leqq C\|u_0\|^2_{L^2(\Omega)},$$

where the constant C depends on T and on the differential operator A. We remind the reader that the Hilbert space structure of $H_0^1(\Omega)$ and, by duality, that of $H^{-1}(\Omega)$, is defined by the inner product on $H_0^1(\Omega)$,

$$(46.14)\qquad ((u, v)) = a(u, v) + \kappa(u, v)_{L^2(\Omega)}$$

where $a(u, v)$ is the sequilinear form associated with A [see (34.16)].

The standard Hölder's inequalities for convolution yield

$$\int_0^T \|v(\cdot, t)\|^2_{H_0{}^1(\Omega)}\, dt = \sum (\kappa + \lambda_j) \int_0^T \left\{\int_0^t \exp[-\lambda_j(t - s)]f_j(s)\, ds\right\}^2 dt$$
$$\leqq \sum (\kappa + \lambda_j)\left\{\int_0^T \exp(-\lambda_j t)\, dt\right\}^2 \int_0^T |f_j(t)|^2\, dt$$
$$= \sum \left\{\frac{\kappa + \lambda_j}{\lambda_j}(1 - \exp(-\lambda_j T))\right\}^2 (\kappa + \lambda_j)^{-1} \int_0^T |f_j(t)|^2\, dt,$$

whence, with a suitable constant $C_1 > 0$, depending on T and on A like C,

$$(46.15)\qquad \int_0^T \|v(\cdot, t)\|^2_{H_0{}^1(\Omega)}\, dt \leqq C_1 \int_0^T \|f(\cdot, t)\|^2_{H^{-1}(\Omega)}\, dt,$$

by applying Proposition 34.2 once more. If we combine (46.13) and (46.15), we obtain

$$(46.16)\qquad \int_0^T \|u(\cdot, t)\|^2_{H_0{}^1(\Omega)}\, dt \leqq C_2^2\left\{\|u_0\|^2_{L^2(\Omega)} + \int_0^T \|f(\cdot, t)\|^2_{H^{-1}(\Omega)}\, dt\right\}.$$

Of course, these existence results, and the related estimates, are nothing new to us: They are particular cases of the theorems, such as Theorem 40.1, obtained by the energy inequalities—or of those obtained by Laplace transformation (Sects. 43 and 44), or by the continuous semigroups theory (Sect. 45). But there are advantages to the eigenfunction expansion approach: In many instances, it provides us with rather concrete, and computable, approximations of the solution. It is immediately apparent that it can be extended to equations of other types than parabolic, for instance to an important class of hyperbolic equations, such as the wave equation, and also to equations, like Schrödinger's, which are neither parabolic nor hyperbolic (see Exercises 46.5 and 46.6).

We shall discuss here the case of hyperbolic equations, as a transition to the results of Sect. 47. The differential equations under study will now be of order 2 with respect to t, specifically [A having the same meaning as in (46.2)]:

$$\frac{\partial^2 u}{\partial t^2} + Au = f \quad \text{in } \Omega \times]0, T[. \tag{46.17}$$

As for the initial conditions, they are now

$$u(\cdot, 0) = u_0, \qquad \frac{\partial u}{\partial t}(\cdot, 0) = u_1. \tag{46.18}$$

The boundary conditions will once again be implicit in the fact that we seek a solution u with values in $H_0^1(\Omega)$. We shall use the series representations of f, u_0, and u in terms of the E_j, as in our treatment in this section of the parabolic problem (46.2)–(46.3). Here we also need a series representation for u_1:

$$u_1 = \sum u_{1j} E_j.$$

We are going to prove the following theorem:

THEOREM 46.1. *Suppose that the open set Ω is bounded and that the differential operator $A(x, \partial/\partial x)$ in Ω is formally self-adjoint and strongly elliptic* [*i.e., satisfies* (34.2) *and* (34.6)].

Then, to every set of data,

$$u_0 \in H_0^1(\Omega), \qquad u_1 \in L^2(\Omega), \qquad f \in L^1(0, T; L^2(\Omega)), \tag{46.19}$$

there is a unique function u such that

$$u \in C^0([0, T]; H_0^1(\Omega)), \qquad \frac{\partial u}{\partial t} \in C^0([0, T]; L^2(\Omega)), \tag{46.20}$$

which satisfies (46.17)–(46.18).

Proof. The coefficients u_j in the series representation (46.7) must satisfy the second-order differential equation

$$u_j'' + \lambda_j u_j = f_j, \qquad 0 < t < T, \qquad [\text{cf. } (46.8)_j], \tag{46.21$_j$}$$

and the initial conditions [cf. $(46.9)_j$]

$$(46.22)_j \qquad u_j(0) = u_{0j}, \qquad u_j'(0) = u_{1j}.$$

The unique solution of $(46.21)_j$–$(46.22)_j$ is given by

$$(46.23) \qquad u_j(t) = u_{0j}\cos(t\sqrt{\lambda_j}) + u_{1j}\frac{\sin(t\sqrt{\lambda_j})}{\sqrt{\lambda_j}} + \int_0^t f_j(s)\frac{\sin\{(t-s)\sqrt{\lambda_j}\}}{\sqrt{\lambda_j}}\,ds.$$

The uniqueness of the solution $u(x, t)$ follows at once from Proposition 34.2 and from the uniqueness of the solutions u_j of $(46.21)_j$–$(46.22)_j$. The fact that (46.20) holds will now be derived from the expressions (46.23). Incidentally, let us point out that the reader should not be bothered by the fact that the square roots of the numbers λ_j (not all of which are necessarily positive) appear in (46.23): It is seen at once that the functions of those square roots $\sqrt{\lambda_j}$ are actually functions of $(\sqrt{\lambda_j})^2 = \lambda_j$. We continue to assume that $H_0^1(\Omega)$ is equipped with the inner product (46.14) and $H^{-1}(\Omega)$ is equipped with the dual Hilbert space structure; thus we are free to apply Proposition 34.2. We have

$$(46.24) \quad \{\sum (\kappa + \lambda_j)|u_j(t)|^2\}^{1/2}$$
$$\leqq \{\sum (\kappa + \lambda_j)|u_{0j}|^2\}^{1/2} + \left\{\sum \left|\frac{\kappa + \lambda_j}{\lambda_j}\right| \{\sin(t\sqrt{\lambda_j})\}^2 |u_{1j}|^2\right\}^{1/2}$$
$$+ \int_0^t \left\{\sum |f_j(s)|^2 \frac{\kappa + \lambda_j}{\lambda_j} \sin\{(t-s)\sqrt{\lambda_j}\}^2\right\}^{1/2} ds.$$

We use the fact that, for a suitable constant $C > 0$ and for all $j = 1, 2, \ldots,$

$$(46.25) \qquad \left|\frac{\kappa + \lambda_j}{\lambda_j}\right| \sin(t\sqrt{\lambda_j})^2 \leqq C.$$

In view of (46.25), we derive from (46.24):

$$(46.26) \quad \|u(\cdot, t)\|_{H_0^1(\Omega)} \leqq \|u_0\|_{H_0^1(\Omega)} + C\|u_1\|_{L^2(\Omega)} + C\int_0^t \|f(\cdot, t)\|_{L^2(\Omega)}\,dt,$$

which proves that $u \in L^\infty(0, T; H_0^1(\Omega))$. Next we show that u is continuous in the closed interval $[0, T]$ with values in $H_0^1(\Omega)$. In order to see this it suffices to derive from the expressions (46.23) an inequality slightly more precise than (46.24), namely

$$(46.27)\quad \{\sum (\kappa + \lambda_j)|u_j(t) - u_j(t')|^2\}^{1/2}$$
$$\leqq \{\sum (\kappa + \lambda_j)|u_{0j}|^2 H_j^2(t, t')\}^{1/2} + \{\sum |u_{1j}|^2 K_j^2(t, t')\}^{1/2}$$
$$+ \left\{\sum \left(\int_{t'}^{t} |f_j(s)| \left|\left[\frac{\kappa + \lambda_j}{\lambda_j}\right]^{1/2} \sin((t' - s)\sqrt{\lambda_j})\right| ds\right)^2\right\}^{1/2}$$
$$+ \left\{\sum \left(\int_0^t |f_j(s)|\,|K_j(t - s, t' - s)|\, ds\right)^2\right\}^{1/2},$$

where we have used the notation

$$(46.28)\qquad H_j(t, t') = \cos(t\sqrt{\lambda_j}) - \cos(t'\sqrt{\lambda_j}),$$

$$(46.29)\qquad K_j(t, t') = \left[\frac{\kappa + \lambda_j}{\lambda_j}\right]^{1/2} [\sin(t\sqrt{\lambda_j}) - \sin(t'\sqrt{\lambda_j})].$$

We have, for a suitable $C > 0$ and all $j = 1, 2, \ldots$, all $t, t' \in [0, T]$,

$$|H_j(t, t')| \leqq 2, \qquad |K_j(t, t')| \leqq 2C.$$

If we fix j, $H_j(t, t')$ and $K_j(t, t')$ converge to zero as t' converges to t. Hence, by the dominated convergence theorem (applied to series), the first two terms on the right-hand side of (46.27) tend to zero as t' tends to t.

In view of the fact that the norm of the integral is not greater than the integral of the norm, we see that the third term is not greater than

$$\int_{t'}^{t} \|f(\cdot, s)\|_{L^2(\Omega)}\, ds,$$

and therefore converges to zero when $t' \to t$. Finally, consider the fourth and last term on the right-hand side of (46.27). To every $\varepsilon > 0$ there is an integer $N_\varepsilon > 0$ such that

$$\left\{\sum_{j > N_\varepsilon} \left(\int_0^T |f_j(s)|\, ds\right)^2\right\}^{1/2} \leqq \varepsilon.$$

On the other hand, there is $\eta > 0$ such that $|t - t'| \leqq \eta$ implies

$$|K_j(t - s, t' - s)| \leqq \varepsilon$$

for every $j = 1, \ldots, N_\varepsilon$, every $s, t, t' \in [0, T]$. From these two facts it follows easily that the last term also tends to zero when $t' \to t$.

A similar argument applies to the t derivative of $u(x, t)$,

$$u_t(x, t) = \sum u_j'(t) E_j(x).$$

It suffices to observe that

$$u_j'(t) = -\sqrt{\lambda_j}\, u_{0j} \sin t(\sqrt{\lambda_j}) + u_{1j} \cos(t\sqrt{\lambda_j}) + \int_0^t f_j(s) \cos\{(t-s)\sqrt{\lambda_j}\}\, ds. \tag{46.30}$$

We leave the details to the student: One must prove that

$$\{\sum |u_j'(t) - u_j'(t')|^2\}^{1/2}$$

converges to zero as t' tends to t.

Remark 46.1. If, in addition to the assumptions in Theorem 46.1, we make the hypothesis that $f \in C^0([0, T]; H^{-1}(\Omega))$, we may derive from the conclusion in Theorem 46.1 and from the equation (46.17) that $\partial^2 u/\partial t^2$ belongs to $C^0([0, T]; H^{-1}(\Omega))$.

Remark 46.2. An essential difference between hyperbolic mixed problems and parabolic ones is that, in the former, time is revertible: By assuming that the right-hand side f in Eq. (46.17) was defined for $-T \leqq t \leqq 0$, we could have solved the *backward* mixed problem, i.e., the same problem as (46.17)–(46.18) except that (46.17) would have to be verified for $(x, t) \in \Omega \times]-T, 0[$. This is not so in the case of a parabolic mixed problem!

Remark 46.3. The reader should compare formula (46.23) with formula (13.10). One may say that the latter is the "global analog" of the former: It is the analog of the former when $\Omega = \mathbf{R}^n$ (and $T = +\infty$). Here again, as in Remark 34.1, we see that the series representations in terms of the eigenfunctions $E_\nu(x)$ have a strong (and deep!) similarity with the Fourier inversion formula. The analogy would be even more obvious if we had used Fourier series instead of Fourier integrals (i.e., if we had dealt with functions on the torus $\mathbf{T}^n$ rather than on $\mathbf{R}^n$; cf. Exercise 13.2).

Exercises

46.1. Let Ω be a bounded interval $a < x < b$ in the real line. Solve, by using the eigenfunction expansion described in Example 34.1, the mixed problem

$$\frac{\partial u}{\partial t} = \frac{\partial^2 u}{\partial x^2}, \qquad a < x < b, \quad t > 0, \tag{46.31}$$

$$u(x, 0) = u_0(x) \in L^2(a, b), \tag{46.32}$$

$$u(a, t) = g_a(t), \qquad u(b, t) = g_b(t), \qquad t > 0. \tag{46.33}$$

Compare with the results in Example 44.2.

46.2. Here we ask the same question as in Exercise 46.1 but with (46.31) replaced by

$$\frac{\partial^2 u}{\partial t^2} = \frac{\partial^2 u}{\partial x^2}, \qquad a < x < b, \quad t > 0, \tag{46.34}$$

and (46.32) replaced by

$$u(x, 0) = u_0(x) \in H_0^1(\Omega), \qquad \frac{\partial u}{\partial t}(x, 0) = u_1(x) \in L^2(\Omega). \tag{46.35}$$

Compare with Exercise 44.7.

46.3. Let Ω be the disk $x^2 + y^2 < R^2$ $(R > 0)$ in the plane. By using the eigenfunction expansion described in Example 34.2, solve the mixed problem

$$\frac{\partial u}{\partial t} = \Delta u, \qquad (x, y) \in \Omega, \qquad t > 0 \qquad \left[\Delta = \left(\frac{\partial}{\partial x}\right)^2 + \left(\frac{\partial}{\partial y}\right)^2\right], \tag{46.36}$$

$$u|_{t=0} = u_0 \in L^2(\Omega) \tag{46.37}$$

$$u(\cdot, t) \in H_0^1(\Omega), \qquad t > 0. \tag{46.38}$$

46.4. Here we ask the same question as in Exercise 46.3 but for the wave equation

$$\frac{\partial^2 u}{\partial t^2} = \Delta u, \qquad (x, y) \in \Omega, \qquad t > 0, \tag{46.39}$$

and with the initial conditions,

$$u\Big|_{t=0} = u_0 \in H_0^1(\Omega), \qquad \frac{\partial u}{\partial t}\Big|_{t=0} = u_1 \in L^2(\Omega). \tag{46.40}$$

46.5. Let A be the same operator as in Eq. (46.2) but consider now the "Schrödinger-like" equation

$$\frac{\partial u}{\partial t} + iAu = f \quad \text{in } \Omega \times]0, T[\qquad (i = \sqrt{-1}), \tag{46.41}$$

with initial condition

$$u(\cdot, 0) = u_0 \in L^2(\Omega). \tag{46.42}$$

Assume that $f \in L^2(0, T; H^{-1}(\Omega))$ and u_0 are given as in (46.6). Determine the coefficients $u_j(t)$ in the expansion (46.7) of the solution u of (46.41)–(46.42). Show that the expansion (46.7) does converge in $L^2(0, T; H_0^1(\Omega))$. What happens when $T \to +\infty$?

46.6. This is the same question as in Exercise 46.1 but with the Schrödinger equation

$$\frac{\partial u}{\partial t} = \sqrt{-1}\,\frac{\partial^2 u}{\partial x^2}, \qquad a < x < b, \quad 0 \leqq t \leqq T, \tag{46.43}$$

substituted for the heat equation (46.31) [also, the side conditions (46.33) are imposed only up to time T].

47

An Abstract Existence and Uniqueness Theorem for a Class of Hyperbolic Mixed Problems. Energy Inequalities

In Section 46 we obtained a result (Theorem 46.1) on the existence and uniqueness of solutions to certain hyperbolic mixed problems. Actually that result also gave some interesting eigenfunction expansions for the solution. But the class of equations and problems to which it applies is by far too restricted. We must be able to handle problems other than the Dirichlet one: for instance the Neumann problem (§37.1), the "hybrid" problem of §37.2, radiation and oblique-derivative problems such as those considered in §37.5, Dirichlet and Neumann problems for higher order strongly elliptic operators (Sects. 36 and 38), and so on. Generalization to include these problems would be desirable also for the parabolic mixed problems; indications on how it could be carried out have been given in a series of exercises—40.1, 41.1, and 42.1 to 42.3 (for applications, see also Exercises 41.4, 41.5, and 42.4). It is for the sake of generality that in this section we state and prove an abstract theorem, susceptible to applications in a wide range of problems of the kind mentioned above. Nevertheless, it might be convenient to think of concrete cases, such as mixed problems for a second-order differential operator of the kind $(\partial/\partial t)^2 + A(x, t, \partial/\partial x)$, where

$$A\left(x, t, \frac{\partial}{\partial x}\right) = -\sum_{j,k=1}^{n} \frac{\partial}{\partial x^j} a^{jk}(x, t) \frac{\partial}{\partial x^k} + \sum_{j=1}^{n} b^j(x, t) \frac{\partial}{\partial x^j} + c(x, t)$$

satisfies the usual strong ellipticity assumptions (40.17) [and also satisfies the boundedness assumption (40.16)]. In contrast to the operators considered in Sect. 47, the ones we deal with here are allowed to have coefficients which vary with t. The fact that $A(x, t, \partial/\partial x)$ is written in the variational form enables us to introduce the usual sesquilinear form, here depending on t,

$$a(t; u, v) = \sum_{j,k=1}^{n} \int_\Omega a^{jk}(x, t) \frac{\partial u}{\partial x^k} \frac{\partial \bar{v}}{\partial x^j}\, dx + \sum_{j=1}^{n} \int_\Omega b^j(x, t) \frac{\partial u}{\partial x^j} \bar{v}\, dx$$

$$+ \int_\Omega c(x, t) u\bar{v}\, dx.$$

In the manner described in some detail in Sect. 37 such a form defines a bounded linear operator (now depending on t, $0 \leqq t \leqq T$) from a Hilbert space $\mathbf{V}$ into its antidual $\overline{\mathbf{V}}'$. We recall that the antidual $\overline{\mathbf{V}}'$ of $\mathbf{V}$ is the Banach space of continuous antilinear functionals on $\mathbf{V}$ (we shall systematically denote by $\langle\ ,\ \rangle^-$ the bracket of the antiduality between $\mathbf{V}$ and $\overline{\mathbf{V}}'$). In the case of the differential operator $A(x, t, \partial/\partial x)$ above, $\mathbf{V}$ might be $H_0^1(\Omega)$ if the boundary conditions in the problem under study are the Dirichlet ones, or else some other function space, most often an intermediary one between $H_0^1(\Omega)$ and $H^1(\Omega)$ (see Sect. 37). If, instead of $A(x, t, \partial/\partial x)$, we were to study a strongly elliptic operator of order $2m$ (with coefficients depending on time), the space $\mathbf{V}$ could be $H_0^m(\Omega)$, or some intermediary space between $H_0^m(\Omega)$ and $H^m(\Omega)$.

At any rate, we have a continuous sesquilinear form on $\mathbf{V} \times \mathbf{V}$, $a(t; \mathbf{u}, \mathbf{v})$, depending on t, and for every such $t \in [0, T]$, it defines a bounded linear operator $A(t) : \mathbf{V} \to \overline{\mathbf{V}}'$, such that

$$a(t; \mathbf{u}, \mathbf{v}) = \langle A(t)\mathbf{u}, \mathbf{v}\rangle^-, \qquad \mathbf{u}, \mathbf{v} \in \mathbf{V}. \tag{47.1}$$

But the reader should be careful not to identify this "abstract" operator $A(t)$ with the restriction to $\mathbf{V}$ of the differential operator $A(x, t, \partial/\partial x)$: $A(t)$ carries information which, in general, $A(x, t, \partial/\partial x)$ does not. Indeed, the choice itself of $\mathbf{V}$ implies certain boundary conditions on the solution, as explained in Sect. 37. In the case where $\mathbf{V} = H_0^1(\Omega)$ we may, in a sense, identify $A(t)$ with $A(x, t, \partial/\partial x)$, viewed as a bounded linear operator $H_0^1(\Omega) \to H^{-1}(\Omega)$—but otherwise this is not so. We refer to Sect. 37 for the more general interpretation. Thus, in the abstract setup, we shall start from the operator $A(t) : \mathbf{V} \to \overline{\mathbf{V}}'$, or from the sesquilinear functional $a(t; \mathbf{u}, \mathbf{v})$.

Some of the limitations encountered in the special situation of Sect. 46 carry over to the present, more general situation: We shall assume that $A(t)$ is formally self-adjoint, in the sense that

$$\langle A(t)\mathbf{u}, \mathbf{v}\rangle^- = \langle \mathbf{u}, A(t)\mathbf{v}\rangle^-, \qquad 0 \leqq t \leqq T, \qquad \mathbf{u}, \mathbf{v} \in \mathbf{V}, \tag{47.2}$$

or equivalently, for the same $\mathbf{u}$, $\mathbf{v}$, t,

$$a(t; \mathbf{u}, \mathbf{v}) = \overline{a(t; \mathbf{v}, \mathbf{u})}. \tag{47.2'}$$

We shall make the usual coerciveness hypothesis:

(47.3) *there are constants* $c_0 > 0$, λ_0 *real, such that, for all* $\lambda \geqq \lambda_0$, *all* $\mathbf{u} \in \mathbf{V}$ *and all* t, $0 \leqq t \leqq T$,

$$\langle(\lambda + A(t))\mathbf{u}, \mathbf{u}\rangle^{-} \geqq c_0 \|\mathbf{u}\|_{\mathbf{V}}^2.$$

We must also specify our requirements on the regularity of $A(t)$, or of $a(t; \mathbf{u}, \mathbf{v})$ with respect to t. We are going to assume that

(47.4) *for each pair of elements* $\mathbf{u}$, $\mathbf{v}$ *of* $\mathbf{V}$, $t \mapsto a(t; \mathbf{u}, \mathbf{v})$ *is a* C^1 *function in the closed interval* $[0, T]$.

The initial value problem we wish to study is the following:

$$\mathbf{u}_{tt}(t) + A(t)\mathbf{u}(t) = \mathbf{f}(t), \qquad 0 < t < T, \tag{47.5}$$

$$\mathbf{u}(0) = \mathbf{u}_0, \qquad \mathbf{u}_t(0) = \mathbf{u}_1, \tag{47.6}$$

where the t subscripts mean differentiation with respect to t. We are going to prove the following result:

THEOREM 47.1. *Suppose that* $A(t)$ *is formally self-adjoint, i.e., that* (47.2) *holds, and that the coerciveness hypothesis* (47.3) *is satisfied. Suppose furthermore that* $a(t; \mathbf{u}, \mathbf{v}) \in C^1([0, T])$ *whatever* $\mathbf{u}, \mathbf{v} \in \mathbf{V}$, *i.e., that* (47.4) *holds.*

Then, to every $\mathbf{u}_0 \in \mathbf{V}$, $\mathbf{u}_1 \in \mathbf{H}$, $\mathbf{f} \in L^1(0, T; \mathbf{H})$, *there is a unique function* $\mathbf{u} \in L^\infty(0, T; \mathbf{V})$ *such that* $\mathbf{u}_t \in L^\infty(0, T; \mathbf{H})$ *and such that* (47.5)–(47.6) *holds.*

Furthermore, $\mathbf{u}$ *is a continuous function in the closed interval* $[0, T]$ *valued in* $\mathbf{V}_\sigma$, *the space* $\mathbf{V}$ *equipped with its weak topology, and* $\mathbf{v}_t$ *is a continuous function in* $[0, T]$ *valued in* $\mathbf{H}_\sigma$.† *There is a constant* $C > 0$, *independent of* $\mathbf{f}$, $\mathbf{u}_0$, $\mathbf{u}_1$, *and of* t, $0 \leqq t \leqq T$, *such that, for these* t*'s,*

$$\|\mathbf{u}(t)\|_{\mathbf{V}} + \|\mathbf{u}_t(t)\|_{\mathbf{H}} \leqq C\left\{\|\mathbf{u}_0\|_{\mathbf{V}} + \|\mathbf{u}_1\|_{\mathbf{H}} + \int_0^t \|\mathbf{f}(t')\|_{\mathbf{H}}\, dt'\right\}. \tag{47.7}$$

Proof. The proof consists of three steps: First, we prove some identities and inequalities of the "energy" type; next we prove the existence of the solution, by the so-called Galerkin method (Example 35.1), using (but only in a finite-dimensional context!) the energy estimates established in part I. Finally we prove the uniqueness of the solution, essentially by a modification of the energy estimates. There are many similarities with the proofs of Theorems 40.1 and 41.1 (which apply to the parabolic evolution equations) but there are also marked differences, in particular in the regularity requirements with respect to the variable t, and also in the fact that we must take $\mathbf{f}$ valued in $\mathbf{H}$

† As usual, since we regard functions as distributions, this means that $\mathbf{u}$ and $\mathbf{u}_t$ might have to be modified on a set of measure zero.

and not in $\overline{\mathbf{V}}'$. Technically, this is due to the fact that, in the derivation of the energy estimates, we are forced to take the inner product of $\mathbf{f}(t)$ with $\mathbf{u}_t(t)$—the latter belongs to $\mathbf{H}$, not to $\mathbf{V}$, and the inner product is therefore to be computed in $\mathbf{H}$.

I Energy Inequalities

Let φ be an arbitrary element of $C^1([0, T]; \mathbf{V})$ such that $\psi = \varphi_{tt} - A(t)\varphi \in L^1(0, T; \mathbf{H})$ [as a consequence, $\varphi_{tt} \in L^1(0, T; \overline{\mathbf{V}}')$]. We have

$$(\varphi_{tt}, \varphi_t)_{\mathbf{H}} + a(t; \varphi, \varphi_t) = (\psi, \varphi_t)_{\mathbf{H}}. \tag{47.8}$$

Let us write, for any $\mathbf{v}$, $\mathbf{w}$ in $\mathbf{V}$, $a_t(t; \mathbf{v}, \mathbf{w}) = (d/dt)a(t; \mathbf{v}, \mathbf{w})$, and take twice the real part of both sides in (47.8). It yields

$$\frac{d}{dt}\{\|\varphi_t\|_{\mathbf{H}}^2 + a(t; \varphi, \varphi)\} = (\psi, \varphi_t)_{\mathbf{H}} + a_t(t; \varphi, \varphi). \tag{47.9}$$

We have used the property (47.2′), i.e., the fact that $a(t; \mathbf{v}, \mathbf{w})$ is Hermitian. We integrate both sides in (47.9) from 0 to t. We obtain

$$\begin{aligned}\|\varphi_t(t)\|_{\mathbf{H}}^2 + a(t; \varphi(t), \varphi(t)) &= \|\varphi_1\|_{\mathbf{H}}^2 + a(0; \varphi_0, \varphi_0)\\ &\quad + \int_0^t \{(\psi(t'), \varphi_t(t'))_{\mathbf{H}} + a_t(t'; \varphi(t'), \varphi(t'))\}\, dt'.\end{aligned} \tag{47.10}$$

Let us now avail ourselves of the coercivity property (47.3). We see that there is a constant $C > 0$, independent of φ (hence of φ_0, φ_1, ψ) and of t such that

$$\begin{aligned}\|\varphi(t)\|_{\mathbf{V}}^2 + \|\varphi_t(t)\|_{\mathbf{H}}^2 \leqq C\Big\{&\|\varphi_0\|_{\mathbf{V}}^2 + \|\varphi_1\|_{\mathbf{H}}^2 + \|\varphi(t)\|_{\mathbf{H}}^2\\ &+ \int_0^t \|\varphi(t')\|_{\mathbf{V}}^2\, dt' + \sup_{0 \leqq t' \leqq t} \|\varphi_t(t')\|_{\mathbf{H}} \int_0^t \|\psi(t')\|_{\mathbf{H}}\, dt'\Big\}.\end{aligned}$$

Let us introduce the following norm:

$$N(t; \varphi) = \sup_{0 \leqq t' \leqq t} \{\|\varphi(t')\|_{\mathbf{V}}^2 + \|\varphi_t(t')\|_{\mathbf{H}}^2\}^{1/2}.$$

Possibly after increasing the constant C, we derive from the preceding inequality:

$$N(t; \varphi) \leqq C\Big\{N(0; \varphi) + \int_0^t \|\psi(t')\|_{\mathbf{H}}\, dt' + \int_0^t N(t'; \varphi)\, dt'\Big\}. \tag{47.11}$$

The classical Gronwall inequality (see Exercise 11.10) enables us to derive from (47.11):

$$N(t;\boldsymbol{\varphi}) \leqq Ce^{Ct}\left\{N(0;\boldsymbol{\varphi}) + \int_0^t \|\boldsymbol{\psi}(t')\|_{\mathbf{H}}\,dt'\right\}, \qquad 0 \leqq t \leqq T, \tag{47.12}$$

in other words, and possibly after some increasing of C,

$$\|\boldsymbol{\varphi}(t)\|_{\mathbf{V}} + \|\boldsymbol{\varphi}_t(t)\|_{\mathbf{H}} \leqq Ce^{CT}\left\{\|\boldsymbol{\varphi}_0\|_{\mathbf{V}} + \|\boldsymbol{\varphi}_1\|_{\mathbf{H}} + \int_0^t \|\boldsymbol{\psi}(t')\|_{\mathbf{H}}\,dt'\right\}. \tag{47.13}$$

II Existence of the Solution

We are going to apply a simplified version of *Galerkin's method* (Example 35.1). The idea of this method is simple (see Sect. 35): In view of the separability of $\mathbf{V}$, we may represent $\mathbf{V}$ as the closure of the union of a strictly increasing sequence of finite-dimensional linear subspaces $\mathbf{V}_J$. By the density of $\mathbf{V}$ in $\mathbf{H}$ we see that $\mathbf{H}$ is also the closure (in $\mathbf{H}$!) of the union of the $\mathbf{V}_J$. By using suitable projections of $\mathbf{V}$ and $\mathbf{H}$ onto the $\mathbf{V}_J$ we transform our infinite-dimensional problem into a system of ordinary differential equations in $\mathbf{V}_J$ (i.e., data and unknown are valued in $\mathbf{V}_J$). The latter yield a unique solution (for suitably chosen initial conditions), $\mathbf{u}_J$, which are then shown to converge, in an appropriate manner, to the sought solution $\mathbf{u}$ of problem (47.5)–(47.6).

In the present situation we shall use *a sequence* $\mathbf{w}_1, \ldots, \mathbf{w}_j, \ldots$ in $\mathbf{V}$ *which forms a complete orthonormal system in* $\mathbf{H}$. Obviously such sequences do exist. The finite-dimensional linear subspace $\mathbf{V}_J$ will be, in our case, the linear span of $\mathbf{w}_1, \ldots, \mathbf{w}_J$. We may select complex numbers $\mathbf{u}_{0J}^j$ such that the partial sums

$$\mathbf{u}_{0J} = \sum_{j=1}^{J} u_{0J}^j\,\mathbf{w}_j$$

converge to $\mathbf{u}_0$ in $\mathbf{V}$. We then define

$$\mathbf{u}_J(t) = \sum_{j=1}^{J} u_J^j(t)\mathbf{w}_j, \qquad J = 1, 2, \ldots,$$

as the solution of the initial value problem:

$$(u_J^j)_{tt}(t) + a(t;\mathbf{u}_J(t), \mathbf{w}_j) = (\mathbf{f}(t), \mathbf{w}_j)_H, \qquad 1 \leqq j \leqq J, \tag{47.14}$$

$$u_J^j(0) = u_{0J}^j, \qquad (u_J^j)_t(0) = (\mathbf{u}_1, \mathbf{w}_j)_H, \qquad 1 \leqq j \leqq J. \tag{47.15}$$

This is a system of J ordinary second-order linear differential equations with C^1 coefficients in $[0, T]$. Indeed,

$$a(t;\mathbf{u}_J(t), \mathbf{w}_j) = \sum_{i=1}^{J} a(t;\mathbf{w}_i, \mathbf{w}_j)u_J^j(t),$$

and its suffices to apply (47.4). From our choice of $\mathbf{f}$ the right-hand sides of (47.14) are integrable functions on $[0, T]$; hence the (unique) solutions u_J^j are C^1 functions and their second derivatives are L^1 functions in $[0, T]$. As a consequence, we have the right to apply the energy inequality (47.13) with $\boldsymbol{\varphi} = \mathbf{u}_J$. We see that there is a positive constant C_1, depending only on the operator $A(t)$ and on T, such that

$$\|\mathbf{u}_J(t)\|_{\mathbf{V}}^2 + \|(\mathbf{u}_J)_t(t)\|_{\mathbf{H}}^2 \leqq C_1\left\{\|\mathbf{u}_0\|_{\mathbf{V}}^2 + \|\mathbf{u}_1\|_{\mathbf{H}}^2 + \left(\int_0^t \|\mathbf{f}(t')\|_{\mathbf{H}}\, dt'\right)^2\right\}. \tag{47.16}$$

We integrate both sides of (47.16) from 0 to T; we reach the conclusion that the $\mathbf{u}_J$ [resp. the $(\mathbf{u}_J)_t$] form a bounded sequence in $L^2(0, T; \mathbf{V})$ [resp. in $L^2(0, T; \mathbf{H})$]. One can therefore extract a subsequence $(\mathbf{u}_{J_\nu})$ which converges weakly in $L^2(0, T; \mathbf{V})$ to a function $\mathbf{u}$, while the $(\mathbf{u}_{J_\nu})_t$ converge weakly in $L^2(0, T; \mathbf{H})$, to $\mathbf{u}_t$. It is easily seen that the limit $\mathbf{u}$ satisfies (47.5)–(47.6) and also, by virtue of (47.16), the energy inequality (47.7). Thus $\mathbf{u} \in L^\infty(0, T; \mathbf{V})$, $\mathbf{u}_t \in L^\infty(0, T; \mathbf{H})$. Observe that it follows from this, from Eq. (47.5), and from our hypotheses on $a(t; \mathbf{v}, \mathbf{w})$ and on $\mathbf{f}$, that $\mathbf{u}_{tt} \in L^1(0, T; \overline{\mathbf{V}}')$. We derive that

$$\mathbf{u} \in C^0([0, T]; \mathbf{H}), \qquad \mathbf{u}_t \in C^0([0, T]; \overline{\mathbf{V}}'). \tag{47.17}$$

This presumes that we have possibly modified $\mathbf{u}$ and $\mathbf{u}_t$ on a set of measure zero; in particular, we may assume that they are bounded functions in $[0, T]$ with values in $\mathbf{V}$ and $\mathbf{H}$, respectively.

Let $\mathbf{v}'$ (resp. $\mathbf{h}$) be an arbitrary element of $\overline{\mathbf{V}}'$ (resp. $\mathbf{H}$), t_0 an arbitrary point, $\{t_j\}$ a sequence converging to t_0 in the closed interval $[0, T]$. We know that $\mathbf{u}(t_j)$ [resp. $\mathbf{u}_t(t_j)$], $j = 0, 1, \ldots$, remains in a bounded subset of $\mathbf{V}$ (resp. $\mathbf{H}$), and hence there is a subsequence $\{j_\nu\}$ such that

$$\langle \mathbf{v}', \mathbf{u}(t_{j_\nu})\rangle^- \qquad [\text{resp. } (\mathbf{u}(t_{j_\nu}), \mathbf{h})_{\mathbf{H}}]$$

converges. By virtue of (47.17) it can only be to

$$\langle \mathbf{v}', \mathbf{u}(t_0)\rangle^- \qquad [\text{resp. } (\mathbf{u}_t(t_0), \mathbf{h})_{\mathbf{H}}]. \tag{47.18}$$

By an elementary argument this means that

$$\langle \mathbf{v}', \mathbf{u}(t_j)\rangle^- \qquad [\text{resp. } (\mathbf{u}_t(t_j), \mathbf{h})_{\mathbf{H}}]$$

converges to (47.18). This proves the continuity properties of $\mathbf{u}$ and $\mathbf{u}_t$ when we regard them as valued in $\mathbf{V}_\sigma$ and $\mathbf{H}_\sigma$, respectively.

III Uniqueness of the Solution

We consider a solution $\mathbf{u} \in L^\infty(0, T; \mathbf{V})$ of the homogeneous equation

$$\mathbf{u}_{tt} + A(t)\mathbf{u} = 0, \qquad 0 < t < T, \tag{47.19}$$

with $\mathbf{u}_t \in L^\infty(0, T; \mathbf{H})$ and $\mathbf{u}(0) = \mathbf{u}_t(0) = 0$. Let us set

$$\mathbf{U}(t) = \int_0^t \mathbf{u}(t')\,dt'$$

and integrate (47.19) from 0 to t'. We get

$$\mathbf{U}_{tt}(t') + \int_0^{t'} A(t'')\mathbf{u}(t'')\,dt'' = \mathbf{U}_{tt}(t') + A(t')\mathbf{U}(t') - \int_0^{t'} A_t(t'')\mathbf{U}(t'')\,dt'' = 0.$$

We exploit now the fact that $\mathbf{U}_t \in L^\infty(0, T; \mathbf{V})$, $\mathbf{U}_{tt} = \mathbf{u}_t \in L^\infty(0, T; \mathbf{H})$ and $A(t)\mathbf{U}(t) \in C^0([0, T]; \overline{\mathbf{V}}')$. We may write

$$(47.20)\quad (\mathbf{U}_{tt}(t'), \mathbf{U}_t(t'))_{\mathbf{H}} + a(t'; \mathbf{U}(t'), \mathbf{U}_t(t')) - \int_0^{t'} a_t(t''; \mathbf{U}(t''), \mathbf{U}_t(t'))\,dt'' = 0.$$

We take twice the real part of the left-hand side of (47.20) and integrate from 0 to t. We obtain

$$(47.21)\quad \|\mathbf{U}_t(t)\|_{\mathbf{H}}^2 + a(t; \mathbf{U}(t), \mathbf{U}(t))$$
$$= \int_0^t a_t(t'; \mathbf{U}(t'), \mathbf{U}(t'))\,dt' + 2\,\mathrm{Re}\int_0^t a_t(t'; \mathbf{U}(t'), \mathbf{U}(t) - \mathbf{U}(t'))\,dt'.$$

We reason as in part I, using the coercivity hypothesis (47.3). We derive from (47.21), for a suitable constant $C > 0$,

$$N(t; \mathbf{U})^2 \leqq C \int_0^t N(t'; \mathbf{U})^2\,dt' + \int_0^t \|\mathbf{U}(t) - \mathbf{U}(t')\|_{\mathbf{V}}^2\,dt' \leqq 5CtN(t; \mathbf{U})^2,$$

which implies $\mathbf{U}(t) = 0$ if $0 \leqq t \leqq t_0 = (5C)^{-1}$. As usual in such a situation, we may now repeat the argument in the interval $[t_0, T]$ instead of the interval $[0, T]$. We easily reach the desired conclusion, that $\mathbf{U}(t)$, and therefore also $\mathbf{u}(t)$, must vanish identically for $0 \leqq t \leqq T$. Q.E.D.

Remark 47.1. It can be shown that the solution $\mathbf{u}$ of (47.5)–(47.6), whose existence and uniqueness are stated in Theorem 47.1, enjoys more regularity with respect to t than what is stated in this theorem. It can be shown, in fact, that $\mathbf{u}$ (resp. $\mathbf{u}_t$) is a continuous function of t, $0 \leqq t \leqq T$, valued in $\mathbf{V}$ (resp. in $\mathbf{H}$) and not only in $\mathbf{V}_\sigma$ (resp. in $\mathbf{H}_\sigma$). But the proof of this fact is considerably more involved than the proof of the weaker statement in Theorem 47.1; it can be found in [LM, Vol. 1, Chap, 3, §8.4], from which we have adapted the argument in this section.

BIBLIOGRAPHY

(References and Further Reading)

On the Subject of Functional Analysis and Distribution Theory

Gelfand, I. M. and Silov, G., "Generalized Functions," Academic Press, New York, 1964 (English translation; contains a wealth of information on and explicit computation of special functions and distributions, fundamental solutions, Fourier transforms, etc.).

[TD] Schwartz, L., "Théorie des Distributions," Hermann, Paris, 1966 (still the best and most comprehensive exposition of the theory).

[TVS, D & K] Treves, F., "Topological Vector Spaces, Distributions and Kernels," Academic Press, New York, 1967.

[Y] Yosida, K., "Functional Analysis," Springer-Verlag, Berlin and New York, 1968 (2nd ed.; a basic reference on the subject of functional analysis, containing extensive material and describing most aspects of the subject).

On the Subject of Linear Partial Differential Equations (General)

Bers, L., John, F. and Schechter, M., "Partial Differential Equations," Wiley (Interscience), New York, 1964 (covers the basic material from a viewpoint somewhat different from the present book).

Ehrenpreis, L., "Fourier Analysis in Several Complex Variables," Wiley (Interscience), New York, 1970 (the theory of overdetermined systems of linear PDEs with constant coefficients, and a rather unorthodox view of its ramifications, by one of the architects of the theory).

[LPDO] Hörmander, L., "Linear Partial Differential Operators," Springer-Verlag, Berlin and New York, 1963 (the general theory of linear PDEs as it was during the 1960s; difficult but very rewarding reading).

Palamodov, V. P., "Linear Differential Operators with Constant Coefficients," Springer-Verlag, Berlin and New York, 1970 (English translation; an exhaustive, and somewhat overwhelming, exposition of the theory of overdetermined systems of linear PDEs with constant coefficients—more rigorous but less diverse than Ehrenpreis' book).

Treves, F., "Linear Partial Differential Equations with Constant Coefficients," Gordon & Breach, New York, 1967.

On the Subject of Linear Partial Differential Equations of Special Type

Agmon, S., "Lectures on Elliptic Boundary Value Problems," Van Nostrand-Reinhold, Mathematical Studies, Princeton, New Jersey, 1965.

Friedman, A., "Generalized Functions and Partial Differential Equations," Prentice-Hall, Englewood Cliffs, New Jersey, 1963.

Friedman, A., "Partial Differential Equations of Parabolic Type," Prentice-Hall, Englewood Cliffs, New Jersey, 1964.

[F] Friedman, A., "Partial Differential Equations," Holt, New York, 1969.

Lions, J. L., "Equations différentielles opérationelles et problèmes aux limites," Springer-Verlag, Berlin and New York, 1961.

[L M] Lions, J. L. and Magenes, E., "Non-homogeneous Boundary Value Problems and Applications," Springer-Verlag, Berlin and New York, 1972.

On Some of the Special Topics Touched upon in the Book

[A] Aubin, J. P., "Approximation of Elliptic Boundary Value Problems," Wiley (Interscience), New York, 1972.

Dinkin, E. B. and Yushkevich, A. A., "Markov Processes," Plenum, New York, 1969 (English translation).

[F S] Fix, G. and Strang, G., "An Analysis of the Finite Element Method," Prentice Hall, Englewood Cliffs, New Jersey, 1973.

[H] Helms, L. L., "Introduction to Potential Theory," Wiley (Interscience), New York, 1969.

Hobson, E. W., "The theory of Spherical Harmonics," Cambridge Univ. Press, London and New York, 1931.

Lavoine, J., "Calcul symbolique des distributions et des pseudo-fonctions," CNRS, Paris, 1959.

[W] Watson, G. N., "Theory of Bessel Functions," 2nd ed., Macmillan, New York, 1944.

Widder, D. V., "Laplace Transform," Princeton Univ. Press, Princeton, New Jersey, 1946.

Bibliographical References

[1] Séminaire Schwartz, "Equations aux dérivées partielles," Institut Henri Poincaré, Paris, 1955.

[2] Stampacchia, G., "Equations elliptiques du second ordre à coefficients discontinus," Séminaire Eq. Dér. Part., Collège de France Paris, 1963/64.

Index

T

U

V

W

Z

B 6
C 7
D 8
E 9
F 0
G 1
H 2
I 3
J 4